D0919787

Number Theory, Trace Formulas and Discrete Groups

Symposium in Honor of Atle Selberg
Oslo, Norway, July 14–21, 1987

Atle Selberg

Number Theory, Trace Formulas and Discrete Groups

Symposium in Honor of Atle Selberg
Oslo, Norway, July 14–21, 1987

Edited by

Karl Egil Aubert

Universitetet I Oslo
Matematisk Institutt
Oslo, Norway

Enrico Bombieri

School of Mathematics
The Institute for Advanced Study
Princeton, New Jersey

Dorian Goldfeld

Department of Mathematics
Columbia University
New York, New York

ACADEMIC PRESS, INC.
Harcourt Brace Jovanovich, Publishers
Boston San Diego New York
Berkeley London Sydney
Tokyo Toronto

ACADEMIC PRESS, INC.
1250 Sixth Avenue, San Diego, CA 92101

United Kingdom Edition published by
ACADEMIC PRESS INC. (LONDON) LTD.
24-28 Oval Road, London NW1 7DX

Library of Congress Cataloging-in-Publication Data
Number theory, trace formulas, and discrete groups: symposium in
honor of Atle Selberg, Oslo, Norway, July 14–21, 1987/edited by
Karl Egil Aubert, Enrico Bombieri, Dorian Goldfeld.
p. cm.
Bibliography: p.
ISBN 0-12-067570-6
1. Selberg trace formula—Congresses. 2. Numbers, Theory of—
Congresses. 3. Discrete groups—Congresses. 4. Selberg, Atle—
Congresses. I. Selberg, Atle. II. Aubert, Karl Egil.
III. Bombieri, Enrico, Date– . IV. Goldfeld, D.
QA241.N877 1988
512′.7—dc 19 88-18096
CIP

Printed in the United States of America
89 90 91 92 9 8 7 6 5 4 3 2 1

Contents

Contributors

The numbers in parentheses refer to the pages on which the authors' contributions begin.

James Arthur (11), *Department of Mathematics, University of Toronto, Toronto, Ontario M5S 1A1, Canada*

Enrico Bombieri (29), *School of Mathematics, The Institute for Advanced Study, Princeton, New Jersey 08540*

Daniel Bump (49), *Department of Mathematics, Stanford University, Stanford, California 94305*

J. B. Conrey (185), *Department of Mathematics, Oklahoma State University, Stillwater, Oklahoma 74078-0613*

Yuval Z. Flicker (201), *Department of Mathematics, Harvard University, Science Center, 1 Oxford Street, Cambridge, Massachusetts 02138*

John B. Friedlander (219), *Physical Sciences Division, Scarborough College, University of Toronto, Scarborough, Ontario M1C 1A4, Canada*

P. X. Gallagher (229), *Department of Mathematics, Columbia University, New York, New York 10027*

Paul B. Garrett (241), *School of Mathematics, University of Minnesota, Twin Cities, Vincent Hall, Church Street SE, Minneapolis, Minnesota 55455*

S. Gelbart (265), *Department of Theoretical Mathematics, The Weizmann Institute of Science, Rehovot 76100, Israel*

A. Ghosh (185), *Department of Mathematics, Oklahoma State University, Stillwater, Oklahoma 74078-0613*

Dorian Goldfeld (281), *Department of Mathematics, Columbia University, New York, New York 10027*

S. M. Gonek (185), *Department of Mathematics, University of Rochester, Rochester, New York 14627*

George Greaves (289), *Department of Pure Mathematics, University College, PO Box 78, Cardiff CF1 1XL, Wales, United Kingdom*

James Lee Hafner (309), *Department K53/801, IBM Research Center, 650 Harry Road, San Jose, California 95120-6099*

H. Halberstam (331), *Department of Mathematics, University of Illinois at Urbana-Champaign, 1409 West Green Street, Urbana, Illinois 61801*

Dennis A. Hejhal (343), *University of Minnesota, School of Mathematics, Minneapolis, Minnesota 55455*

H. Iwaniec (371), *Department of Mathematics, Rutgers University, New Brunswick, New Jersey 08903*

Hervé Jacquet (111), *Department of Mathematics, Columbia University, New York, New York 10027*

David A. Kazhdan (201), *Department of Mathematics, Harvard University, Science Center, 1 Oxford Street, Cambridge, Massachusetts 02138*

R. P. Langlands (125), *School of Mathematics, The Institute for Advanced Study, Princeton, New Jersey 08540*

G. A. Margulis (377), *Institute of Problems of Information Transmission, ul Ermolova 19, Moskva 103051, USSR*

Hugh L. Montgomery (157), *Department of Mathematics, The University of Michigan, Ann Arbor, Michigan 48109-1003*

G. D. Mostow (169), *Department of Mathematics, Yale University, Box 2155, Yale Station, New Haven, Connecticut 06520*

Julia Mueller (399), *Department of Mathematics, Fordham University, Bronx, New York 10458*

S. J. Patterson (409), *Mathematisches Institut der Georg-August-Universitat, Bunsenstrasse 3-5, D-3400 Gottingen, Federal Republic of Germany*

I. Piatetski-Shapiro (443), *School of Mathematical Sciences, Tel-Aviv University, Ramat-Aviv, 69978 Tel-Aviv, Israel*

Yao Qi (331), *Department of Mathematics, University of Illinois at Urbana-Champaign, 1409 West Green Street, Urbana, Illinois 61801*

H.-E. Richert (445), *Department of Mathematics, University of Illinois, Urbana, Illinois 61801*

P. Sarnak (457), *Department of Mathematics, Stanford University, Stanford, California 94305*

Atle Selberg (467), *School of Mathematics, The Institute for Advanced Study, Princeton, New Jersey 08540*

Lou Shituo (331), *Department of Mathematics, University of Illinois at Urbana-Champaign, 1409 West Green Street, Urbana, Illinois 61801*

Kai-man Tsang (485), *Department of Mathematics, University of Hong Kong, Pokfulam Road, Hong Kong*

R. C. Vaughan (503), *Department of Mathematics, Imperial College of Science and Technology, Huxley Building, Queen's Gate, London SW7 2BZ, England, United Kingdom*

Andre Weil (1), *School of Mathematics, The Institute for Advanced Study, Princeton, New Jersey 08540*

Participants in the Selberg Symposium, 1987

D. Andreoli
J. Arthur
K. E. Aubert
K. Barner
P. T. Bateman
E. Bombieri
R. Bruggeman
S. Böcherer
D. Bump
N. Baas
L. Carleson
J. W. S. Cassels
P. Cohen
B. Conrey
A. Deitmar
G. Dirdal
R. Dvornicich
J. Elstrodt
Y. Flicker
S. Friedberg
J. Friedlander
A. Fujii
P. X. Gallagher
S. Gelbart
D. Goldfeld
G. Greaves
L. Gårding
J. L. Hafner
J. Hakim
H. Halberstam
K. Hashimoto
D. Hejhal
A. Holme
C. Hooley
H. Iwaniec
H. Jacquet
E. Kallin
K. Kueh
N. V. Kuznetsov
M. B. Landstad

R. P. Langlands
R. Ledgard
O. Lehto
A. Lubotzky
L. Mantini
G. A. Margulis
H. Martens
O. Martinussen
G. Mostow
J. Mueller
R. Murty
A. Odlyzko
S. J. Patterson
A. Perelli
K. Peters
I. Piatetski-Shapiro
H. Pohlmann
M. S. Raghunathan
F. Recillas-Juarez
H. E. Richert
Ø. Rødseth
P. Sarnak
W. M. Schmidt
R. Schulze-Pillot
A. Selberg
S. Selberg
E. S. Selmer
V. Sos
E. Stade
A. Stadler
A. Stray
S. A. Strømme
T. Suzuki
J. Szmidt
R. Tchangang
K. Tsang
H. Tverberg
R. C. Vaughan
J. Vaughn
A. Voros
M. Walter
A. Weil
A. Winkler
D. Wright
U. Zannier
J. Aarnes

The Organizers of the Selberg Symposium
hereby express their sincere thanks to
those donors who have made this symposium possible.

The Mathematical Trust, Donors:
Landon Clay, The James Vaughn, Jr., Foundation Trust

The Norwegian Research Council for Science and the Humanities

The Royal Norwegian Council for Scientific and Industrial Research

The Norwegian Academy of Sciences and Letters

The Norwegian Mathematical Society

Preface

By a combined Norwegian–American initiative, some 100 mathematicians from different parts of the world gathered at the University of Oslo during the week of June 14–20, 1987, to celebrate the 70th birthday of Atle Selberg. The present volume contains the 30 invited lectures given at this symposium. From the content of these lectures, the reader will gather that the scientific program of this "Selberg symposium" revolved around the topics and themes to which Selberg has contributed most significantly over a period of 50 years: modular forms, the Riemann zeta function (as well as other zeta functions), analytic number theory, sieve methods, discrete groups, and trace formulae. The present volume certainly testifies to the great impact that Selberg's work has had on the present generation of mathematicians and on the development of a wide variety of topics. His many-sided achievements place him squarely as one of the truly great mathematicians of the 20th century. In fact, in one respect Selberg represents something quite out of the ordinary even among great mathematicians. During the first 15 years of his mathematical career, 1935–1950, Selberg certainly appeared more as a mathematician interested in the special than in the general. His first inspiration in this direction came from the reading of Ramanujan and thereafter from hard and unsolved problems in prime number theory, including the Riemann hypothesis. It is a very rare phenomenon that a mathematician with such a pronounced leaning towards the special, as exhibited in Selberg's early work, later in life appears as a theory builder by producing such a many-sided and far reaching result as the Selberg trace formula, a result in harmonic analysis with applications ranging from number theory to physics.

Some words of thanks are also highly appropriate. Without the generous support offered by The Mathematical Trust, represented by Mr. and Mrs. James Vaughn, Jr., this symposium could not have been carried through in the way it was. Important support was also given by the two major Norwegian research councils, as well as from the Norwegian Academy of Sciences and Letters. However, the interest and the success of the Selberg Symposium was above all achieved by the friendly presence of so many first-rate mathematicians.

—Karl Egil Aubert

Foreword

Number Theory, Trace Formulas and Discrete Groups: By the title alone, any mathematician will recognize a symposium dedicated to areas on which Atle Selberg has left an indelible imprint. However, the purpose of this symposium went well beyond that of honoring Selberg on the occasion of his 70th birthday; the purpose was also to look at the progress done in the fields opened by Selberg's pioneering work and to point out new directions for the future. Thus the reader will find essentially two distinct types of papers presented in this volume, namely survey lectures, sometimes more than one in the same field, and original contributions, including a remarkable one by Selberg himself.

Mathematicians are familiar with Selberg's contributions to sieve methods, analytic number theory, the Selberg trace formula and the associated harmonic analysis, the theory of discrete groups. They all belong to the history of mathematics, and they all have blossomed into full-fledged theories, in fact in some cases to points that even their creator could not foresee. There is, however, another face to Selberg's contributions to the mathematical world that goes beyond the published papers and hard theorems. In fact, Atle Selberg has also been a great teacher, and many of the participants in this symposium were, at one time or another, his students. Selberg's teaching has been done not in huge and crowded classrooms but in the quiet of his study in Princeton, where he freely shares his ideas and deep knowledge with generations of scholars.

We believe that the papers in this volume will present a vivid picture of areas of mathematics close to Selberg's heart, and we wish to extend our sincere thanks to all contributors and to the sponsors, without whose help this meeting could not have taken place.

—Enrico Bombieri

HISTORICAL INTRODUCTION

1 Prehistory of the Zeta-Function*

ANDRÉ WEIL

In substance at least, the fact that an *infinite* series can have a *finite* sum was known to the Greeks, at first as a philosophical paradox ("Achilles and the tortoise"), then, with Archimedes if not earlier, as a mathematical result used by him in his *Quadrature of the Parabola* (v. [1], pp. 312–315, prop. XXIV); there, in substance, he sums the infinite geometric progression with the ratio $\frac{1}{4}$. With equal ease, using *Eucl.* IX.35, he could have summed any progression of ratio < 1; the case of the ratio $\frac{1}{2}$ is already implicit in *Eucl.* X.1.

In 1644, obviously inspired by Archimedes' tract, Torricelli published in Florence his *De dimensione parabolae* ("the tritest of all trite subjects," he calls it), covering the same ground and much more ([2] pp. 89–162). There, not content with geometric progressions, he has ([2], pp. 149–150) a scholium stating, for any "finite or infinite" decreasing

* ***N.B.*** In this paper, the notation $\zeta(m)$, with m real, will be used as an abbreviation for the series (convergent or not) that, in modern notation, would be written as $\sum_{n=1}^{\infty} n^{-m}$ and that earlier writers (including Euler, but not yet Mengoli) wrote as

$$\frac{1}{1^m} + \frac{1}{2^m} + \frac{1}{3^m} + \frac{1}{4^m} + \text{etc.}$$

NUMBER THEORY,
TRACE FORMULAS and
DISCRETE GROUPS

ISBN 0-12-067570-6

sequence of positive numbers, the identity that we would write, in the case of a finite sequence $(a_0, a_1, \ldots, a_n)$, as

$$a_0 = \sum_{i=0}^{n-1} (a_i - a_{i+1}) + a_n;$$

here a_n is "the last number" of the sequence, with the understanding that, for an infinite geometric progression, this "last number" is "nothing or a point"; in our language, of course, one would say that a_n is to be replaced by the limit of the sequence. From this Torricelli derives an elegant proof for the sum of an infinite geometric progression of ratio < 1.

In 1650, quite possibly under the impression of that scholium, a young professor in Bologna, Pietro Mengoli, the successor of the great Cavalieri, published a book ([3]) entirely dedicated to the theory of infinite series. Its title, *Novae Quadraturae Arithmeticae Seu De Additione Fractionum*, seems to have been meant as a reference to Archimedes and Torricelli; actually no "quadrature" (i.e., no calculation of areas) occurs in the book. Even for its contemporaries, the almost complete lack of an algebraic notation must have made it hard reading (cf. [4]). With two exceptions (to be dealt with presently), it is exclusively concerned with series in which the summation can be carried out explicitly by elementary means, indeed as an application of Torricelli's identity. The first example, and a typical one, is that of the sum of the reciprocals of the "triangular numbers" $n(n + 1)/2$, i.e., of the series

$$\frac{1}{3} + \frac{1}{6} + \frac{1}{10} + \frac{1}{15} + \text{etc.}$$

Here, of course, the m-th term is

$$\frac{2}{(m + 1)(m + 2)} = \frac{2}{m + 1} - \frac{2}{m + 2},$$

so that the sum of the first m terms is $1 - 2/m + 2$, and the sum of the series is 1.

More important for our present purpose is the fact that Mengoli proves the divergence of the "harmonic series" $\zeta(1)$ and raises, also for the first time, the question of the summation of the series $\zeta(2)$. About the latter, he expresses (as well he might) his wonderment at the fact that the reciprocals of the "triangular numbers" can be summed, but not those of the "square numbers;" "this," he writes, "requires the help of a richer intellect [than mine]". For the divergence of the harmonic series, however, he brings a clever proof, based on the inequality

$$\frac{1}{a-1}+\frac{1}{a}+\frac{1}{a+1}>\frac{3}{a},$$

which can be expressed by the diagram

$$\underbrace{\frac{1}{1}}+\underbrace{\frac{1}{2}+\frac{1}{3}+\frac{1}{4}}_{>\frac{3}{3}=1}+\underbrace{\underbrace{\frac{1}{5}+\frac{1}{6}+\frac{1}{7}}_{>\frac{3}{6}=\frac{1}{2}}+\underbrace{\frac{1}{8}+\frac{1}{9}+\frac{1}{10}}_{>\frac{3}{9}=\frac{1}{3}}+\underbrace{\frac{1}{11}+\frac{1}{12}+\frac{1}{13}}_{>\frac{3}{12}=\frac{1}{4}}}_{>\frac{3}{3}=1,}+\text{ etc.}$$

showing that the series can be split up into infinitely many partial sums (with, respectively, 1, 3, 9, 27 ... terms), each of which is > 1.

Mengoli's *Novae Quadraturae* seem to have remained almost entirely unknown; the only contemporary reference to it appears in a communication from Collins, contained in a letter of 1673 from Oldenburg to Leibniz ([5a], Vol. I-1, p. 39; [5b], p. 85), where Collins quotes some of Mengoli's results and repeats Mengoli's query about the sum of $\zeta(2)$. In the meantime, however, some notable progress had been made. Firstly, in 1668, N. Kauffman, better known as Mercator, had effected "the quadrature of the hyperbola." ([6]) His method, which amounted to the termwise integration of the series

$$\frac{1}{1+x}=1-x+x^2-x^3+\cdots,$$

gave him for $\log(1+x)$ the power series

$$\log(1+x)=\frac{x}{1}-\frac{x^2}{2}+\frac{x^3}{3}-\cdots,$$

with $0<x<1$. Also in the same year, by means of a careful analysis, Brouncker established the validity of the above formula for $x=1$ ([7]), i.e.,

$$\log 2=1-\frac{1}{2}+\frac{1}{3}+\cdots$$

(a series closely related to $\zeta(1)$, as Jacob Bernoulli was later to observe). Then Newton, in his *De Analysi per Aequationes Infinitas* ([8]), had raised power-series expansions to the status of a universal tool for analysis. Soon Leibniz was to effect "the quadrature of the circle", meaning by this the formula

$$\frac{\pi}{4} = 1 - \frac{1}{3} + \frac{1}{5} - \cdots,$$

which he proved in substance by integrating the series

$$\frac{1}{1+x^2} = 1 - x^2 + x^4 - \cdots.$$

Leibniz's discovery did not appear in print until 1682, nor did Newton's until much later. In 1682 Leibniz published his formula for $\pi/4$, also mentioning some others of the above series, in an article ([5a], Vol. II-1, pp. 118–122) in one of the very first numbers of the *Acta Eruditorum*, newly started in Leipzig under his auspices. Next, in 1689, a mathematician in Basel, still little known at the time, Jacob Bernoulli, inaugurated, under the title *Positiones de Seriebus Infinitis*, a sequence of dissertations to be "defended" by his students at the university, which he eventually built up into a systematic treatise on infinite series. In this first instalment ([9], pp. 375–402) he rediscovered, among other results that actually had been known to Mengoli, the divergence of the series $\zeta(1)$, and, just as Mengoli, he expressed his puzzlement at the difficulty offered by $\zeta(2)$; "it is finite," he writes, "as appears from the comparison with another which obviously majorizes it," but "its summation is more difficult than one would have expected, and whoever will obtain it and communicate it to us will earn our deep gratitude."

So far, of all the series $\zeta(m)$, only $\zeta(1)$, $\zeta(2)$, and once, incidentally, $\zeta(3)$ had been mentioned. This changes with Jacob Bernoulli's second *Positiones* of 1692 ([9], pp. 517–542). As he had not yet at his disposal the exponential function a^x for all real x, he takes for m an arbitrary rational number (presumably a positive one). Clearly he knows that $\zeta(m)$ converges for $m \geq 2$ and diverges for $m \leq 1$, since it does so for $m = 2$ and $m = 1$. It is not clear whether he knows that it converges for $1 < m < 2$, but, anyway, his treatment at this point (*Pos.* XXIV, *Schol.*; [9] pp. 529–533) is purely formal. Splitting the series $\zeta(m)$ into the series

$$\phi(m) = \frac{1}{2^m} + \frac{1}{4^m} + \frac{1}{6^m} + \text{etc.}, \qquad \psi(m) = 1 + \frac{1}{3^m} + \frac{1}{5^m} + \text{etc.},$$

he obtains

$$\zeta(m) = \phi(m) + \psi(m), \qquad \phi(m) = 2^{-m}\zeta(m),$$

and therefore

$$\frac{\psi(m)}{\phi(m)} = 2^m - 1,$$

a result that he finds paradoxical for $m = \frac{1}{2}$, since in that case it gives $\psi(m) < \phi(m)$, while each term in the series $\psi(m)$ is greater than the corresponding term in $\phi(m)$. "It seems," he adds, "that a finite mind cannot comprehend the infinite"! Had he, instead of this naive remark, rewritten his result in the form

$$\zeta(m) = (1 - 2^{-m})^{-1}\psi(m),$$

he would have taken the first step towards the expression of $\zeta(m)$ as an "Eulerian" product.

No mention has yet been made of the numerical evaluation of the series $\zeta(m)$, a serious problem in view of their slow convergence. One can hardly regard the lengthy but inconclusive discussions, in the correspondence between Leibniz and Jacob Bernoulli ([5a], Vol. II-3, pp. 25–27, 32–34, 44–45, 49) on the evaluation of the partial sums of $\zeta(1)$ as a contribution to that subject. The calculation of $\zeta(2)$ was undertaken by Daniel Bernoulli in 1728 and by Goldbach in 1729 ([10], Vol. II, pp. 263 and 281–282), with preliminary results, soon to be improved upon by Euler.

This seems to have given Euler his first opportunity for a contact with the zeta-function. As with most of the questions that ever attracted his attention, he never abandoned it, soon making a number of fundamental contributions to it (cf. [11], pp. 257–276). First, he discovered the so-called Euler–MacLaurin formula, which enabled him to calculate, with a high degree of approximation, the sums $\zeta(m)$ for $m \geq 2$ and the partial sums of $\zeta(1)$ ([12], t. 14, pp. 119–122). This, however, threw no light on the theory of the zeta-function, except inasmuch as it introduced for the first time the Bernoulli numbers into the subject, but merely as coefficients in the Euler–MacLaurin formula.

Next came, in 1735, the sensational discovery of the formula

$$\zeta(2) = \frac{\pi^2}{6},$$

based on a bold application of the theory of algebraic equations to the transcendental equation $0 = 1 - \sin x$ ([12], t. 14, pp. 73–74). This was soon followed by the calculation of $\zeta(m)$ for $m = 4, 6$, etc. by the same method, and by a number of contributions to the theory of trigonometric functions that eventually legitimized the results on $\zeta(2)$, $\zeta(4)$, etc. Another paper, in 1737 ([12], t. 14, pp. 216–244), established the "Eulerian product" for $\zeta(m)$ and various allied series, including

$$L(m) = 1 - \frac{1}{3^m} + \frac{1}{5^m} - \frac{1}{7^m} + \text{etc.}$$

He devoted a whole chapter (Chapter XV) of his great *Introductio in Analysin Infinitorum* of 1745 ([12], t. 8) to this same subject. From this he derived, or thought he could derive, the infinite product for $L(1)$ ([12], t. 14, p. 233)

$$L(1) = \prod \frac{p}{p \pm 1},$$

where the product is taken over all odd primes arranged in increasing order and the sign is given by $p \pm 1 \equiv 0 \pmod 4$.

Equally momentous in its ultimate consequences, but no less neglected for over a century, was Euler's discovery of functional equations for the series in question. This began in 1739 ([12], t. 14, p. 443) with the relation

$$1 - 2^m + 3^m - 4^m + \text{etc.} = \frac{\pm 2.1.2.3 \cdots m}{\pi^{m+1}} \left(1 + \frac{1}{3^{m+1}} + \frac{1}{5^{m+1}} + \text{etc.}\right)$$

for $m = 1, 3, 5, 7$. The left-hand side is to be understood here according to Euler's views on divergent series, i.e., in this case by what is now known as the Abel summation. Clearly this is equivalent to the functional equation for $\zeta(s)$ when $s = 2, 4, 6, 8$, or, rather, as Euler suggests at once, for all positive even values of s. In 1749, under the title "Remarques sur un beau rapport entre les séries de puissances tant directes que réciproques" ([12], t. 15, pp. 70–90), Euler not only gave a tentative proof for the above formula, but wrote down conjecturally the correct functional equation for the zeta-function for arbitrary values of the argument (or, rather, what amounts to the same, for the closely related series

$$(1 - 2^{1-n})\zeta(n) = 1 - \frac{1}{2^n} + \frac{1}{3^n} - \frac{1}{4^n} + \text{etc.})$$

and also for the series $L(n)$ given as above, adding that the latter was just as certain as the former and might be easier to prove, "thus throwing much light on a number of similar investigations."

Euler's last paper on the subject, based on investigations carried out in 1752, was written in 1775 and published posthumously in 1785 ([12], t. 4, pp. 146–153). It deals with the series $\sum \pm 1/p$, where the sum is taken over all odd primes (in increasing order), and the sign, just as in the infinite product for $L(1)$, is given by $p \pm 1 \equiv 0 \pmod 4$. Here he takes as his starting point his infinite products for $\zeta(2)$ and $L(1)$, whose values $\zeta(2) = \pi^2/6$ and $L(1) = \pi/4$ are known to him, giving the formula

$$\frac{3\zeta(2)}{4L(1)^2} = 2 = \prod \frac{p \pm 1}{p \mp 1},$$

and therefore

$$\frac{1}{2} \log 2 = 0{,}3465735902 \cdots = \sum \pm \frac{1}{p} + \frac{1}{3} \sum \pm \frac{1}{p^3} + \frac{1}{5} \sum \pm \frac{1}{p^5} + \text{etc.}$$

In the right-hand side, the first series is the one to be calculated; all others are absolutely (but slowly) convergent. Euler evaluates them numerically by comparison with the known values of $L(3)$, $L(5)$, etc., and obtains finally

$$\sum \pm \frac{1}{p} = 0{,}3349816 \cdots$$

(whereas, in 1752, he had only obtained for this the value $0{,}334980 \cdots$). In view of the fact, also a consequence of his earlier results, that $\sum 1/p$ is infinite, this enables him to conclude that there are infinitely many primes of the form $4n + 1$, and infinitely many of the form $4n - 1$, thus preluding to Dirichlet's famous articles of 1837 on the arithmetic progression ([13], Bd. I, pp. 307–343).

Dirichlet's papers, as outstanding for mathematical rigor as Euler had been careless about such matters, are too well known to require detailed comments here. It will be enough to point out that, based as they are on Euler's *Introductio* of 1745 and on Gauss' treatment of the multiplicative group of integers prime to N modulo N for any N, they introduce for the first time the "Dirichlet series" $L_\chi(s)$ attached to the characters of such groups, but only for real $s > 1$ when they are absolutely convergent; in fact, they deal exclusively with the properties of such series for s tending to 1. It was still a far cry from this to their analytic continuation and their functional equation.

That last step was taken by Riemann in 1859, at least for the zeta-function itself; but it had been preceded by three proofs for the functional equation for the series

$$L(s) = 1 - \frac{1}{3^s} + \frac{1}{5^s} - \frac{1}{7^s} + \text{etc.}$$

in the range $0 < s < 1$, where it is conditionally convergent. The Swedish mathematician Malmstén included his proof (mentioning that a similar proof could be given for the function

$$(1 - 2^{1-s})\zeta(s) = 1 - \frac{1}{2^s} + \frac{1}{3^s} - \frac{1}{4^s} + \text{etc.}$$

and recalling that both results had been announced by Euler) in a long paper ([14]) written in May 1846 and published in *Crelle's Journal* in 1849. Schlömilch, obviously unaware of Malmstén's paper and apparently also of Euler's priority, announced the same result as an exercise in *Grunert's Archiv*, also in 1849 ([15]); Clausen published a solution to this "exercise" in the same *Archiv* in 1858 ([16]), and Schlömilch then published his own proof ([17]).

It has sometimes been suggested that these publications gave Riemann the incentive for his research on the zeta-function; but an even more likely source has recently turned up ([18]) in the form of Eisenstein's own copy of Gauss's *Disquisitiones* in the French translation by Poullet-Delisle (Paris, 1807). On the last blank page of that volume, Eisenstein jotted down still another proof for the functional equation of $L(s)$; essentially it consists of a straightforward application, without any remark or reference to justify it, of Poisson's summation to the series in question, coupled with the formula

$$\int_0^\infty e^{\sigma\psi i}\psi^{q-1}\, d\psi = \frac{\Gamma(q)}{\sigma^q}\, e^{q\pi i/2} \qquad (0 < q < 1,\ \sigma > 0),$$

for which Eisenstein quotes Dirichlet ([13], Bd. I, p. 401). It gives the Fourier transform of the function equal to 0 for $x < 0$ and to x^{q-1} for $x > 0$.

Eisenstein's proof is dated with the words "Scripsi 7 April 1849." As he does not claim the result as his own, he could conceivably have gotten it either from Malmstén or from Schlömilch. Of particular interest, however, is the fact that April 1849 is precisely the time when Riemann finally left Berlin for Göttingen. He and Eisenstein had been

intimate friends. Thus it is not only possible, but even quite likely, that Eisenstein had discussed his proof of 1849 with Riemann before the latter's departure. If so, this might have been the origin of Riemann's paper of 1859.

References

[1] Archimedes. *Opera Omnia* ..., Vol. II. J. L. Heiberg, ed., Lipsiae, (1913).

[2] Torricelli. *Opere di Evangelista Torricelli*, Vol. I, Part 1. G. Loria and G. Vassura, eds., Faenza (1919).

[3] Mengoli, Pietro. *Novae Qudraturae Arithmeticae seu De Additione Fractionum*, Bononiae (1650).

[4] Eneström, G. "Zur Geschichte der unendlichen Reihen um die Mitte des siebzehnten Jahrhunderts," *Bibl. Math.* (III)12:135–148 (1911–12).

[5a] Gerhardt, C. I., ed. *Leibnizens mathematische Schriften*. 6 vols., Halle (1849–63).

[5b] Gerhardt, C. I., ed. *Der Briefwechsel von Gottfried Wilhelm Leibnitz*. Berlin (1899); G. Olm, ed., Hildesheim (1962).

[6] Mercator, Auctore Nicolao Mercatore. *Logarithmotechnia ··· accedit vera Quadratura Hyperbolae* ..., Londini (MDCLXVIII).

[7] Brouncker, Lord Viscount. "The Squaring of the Hyperbola by an infinite series of rational Numbers ··· by that eminent Mathematician the right Honourable the Lord Viscount Brouncker," *Phil. Trans.* (1668).

[8] Whiteside, D. T., ed. *The Mathematical Papers of Isaac Newton*, Vol. II, Cambridge (1968).

[9] Bernoulli, Jacob. *Opera*, Tomus Primus, Genevae (1744).

[10] Fuss, P.-H. *Correspondance mathématique et physique* ..., Vol. II, St. Petersbourg (1843), Johnson Reprint Corp. (1968).

[11] Weil, A. *Number Theory: An Approach through History*, Birkhäuser Boston (1983).

[12] Euler, Leonhard. *Opera Omnia, Series Prima*, 27 vols., Leipzig, Zurich (1911–56).

[13] Dirichlet. *G. Lejeune Dirichlet's Werke*, Bd. I, Berlin (1889).

[14] Malmstén, C. J. "De integralibus quibusdam definitis seriebusque infinitis," *J. fur reine u. ang. Math.* (Crelle's Journal), **38**:1–39 (1849).

[15] Schlömilch, O. "Uebungsaufgaben für Schüler, Lehrsatz von dem Herm Prof. Dr. Schlömilch," *Archiv der Math. u. Phys.* (Grunert's Archiv), 12:415 (1849).

[16] Clausen, T. "Beweis des von Schlömilch ... aufgestellten Lehrsatzes," *Archiv der Math. u. Phys.* (Grunert's Archiv), **30**:166–169 (1858).

[17] Schlömilch, O. "Ueber eine Eigenschaft gewisser Reihen," *Zeitschr. fur Math. u. Phys.* **3**:130–132 (1858).

[18] Weil, A. *On Eisenstein's copy of the Disquisitiones*, to appear.

SURVEY LECTURES ON SELBERG'S WORK

2 The Trace Formula and Hecke Operators

JAMES ARTHUR

This lecture is intended as a general introduction to the trace formula. We shall describe a formula that is a natural generalization of the Selberg trace formula for compact quotient. Selberg also established trace formulas for noncompact quotients of rank 1, and our formula can be regarded as an analogue for general rank of these. As an application, we shall look at the "finite case" of the trace formula. We shall describe a finite closed formula for the traces of Hecke operators on certain eingenspaces.

A short introduction of this nature will by necessity be rather superficial. The details of the trace formula are in [1(e)] (and the references there), while the formula for the traces of Hecke operators is proved in [1(f)]. There are also other survey articles [1(c)], [1(d)], [5], and [1(g)], where some of the topics in this paper are discussed in more detail and others are treated from a different point of view.

1

Suppose that G is a locally compact group that is unimodular and that Γ is a discrete subgroup of G. There is a right G-invariant measure on the coset space $\Gamma\backslash G$ that is uniquely determined up to a constant

NUMBER THEORY,
TRACE FORMULAS and
DISCRETE GROUPS

ISBN 0-12-067570-6

multiple. We can therefore take the Hilbert space $L^2(\Gamma\backslash G)$ of square integrable functions on $\Gamma\backslash G$. Define

$$(R(y)\phi)(x) = \phi(xy), \qquad \phi \in L^2(\Gamma\backslash G),\ x, y \in G.$$

Then R is a unitary representation of G on $L^2(\Gamma\backslash G)$.

One would like to obtain information on the decomposition of R into irreducible representations. Selberg approached the problem by studying the operators

$$R(f) = \int_G f(y)R(y)\,dy, \qquad f \in C_c(G),$$

on $L^2(\Gamma\backslash G)$. If ϕ belongs to $L^2(\Gamma\backslash G)$, one can write

$$\begin{aligned}(R(f)\phi)(x) &= \int_G f(y)\phi(xy)\,dy \\ &= \int_G f(x^{-1}y)\phi(y)\,dy \\ &= \int_{\Gamma\backslash G} \sum_{\gamma\in\Gamma} f(x^{-1}\gamma y)\phi(\gamma y)\,dy \\ &= \int_{\Gamma\backslash G} \Big(\sum_{\gamma\in\Gamma} f(x^{-1}\gamma y)\Big)\phi(y)\,dy.\end{aligned}$$

Therefore, $R(f)$ is an integral operator with kernel

$$K(x, y) = \sum_{\gamma\in\Gamma} f(x^{-1}\gamma y). \tag{1.1}$$

Suppose first that $\Gamma\backslash G$ is compact. Then under some mild restriction on f, the operator $R(f)$ is of trace class, and its trace is equal to the integral of the kernel on the diagonal. This is so, for example, if G is a Lie group and f is smooth. One can then write

$$\begin{aligned}\operatorname{tr} R(f) &= \int_{\Gamma\backslash G} \sum_{\gamma\in\Gamma} f(x^{-1}\gamma x)\,dx \\ &= \int_{\Gamma\backslash G} \sum_{\gamma\in\{\Gamma\}} \sum_{\delta\in\Gamma_\gamma\backslash\Gamma} f(x^{-1}\delta^{-1}\gamma\delta x)\,dx \\ &= \sum_{\gamma\in\{\Gamma\}} \int_{\Gamma_\gamma\backslash G} f(x^{-1}\gamma x)\,dx \\ &= \sum_{\gamma\in\{\Gamma\}} \int_{G_\gamma\backslash G} \int_{\Gamma_\gamma\backslash G_\gamma} f(x^{-1}u^{-1}\gamma ux)\,du\,dx \\ &= \sum_{\gamma\in\{\Gamma\}} a_\Gamma^G(\gamma) \int_{G_\gamma\backslash G} f(x^{-1}\gamma x)\,dx,\end{aligned}$$

where

$$a_\Gamma^G(\gamma) = \text{volume}(\Gamma_\gamma \backslash G_\gamma).$$

Here, $\{\Gamma\}$ is a set of representatives of conjugacy classes in Γ, and Γ_γ and G_γ denote the centralizers of γ in Γ and G. Implicit in the discussion is the absolute convergence of the various sums and integrals. Now it can also be seen that R decomposes into a direct sum of irreducible unitary representations with finite multiplicities. It follows that

$$\text{tr } R(f) = \sum_{\pi \in \Pi(G)} a_\Gamma^G(\pi) \text{ tr } \pi(f),$$

where $\Pi(G)$ is a set of equivalence classes of irreducible representations of G, and $a_\Gamma^G(\pi)$ is a positive integer. We can therefore write

$$\sum_{\gamma \in \{\Gamma\}} a_\Gamma^G(\gamma) I_G(\gamma, f) = \sum_{\pi \in \Pi(G)} a_\Gamma^G(\pi) I_G(\pi, f), \tag{1.2}$$

where

$$I_G(\gamma, f) = \int_{G_\gamma \backslash G} f(x^{-1}\gamma x)\, dx$$

and

$$I_G(\pi, f) = \text{tr } \pi(f).$$

This is the Selberg trace formula for compact quotient, introduced in [6(a)]. (Selberg's original formula actually took a slightly different form. The present form is due to Tamagawa [9].)

Example 1. Suppose that $G = \mathbb{R}$ and $\Gamma = \mathbb{Z}$. Then (1.2) becomes

$$\sum_{n \in \mathbb{Z}} f(n) = \sum_{n \in \mathbb{Z}} \hat{f}(2\pi n), \qquad f \in C_c^\infty(\mathbb{R}),$$

the Poisson summation formula.

Example 2. Suppose that G is a finite group and that $\Gamma \subset G$ is an arbitrary subgroup. Let π be an irreducible unitary representation of G and set

$$f(x) = \text{tr } \pi(x^{-1}).$$

Writing the left-hand side of the trace formula as

$$\sum_{x \in \Gamma \backslash G} \sum_{\gamma \in \Gamma} \text{tr } \pi((x^{-1}\gamma x)^{-1}) = |G|\,|\Gamma|^{-1} \sum_{\gamma \in \Gamma} \text{tr } \pi(\gamma^{-1}),$$

and applying the theory of characters, we find that $a_\Gamma^G(\pi)$ equals the multiplicity of the trivial representation in the restriction of π to Γ. But $a_\Gamma^G(\pi)$ is, by definition, the multiplicity of π in the representation of G induced from the trivial representation of Γ. The equality of these two multiplicities is just Frobenius reciprocity for finite groups. Frobenius reciprocity applies more generally to an arbitrary irreducible unitary representation of Γ, but so in fact does the Selberg trace formula. The arguments above apply equally well to spaces of square integrable, Γ-equivariant sections on G. The Selberg trace formula is therefore a generalization of Frobenius reciprocity.

We have chosen the notation $I_G(\gamma, f)$ and $I_G(\pi, f)$ in (1.2) to emphasize that as distributions in f, these functions are invariant. They remain unchanged if f is replaced by a conjugate

$$f^y(x) = f(yxy^{-1}).$$

The importance of such distributions is that they are completely determined from only partial information on f. One could expect to be able to evaluate any invariant distribution only knowing the orbital integrals of f on the conjugacy classes of G, or alternatively, the values of the characters at f of the irreducible unitary representation of G.

Consider, for example, the special case that $G = \mathrm{SL}(2, \mathbb{R})$. Assume that f is smooth and bi-invariant under the maximal compact subgroup $\mathrm{SO}(2, \mathbb{R})$. This was the case Selberg treated in greatest detail. The value at f of any invariant distribution depends only on the symmetric function

$$g(u) = |e^{u/2} - e^{-u/2}| \int_{A\backslash G} f\left(x^{-1}\begin{bmatrix} e^{u/2} & 0 \\ 0 & e^{-u/2}\end{bmatrix}x\right) dx, \qquad u \in \mathbb{R},\ u \neq 0,$$

where A denotes the subgroup of diagonal matrices in $\mathrm{SL}(2, \mathbb{R})$. It could equally well be expressed in terms of the function

$$h(r) = \int_{-\infty}^{\infty} e^{iru} g(u)\, du = \mathrm{tr}\ \pi_r(f), \qquad r \in \mathbb{R},$$

in which $\{\pi_r\}$ is the principal series of induced representations. Written in terms of g and h, (1.2) becomes the more concrete formula given on page 74 of [8(a)]. Selberg noticed a remarkable similarity between this formula and the "explicit formulas" of algebraic number theory. The analogue of the numbers

$$\{\log (p^n) : p \text{ prime},\ n \geq 1\}$$

is the set of lengths of closed geodesics on the Riemann surface

$$X_\Gamma = \Gamma\backslash \mathrm{SL}(2, \mathbb{R})/\mathrm{SO}(2, \mathbb{R}),$$

while the role of the zeros of the Riemann zeta-function is played by the points

$$\left\{\frac{1}{2} + \sqrt{-1}\, r : \pi_r \in \Pi(G)\right\}.$$

By choosing g suitably, Selberg obtained a sharp asymptotic estimate for the number of closed geodesics of length less than a given number. By varying h instead, he established an asymptotic formula for the distribution of the eigenvalues of the Laplacian of X_Γ.

Now, suppose only that $\Gamma\backslash G$ has finite invariant volume. For example, Γ could be a congruence subgroup

$$\{\gamma \in \mathrm{SL}(2, \mathbb{Z}) : \gamma \equiv 1 (\mathrm{mod}\ N)\}$$

of $\mathrm{SL}(2, \mathbb{R})$. Then everything becomes much more difficult. For $G = \mathrm{SL}(2, \mathbb{R})$, Selberg derived a trace formula in detail. Among other things, he gave a finite closed formula for the trace of the Hecke operators on the space of modular forms of weight $2k$, for $k > 1$. (See [8(a), p. 85].) In the later paper [8(b)], Selberg outlined an argument for establishing a trace formula for noncompact quotient when G has real rank 1. He also emphasized the importance of establishing such a result in general.

In this lecture we shall describe a general trace formula. It will be valid if G is an arbitrary reductive Lie group, and Γ is any arithmetic subgroup that is defined by congruence conditions.

2

In dealing with congruence subgroups, it is most efficient to work over the adèles. Therefore, we change notation slightly and take G to be a reductive algebraic group over $\mathbb{Q}$. The adèle ring

$$\mathbb{A} = \mathbb{R} \times \mathbb{A}_0 = \mathbb{R} \times \mathbb{Q}_2 \times \mathbb{Q}_3 \times \mathbb{Q}_5 \times \cdots$$

is locally compact ring that contains $\mathbb{Q}$ diagonally as a discrete subring. Moreover, $G(\mathbb{A})$ is a locally compact group that contains $G(\mathbb{Q})$ as a discrete subgroup. At first glance, $G(\mathbb{Q})\backslash G(\mathbb{A})$ might seem an ungainly substitute for the quotient of $G(\mathbb{R})$ by a congruence subgroup. However, the study of the two are equivalent. We shall assume for simplicity that G is semisimple and simply connected, and that $G(\mathbb{R})$

has no compact factors. Let K_0 be an open compact subgroup of the group $G(\mathbb{A}_0)$ of finite adèles. The strong approximation theorem asserts that

$$G(\mathbb{A}) = G(\mathbb{Q})K_0 G(\mathbb{R}). \tag{2.1}$$

If follows that there is a $G(\mathbb{R})$-equivariant homeomorphism

$$\Gamma \backslash G(\mathbb{R}) \xrightarrow{\sim} G(\mathbb{Q}) \backslash G(\mathbb{A})/K_0,$$

where

$$\Gamma = G(\mathbb{Q})K_0 \cap G(\mathbb{R})$$

is a congruence subgroup of $G(\mathbb{R})$. Conversely, any congruence subgroup can be obtained in this way. Thus, instead of working with $L^2(\Gamma \backslash G(\mathbb{R}))$, one can work with the $G(\mathbb{R})$-invariant subspace of functions in $L^2(G(\mathbb{Q}) \backslash G(\mathbb{A})$ that are right invariant under K_0. The advantage of the adèlic picture is that the conjugacy classes $G(\mathbb{Q})$ are much easier to deal with than those of Γ.

Suppose that f is a function in $C_c^\infty(G(\mathbb{A}))$. This means that f is a finite linear combination of functions $f_{\mathbb{R}} f_0$, where $f_{\mathbb{R}}$ belongs to $C_c^\infty(G(\mathbb{R}))$ and f_0 is a locally constant function of compact support on $G(\mathbb{A}_0)$. For example, if one is interested in the action of a function $f_{\mathbb{R}}$ on $L^2(\Gamma \backslash G(\mathbb{R}))$, one could take f to be the product of $f_{\mathbb{R}}$ with the unit function 1_{K_0}. (By definition, 1_{K_0} is the characteristic function of K_0 divided by the volume of K_0). Consider the values

$$K(x, x) = \sum_{\gamma \in G(\mathbb{Q})} f(x^{-1}\gamma x), \qquad x \in G(\mathbb{Q}) \backslash G(\mathbb{A}),$$

of the kernel on the diagonal. Since $G(\mathbb{Q}) \backslash G(\mathbb{A})$ is not in general compact, this function is not generally integrable. What causes the integral to diverge? Experiments with examples suggest that the contribution to the integral of a conjugacy class in $G(\mathbb{Q})$ diverges when the conjugacy class intersects a proper parabolic subgroup defined over $\mathbb{Q}$. The more parabolics it meets, the worse will generally be the divergence of the integral. It turns out that by adding a correction term for each standard parabolic subgroup of G, one can truncate $K(x, x)$ in a uniform way so that its integral converges. Let us briefly describe this process in the special case that $G = \mathrm{SL}(n)$. (For a fuller illustration of the case of $\mathrm{SL}(n)$, see the survey [1(c)].)

The standard parabolic subgroups P of SL(n) are parametrized by partitions $(n_1, \ldots, n_k)$ of n. To each such partition corresponds subgroups

$$P = \left\{ \begin{pmatrix} \boxed{n_1} & * & \cdots & * \\ & \boxed{n_2} & & \vdots \\ & & \ddots & \vdots \\ 0 & & & \boxed{n_k} \end{pmatrix} \right\},$$

$$M = \left\{ \begin{pmatrix} \boxed{n_1} & & & 0 \\ & \boxed{n_2} & & \\ & & \ddots & \\ 0 & & & \boxed{*} \end{pmatrix} \right\},$$

$$N = \left\{ \begin{pmatrix} I & \cdots & \cdots & * \\ & I & & \vdots \\ & & \ddots & \vdots \\ 0 & & & I \end{pmatrix} \right\},$$

and

$$A_M = A_P = \left\{ \begin{pmatrix} t_1 I & & & 0 \\ & t_2 I & & \\ & & \ddots & \\ 0 & & & t_k I \end{pmatrix} \right\}$$

of SL(n). One has the decomposition

$$\mathrm{SL}(n, \mathbb{A}) = P(\mathbb{A})K = N(\mathbb{A})M(\mathbb{A})K,$$

where

$$K = \mathrm{SO}_n(\mathbb{R}) \times \left(\prod_p \mathrm{SL}_n(\mathbb{Z}_p) \right)$$

is the maximal compact subgroup. For any point

$$x = n \begin{pmatrix} m_1 & & \\ & \ddots & \\ & & m_k \end{pmatrix} k, \qquad n \in N(\mathbb{A}),\ m_i \in GL(n_i, \mathbb{A}),\ k \in K,$$

in SL(n, $\mathbb{A}$), set

$$H_P(x) = (\log |\det m_1|, \ldots, \log |\det m_k|).$$

Then $H_P(x)$ takes values in the real vector space

$$\mathbf{a}_M = \mathbf{a}_P = \{(u_1, \ldots, u_k): \sum u_i = 0\}.$$

The truncation of $K(x, x)$ will depend on a point

$$T = (t_1, \ldots, t_n), \qquad t_i \in \mathbb{R}, \sum t_i = 0,$$

which is suitably regular, in the sense that t_i is much larger than t_{i+1} for every i. For any P, write

$$T_P = (t_1 + \cdots + t_{n_1}, t_{n_1+1} + \cdots + t_{n_2}, \ldots)$$

for the corresponding point in $\mathbf{a}_P$, and let $\hat{\tau}_P$ be the characteristic function of

$$\{(u_1, \ldots, u_k) \in \mathbf{a}_P : u_1 + \cdots + u_i > u_{i+1} + \cdots + u_k, \qquad 1 \leq i \leq k\}.$$

Example 3. Suppose that $G = \mathrm{SL}(3)$ and that P corresponds to the minimal partition (1, 1, 1). Then $\mathbf{a}_P$ is a two-dimensional space with six chambers. The function

$$H \to \hat{\tau}_P(H - T_P), \qquad H \in \mathbf{a}_P,$$

is the characteristic function of the convex shaded dual chamber.

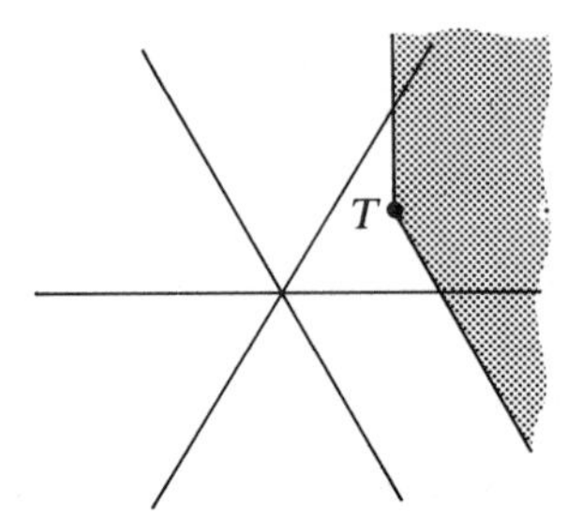

The notion of a standard parabolic subgroup exists, of course, for arbitrary G, as do the other objects described for $\mathrm{SL}(n)$. For any P, we have a kernel

$$K_P(x, y) = \int_{N(\mathbb{A})} \sum_{\gamma \in M(\mathbb{Q})} f(x^{-1}\gamma n y)\, dn,$$

for the right convolution operator of f on $L^2(N(\mathbb{A})M(\mathbb{Q})\backslash G(\mathbb{A}))$. We can now write down the truncated kernel. It is an expression

$$k^T(x, f) = \sum_P (-1)^{\dim A_P} \sum_{\delta \in P(\mathbb{Q})\backslash G(\mathbb{Q})} K_P(\delta x, \delta x)\hat{\tau}_P(H_P(\delta x) - T_P), \quad (2.2)$$

defined by a finite alternating sum over standard parabolic subgroups P of G. The inner sum over δ may also be taken over a finite set, but this depends upon x. The purpose of this inner sum is to make the corresponding term left $G(\mathbb{Q})$-invariant. Observe that the term corresponding to $P = G$ is just $K(x, x)$. A term corresponding to $P \neq G$ can be regarded as a function that is supported on a neighborhood of infinity in $G(\mathbb{Q})\backslash G(\mathbb{A})$.

Theorem 2.1.
(a) *The function $k^T(x, f)$ is integrable over $G(\mathbb{Q})\backslash G(\mathbb{A})$.*
(b) *The function*

$$J^T(f) = \int_{G(\mathbb{Q})\backslash G(\mathbb{A})} k^T(x, f)\, dx$$

is a polynomial, for values of T that are suitably regular.

(See [1(a), Theorem 7.1] and [1(b), Proposition 2.3].)

As a polynomial, $J^T(f)$ can be extended to all values of T, even though it is defined as a convergent integral only for T in some chamber. Set

$$J(f) = J^0(f).$$

Then J is a distribution on $C_c^\infty(G(\mathbb{A}))$. An obvious question is how to evaluate it more explicitly.

3

Theorem 2.1 is just the first of a number of steps. We have described it in order to give some flavor of what is involved. The remaining steps are more elaborate, and we shall discuss them in only the most cursory manner.

Theorem 2.1 provides only a definition of a distribution $J(f)$. There is not yet any trace formula. For this, one needs to look at the representation theoretic expansion of $K(x, x)$. Since $G(\mathbb{Q})\backslash G(\mathbb{A})$ is generally noncompact, R is no longer a direct sum of irreducible representations. Rather, we have

$$R = R_{\mathrm{disc}} \oplus R_{\mathrm{cont}},$$

where R_{disc} is a direct sum of irreducible representations, and R_{cont} decomposes continuously. The decomposition of R_{cont} can be described in terms of the constituents of the analogues of R_{disc} for Levi components of standard parabolic subgroups. This description is part of the theory of Eisenstein series initiated by Selberg [8(a)] and established for general groups by Langlands [6(b)]. We shall say here only that the theory of Eisenstein series provides a second formula for the kernel. One obtains

$$K(x, y) = \sum_{P_1} \sum_{\phi} \int_{i\mathfrak{a}^*_{P_1}} E(x, \boldsymbol{I}_{P_1}(\lambda, f)\phi, \lambda)\overline{E(y, \phi, \lambda)}\, d\lambda, \tag{3.1}$$

where P_1 is summed over standard parabolic subgroups, ϕ is summed over an orthonormal basis of $L^2_{\mathrm{disc}}(P_1(\mathbb{Q})A_{P_1}(\mathbb{R})^0N_{P_1}(\mathbb{A})\backslash G(\mathbb{A}))$ (the subspace that decomposes discretely under the action of $G(\mathbb{A})$), $\boldsymbol{I}_{P_1}(\lambda)$ is an induced representation, and $E(\cdot, \cdot, \cdot)$ is the Eisenstein series (or, rather, its analytic continuation to imaginary λ). Notice that the term corresponding to $P_1 = G$ is just the kernel of the operator $R_{\mathrm{disc}}(f)$. More generally, for any P, Eisenstein series give a second formula for the kernel $K_P(x, y)$. One has only to restrict the sum in (3.1) to those P_1 that are contained in P and to take partial Eisenstein series from P_1 to P. Substituted into (2.2), these formulas provide a second expression for $k^T(x, f)$.

Thus, we have two distinct expressions for the integrable function $k^T(x, f)$. One is a geometric expansion related to conjugacy classes, which originates with the formula (1.1), and the other is a spectral expansion related to representation theory, which originates with the formula (3.1). We therefore obtain two expressions for the integral

$$J^T(f) = \int_{G(\mathbb{Q})\backslash G(\mathbb{A})} k^T(x, f)\, dx.$$

At this stage, the two expressions are too abstract to be of much value. Nevertheless, it turns out that each can be rewritten in a rather explicit form. This is by far the most difficult part of the process. In the end, however, one obtains two different formulas for the polynomial $J^T(f)$. In particular, by specializing T one obtains two different formulas for the distribution $J(f)$. One is a geometric expansion

$$J(f) = \sum_{M} |\mathrm{ch}(\mathbf{a}_M)|^{-1} \sum_{\gamma \in \{M(\mathbb{Q})\}} a^M(\gamma)J_M(\gamma, f), \tag{3.2}$$

and the other is a spectral expansion

$$J(f) = \sum_{M} |\mathrm{ch}(\mathbf{a}_M)|^{-1} \int_{\Pi(M)} a^M(\pi)J_M(\pi, f)\, d\pi. \tag{3.3}$$

The trace formula can be regarded as the equality of the two.

We shall have to be content to say only a few words about the terms in (3.2) and (3.3). In each case, M is summed over Levi components of standard parabolic subgroups P of G, and $|\mathrm{ch}(\mathbf{a}_M)|$ denotes the number of chambers in the vector space $\mathbf{a}_M$. In (3.2), $\{M(\mathbb{Q})\}$ stands for the conjugacy classes in $M(\mathbb{Q})$. In (3.3), $\Pi(M)$ is a set of irreducible unitary representations of the group

$$M(\mathbb{A})^1 = \{x \in M(\mathbb{A}) : H_P(x) = 0\},$$

equipped with a certain measure $d\pi$. The functions $a^M(\gamma)$ and $a^M(\pi)$ are global in nature and depend only on the subgroup M. If $M(\mathbb{Q})\backslash M(\mathbb{A})^1$ is compact, they are equal to the coefficients $a^{M(\mathbb{A})^1}_{M(\mathbb{Q})}(\gamma)$ and $a^{M(\mathbb{A})^1}_{M(\mathbb{Q})}(\pi)$ that occur in (1.2). The distributions $J_M(\gamma, f)$ and $J_M(\pi, f)$ are local in nature. If $M = G$, they equal the distributions $I_{G(\mathbb{A})}(\gamma, f)$ and $I_{G(\mathbb{A})}(\pi, f)$ in (1.2). For $M \neq G$, however, they are more complicated. For example, $J_M(\gamma, f)$ is the orbital integral of f over the conjugacy class of γ, but not with respect to the invariant measure. The invariant measure has instead to be weighted by the volume of a certain convex hull, which depends on x.

Example 3 (continued). Suppose that $G = \mathrm{SL}(3)$, that M corresponds to the minimal partition $(1, 1, 1)$, and that γ is a diagonal element with distinct eigenvalues. Then

$$J_M(\gamma, f) = \int_{A_M(\mathbb{A})\backslash G(\mathbb{A})} f(x^{-1}\gamma x) v_M(x)\, dx,$$

where $v_M(x)$ is the volume of the convex hull of the set

$$\{s^{-1} H_P(w_s x) : s \in W(\mathbf{a}_M)\}.$$

The Weyl group $W(\mathbf{a}_M)$ here is isomorphic to S_3, and for every element s, w_s is the associated permutation matrix. The convex hull is represented by the shaded region

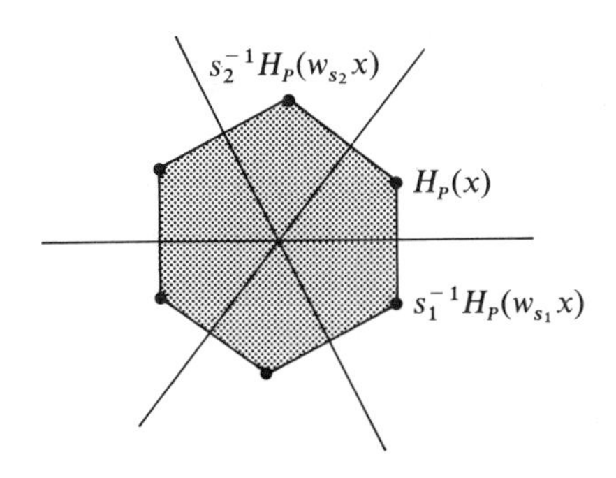

The example illustrates a weakness of the identity obtained from (3.2) and (3.3). Unlike the formula (1.2) for compact quotient, the individual terms are not invariant distributions. They depend on more than just the irreducible characters of f. Fortunately, there is a way to rectify this. For technical reasons, one must insist that f be K-finite, but this is of no great consequence. Then there is natural process that associates an invariant linear functional $I(f)$ to $J(f)$. The same process attaches invariant linear functions $I_M(\gamma, f)$ and $I_M(\pi, f)$ to $J_M(\gamma, f)$ and $J_M(\pi, f)$ such that

$$I(f) = \sum_M |\mathrm{ch}(\mathbf{a}_M)|^{-1} \sum_{\gamma \in \{M(\mathbb{Q})\}} a^M(\gamma) I_M(\gamma, f)$$

and

$$I(f) = \sum_M |\mathrm{ch}(\mathbf{a}_M)|^{-1} \int_{\Pi(M)} a^M(\pi) I_M(\pi, f)\, d\pi.$$

The invariant trace formula is just the identity

$$\sum_M |\mathrm{ch}(\mathbf{a}_M)|^{-1} \sum_{\gamma \in \{M(\mathbb{Q})\}} a^M(\gamma) I_M(\gamma, f)$$
$$= \sum_M |\mathrm{ch}(\mathbf{a}_M)|^{-1} \int_{\Pi(M)} a^M(\pi) I_M(\pi, f)\, d\pi. \quad (3.4)$$

(The details of the construction are contained in [1(e), Sections I.2–I.3, Sections II.2–II,4]. For a general idea of how it works, see the introduction to [1(b)].) If $M = G$, the distributions $J_G(\gamma, f)$ and $J_G(\pi, f)$ are already invariant, and the process does not alter them. Consider the special case that G is anisotropic over $\mathbb{Q}$. Then there are no proper (rational) parabolic subgroups, and the only summands come from $M = G$. The formula (3.4) reduces to (1.2), which is to be expected since $G(\mathbb{Q})\backslash G(\mathbb{A})$ is compact. Thus, (3.4) is a generalization of (1.2) in which the additional terms are contributed by the proper Levi components M.

In the interests of simplicity, we have passed over two technical complications. On the left-hand side of (3.4) (as well as in (3.2)), the notation should actually include a large finite set S of valuations of $\mathbb{Q}$ (that depends in a simple way on f). For if $\gamma \in M(\mathbb{Q})$ is unipotent, the orbital integral on $G(\mathbb{A})$ at γ diverges. It must instead be taken over $G(\mathbb{Q}_S)$. The functions $a^M(\gamma)$ and $I_M(\gamma, f)$, and also the conjugacy relation $\{M(\mathbb{Q})\}$, really depend implicitly on S. The other point is that the integrals over $\Pi(M)$ in (3.3) and (3.4) are not known to converge. This is tied up with the fact that the operator $R_{\mathrm{disc}}(f)$ is not known to be of trace class. However, there is a way to group the terms in the sum-integral over M and $\Pi(M)$ to make them converge. This is supposed to be implicit in (3.4). These complications are serious if one wants to derive the kind of asymptotic formulas available for compact quotient. However, they do not seem to be of any consequence in applications such as base change, that entail a comparison of two trace formulas.

The expression (3.4) is certainly a formula, but the reader is perhaps wondering where the trace is. It is buried in the term corresponding to $M = G$, on the spectral side of (3.4). We have not described $a^G(\pi)$ in general, but this function is actually defined explicitly as a finite sum of terms, one of which is the multiplicity with which π occurs discretely in $L^2(G(\mathbb{Q})\backslash G(\mathbb{A})^1)$. If we were able to transfer everything but these multiplicities to the left-hand side of (3.4), we would be left with a formula for the trace of $R_{\mathrm{disc}}(f)$. This, however, is not allowed, since we don't know at the moment that $R_{\mathrm{disc}}(f)$ is of trace class. What is known to be of trace class is $R_{\mathrm{cusp}}(f)$, the restriction of $R_{\mathrm{disc}}(f)$ to the space of cusp forms. One can always rewrite (3.4) as a formula for the trace of $R_{\mathrm{cusp}}(f)$. It simply entails a convergent grouping of the terms that would otherwise be the formula for the trace of $R_{\mathrm{disc}}(f)$.

4

Suppose that K_0 is an open compact subgroup of $G(\mathbb{A}_0)$. The Hecke algebra

$$\boldsymbol{H}_{K_0} = C_c(K_0\backslash G(\mathbb{A}_0)/K_0)$$

acts on $L^2(G(\mathbb{Q})\backslash G(\mathbb{A})/K_0)$ on the right by convolution. If

$$\Gamma = G(\mathbb{Q})K_0 \cap G(\mathbb{R})$$

as before, we have a $G(\mathbb{R})$-isomorphism

$$L^2(G(\mathbb{Q})\backslash G(\mathbb{A})/K_0) \xrightarrow{\sim} L^2(\Gamma\backslash G(\mathbb{R})).$$

It is easy to describe the action of $\boldsymbol{H}_{K_0}$ in terms of the space on the right. Writing

$$g = g_{\mathbb{R}} g_0, \qquad g_{\mathbb{R}} \in G(\mathbb{R}),\ g_0 \in G(\mathbb{A}_0),$$

for any element $g \in G(\mathbb{Q})$, we first observe from (2.1) that

$$\{g_0 : g \in G(\mathbb{Q})\}$$

is dense in $G(\mathbb{A}_0)$. We can therefore assume that $h \in \boldsymbol{H}_{K_0}$ is the product of $\mathrm{vol}(K_0)^{-1}$ with the characteristic function of

$$K_0 g K_0, \qquad g \in G(\mathbb{Q}),$$

since any element in $\boldsymbol{H}_{K_0}$ is a linear combination of such functions. Let ϕ be a function in the Hilbert space, and take $x \in G(\mathbb{R})$. Then

$$\begin{aligned}(\phi * h)(x) &= \int_{G(\mathbb{A}_0)} \phi(xy)h(y)\,dy \\ &= \sum_{k \in K_0/K_0 \cap g_0 K_0 g_0^{-1}} \phi(xkg_0) \\ &= \sum_{\gamma \in \Gamma_{\mathrm{diag}}/\Gamma_{\mathrm{diag}} \cap g \Gamma_{\mathrm{diag}} g^{-1}} \phi(\gamma_0 g_0 x) \\ &= \sum_{\gamma_{\mathbb{R}} \in \Gamma/\Gamma \cap g_{\mathbb{R}} \Gamma g_{\mathbb{R}}^{-1}} \phi(g_{\mathbb{R}}^{-1}\gamma_{\mathbb{R}}^{-1}x),\end{aligned}$$

where

$$\Gamma_{\mathrm{diag}} = G(\mathbb{R})K_0 \cap G(\mathbb{Q}).$$

This is closer to the classical definition. Actually, in the special case that $G = \mathrm{SL}(2)$ and $\Gamma = \mathrm{SL}(2, \mathbb{Z})$, the prescription above gives only the classical Hecke operators

$$T(n), \qquad n \in \mathbb{N},$$

in which n is a square. To get them all, one would need to take GL(2), a nonsemisimple group that we excluded with our original simplifying assumption.

Suppose that $\pi_{\mathbb{R}}$ is an irreducible unitary representation of $G(\mathbb{R})$. Let $m_{\mathrm{disc}}(\pi_{\mathbb{R}}, K_0)$ be the multiplicity with which $\pi_{\mathbb{R}}$ occurs discretely in $L^2(G(\mathbb{Q})\backslash G(\mathbb{A})/K_0)$. If h belongs to $\boldsymbol{H}_{K_0}$, let $R_{\mathrm{disc}}(\pi_{\mathbb{R}}, h)$ be the operator obtained by restricting h to the $\pi_{\mathbb{R}}$-isotypical subspace of $L^2(G(\mathbb{Q})\backslash G(\mathbb{A})/K_0)$. It can be identified with an $m_{\mathrm{disc}}(\pi_{\mathbb{R}}, K_0) \times m_{\mathrm{disc}}(\pi_{\mathbb{R}}, K_0)$-matrix. One would like an explicit formula for $m_{\mathrm{disc}}(\pi_{\mathbb{R}}, K_0)$ or, more generally, a formula for the trace of $R_{\mathrm{disc}}(\pi_{\mathbb{R}}, h)$. This, of course, is too much to ask in general. However, it is reasonable to ask the question when $\pi_{\mathbb{R}}$ belongs to $\Pi_{\mathrm{disc}}(G(\mathbb{R}))$, the discrete series of $G(\mathbb{R})$. This is essentially what Selberg's formula [8(a), p. 85] gives in the special case that $G = \mathrm{SL}(2)$, $\Gamma = \mathrm{SL}(2, \mathbb{Z})$, and $\pi_{\mathbb{R}}$ is any but the lowest discrete series.

Recall that $G(\mathbb{R})$ has a discrete series if and only if it has a compact Cartan subgroup. Assume that this is the case. Then $\Pi_{\mathrm{disc}}(G(\mathbb{R}))$ is disjoint union of finite sets $\Pi_{\text{disc}}(\mu)$, parametrized by the irreducible finite dimensional representations μ of G. The number $w(G)$ of elements in a set $\Pi_{\text{disc}}(\mu)$ equals the order of a quotient of Weyl groups and is independent of μ; the set itself contains all the representations in the discrete series with the same infinitesimal character as μ. These facts are, of course, part of the classification [4(b)] of Harish-Chandra.

We shall describe a formula for the sum over $\pi_{\mathbb{R}} \in \Pi_{\text{disc}}(\mu)$ of the traces of the Hecke operators. We must first define the terms that appear. Suppose that M is a Levi component of a standard parabolic subgroup of G over $\mathbb{Q}$. Let $\Phi'_M(\gamma, \mu)$ be the function on $M(\mathbb{R})$ which equals

$$|\det(1 - \mathrm{Ad}(\gamma))_{\mathbf{g}/\mathbf{m}}|^{1/2} \sum_{\pi_{\mathbb{R}} \in \Pi_{\text{disc}}(\mu)} \Theta_{\pi_{\mathbb{R}}}(\gamma)$$

if $\gamma \in M(\mathbb{R})$ is $\mathbb{R}$-elliptic (i.e., belongs to a Cartan subgroup that is compact modulo $A_M(\mathbb{R})$), and which equals 0 otherwise. Here $\Theta_{\pi_{\mathbb{R}}}$ stands for the character $\pi_{\mathbb{R}}$, and $\mathbf{g}$ and $\mathbf{m}$ are the Lie algebras of G and M. One can express $\Phi'_M(\gamma, \mu)$ in terms of formulas of Harish-Chandra [4(a)], which are reminiscent of the Weyl character formula. Observe that $\Phi'_M(\gamma, \mu)$ vanishes unless γ is semisimple. Now, suppose that γ belongs to $M(\mathbb{Q})$ and is semisimple. Write

$$h_M(\gamma) = \delta_P(\gamma_0)^{1/2} \int_{K_{0,\max}} \int_{N_P(\mathbb{A}_0)} \int_{M_\gamma(\mathbb{A}_0)\backslash M(\mathbb{A}_0)} h(k^{-1}m^{-1}\gamma m n k)\, dm\, dn\, dk,$$

where $\delta_P(\gamma_0)$ is the modular function of P, evaluated at the finite adèlic component of γ, and $K_{0,\max}$ is a suitable maximal compact subgroup of $G(\mathbb{A}_0)$. This is essentially an invariant p-adic orbital integral and is no more complicated than the distributions in the trace formula for compact quotient. Finally, there is a constant $\chi(M_\gamma)$ which is defined if $\gamma \in M(\mathbb{Q})$ is $\mathbb{R}$-elliptic. If G has no factors of type E_8,

$$\chi(M_\gamma) = (-1)^{q(M_\gamma)} \operatorname{vol}(\overline{M}_\gamma(\mathbb{Q})\backslash \overline{M}_\gamma(\mathbb{A}_0)) w(M_\gamma)^{-1},$$

where $q(M_\gamma)$ is the dimension of the symmetric space attached to M_γ and $\overline{M}_\gamma$ is any inner twist of M_γ such that $\overline{M}_\gamma(\mathbb{R})/A_M(\mathbb{R})^0$ is compact. This relies on a theorem of Kottwitz [5] that requires the Hasse principle. Otherwise, $\chi(M_\gamma)$ must be given by a slightly more complicated formula.

Theorem 4.1. *Suppose the μ is an irreducible finite dimensional representation of G whose highest weight is nonsingular. Then*

$$\sum_{\pi_{\mathbb{R}} \in \Pi_{\text{disc}}(\mu)} \operatorname{tr}(R_{\text{disc}}(\pi_{\mathbb{R}}, h))$$

equals

$$\sum_{M} (-1)^{\dim A_M} |\mathrm{ch}\,(\mathbf{a}_M)|^{-1} \sum_{\gamma \in \{M(\mathbb{Q})\}} \chi(M_\gamma)\Phi'_M(\gamma, \mu)h_M(\gamma).$$

This theorem is proved in [1(f), Corollary 6.2]. It expresses the trace of Hecke operators as a finite closed formula. The reader might want to compare the formula with those in [8(a), p. 85] (for $G = \mathrm{SL}(2)$), [2, p. 283] ($G = P\mathrm{GL}(2)$), and [7, p. 307] (G of real rank 1). The main step would be to convert the p-adic orbital integrals into suitable finite sums. However, for future applications to Shimura varieties, the formula is best left in adèlic form. This is more natural for comparison with the Lefshetz fixed point formula in characteristic p, as one can see from the formulas for GL(2) in [6(a), Sections 5–6].

References

[1.] Arthur, J.

(a) "A trace formula for reductive groups I: Terms associated to classes in G($\mathbb{Q}$)," *Duke Math. J.* **45** (1978), 911–952.

(b) "The trace formula in invariant form," *Ann. of Math.* **114** (1981), 1–74.

(c) "The trace formula for reductive groups," In: *Journées Automorphes*, Publications Mathématiques de l'Université de Paris VII, 1–41.

(d) "The trace formula for noncompact quotient," *Proceedings of the International Congress of Mathematicians, Warsaw, 1983,* **2** (1983), 849–859.

(e) "The invariant trace formula I. Local theory," *J. Amer. Math. Soc.* **1** (1988), 323–383; "II. Global theory," *J. Amer. Math. Soc.* **1** (1988), 501–554. Update.

(f) "The L^2-Lefschetz numbers of Hecke operators," preprint.

(g) "Characters, harmonic analysis, and an L^2-Lefschetz formula," update to appear in *Proc. Sympos. Pure. Math.* (Weyl Symposium), A.M.S.

[2.] Duflo, M., and Labesse, J.-P., "Sur la formule des traces de Selberg," *Ann. Scient. Éc. Norm. Sup.*, **4** (1971), 193–284.

[3.] Harish-Chandra

(a) "Discrete series for semisimple groups I," Acta Math. **113** (1965), 241–318; also *Collected Papers*, Vol. **III**, 403–481.

(b) "Discrete series for semisimple groups II," *Acta. Math.* **116** (1966), 1–111; also *Collected Papers*, Vol. **III**, 537–648.

[4.] Kottwitz, R., "Tamagawa numbers," *Ann. of Math.*, **127** (1988), 629–646. Update.

[5.] Labesse, J.-P., "La formule des traces d'Arthur-Selberg," (Séminaire Bourbaki, 636), *Astérisque* **133** (1986), 73–88.

[6.] Langlands, R. P.

(a) "Modular forms and *l*-adic representations," *Lecture Notes in Mathematics.* **349** (1973), 361–500.

(b) "On the Functional Equations Satisfied by Eisenstein Series," Lecture Notes in Mathematics. **544** (1976).

[7.] Osborne, M. S., and Warner, G. "Multiplicities of the integrable discrete series," *J. Funct. Anal.* **30** (1978), 287–310.

[8.] Selberg, A.

(a) "Harmonic analysis and discontinuous groups in weakly symmetric Riemannian spaces with applications to Dirichlet series," *J. Indian Math. Sol.* **20** 1956), 47–87.

(b) "Discontinuous groups and harmonic analysis," *Proceedings of the International Congress of Mathematicians, Stockholm.* (1962), 177–189.

[9.] Tamagawa, T. "On Selberg's trace formula," J. Fac. Sci. Univ. Tokyo, Sect. I. **8** (1960), 363–386.

Note Added in Proof.

The trace class problem has been recently solved by W. Müller ("The trace class conjecture in the theory of automorphic forms," preprint). In particular, (3.4) can now be written as a formula for the trace of $R_{\mathrm{disc}}(f)$. It would be interesting to investigate the convergence properties of the other terms on the right-hand side of (3.4).

3 Selberg's Sieve and Its Applications

ENRICO BOMBIERI

1

The theory of sieves is one of the fields in which, through his ideas and innovations, Selberg has left a very clear mark. In this lecture I will attempt to explain some of Selberg's contributions, and I will try to comment on some of the recent work in sieve theory. Selberg himself has written on the general subject of sieve methods [S1–S4] and a thorough study of these papers is essential for anyone who wants to understand how a sieve works.

What is a sieve? Let us review quickly the most classical instances of a sieve.

1.1. The Sieve of Eratosthenes

Let us consider the prime numbers $p \leq z$, and from a table of positive integers $2 \leq n \leq z^2$ let us cross out all multiples of p except p itself. Then what is left are the primes $p \leq z^2$.

Although this sieve can be used fairly effectively for the construction of numerical tables, it is not well suited for theoretical investigations.

NUMBER THEORY,
TRACE FORMULAS and
DISCRETE GROUPS

ISBN 0-12-067570-6

1.2. The Legendre Formula

The number of integers $n \le x$, $n \equiv 0 \pmod d$ is clearly equal to $[x/d]$. In the sieve of Eratosthenes we start with $[x]$ integers, then we strike out the multiples of 2, 3, ..., p, ...; this lead us to

$$[x] - \sum_p \left[\frac{x}{p}\right];$$

in doing this, we have struck out twice the integers divisible by two primes, say p, q, which leads us to consider

$$[x] - \sum_p \left[\frac{x}{p}\right] + \sum_{pq} \left[\frac{x}{pq}\right];$$

and so on. We see that:

Legendre's Formula. *Let* $P(z) = \prod_{p \le z} p$. *If* $z = [\sqrt{x}]$ *then*

$$\pi(x) - \pi(z) + 1 - \sum_{d \mid P(z)} \mu(d)\left[\frac{x}{d}\right]. \tag{1.1}$$

Moreover, for $r = 0, 1, 2, \ldots$ *we have:**

$$\pi(x) - \pi(z) + 1 \le \sum_{\substack{d \mid P(z) \\ \nu(d) \le 2r}} \mu(d)\left[\frac{x}{d}\right], \tag{1.2}$$

$$\pi(x) - \pi(z) + 1 \ge \sum_{\substack{d \mid P(z) \\ \nu(d) \le 2r+1}} \mu(d)\left[\frac{x}{d}\right]. \tag{1.3}$$

This is a special case of the inclusion-exclusion principle:

Theorem 1. *Let N be a finite set, let $\mathscr{A}$ be a finite set of properties associated to elements $n \in N$, and for each subset $A \subset \mathscr{A}$ let N_A be the subset of elements of N that have each one of the properties $\alpha \in A$. Let $|A|$ denote the cardinality of a set A.*

* Here $\nu(d)$ is the number of prime factors of d.

If N is the subset of elements of N that have none of the properties $\alpha \in \mathscr{A}$, we have:

$$|N_0| \leq \sum_{j=0}^{2r} (-1)^j \sum_{|A|=j} |N_A| \tag{1.4}$$

$$|N_0| \leq \sum_{j=0}^{2r+1} (-1)^j \sum_{|A|=j} |N_A|. \tag{1.5}$$

Proof. Let $\phi_A : N \to \{0, 1\}$ be the characteristic function of the set N_A. Clearly $\phi_{A \cup B} = \phi_A \phi_B$. The characteristic function of N_0 is simply

$$\prod_{\alpha \in \mathscr{A}} (1 - \phi_\alpha),$$

and now the inequalities (here $0 \leq z_\alpha \leq 1$)

$$\prod_\alpha (1 - z_\alpha) \leq 1 \tag{1.6$_0$}$$

$$\prod_\alpha (1 - z_\alpha) \geq 1 - \sum z_\alpha \tag{1.6$_1$}$$

$$\prod_\alpha (1 - z_\alpha) \leq 1 - \sum z_\alpha + \sum z_\alpha z_\beta \tag{1.6$_2$}$$

prove

$$\prod_{\alpha \in \mathscr{A}} (1 - \phi_\alpha) \leq \sum_{j=0}^{2r} (-1)^j \sum_{|A|=j} \phi_A \tag{1.7}$$

$$\prod_{\alpha \in \mathscr{A}} (1 - \phi_\alpha) \leq \sum_{j=0}^{2r-1} (-1)^j \sum_{|A|=j} \phi_A. \tag{1.8}$$

We may think of inequalities (1.7) and (1.8) as being the basic building blocks of *Brun's sieve*.

1.3. Sieves

Let $\mathscr{P}$ be a finite set of primes and let $P(z) = P = \prod_{\substack{p \leq z \\ p \in \mathscr{P}}} p$. Let us consider a sequence a_n, $n = 1, 2, \ldots$. We want to obtain information on

$$\sum_{(n,P)=1} a_n = S(\mathscr{A}; \mathscr{P}, z)$$

from knowledge about

$$\sum_{n \equiv 0 (\mathrm{mod}\, d)} a_n = |\mathscr{A}_d|$$

and

$$\sum a_n = |\mathscr{A}|.$$

If (P) is the set of square-free integers with prime factors dividing P, we see that

$$\begin{aligned} S(\mathscr{A}; \mathscr{P}, z) &= \sum a_n \left(\sum_{\substack{d \mid n \\ d \mid P}} \mu(d) \right) \\ &= \sum_{d \mid P} \mu(d) \sum_{n \equiv 0 (\mathrm{mod}\, d)} a_n \\ &= \sum_{d \mid P} \mu(d) |\mathscr{A}_d|. \end{aligned}$$

which is the generalization of Legendre formula. More generally, instead of considering $\sum_{(n,P)=1} a_n$, we may deal with the following situation: Let $w(n)$ be a function depending on the factorization of n. What can we say about $\sum a_n w(n)$ from knowledge of sums of the kind $\sum_{n \equiv 0 (\mathrm{mod}\, d)} a_n$? We shall refer to $w(n)$ as a *weight* and $S(\mathscr{A}; w) = \sum a_n w(n)$ as the weighted sifted function.

For a reason that will be clear in a moment, elements in the sequence are usually taken with "multiplicity" $a_n \geq 0$. The trouble is that we have a control of $S(\mathscr{A}; w)$ only for rather special w, and one of the main points of a sieve consists in the remark that *if* $a_n \geq 0$ *and if* $w_- \leq w \leq w_+$ *then*

$$S(\mathscr{A}; w_-) \leq S(\mathscr{A}; w) \leq S(\mathscr{A}; w_+).$$

A more classical form of a sieve is as follows. Let a sequence $\mathscr{A}_0$, a set $\mathscr{P}$ of primes be given, and for every $p \in \mathscr{P}$ let Ω_p be a set of residue classes mod p. We want to study the sifted function

$$S(\mathscr{A}_0; \mathscr{P}, \Omega_p) = \sum_{\substack{n \not\equiv \Omega_p \mathrm{mod}\, p \\ \mathrm{all}\, p \in \mathscr{P}}} a_n^0.$$

In the special case $|\Omega_p| \leq k$, this can be easily reduced to the situation described before. Indeed, by the Chinese remainder theorem, we can find $b_1, b_2, \ldots, b_k$ so that

$$\Omega_p = b_1 b_2 \ldots b_k \bmod p.$$

Now let $\mathscr{A}$ be the sequence $(m - b_1)\cdots(m - b_k)$, $m \in \mathscr{A}_0$, i.e.,

$$a_n = \begin{cases} a_m^0 & \text{if } n = (m - b_1)\ldots(m - b_k) \\ 0 & \text{otherwise.} \end{cases}$$

Then

$$S(\mathscr{A}_0; \mathscr{P}, \Omega_p) = S(\mathscr{A}; \mathscr{P}, \{0\}).$$

1.4. *The Brun Sieve and the Fundamental Lemma*

Suppose we have two functions on (P), say $\chi_+(d)$ and $\chi_-(d)$, such that

$$\sum_{\substack{d|n \\ d\in(P)}} \mu(d) \le \sum_{\substack{d|n \\ d\in(P)}} \mu(d)\chi_+(d), \tag{1.9}$$

$$\sum_{\substack{d|n \\ d\in(P)}} \mu(d) \ge \sum_{\substack{d|n \\ d\in(P)}} \mu(d)\chi_-(d). \tag{1.10}$$

Let $\mathscr{A}$ be a positive sequence (i.e., $a_n \ge 0$) and let us consider $S(\mathscr{A}; \mathscr{P}, z)$. We have

$$S(\mathscr{A}; \mathscr{P}, z) \le \sum_{d\in(P)} \mu(d)\chi_+(d)|\mathscr{A}_d|, \tag{1.11}$$

$$S(\mathscr{A}; \mathscr{P}, z) \ge \sum_{d\in(P)} \mu(d)\chi_-(d)|\mathscr{A}_d|, \tag{1.12}$$

and we have seen that

$$\chi_+(d) = \begin{cases} 1 & \text{if } d = p_1\ldots p_j, \quad j \le 2r \\ 0 & \text{otherwise} \end{cases} \tag{1.13}$$

$$\chi_-(d) = \begin{cases} 1 & \text{if } d = p_1\ldots p_j, \quad j \le 2r+1 \\ 0 & \text{otherwise} \end{cases} \tag{1.14}$$

are indeed *admissible* χ_+, χ_-.

Lemma. *Let $q(d)$ denote the smallest prime factor of d. Let X_+, χ_- satisfy*

$$\mu(d)\{\chi_+(d) - \chi_+(pd)\} \ge 0,$$

$$\mu(d)\{\chi_-(d) - \chi_-(pd)\} \le 0$$

whenever $pd \in (P)$ and χ_- are admissible.

Proof. Let $p = q(n)$. We have

$$\sum_{d|n} \mu(d)\chi(d) = \sum_{d|n/p} \mu(d)\chi(d) + \sum_{d|n/p} \mu(pd)\chi(pd)$$
$$= \sum_{d|n/p} \mu(d)\{\chi(d) - \chi(pd)\};$$

the result follows.

We use this principle to illustrate Brun's sieve. We assume

$$|\mathscr{A}_d| = \frac{1}{f(d)} X + R_d,$$

where R_d is a remainder term. This yields

$$|\mathscr{A}| \leq X \sum \frac{\chi_+(d)\mu(d)}{f(d)} + \sum |\chi_+(d)||R_d|, \qquad (1.15)_+$$

$$|\mathscr{A}| \geq X \sum \frac{\chi_-(d)\mu(d)}{f(d)} - \sum |\chi_-(d)||R_d|. \qquad (1.15)_-$$

Now we have two main problems. One is the control of

$$\sum \frac{\mu(d)\chi_\pm(d)}{f(d)},$$

and the other that of $\sum |\chi_\pm(d)||R_d|$. Of course, we do not want $\sum |\chi_\pm(d)||R_d|$ to become too large. This will put some restriction on how big is the set of d's for which $\chi_\pm(d) \neq 0$.

In most cases, $1/f(d)$ will be a multiplicative function on the set $d \in (P)$. This is a natural condition, because it means that the distribution of $\mathscr{A}$ on $n = d_1 d_2 m$ depends only on asking $n \equiv 0 \pmod{d_1}$, $n \equiv 0 \pmod{d_2}$ separately, provided $(d_1, d_2) = 1$; no additional condition comes from asking a simultaneous congruence.

The functions χ_+, χ_- that determine the sieve are in practice never big, usually between 0 and 1. We shall suppose

$$|\chi_+(d)|, \qquad |\chi_-(d)| \ll 1$$

and, more precisely, we choose

$$2 = z_m < z_{m-1} < \cdots < z_1 < z_0 = z; \qquad (1.16)$$

$\chi_+(d) = 1$ *if the number of prime factors of* d, $z_n < p \leq z$ *is* $2b_n$ *and* 0 *otherwise. Here* b_n *is a positive integer*. We claim that χ_+ satisfies the condition of the lemma. Indeed, if $\chi_+(d) = 0$ then *a fortiori* $\chi_+(pd) = 0$; if $\mu(d) = 1$, again the condition is verified. Hence we must assume $\mu(d) = -1$ and $\chi_+(d) = 1$. Let $z_n < q(d) \le z_{n-1}$, so that the number of prime factors of d is $\le 2b_n - 1$ in the range $(z_n, z]$ and $\le 2b_r$ in the ranges $(z_n, z]$ if $r < n$. If $p \le z_n < q(d)$ then $\chi_+(pd) = 1$, by definition. If $z_n < p < q(d)$ then $\chi_+(pd) = 1$ again because d had $\le 2b_n - 1$ factors in $(z_n, z]$, hence pd has still $\le 2b_n$ factors.

Similarly, we take:

$\chi_-(d) = 1$ *if the number of prime factors of* d, $z_n < p \le z$ *is* $2b_n - 1$

and 0 *otherwise.*

Now we have to control $\sum \chi_\pm(d)|R_d|$ and $\sum \mu(d)\chi_\pm(d)/f(d)$. Here one must assume some mild regularity conditions on $f(d)$ of the type

$$0 \le \frac{1}{f(d)} \le 1 - \frac{1}{A_1}; \tag{Ω_1}$$

$$\sum_{w < p \le z} \frac{\log p}{f(p)} \le k \log \frac{z}{w} + A_2, \tag{$\Omega_2(k)$}$$

for some constants A_1, A_2. The points $z_m, \ldots, z_0$ in Brun's sieve are chosen as

$$z_r = z^{e^{-u_r}} \tag{1.17}$$

with $0 = u_0 < u_1 < \cdots < u_{m-1} < \infty$, $z_m = \frac{3}{2}$, and using (Ω_1) and $(\Omega_2(k))$ one obtains rather complicated estimates of the kind

$$\sum \frac{\mu(d)\chi_+(d)}{f(d)} \lesssim W(z) \sum_{r=1}^{\infty} e^{ku_{r-1}} \sum{}' \left[\int_0^{k(u_r - u_{r-1})} \frac{(ku)^{\beta_r}}{\beta_r!} e^u \, du \right] \cdot \prod_{s=1}^{r-1} \frac{(k(u_s - u_{s-1}))^{\beta_s}}{\beta_s!}, \tag{1.18}$$

where

$$W(z) = \prod_{\substack{p \le z \\ p \in \mathcal{P}}} \left[1 - \frac{1}{f(p)} \right] \tag{1.19}$$

and where $\sum'$ runs over sets $\beta_1, \ldots, \beta_r$ of integers satisfying

$$\beta_1 + \cdots + \beta_r = 2b_r,$$

$$\beta_1 + \cdots + \beta_s = 2b_s,$$

for $s = 1, \ldots, r-1$ and

$$\sum_{r=1}^{\infty} \beta_r e^{-u_{r-1}} \le \frac{\log X}{\log z}.$$

Inequality (1.18) leads to upper bounds for the cardinality of the sifted set $\mathscr{A}$; lower bounds are obtained by dealing with $\sum \mu(d)\chi_-(d)/f(d)$. However, in this case the lower bound is useful only if it is positive, while upper bounds always carry some nontrivial information. This leads to the problem of estimating the right-hand side of inequalities such as (1.18).

2. Selberg's Sieve

In 1946 Selberg discovered a new approach to the problem of determining admissible $\chi_\pm$ sets. It is useful to generalize the situation a bit; what follows is essentially Selberg's own description of the method in [S4].

We abandon the notation of the preceding section in favor of the following simpler one. We denote:

S = finite set of weighted integers n, with weight w_n

P = finite set of primes

(P) = the set of square-free numbers generated by the primes in P

$\langle P \rangle$ = the set of all integers generated by the primes in P

$$N_d = \sum_{\substack{n \in S \\ d|n}} w_n$$

$$S_d = \{n \in S;\ d|n\}$$

$$M(S, P) = \sum_{(n,P)=1} w_n.$$

2.1 The Λ Method

Let $\Lambda = \{\lambda_d\}$ be a collection of real numbers λ_d for $d \in (P)$, with the property that

$$\sum_{\substack{d|n \\ d \in (P)}} \mu(d) \le \sum_{\substack{d|n \\ d \in (P)}} \lambda_d$$

for every $n \in S$. The exclusion-inclusion principle yields

$$M(S, P) = \sum_{n \in S} \Bigg[\sum_{\substack{d|n \\ d \in (P)}} \mu(d) \Bigg] w_n;$$

hence if $w_n \geq 0$ we have

$$M(S, P) \leq \sum_{d} \lambda_d N_d; \tag{2.1}$$

we then say that Λ forms a $\Lambda^+(P)$ method. If instead

$$\sum_{\substack{d|n \\ d\in(P)}} \mu(d) \geq \sum_{\substack{d|n \\ d\in(P)}} \lambda_d, \tag{2.2}$$

we have a $\Lambda^-(P)$ method. The basic problem is that of estimating

$$M^+(S, P) = \min_{\Lambda^+(P)} \sum \lambda_d N_d$$

$$M^-(S, P) = \max_{\Lambda^-(P)} \sum \lambda_d N_d.$$

It is clear that an optimal sieve has

$$\lambda_1 = 1 \tag{2.3}$$

for a Λ^+ sieve and for a Λ^- sieve that yields a positive lower bound. Rather than entering into the intricacies of constructing explicit sieves Λ (the Brun sieve shows how things can become horrendously complicated), Selberg had the idea of looking first at the set of all sieves. If

$$\Lambda^{(i)} = \{\lambda_d^{(i)}\}, \qquad i = 1, 2$$

are two sieves, then the formula

$$\lambda_d = \sum_{[d_1, d_2] = d} \lambda_{d_1}^{(1)} \lambda_{d_2}^{(2)}, \tag{2.4}$$

where $[d_1, d_2]$ is the least common multiple of d_1, d_2, defines a sieve that is the *product* of $\Lambda^{(1)}$ and $\Lambda^{(2)}$:

$$\sum_{d|n} \lambda_d = \left[\sum_{d_1|n} \lambda_{d_1}^{(1)}\right]\left[\sum_{d_2|n} \lambda_{d_2}^{(2)}\right]. \tag{2.5}$$

One can denote it by $\Lambda^{(1)}\Lambda^{(2)}$. It is now clear that

$$\Lambda^2 \text{ is always a } \Lambda^+ \tag{2.6}$$

$$\Lambda^-\Lambda^+ \text{ is always a } \Lambda^-, \tag{2.7}$$

hence

$$\Lambda^-\Lambda^2 \text{ is always a } \Lambda^-. \tag{2.8}$$

This simple remark produces at a stroke a large and flexible class of $\Lambda^{\pm}$ sieves, and Selberg's next idea is to optimize the sieves so obtained relative to the problem at hand. Thus, rather than constructing a sieve good for all purposes, one finds the best sieve in relation to the problem. This idea, as well as the use of the Λ^2 sieve, appears to find its origin in Selberg's work on the analytic theory of the Riemann zeta-function, where he introduced so-called mollifiers by the Λ^2 method.

In practice, S is a set S_x depending on a parameter x, $P = \{p \leq Z\}\backslash$(finite set), $N_d = x/f(d) + R_d$ where R_d is a remainder that is assumed to be small in a suitable sense and $f(d) = d/n(d)$ is multiplicative. It should be noted that Selberg, in his work on the Riemann zeta-function, encountered more general N_d's, with a main term of the type $x/f(d) \log x/d$ and dealt successfully with them.

The function $u(d)$ is typically bounded on average, satisfying conditions of the type

$$\sum_{p<\xi} \frac{u(p)}{p} \log p \sim \kappa \log \xi \tag{2.9}$$

$$\sum_{p<\xi} \frac{u(p)}{p} \log p = \kappa \log \xi + O(1) \tag{2.9$'$}$$

$$\sum_{p<\xi} \frac{u(p)}{p} \log p \leq \kappa \log \xi + O(1). \tag{2.10}$$

In a classical sieve, $u(p)$ is the number of residue classes $(\bmod p)$ that are removed from S, and therefore one refers to such a sieve as being a κ-residue sieve. In the literature, one often refers to κ as being the dimension of a sieve, $\kappa = 1$ corresponding to a linear sieve. We find this terminology rather unfortunate and somewhat misleading, and we shall speak instead of κ-residue sieves and refer to κ as being the density of the sieve.

The upper bound for $M(S, P)$ now becomes

$$M(S, P) \leq x \sum \frac{\lambda_d}{f(d)} + \sum \lambda_d R_d$$

for any Λ^+ sieve. If

$$\theta_d = \sum_{d'|d} \lambda_{d'} \tag{2.11}$$

then

$$\sum_{(P)} \frac{\lambda_d}{f(d)} = W_n(P)^{-1} \sum_{(P)} \frac{\theta_d}{f'(d)}, \tag{2.12}$$

where f' is the multiplicative function generated by $f(p) - 1$ (the so-called arithmetic derivative of f) and where

$$\frac{1}{W_u(P)} = \prod_P \left[1 - \frac{u(p)}{p}\right]. \tag{2.13}$$

In most applications, assumptions such as (2.9) yield $W_u(P) \sim C_\kappa(\log z)^\kappa$ and $\sum_{(P)} \theta_d/f'(d)$ is more or less bounded so that the main term in the estimate for $M(S, P)$ is of order $x(\log z)^{-\kappa}$. Thus we need

$$\sum \lambda_d R_d \ll x^{1-\varepsilon} \tag{2.14}$$

for $\sum \lambda_d R_d$ to play the role of an irrelevant error term. A simple way of ensuring (2.14) consists in choosing $\lambda_d = 0$ for $d > \xi$, with $\xi = x^\alpha$ for some α, and estimating R_d either directly or on average. If we can obtain a useful bound for $\sum \lambda_d R_d$ we say that S is well distributed up to level ξ.

Of course, entirely similar considerations are valid when we consider lower bounds.

2.2 *A Theorem of Selberg and the Sieving Limit*

Let us consider now the case in which we can choose $\xi = x^{1-\varepsilon}$ for every $\varepsilon > 0$, and let $P = \{p \le z\}$ with $z = x^\alpha$, $0 < \alpha \le 1$. For a given sieve Λ, let

$$T_u(\Lambda) = \sum_{(P)} \frac{\theta_d}{f'(d)}$$

and let us look at $\inf T_u(\Lambda^+)$ and $\sup T_u(\Lambda^-)$. It is not difficult to see that the above quantities, up to $o(1)$ terms as $x \to \infty$, depend only on κ, α; if we denote them by $T_\kappa^\pm(\alpha)$ we can write

$$T_\kappa^-(\alpha)\frac{x}{W_u(P)} \lesssim M(S, P) \lesssim T_\kappa^+(\alpha)\frac{x}{W_u(P)}. \tag{2.15}$$

Inequality (2.15) represents the most one can extract from a sieve, in the sense that by choosing S appropriately one can asymptotically approach both sides of (2.15). It is clearly of primary importance to be able to compute $T_\kappa^\pm(\alpha)$, and much work has been devoted to finding practical estimates for these quantities. A theoretical (but not practical) solution has been found by Selberg [S4].

Let F^+ be the class of functions $F(u_1, \ldots, u_r)$, $r = 1, 2, \ldots$ such that

$$F(u_1, \ldots, u_r) = 0 \qquad \text{if } u_1 + \cdots + u_r \geq 1,$$

and

$$\theta(u_1, \ldots, u_r) = 1 - \sum F(u_i) + \sum_{i \leq j} F(u_i, u_j) - \cdots + (-1)^r F(u_1, \ldots, u_r)$$

satisfies

$$\theta \geq 0 \qquad \text{for every } r;$$

similarly, let F^- be the class for which

$$\theta \leq 0 \qquad \text{for every } r.$$

Theorem (Selberg). *We have*

$$T_\kappa^+(\alpha) = \underset{F^+}{\text{l.u.b.}} \left\{1 + \kappa \int_0^\alpha \frac{\theta(u_1)}{u_1}\, du_1 + \frac{\kappa^2}{2!} \int_0^\alpha \int_0^\alpha \frac{\theta(u_1, u_2)}{u_1 u_2}\, du_1\, du_2 + \cdots\right\}$$

and

$$T_\kappa^-(\alpha) = \underset{F^-}{\text{g.l.b.}} \left\{1 + \kappa \int_0^\alpha \frac{\theta(u_1)}{u_1}\, du_1 + \frac{\kappa^2}{2!} \int_0^\alpha \int_0^\alpha \frac{\theta(u_1, u_2)}{u_1 u_2}\, du_1\, du_2 + \cdots\right\}.$$

Moreover, Selberg has shown how to approximate the above quantities by means of a finite problem. Then Selberg proves that

$$T_\kappa^+(\eta, \alpha) \leq T_\kappa^+(\alpha) \leq (1 - \sqrt{\eta})^{-\kappa-1} T_\kappa^+(\eta, \alpha)$$

whenever $\eta < \min(\alpha, 10^{-5}, 1/200\,\kappa^2)$, with

$$T_\kappa^+(\eta, \alpha) = \min_{F^+} \left(\frac{\alpha}{\eta}\right)^\kappa \left\{1 - \kappa \int_\eta^\alpha F(u_1) \frac{du_1}{u_1} + \frac{\kappa^2}{2!} \int_\eta^\alpha \int_\eta^\alpha F(u_1, u_2) \frac{du_1\, du_2}{u_1\ \ u_2} - \cdots\right\}.$$

A similar result holds for $T_\kappa^-(\alpha)$. For each $\eta > 0$, $T_\kappa^\pm(\eta, \alpha)$ involves only a sum with at most $1/\eta$ terms, and thus the determination of $T_\kappa^\pm(\eta, \alpha)$ is a problem in linear programming; in this sense, Selberg's solution is constructive.

The *sieving limit* $\alpha(\kappa)$ is defined by

$$T_\kappa^-(\alpha) > 0 \qquad \text{if } 0 \le \alpha < \alpha(\kappa),$$
$$T_\kappa^-(\alpha(\kappa)) = 0 \qquad \text{if } \kappa \ge \tfrac{1}{2}.$$

We know

$$\alpha(\tfrac{1}{2}) = 1$$
$$\alpha(1) = \tfrac{1}{2},$$

but the exact value of the important quantity $\alpha(2)$ is still unknown. Theoretically it could be determined with arbitrary precision using Selberg's result.

2.3 *Constructing Sieves*

There are many ways of constructing explicit $\Lambda^\pm$ sieves. For example, the Λ^2 sieve of Selberg is the most elegant construction of a Λ^+ sieve; its optimization leads to the determination of the minimum of a positive definite quadratic form, which Selberg does elegantly in [S1] by transforming it in a sum of squares.

Surprisingly enough, the lower bound sieve $\Lambda^-\Lambda^2$ has not been exploited as it deserves. (See, however, the important comments by Selberg in [S4] and [B, Sections 8–9], for an explicit treatment of a special case.) The celebrated *Buchstab iteration* (1927) based on the relation

$$\sum_{(n,P(z))=1} = \sum_{(n,P(y))=1} - \sum_{y<p\le z} \sum_{\substack{(n,P(p))=1 \\ n\equiv 0 \,(\mathrm{mod}\, p)}}$$

leads to the following procedure.

Suppose we start with two sieves, a $\Lambda^\pm(P(z))$ and $\Lambda^\mp(P(y))$, with $y < z$. Then we can extend $\Lambda(P(y))$ to a $\Lambda^+(P(z))$ by means of

$$\Lambda^\mp(P(z))S = \Lambda^\mp(P(y))S - \sum_{y<p\le z} \Lambda^\pm(P(p))(S_p). \tag{2.16}$$

One starts with z, y very small and, for example, a Brun $\Lambda^\pm$ (the Fundamental Lemma) and proceeds to extend the ranges of the sieves by using (2.16). From (2.16) one can show the following. Let $1 \le u < v$, and let $T_\kappa^+(1/u) \le f_\kappa^+(u)$, $T_\kappa^-(1/u) \ge f_\kappa^-(u)$. Then

$$v^\kappa f_\kappa^\pm(v) = \int_u^\infty f_\kappa^\mp(t-1)\, dt^\kappa \tag{2.17}$$

form a new set of functions $u^\kappa f_\kappa^\mp(u)$. The origin of (2.17) is clear from (2.16) if we write $z = x^{1/u}$, $y = x^{1/v}$, and replace the sum by an integral (the argument $t - 1$ for $f_\kappa^\mp$ in the integral comes from replacing S by the set $S_p = \{n \in S, n \equiv 0 \pmod p\}$, while the powers u^κ, v^κ, t^κ come from the term $1/W_u(P)$ in (2.12)).

If this procedure has a limit, by letting $v \to u$ in (2.17) we obtain the Buchstab differential-difference equation

$$(u^\kappa f_\kappa^\mp(u))' = \kappa u^{\kappa-1} f_\kappa^\mp(u-1) \tag{2.18}$$

with initial condition

$$f_\kappa^\pm(u) = 1 + 0(e^{-u}) \tag{2.19}$$

as $u \to \infty$, which corresponds to starting with Brun's Fundamental Lemma.

An independent combinatorial approach, which leads to the limit of Buchstab's iteration, was found by B. Rosser in the 1950s and independently by Iwaniec [I], who was also able to give the remainder term $\sum \lambda_d R_d$ a "bilinear" structure, which turns out to be very important for applications.

The Buchstab–Rosser–Iwaniec sieve turns out to be optimal in two important cases, $\kappa = \frac{1}{2}$ and $\kappa = 1$. Again, we owe to Selberg the extremal examples. If $\kappa = 1$, one defines

$$S_1^+ = \{n, \Omega(n) \equiv 1 \pmod 2\},$$
$$S_1^- = \{n, \Omega(n) \equiv 0 \pmod{\ }\},$$

where $\Omega(n)$ is the total number of prime factors of n, and for $\kappa = \frac{1}{2}$ one takes

$$S_{1/2}^+ = \{n \equiv 1 \pmod 4, p|n \Rightarrow p \equiv 3 \pmod 4\}$$
$$S_{1/2}^- = \{n \equiv 3 \pmod 4, p|n \Rightarrow p \equiv 3 \pmod 4\}.$$

No extremal sieves are known if $\kappa \neq \frac{1}{2}, 1$.

2.4. *The Λ^2 Sieve*

Although the Buchstab–Rosser–Iwaniec sieve is nearly optimal for small densities κ, it turns out that Selberg's Λ^2 sieve is far superior for large κ. If S is well distributed up to level ξ, then choosing $\lambda_d = 0$ for $d > \xi^{1/2-\varepsilon}$ ensures that the remainder $\sum \lambda_{d_1}\lambda_{d_2} R_{[d_1,d_2]}$ in the Λ^2 sieve is small. If we minimize the quadratic form $\sum_{\substack{d \in (P) \\ d < \xi}} \theta(d)/f'(d)$ we obtain a main term

$$x\Big/\Bigg[\sum_{\substack{d \in (P) \\ d < \xi}} \frac{1}{f'(d)}\Bigg] \sim \sigma_\kappa^+(u)^{-1} W_\kappa(z)^{-1} x, \tag{2.20}$$

where $\sigma_\kappa^+(u)$, $u = \log \xi/\log z$, satisfies a certain differential-difference equation, not unlike Buchstab's. Comparison of $\sigma_\kappa^+(u)^{-1}$ with Buchstab's $f_\kappa^+(u)$ shows that for large κ

$$\sigma_\kappa^+(u)^{-1} < f_\kappa^+(u).$$

This is typical of Selberg's style: simplicity and elegance of method, powerful results.

Obtaining a good bound for a lower-bound sieve and, in particular, for the sieving limit $\alpha(\kappa)$ is not a simple matter. Ankeny and Onishi [A–O] have worked out the effect of a Buchstab iteration on Selberg's upper-bound method. A second iteration was carried out by Porter [P], and the limit was obtained by Iwaniec, van de Lune, and te Riele [ILR]. The value for $\alpha(\kappa)$ in the Iwaniec-van de Lune and te Riele sieve is asymptotic to $1/c\kappa$, with $c > 2$. However, here again Selberg has shown that $\alpha(\kappa) \gtrsim 1/2\kappa$ can be obtained by going back to his original remark that every $\Lambda^-\Lambda^2$ sieve is a Λ^- sieve.

We shall not comment any further on this point, since it will be covered by Selberg's talk, and only mention that it is not unlikely that $\alpha(\kappa) \sim 1/2\kappa$ as $\kappa \to \infty$.

3. Further Developments

3.1. Going Beyond the Level of Distribution

In many applications of a sieve, determining the level of distribution ξ of S can be an exceedingly difficult task. For example, if $S = \{p + 2\}$ is the set of translates of primes $p \leq x$, it is known that S has level of distribution $x^{1/2-\varepsilon}$, and it is a conjecture of Elliott and Halberstam that one can take $x^{1-\varepsilon}$ in place of $x^{1/2-\varepsilon}$. On the other hand, the remainder terms appearing in a sieve often have a special structure that can be used with advantage. Selberg's Λ^2 sieve is a good example. Here we get

$$M(S, P) \leq x \sum \frac{\lambda_d \lambda_{d'}}{f([(d, d'])} + \sum R_{[d,d']} \lambda_d \lambda_{d'} \tag{3.1}$$

and the remainder $\sum R_{[d,d']}\lambda_d\lambda_{d'}$ is a *quadratic form* Q in the λ's. Thus not only do we have the obvious bound $|Q| \le \sum |R_{[d,d']}\lambda_d\lambda_{d'}|$, but in fact

$$|Q| \le \|Q\| \sum \lambda_d^2, \tag{3.2}$$

where $\|Q\|$ is the largest eigenvalue of Q. In several cases, it is possible to control $\|Q\|$ nontrivially and thus obtain results that go beyond the reach of the general sieve for arbitrary sequences, in particular beyond the sieving limit $\alpha(\kappa)$. The Buchstab–Rosser–Iwaniec sieve does not lend itself immediately to such a treatement, and it was an important achievement of Iwaniec [I] to show how to estimate the remainder terms in terms of bilinear forms. The Iwaniec sieve has become one of the most important and flexible tools in sieve theory.

How does one apply these ideas in concrete situations? It suffices here to mention one example: If S is an interval, R_d can be expanded in a Fourier series (by means of Poisson summation formula), allowing situations to be treated in which $d > |S|$, where the remainder term oscillates and is larger than the main term. Important results have been obtained by several authors using these ideas, and we refer to the books by Hooley [H], Halberstam and Richert [H–R], Richert [R], and to the Bourbaki Seminar [D] for further information.

3.2. Weights

Another important aspect in sieve theory consists of the use of weights. The use of sieves in a sequence often is to show that S contains r-almost primes, i.e., elements with $\Omega(n) \le r$. The "pure" sieve described before proves that (say, $R_d = 0(x^{\varepsilon})$) $r < 1/\alpha(\kappa) + 1$ and, more precisely, that S has r-almost primes *all* of whose factors are $> x^{\alpha(\kappa)-\varepsilon}$. However, for our goal of finding r-almost primes P_r in a sequence, this is too much information. In finding P_r's in a sequence, we may allow elements in S with one large factor and only a few small factors, etc. What we want is their number, not the location. The use of weights obviates this difficulty.

Weights were first introduced by Kuhn [K]; they also appear early in Selberg's work. Selberg's approach is remarkably simple and efficient: If

$$\sum_S w_n \Bigg(\sum_{\substack{d\mid n\\ d<z}} \lambda_d\Bigg)^2 \Bigg(\sum_{\substack{d\mid n\\ d<z_0}} a_d\Bigg) < C \sum_S w_n \Bigg(\sum_{\substack{d\mid n\\ d<z}} \lambda_d\Bigg)^2 \tag{3.3}$$

then

$$\sum_{\substack{d|n \\ d<z}} a_d \leq C \tag{3.4}$$

quite often.

For example, Selberg treats

$$\sum (d(n) + d(n+2)) \left(\sum_{\substack{d|n(n+2) \\ d<z}} \lambda_d \right)^2$$

and proves that

$$d(n) + d(n+2) < 16$$

often, so $n(n+2) = P_5$ (still the best result obtained by elementary means).

Another example, choosing $a_d = 1$ if d is a prime $d = p < z_0$, is treated in [B1]. There one shows that $p + 2 = P_4$ infinitely often, with $\log 2 < 1$ being the only numerical calculation needed in the proof. Unfortunately, more precise results often require quite extensive numerical calculations so that one quickly loses the naturalness and elegance that can be achieved in proving weaker results.

The general theory of weights is in a considerable state of flux, and important results have been obtained by Richert [H–R], Greaves [G], and Laborde. A possible approach to general weights is as follows. An *r-weight* $w(n)$ is an arithmetical function defined on r-almost primes $n = p_1, \ldots, P_s, p_1 \geq \cdots \geq p_s, s \leq r$ of the type

$$w(n) = w_s\left(\frac{\log p_i}{\log x}, \ldots, \frac{\log p_s}{\log x}\right),$$

with w_s a "reasonable" function of $u_1, \ldots, u_s$ on

$$U_s^* = \{u_1 \geq \cdots \geq u_s \geq 0,\ u_1 + \cdots + u_s \leq 1\}.$$

Now we want upper and lower bounds for

$$\sum_{n \in S} w(n).$$

For example, if

$$w_s(u_1, \ldots, u_s) = 0,$$

whenever $u_s < \log z/\log x$, we have

$$\sum_{n \in S} w(n) = \sum_{(n,P)=1} w(n),$$

which is the problem considered so far.

Let us assume that S is "local" in the sense that $n \in S$ is $x^{1-\varepsilon} < n < x^{1+\varepsilon}$. Then we deal with weights

$$w(n) = w_s\left(\frac{\log p_1}{\log n}, \ldots, \frac{\log p_s}{\log n}\right)$$

on

$$U_s = \{u_1 \geq \cdots \geq u_s;\ u_1 + \cdots + u_s = 1\}.$$

We can define a $T_\kappa^\pm(w)$ analogous to $T_\kappa^\pm(\alpha)$ in Section 2. Suppose now that the density κ is 1. Then Selberg's Λ^2 sieve can be used to prove [B2]:

Theorem

$$T_1^+(w) = W^+ + |W^-|$$

$$T_1^-(w) = W_1^+ - |W^-|$$

where

$$W^+ = \sum_{h=1}^{\infty} \int_{U_h} w_h \frac{du_1}{u_1} \cdots \frac{du_{h-1}}{u_{h-1}}$$

$$W^- = \sum_{h=1}^{\infty} (-1)^h \int_{U_h} w_h \frac{du_1}{u_1} \cdots \frac{du_{h-1}}{u_{h-1}}.$$

[Note: $w_1(u) = w_1(1) =$ constant, so $T_1^-(w_1) = 0$, which we knew.]

The proof is in several steps:

(a) reduction to a standard set of weights

(b) reduction to $W^- = 0$ for a proof of

$$T_1^+(w) \leq W^+ + |W^-|$$

$$T_1^-(w) \geq W^+ - |W^-|;$$

(c) proof in case (b) using Selberg's sieve

(d) use of Selberg sequences $S_1^\pm$ to prove that the inequalities are sharp.

As a special case, we recover the Buchstab–Rosser–Jurkat and Richert–Iwaniec result for $\kappa = 1$, taking

$$w_h = \begin{cases} 1 & \text{if } u_h \geq \alpha \\ 0 & \text{if } 0 \leq u_h < \alpha. \end{cases}$$

If $W^- = 0$ we get asymptotics: Choose, for example,

$$\begin{aligned} w_1 &= 1 \\ w_2(u_1, u_2) &= 2u_1u_2 \\ w_h &= 0 \qquad \text{if } h \geq 3 \end{aligned}$$

and, surprise, the result is the celebrated *Selberg formula*

$$\sum_{n \in S} \Lambda(n) \log n + \sum_{mn \in S} \Lambda(m)\Lambda(n) = 2\frac{x}{W_1(x)}(\log x)^2 + o(x \log x)$$

for the set S.

Selberg was led to his celebrated formula by sieve considerations; once discovered, direct proofs could be found. The above result shows again the sieve theoretic reasons for these results.

References

[A-O] Ankeny, N.C. and Onishi, H. "The general sieve," *Acta Arith.* **10** (1964-65), 31–62.

[B1] Bombieri, E. "*Le grand crible dans la théorie analytique des nombres.*" II ed. *Astérisque Soc. Math. France.* **18** (1987).

[B2] Bombieri, E. "The asymptotic sieve." *Mem. Accad. Naz.* **XL** (1976), 243–269.

[D] Deshouillers, J.M. "Progrés récents des petits cribles arithmétiques (d'après Chen et Iwaniec)," *Sém. Bourkbaki.* **520** (1977/78), 1–15.

[G] Greaves, G. "Rosser's sieve with Weights." In: *Recent Progress in Analytic Number Theory*, H. Halberstram and C. Hooley, eds. Academic Press (1981), 61–68.

[H] Hooley, C. "*Applications of sieve methods to the theory of numbers.*" *Cambridge Tracts in Math.* **70**, Cambridge Univ. Press (1976).

[H-R] Halberstam, H. and Richert, H.E. *Sieve Methods.* Academic Press (1974).

[I] Iwaniec, H. "*Rosser's sieve-bilinear forms of the remainder terms —some applications.*" In: *Recent Progress in Analytic Number Theory*, H. Halberstam and C. Hooley, eds. Academic Press (1981), 203–230.

[I-L-R] Iwaniec, H., Lune, van de J., and Riele, H. te, "The limits of Buchstab's iteration sieve, "*Report Dept. of Pure Math. Mathematisch Centrum, Amsterdam* (1979).

[K] Kuhn, P. "Zur Viggo Brunschen Siebmethode," *I. Norske Vid. Selsk. Forh. Trondhjem.* **14** (1941), 145–148.

[R] Richert, H.E. *Lectures on Sieve Methods*. Tata Inst. Bombay (1976).

[S1] Selberg, A. "On an elementary method in the theory of primes," *Norske Vid. Selsk. Forh. Trondhjem.* **19** (1947), 64–67.

[S2] Selberg, A. "On elementary methods in prime number theory and their limitations," *11 Skand. Mat. Kongr. Trondhjem.* (1949), 13–22.

[S3] Selberg, A. "The general sieve method and its place in prime number theory." *Proc. Int. Math. Congr. Cambridge, Mass.* **I** (1950), 286–292.

[S4] Selberg, A. "Sieve methods," *Proc. Symp. Pure Math. AMS.* **20** (1971), 311–351.

4 The Rankin–Selberg Method: A Survey

DANIEL BUMP

1. The Rankin–Selberg Method for SL(2, Z)

1.1. Prehistory: Ramanujan

The original proof by Hadamard and de la Vallée Poussin of the prime number theorem depends on the fact that the Riemann zeta-function $\zeta(1 + it) \neq 0$ for real values of t. An interesting proof of this was given in 1930 by Ingham [36], based on a formula of Ramanujan [75]. This proof is related to a modern proof of this nonvanishing (cf. Section 1.5 below), which uses the Rankin–Selberg method explicitly, and which establishes the corresponding fact for L-series of automorphic forms on GL(n).

Let us review Ramanujan's formula, which is instructive.

Ramanujan proves that if $\sigma_a(n) = \sum d^a$, where the sum is over all divisors d of n, then

$$\sum_{n=1}^{\infty} \frac{\sigma_a(n)\sigma_b(n)}{n^s} = \frac{\zeta(s)\zeta(s-a)\zeta(s-b)\zeta(s-a-b)}{\zeta(2s-a-b)} \tag{1.1.1}$$

NUMBER THEORY,
TRACE FORMULAS and
DISCRETE GROUPS

ISBN 0-12-067570-6

To see this, note that both sides have Euler products, so it is sufficient to prove that

$$\sum_{k=0}^{\infty} \frac{\sigma_a(p^k)\sigma_b(p^k)}{p^{ks}} = \frac{1-p^{-2s+a+b}}{(1-p^{-s})(1-p^{-s+b})(1-p^{-s+b})(1-p^{-s+a+b})}. \tag{1.1.2}$$

This follows from the

Lemma. *If*

$$\sum_{k=0}^{\infty} f(k)x^k = (1-\alpha x)^{-1}(1-\alpha' x)^{-1},$$

$$\sum_{k=0}^{\infty} g(k)x^k = (1-\beta x)^{-1}(1-\beta' x)^{-1},$$

then

$$\sum_{k=0}^{\infty} f(k)g(k)x^k = \frac{1-\alpha\alpha'\beta\beta' x^2}{(1-\alpha\beta x)(1-\alpha'\beta x)(1-\alpha\beta' x)(1-\alpha'\beta' x)}.$$

Indeed, taking $\alpha = 1$, $\alpha' = p^a$, $\beta = 1$, $\beta' = p^b$, we obtain (1.1.2).

Since we wish to make a point, we will give a proof of the lemma. Denoting

$$\sum_{k=0}^{\infty} f(k)x^k = \phi(x), \qquad \sum_{k=0}^{\infty} g(k)x^k = \psi(x),$$

consider the integral

$$\frac{1}{2\pi i}\int \phi(xz)\psi(z^{-1})\,\frac{dz}{z},$$

where the path of integration circles the origin in the positive sense so that the poles of $\phi(xz)$ are outside the path of integration and the poles of $\psi(z^{-1})$ are inside. This is possible if x is sufficiently small. The integral equals

$$\sum_{k,l} f(k)g(l)x^k \times \frac{1}{2\pi i}\int z^{k-l-1}\,dz = \sum_k f(k)g(k)x^k.$$

Thus it is sufficient to show that

$$\frac{1}{2\pi i}\int (1-\alpha xz)^{-1}(1-\alpha' xz)^{-1}(1-\beta z^{-1})^{-1}(1-\beta' z^{-1})^{-1}\frac{dz}{z}$$

$$=\frac{1-\alpha\alpha'\beta\beta' x^2}{(1-\alpha\beta x)(1-\alpha'\beta x)(1-\alpha\beta' x)(1-\alpha'\beta' x)}. \tag{1.1.3}$$

Indeed, the left side is equal to the sum of the poles at $z=\beta$ and $z=\beta'$, and summing these gives the right-hand side.

In the form (1.1.3), we see that Ramanujan's formula is formally very analogous to Barnes' Lemma [6], which states that

$$\frac{1}{2\pi i}\int_{-i\infty}^{i\infty}\Gamma(\alpha+s)\Gamma(\beta+s)\Gamma(\gamma-s)\Gamma(\delta-s)\,ds$$

$$=\frac{\Gamma(\alpha+\gamma)\Gamma(\alpha+\delta)\Gamma(\beta+\gamma)\Gamma(\beta+\delta)}{\Gamma(\alpha+\beta+\delta+\gamma)}. \tag{1.1.4}$$

Here the line of integration is to be taken to the right of the poles of $\Gamma(\alpha+s)\Gamma(\beta+s)$ and to left of the poles of $\Gamma(\gamma-s)\Gamma(\delta-s)$.

Ingham's proof that $\zeta(1+it)\neq 0$ may now be explained. One takes $a=-b=it$ in Ramanujan's formula and sees that

$$\sum\frac{|\sigma_{it}(n)|^2}{n^s}=\frac{\zeta(s)^2\zeta(s-it)\zeta(s+it)}{\zeta(2s)}. \tag{1.1.5}$$

If $\zeta(1+it)=0$, this has no pole at $s=1$. There are now a number of different ways of deriving a contradiction, since this is a Dirichlet series with positive coefficients. The connection of this proof with the Rankin–Selberg method will be explained in Section 1.5 below.

1.2. The Ramanujan Conjecture

The earliest applications of the Rankin–Selberg method were to the estimation of the Fourier coefficients of modular forms. Here again we encounter Ramanujan. An *automorphic form* for $\Gamma=\mathrm{SL}(2,\mathbf{Z})$ is a function ϕ on the upper half-plane $\mathscr{H}$ with the properties that

1. *We have*

$$\phi\left(\frac{az+b}{cz+d}\right)=(cz+d)^k\phi(z) \tag{1.2.1}$$

for $\begin{pmatrix} a & b \\ c & d\end{pmatrix}\in\Gamma$;

2. *$\phi(iy) = O(y^n)$ for some n as $y \to \infty$;*

Moreover, we shall assume either that

3a. *k is an even positive integer, and ϕ is holomorphic;*

or:

3b. *$k = 0$ and ϕ is an eigenfunction of the noneuclidean Laplacian: if $z = x + iy \in \mathscr{H}$,*

$$= -y^2\left(\frac{\partial^2}{\partial x^2} + \frac{\partial^2}{\partial y^2}\right)\phi(z) = \lambda\phi(z). \tag{1.2.2}$$

In case (3a), we call ϕ a *modular form* of weight k, and in the second, we call ϕ a *Maass form* with eigenvalue λ.

In the special case that ϕ satisfies

$$\int_0^1 \phi(x + iy)\, dx = 0,$$

ϕ is called a *cusp form*. In this case, ϕ has a Fourier expansion

$$\phi(z) = \sum_{n \neq 0} |n|^{(k-1)/2} W(nz). \tag{1.2.3}$$

Note: We shall use the notations $a(n)$ and a_n interchangeably. We have used this particular normalization since it will allow us to give a uniform discussion of Maass and modular forms. Here, if ϕ is a modular form, $W(z) = e^{iz}$, while if ϕ is a Maass form,

$$W(z) = \sqrt{|y|}K_{ir}(2\pi|y|)e^{2\pi i x},$$

where K_ν is a Bessel function, and r is either root of the equation $\frac{1}{4} + r^2 = \lambda$. Negative values of n may occur in (1.2.3) only for Maass forms.

Since the Laplacian is self-adjoint, any eigenvalue λ of the Laplacian must be a positive real number, and therefore r is either real or of the form it with t real, $-\frac{1}{2} < t < \frac{1}{2}$. It was proved by Selberg that the second possibility does not occur for Maass forms that are automorphic for $SL(2, \mathbf{Z})$. (The first published proof of this is due to Roelcke.) Selberg further conjectured, around 1956, that this second possibility also cannot occur for congruence subgroups. Since this amounts to the assertion that the eigenvalue $\lambda \geq \frac{1}{4}$, this is known as Selberg's "$\frac{1}{4}$" conjecture.

One may diagonalize the Maass forms with respect to the linear mapping $\phi(z) \to \phi(-\bar{z})$. Thus we are led to consider only Maass forms satisfying $a(n) = a(-n)$ or $a(n) = -a(-n)$. The first type are called *even*, the second *odd*. *We shall only consider even Maass forms here.*

Ramanujan considered the function $\tau(n)$ defined by

$$\sum_{n=1}^{\infty} \tau(n) q^n = q \prod_{k=1}^{\infty} (1 - q^k)^{24}.$$

With the above normalization, $\tilde{\tau}(n) = n^{-11/2}\tau(n)$ are the Fourier coefficients of the modular form of weight 12, which is denoted Δ. Ramanujan conjectured in 1916 (cf. [74]) the Euler product representation

$$L(s, \Delta) = \sum_{n=1}^{\infty} \tilde{\tau}(n) n^{-s} = \prod_{p} (1 - \tilde{\tau}(p) p^{-s} + p^{-2s})^{-1}.$$

This was proved by Mordell [58]. Ramanujan then observed that the "Hecke polynomial" $1 - \tilde{\tau}(p)x + x^2$ that occurs in the p-th Euler factor would have two complex conjugate roots if $\tilde{\tau}(p) < 2$, instead of two real roots, and on the basis of considerable calculation, conjectured that this would be the case. This is the nonarchimedean analog, for this particular modular form, of Selberg's $\frac{1}{4}$ conjecture. It may be seen, by considering $\tilde{\tau}(p^k)$ for large k, that Ramanujan's conjecture is equivalent to the hypothesis that

$$\tilde{\tau}(n) = O(n^{\varepsilon}) \qquad \text{for any } \varepsilon > 0, \tag{1.2.4}$$

and one may naturally make the same conjecture for any cusp form. For modular forms, this was established in 1970 by Deligne [18], as a corollary to the proof of the Weil conjectures, fulfilling a 1965 prediction of Selberg [81], who wrote that

> ... it seems likely that ... a proof of the Ramanujan conjecture is [most] likely to result from future developments in algebraic geometry.

The Ramanujan conjecture and its analog, Selberg's $\frac{1}{4}$ conjecture, remain open for Maass forms.

1.3. The Rankin–Selberg Method: Functional Equations

Hecke [33] introduced operators $T(p)$ on cusp forms for each p that play a role similar to the Laplacian. These operators are self-adjoint and mutually commutative, hence may be simultaneously diagonalized. If ϕ

is an eigenfunction of the Hecke operators, normalized so that the first Fourier coefficient $a(1) = 1$, then the Fourier coefficients are multiplicative. The L-series is

$$L(s, \phi) = \sum_{n=1}^{\infty} a(n)n^{-s} = \prod_{p} (1 - a(p)p^{-s} + p^{-2s})^{-1}. \qquad (1.3.1)$$

It is wise to follow Ramanujan and to factor each Euler factor so that

$$L(s, \phi) = \prod_{p} (1 - \alpha_p p^{-s})^{-1}(1 - \alpha'_p p^{-s})^{-1}. \qquad (1.3.2)$$

This has a functional equation, namely, that

$$\Lambda(s, \phi) = i^k \Lambda(1 - s, \phi), \qquad (1.3.3)$$

where $\Lambda(s, \phi) = L_\infty(s, \phi)$. Here

$$L_\infty(s, \phi) = \begin{cases} (2\pi)^{-s}\Gamma\left(\frac{1}{2}(k-1) + s\right) & \text{if } \phi \text{ is modular;} \\ \pi^{-s}\Gamma\left(\frac{s+ir}{2}\right)\Gamma\left(\frac{s-ir}{2}\right) & \text{if } \phi \text{ is a Maass form.} \end{cases}$$

The functional equation for $L(s, \Delta)$ was first published by Wilton [99].

Rankin [76] and, independently at around the same time, Selberg [80] introduced a new tool into the study of cusp forms, which is known today as the Rankin–Selberg method. This gives the functional equation of a new kind of Euler product, and as a corollary, gives new estimates for the Fourier coefficients.

The paper of Rankin was motivated by some questions posed by G. H. Hardy. In the first part, which does not depend on the Rankin–Selberg method, Rankin proves, using the technique originally applied to the zeta-function by Hadamard and de la Vallée Poussin, that $L(s, \Delta)$ has no zeros on the line $1 + it$. This depends on the estimate

$$Ax < \sum_{n \le x} \tilde{\tau}(n)^2 < Bx,$$

which Hardy proved as early as 1918 by an ad hoc method depending on the infinite product for Ramanujan's discriminant function. In the second part, Rankin introduces the Rankin–Selberg method, and from it he derives the more refined estimates for the Fourier coefficients of modular forms, to be described below in Section 4.

The paper of Selberg is short—only four pages long. Selberg proves the functional equation and then states (essentially the same) consequences for the Fourier coefficients without proof. Unlike Rankin, Selberg treated the functional equations as the main point of the paper. A later paper [81] of Selberg discusses some more technical points concerning the Rankin–Selberg method.

Let ϕ and ψ be cusp forms with L-series (1.3.2) above, and

$$L(s, \psi) = \prod_p (1 - \beta_p p^{-s})^{-1}(1 - \beta'_\lambda \lambda^{-\nu})^{-1}.$$

The Rankin–Selberg method yields an integral representation of the Euler product

$$L(s, \phi \times \overline{\psi}) = \prod_p (1 - \alpha_p \overline{\beta}_p p^{-s})^{-1}(1 - \alpha'_p \overline{\beta}_p p^{-s})^{-1}$$
$$\times (1 - \alpha_p \overline{\beta}'_\lambda \lambda^{-\nu})^{-1}(1 - \alpha'_p \overline{\beta}'_\lambda \lambda^{-\nu})^{-1}. \qquad (1.3.4)$$

To see this, let us suppose for simplicity that ϕ and ψ have the same weight k—in general one does not make this assumption. We have the Fourier expansions

$$\phi(z) = \sum_{n \neq 0} a(n)|n|^{(k-1)/2} W(nz), \qquad \psi(z) = \sum_{n \neq 0} b(n)|n|^{(k-1)/2} W(nz).$$

We consider the integral

$$\int_{\Gamma \backslash \mathscr{H}} \phi(z)\overline{\psi(z)} y^k E^*(z, s) \frac{dx\, dy}{y^2}, \qquad (1.3.5)$$

where the integral is over a fundamental domain for Γ, and the "Eisenstein series" or Epstein zeta-function

$$E^*(z, s) = \pi^{-s}\Gamma(s)\zeta(2s)E(z, s),$$

$$E(z, s) = \sum_{\gamma \in \Gamma_\infty \backslash \Gamma} y(\gamma z)^s = \frac{1}{2} \sum_{(c, d) = 1} \frac{y^s}{|cz + d|^{2s}} \qquad (1.3.6)$$

is convergent if $\mathrm{re}(s) > 1$. The integral (1.3.5) "unfolds" in the manner that is typical of the Rankin–Selberg method. We may compute this integral in terms of the Fourier expansions, and we see that (1.3.5) is equal to

$$2\pi^{-s}\Gamma(s) \int_0^\infty W(iy)\overline{V(iy)} y^{s+k-1} \frac{dy}{y} \times \zeta(2s) \sum_{n=1}^\infty a(n)\overline{b(n)} n^{-s}. \qquad (1.3.7)$$

Now $E^*(z, s)$ has meromorphic continuation to all values of s with simple poles at $s = 0$ and $s = 1$, and satisfies the functional equation

$$E^*(z, s) = E^*(z, 1 - s). \qquad (1.3.8)$$

This functional equation may be established quite simply by Poisson summation, but for more general Eisenstein series, the analytic continuation and functional equations require the deeper analysis that was introduced by Selberg in [81] and expounded in generality by Langlands [52].

It follows from the lemma of Section 1.1 that the Dirichlet series

$$L(s, \phi \times \overline{\psi}) = \zeta(2s) \sum a(n)\overline{b(n)}n^{-s}.$$

Let

$$\Lambda(s, \phi \times \bar{\psi}) = L_\infty(s, \phi \times \bar{\psi})L(s, \phi \times \bar{\psi}),$$

where

$$L_\infty(s, \phi \times \bar{\psi}) = \begin{cases} (2\pi)^{-2s-k+1}\Gamma(s)\Gamma(s+k-1) & \text{if } \phi \text{ is holomorphic;} \\ \pi^{-2s}\Gamma\left(\frac{s+ir+iq}{2}\right)\Gamma\left(\frac{s-ir+iq}{2}\right)\Gamma\left(\frac{s+ir-iq}{2}\right) & \\ \quad \times \Gamma\left(\frac{s-ir-iq}{2}\right) & \text{if } \phi, \psi \text{ are Maass forms.} \end{cases}$$

In the second case, $\frac{1}{4} + r^2$, $\frac{1}{4} + q^2$ are the eigenvalues of the Laplacian corresponding to ϕ and ψ.

The significance of this is that the integral occuring in (1.3.7) is equal to $L_\infty(s, \phi \times \bar{\psi})$. At least in the second case (Maass forms), the evaluation of the integral may be accomplished by means of Barnes' Lemma (1.1.4).

Thus we obtain from (1.3.8) the functional equation

$$\Lambda(s, \phi \times \bar{\psi}) = \Lambda(1 - s, \phi \times \bar{\psi}). \tag{1.3.9}$$

The residue at $s = 1$ of the Eisenstein series is the constant function. Thus, $L(s, \phi \times \bar{\psi})$ may have a pole at $s = 1$ in which the residue is the inner product of ϕ with ψ. Now, if ϕ and ψ are distinct Hecke eigenforms, they are orthogonal, and hence we see that $L(s, \phi \times \bar{\psi})$ has a pole if and only if $\phi = \psi$.

In [81], Selberg generalizes the integral representation (1.3.7) by replacing the Eisenstein series by a Poincaré series to obtain the meromorphic continuation of the series

$$\sum_{n=1}^{\infty} a(n)\overline{b(n+m)}n^{-s}.$$

The poles of this series are seen to be related to the Eigenvalues of the Laplacian for SL(2, **Z**), for the coefficients in the spectral expansion of the Poincaré series have poles at these locations. A further paper on this topic is Goldfeld [31].

It should be mentioned that frequently one wishes to extend the Rankin–Selberg method to cover forms that are not cuspidal and may, in fact, not decay rapidly at infinity. Zagier [100] has given a convenient formulation for such problems.

1.4. Estimation of Fourier Coefficients

As noted above, the Ramanujan conjecture is the assertion, for Fourier coefficients of an automorphic form, that

$$a(n) = O(n^{\varepsilon}) \tag{1.4.1}$$

for any $\varepsilon > 0$. For modular forms, this was proved by Deligne [18], following on earlier work of Eichler, Shimura, Sato, Ihara, and others, using techniques from algebraic geometry that almost certainly do not generalize to Maass forms (except perhaps those with eigenvalue $\frac{1}{4}$, which, in view of the recent work of Blasius, Clozel, and Ramakrishnan, have perhaps more in common with modular forms than with Maass forms).

The "trivial estimate," which was known to Hardy, is

$$a(n) = O(n^{1/2+\varepsilon}). \tag{1.4.2}$$

The Rankin–Selberg method gives an improvement over this. Simply the fact that the Dirichlet series $\sum |a(n)|^2 n^{-s}$ has its first (simple) pole at $s = 1$ implies, by partial summation, that the Ramanujan conjecture is true "on average," that is, that

$$\sum_{n \le x} a(n) = O(x),$$

which is sufficient for many purposes. On the other hand, a more refined Tauberian argument shows that the functional equation (1.3.9) implies that

$$a(n) = O(n^{3/10+\varepsilon}). \tag{1.4.3}$$

An entirely different method of estimating the Fourier coefficients of automorphic forms comes from the theory of Poincaré series. It might be argued that the fundamental calculation here—the inner product of a Poincaré series with an automorphic form—is an application of the

Rankin–Selberg method, in that the unfolding of the Poincaré series is similar to the unfolding of the Eisenstein series. However, this topic is outside the scope of our discussion. Suffice it to say that the best estimate that comes from this technique is

$$a(n) = O(n^{1/4+\varepsilon}). \tag{1.4.4}$$

This estimate depends on Weil's estimate for Kloosterman sums, and the theory of Poincaré series for Maass forms is another of Selberg's deep contributions. For more information, we refer to Selberg [81] and the references therein. One also obtains from this investigation the analogous estimate $\lambda \geq \frac{3}{16}$ for eigenvalues of the Laplacian on congruence subgroups.

The conjectures of Langlands [54] imply the Ramanujan conjecture. For they imply that the "symmetric power" L-functions

$$L(s, \phi, \vee^k) = \prod_p (1 - \alpha_p^k p^{-s})^{-1}(1 - \alpha_p^{k-1}\alpha'_p p^{-s})^{-1} \cdots (1 - \alpha'^k_p p^{-s})^{-1} \tag{1.4.5}$$

have analytic continuation and functional equation. It was noticed by Serre—see the arguments in the appendix to Chapter I in [83]—that these functional equations imply—for *most* automorphic forms, i.e., for all except those associated with Hecke characters of quadratic fields by means of the Weil representation—not only the Ramanujan conjecture, but the so-called Sato–Tate conjecture, that the angles θ_p defined by

$$2\cos(\theta_p) = \alpha_p + \alpha'_p, \qquad 0 \leq \theta_p \leq \pi,$$

are equidistributed with respect to the measure $\pi^{-1}\sin^2(\theta)\,d\theta$.

For example, the convolution $L(s, \phi \times \bar{\phi})$ equals $\zeta(s)L(s, \phi, \vee^2)$, which shows that the Rankin–Selberg estimate 1.4.3 is within this progression of ideas.

A different method of proving (1.4.4), and the "$\frac{3}{16}$" estimate for eigenvalues of the Laplacian for congruence subgroups, comes from the existence of an automorphic form on GL(3) whose L-function is $L(s, \phi, \vee^2)$. This result, due to Gelbart and Jacquet [23], is discussed in Section 1.8 below as an application of a refinement of the Rankin–Selberg method due to Shimura [88].

The best estimate known for the Fourier coefficients of Maass forms is

$$a(n) = O(n^{1/5}). \tag{1.4.6}$$

This estimate is due to Shahidi (cf. Gelbart and Shahidi [28]). The slightly weaker exponent $\frac{1}{5} + \varepsilon$ was noticed by several people. See also Section 1.8 below.

One may also ask for estimates on the Fourier coefficients of automorphic forms on nonarithmetic Fuchsian groups. Although Selberg [81] obtained a nontrivial estimate for forms on subgroups of finite index in SL(2, **Z**), little was known for arbitrary Fuchsian groups until Good [32] obtained the estimate

$$a(n) = O(n^{1/3+\varepsilon})$$

for holomorphic cusp forms on arbitrary finitely generated Fuchsian groups. This work is based on the Rankin–Selberg method. The essential new idea in this paper was the expansion of $|\phi(z)|^2$ in terms of nonholomorphic automorphic forms and deep careful analysis of the result in terms of Poincaré series.

Phillips and Sarnak [67] showed that for nonarithmetic groups, Maass cusp forms are probably rare. For groups obtained by deforming a given Fuchsian group, they gave a condition for the "destruction" of a given Maass cusp form—the form is destroyed if a special value of the Rankin–Selberg convolution of the form with a certain holomorphic form parametrizing the direction of deformation is nonzero. This is probably almost always the case, so, for example, a group obtained by deforming a congruence subgroup of SL(2, **Z**) would seem unlikely to have many (if any) Maass cusp forms.

1.5. Nonvanishing Theorems

We return to the topic of Section 1.1. It has been noticed by a number of people that the theory of Eisenstein series provides a proof that the Riemann zeta-function does not vanish on the line re(s) = 1. Jacquet and Shalika [45] extended this theorem to a very wide class of Euler products. This proof may be related to the proof of the nonvanishing of Ingham, already mentioned.

The "constant term" of the Eisenstein series,

$$\int_0^1 E^*(x+iy, s)\,dx = \pi^{-s}\Gamma(s)\zeta(2s)y^s + \pi^{s-1}\Gamma(1-s)\zeta(2-2s)y^{1-s}.$$

Supposing that $\zeta(1+it) = 0$, it follows that

$$\int_0^1 E^*\left(x+iy, \frac{1}{2}+\frac{it}{2}\right)dx = 0,$$

so this Eisenstein series is a cusp form. This is a contradiction, because the Eisenstein series is orthogonal to the cusp forms.

To see the connection with Ingham's proof, let us consider the integral

$$\int_{\Gamma\backslash\mathscr{H}} \left| E^*\left(z, \frac{1}{2} + \frac{it}{2}\right) \right|^2 E^*(z, s) \frac{dx\, dy}{y^2}$$

for large s. This integral would be convergent if $\zeta(1 + it) = 0$, owing to the rapid decay of the Bessel functions in the Fourier expansion of $E^*(z, \frac{1}{2} + it/2)$. Computing the integral by the Rankin–Selberg method gives

$$\pi^{-2s}\Gamma\left(\frac{s + it}{2}\right)\Gamma\left(\frac{s}{2}\right)^2 \Gamma\left(\frac{s - it}{2}\right)\zeta(2s) \sum_n |\sigma_{it}(n)|^2 n^{-s}.$$

The residue at $s = 1$ will be the inner product of $E^*(z, \frac{1}{2} + it/2)$ with itself, which is positive, proving again that the Dirichlet series (1.1.5) must have a pole at $s = 1$. Thus we obtain Ingham's proof by taking the residue in the Eisenstein-theoretic proof.

It has been pointed out that this proof may be improved to give a zero-free region.

1.6. The Paper of Doi and Naganuma

The phenomenon known now as base change for modular forms on GL(2) was first discovered empirically by Doi and Naganuma [19]. On a suggestion of Shimura, they were able to establish the phenomenon as an application of the Rankin–Selberg method in [20]. Let $K/\mathbf{Q}$ be a quadratic extension of discriminant D. (Of course one could work over any number field instead of $\mathbf{Q}$.) Given an automorphic form ϕ on GL(2) over the ground field $\mathbf{Q}$, there should be an automorphic form $\hat{\phi}$ on GL(2) over K such that

$$L(s, \hat{\phi}) = L(s, \phi)L\left(s, \phi \times \left(\frac{D}{\cdot}\right)\right). \tag{1.6.1}$$

Here the "twisted" L-function is defined by

$$L\left(s, \phi \times \left(\frac{D}{\cdot}\right)\right) = \sum_n \left(\frac{D}{n}\right) a(n) n^{-s}$$

in terms of the Legendre symbol. This is the precise GL(2) analog (for quadratic extensions) of the familiar factorization of Abelian

L-functions in an extension, from class field theory. Indeed, the conjectures of Langlands [54], which contain class field theory as a special case, show that the factorization of Artin L-functions in an extension should be regarded as a reflection of a similar sort of factorization for L-functions of automorphic forms, of which (1.6.1) would be a special case.

Doi and Naganuma established functional equations of twists of the Dirichlet series defined by (1.6.1) by Hecke characters of K, and from the converse theorem of Weil [97], they concluded that this Dirichlet series is the L-function of an automorphic form. Recall that the converse theorem of Weil asserts that a Dirichlet series, sufficiently many of whose twists by Hecke characters have the right functional equations, is the L-function of an automorphic form. This principle is valid for automorphic forms on GL(2) (cf. Weil [97], and Jacquet and Langlands [39]), and for automorphic forms on GL(3) (cf. Jacquet, Piatetski–Shapiro, and Shalika [42]). Actually, Doi and Naganuma restricted themselves to the case where K is real with class number one, ϕ is holomorphic, and they only considered twists by Hecke characters of conductor one. However, these restrictions were eliminated by Jacquet [37].

In order to prove functional equations of twists of (1.6.1), given a Hecke character χ of K, there exists an automorphic form θ_χ on GL(2) over the ground field $\mathbf{Q}$ with the property that

$$L(s, \theta_\chi) = L(s, \chi).$$

For example, θ_χ may be constructed by means of the Weil representation. It may then be calculated that

$$L(s, \hat{\phi} \times \chi) = L(s, \phi \times \theta_\chi),$$

where the right-hand side is a Rankin–Selberg convolution. Thus the Rankin–Selberg method gives enough functional equations to conclude that (1.6.1) is the L-function of an automorphic form on GL(2).

After this, this method of proving base change went into eclipse due to the successes of Saito [78], Shintani [89], and Langlands [51], who proved base change on GL(2) for cyclic extensions in full generality using the trace formula. However, the Rankin–Selberg method was revived in this context by Jacquet, Piatetski–Shapiro, and Shalika [43], who proved base change by the original method for cubic extensions. For non-normal cubic extensions, this was a new theorem and resulted in the proof of a new case of Artin's conjecture by Tunnell [96].

Automorphic forms for $\Gamma_0(D)$ with the quadratic character as Nebentypus also have lifts which are *a priori* indistinguishable from the lifts from $\mathrm{SL}(2, \mathbf{Z})$. These were considered by Naganuma [59]. Asai [5] showed that the images of the two liftings are orthogonal by showing that the images could be characterized in terms of the poles of certain Rankin–Selberg convolutions. Gelbart and Shahidi [28] give further references for Asai's L-functions.

1.7. *Triple Convolutions*

Given the functional equations of convolutions such as (1.3.4), one is naturally led to consider multiple convolutions. For example, given three cusp forms ϕ, ψ, and ρ, with

$$L(s, \phi) = \prod_p (1 - \alpha_1(p)p^{-s})^{-1}(1 - \alpha_2(p)p^{-s})^{-1},$$

$$L(s, \psi) = \prod_p (1 - \beta_1(p)p^{-s})^{-1}(1 - \beta_2(p)p^{-s})^{-1},$$

and

$$L(s, \rho) = \prod_p (1 - \gamma_1(p)p^{-s})^{-1}(1 - \gamma_2(p)p^{-s})^{-1},$$

consider the convolution

$$L(s, \phi \times \psi \times \rho) = \prod_p \prod_{i,j,k} (1 - \alpha_i(p)\beta_j(p)\gamma_k(p)p^{-s})^{-1}. \tag{1.7.1}$$

There does not seem to be any way of adapting the Rankin–Selberg method of Section 1.3 to prove the functional equation for this. Nevertheless, Garrett [22] has found an interpretation of this Euler product as a Rankin–Selberg integral, but one involving a higher rank group. Specifically, one restricts an Eisenstein series of maximal parabolic type (in the residual spectrum) on $\mathrm{Sp}(6)$ to $\mathrm{SL}(2) \times \mathrm{SL}(2) \times \mathrm{SL}(2)$ and integrates against $\phi \times \psi \times \rho$. This method of restriction of Eisenstein series is closely related to the integrals of Piatetski-Shapiro and Rallis to be considered in Section 3.2. Moreover, Piatetski–Shapiro and Rallis [72] were able to obtain an analog of Asai's L-function [5] for cubic extensions by a modification of Garrett's method.

Moreno has also considered various triple convolutions from a different point of view, namely, the Langlands–Shahidi method. Langlands [53] observed that the functional equations of Eisenstein series on various groups led to the analytic continuation in many instances of what are now known as Langlands L-functions. For example, by use of the exceptional group G_2, he obtained the analytic continuation of the symmetric cube L-function (cf. Section 1.4 above). By making use of an important formula, which is due independently to Kato [46] and to Casselman and Shalika [17], Shahidi [84] refined this method to yield functional equations. Only certain Langlands L-functions may be obtained by this method, of course. Also, there is the problem that one does not obtain complete information on the location of the poles. It should be mentioned that for these Eisenstein series the functional equations must definitely be proved by the methods of spectral theory introduced by Selberg [82], and worked out in detail by Langlands [52], so this is yet another area where the ideas of Selberg are decisive.

Moreno has considered Eisenstein series on groups of type D_n, E_6, E_7, and E_8 that lead to triple convolutions. For example, given the three cusp forms ϕ, ψ, and ρ as above, one finds that there is a maximal parabolic subgroup of spin(8) whose Levi factor is isogenous to three copies of PGL(2). Thus one may use the three forms to construct an Eisenstein series. The formula of Kato–Casselman–Shalika shows that the natural "denominator" of this Eisenstein series is the triple convolution (1.7.1), times a zeta-function, leading to a proof of the functional equation.

However, Garrett's method is superior in that it should in principle yield complete information about the location of the poles of the convolution, and, as Garrett showed, the algebraicity of certain special values, in accordance with a general conjecture of Deligne. The location of the poles is important, because, for example, of the following considerations. It is believed that the convolution (1.3.4) should be the L-function of an automorphic form on GL(4). One might approach this problem by means of the converse theorem. Jacquet, Piatetski–Shapiro, and Shalika believe that they can prove a converse theorem on GL(4) in which the main hypothesis involves twisting the given Dirichlet series by automorphic forms on GL(2). For example, twisting the Dirichlet series (1.3.4) by ρ gives the triple convolution (1.7.1). Thus Garrett's theorem should lead to a proof that (1.3.4) is the L-function of an automorphic form on GL(4). Precise knowledge of the poles is essential in this argument.

1.8. The Symmetric Square

If $\phi = \psi$ in (1.3.4), the convolution is at least formally divisible by

$$\zeta(s) = \prod_p (1 - \alpha_p \alpha'_p p^{-s})^{-1},$$

since $\alpha_p \alpha'_p = 1$. The quotient is the symmetric square

$$L(s, \phi, \vee^2) = \prod_p (1 - \alpha_p^2 p^{-s})^{-1}(1 - \alpha_p \alpha'_p p^{-s})^{-1}(1 - \alpha_p'^2 p^{-s})^{-1}$$

$$= \frac{L(s, \phi \times \bar{\phi})}{\zeta(s)} \tag{1.8.1}$$

This observation proves the functional equation of the symmetric square, but does not show that it is entire, since it is not obvious that the quotient does not have poles at the zeros of $\zeta(s)$. For this, Shimura [88] considered a different sort of Rankin–Selberg convolution. Specifically, he considered a convolution that we will denote as $L(s, \phi \times \theta)$, where

$$\theta(z) = \sum_{-\infty}^{\infty} e^{2\pi i n^2 z}$$

is the classical theta function studied by Jacobi and Riemann. The idea is that the only nonzero coefficients of this theta function are the squares, and so convolving this with ϕ picks off the coefficients of ϕ that are squares, leading to a Dirichlet series that is closely related to the symmetric square. Since Shimura considers holomorphic cusp forms, we shall consider Maass forms here. Suppose ϕ is a Maass form, with eigenvalue $\frac{1}{4} + r^2$. We find that

$$\int_{\Gamma_0(4)\backslash\mathscr{H}} \phi(z)\overline{\theta}(z)\tilde{E}(z, s)\,\frac{dx\,dy}{y^2}$$

$$= \pi^{3/4 - 3s}\Gamma\left(s - \frac{1}{4} + ir\right)\Gamma\left(s - \frac{1}{4}\right)\Gamma\left(s - \frac{1}{4} - ir\right)L\left(2s - \frac{1}{2}, \phi, \vee^2\right), \tag{1.8.2}$$

where the "metaplectic" Eisenstein series

$$\tilde{E}(z, s) = \pi^{1/2 - 2s}\Gamma\left(2s - \frac{1}{2}\right)\zeta(4s - 1)$$

$$\times\, y^{s + 1/4} \sum_{\gamma \in \Gamma_\infty \backslash \Gamma_0(4)} \varepsilon_d \left(\frac{c}{d}\right) \frac{(cz + d)^{-1/2}}{|cz + d|^{2s - 1/2}}.$$

We follow Shimura's notations for the quadratic symbol and the square root function here, and

$$\varepsilon_d = \begin{cases} 1 & \text{if } d \equiv 1 (\text{mod } 4); \\ i & \text{if } d \equiv -1 (\text{mod } 4). \end{cases}$$

This Eisenstein series has analytic continuation, with a simple pole at $s = \frac{3}{4}$ and satisfies a functional equation with respect to $s \to 1 - s$. The precise form of the functional equation is not relevant here, since we already know the functional equation of the symmetric square from (1.8.1). However, the representation (1.8.2) does give us new information, namely, it shows that $L(s, \phi, \vee^2)$ is entire.

Actually, one might well ask whether the symmetric square L-function might not inherit a pole from the pole of the metaplectic Eisenstein series. In fact, since the representation (1.8.1) shows that the $L(s, \phi, \vee^2)$ has no pole at $s = 1$, we see that it does not. The pole of the metaplectic Eisenstein series is, up to a constant, $y^{1/2}\theta(z)$, and so taking residues in (1.8.2) we see that

$$\int_{\Gamma_0(4)\backslash\mathscr{H}} \phi(z)|\theta(z)|^2 y^{1/2} \frac{dx\, dy}{y^2} = 0.$$

Gelbart and Jacquet [23] used this method of Shimura to prove the existence of an automorphic form on GL(3) whose L-function is $L(s, \phi, \vee^2)$, as a consequence of the converse theorem of Jacquet, Piatetski–Shapiro, and Shalika [42]. This has various consequences. It is an immediate consequence that one has the estimate (1.4.4) above—otherwise, the (adelization of the) form on GL(3) would not be an L^2-function, by a classification theorem for unitary representations of $\mathrm{GL}(3, Q_p)$. Indeed, this method is slightly better than Selberg's, since it gives $a(n) = O(n^{1/4})$, a slight improvement over (1.4.4). One also obtains the eigenvalue estimate $\lambda > \frac{3}{16}$ (where Selberg's method gives only $\lambda \geq \frac{3}{16}$) from the classification of unitary representations of $\mathrm{GL}(3, \mathbf{R})$.

Moreover, one may feed the lifted form back into the Rankin–Selberg method, or the method of Langlands and Shahidi (which was discussed in Section 1.7). Thus one obtains the stronger estimate (1.4.6) or the functional equations of $L(s, \phi, \vee^n)$ for $n \leq 5$. For example, $L(s, \phi, \vee^5)$ occurs in the Fourier coefficients of Eisenstein series on either the group F_4 or E_8—the construction of the Eisenstein series involved in either of these examples requires the symmetric square lift to GL(3). These examples were considered by Moreno and Shahidi. Furthermore, the symmetric square lift was an essential ingredient in the proof of certain cases of Artin's conjecture by Langlands [51].

One may ask whether higher symmetric powers may be realized as Rankin–Selberg convolutions. The natural idea would be to construct the L-functions as convolutions with generalized theta series on the higher metaplectic groups. Several people have tried to construct such convolutions without success, but the idea remains a tantalizing one.

2. The Rankin–Selberg Method for the General Linear Groups

Godement and Jacquet [30] extended the Euler products $L(s, \phi)$ and their functional equations (1.3.3) to the case where ϕ is an automorphic form on $\mathrm{GL}(n)$. This possibility was known to Selberg. The Euler product will have the form

$$L(s, \phi) = \prod_p (1 - \alpha_1(p)p^{-s})^{-1} \cdots (1 - \alpha_n(p)p^{-s})^{-1}.$$

There will be some gamma factors to go with this Euler product and a functional equation with respect to $s \to 1 - s$. If now ψ is an automorphic form on $\mathrm{GL}(m)$, with

$$L(s, \psi) = \prod_p (1 - \beta_1(p)p^{-s})^{-1} \cdots (1 - \beta_m(p)p^{-s})^{-1},$$

then the Euler product which generalizes that considered in Section 3 is

$$L(s, \phi \times \psi) = \prod_p \prod_{i,j} (1 - \alpha_i(p)\beta_j(p)p^{-s})^{-1}.$$

Jacquet, Piatetski–Shapiro, and Shalika [41] considered how to generalize the method of Rankin and Selberg to give the functional equation of this Euler product. In this generality, the Rankin–Selberg convolution contains the functional equation of $L(s, \phi)$ as a special case, for, as we have mentioned, an automorphic form on $\mathrm{GL}(1)$ is a Hecke character, and so taking ψ to be the principal character on $\mathrm{GL}(1)$ gives $L(s, \phi)$.

Much consideration must be given to the attempt to classify different methods of proving functional equations of L-functions. Even in the restricted domain of direct integral representations, there would seem to be two different types of integrals. For example, Piatetski–Shapiro

[69] distinguishes between those integral representations that involve Eisenstein series and those that do not. If only the former are entitled to the title "Rankin–Selberg convolution," then the integrals under consideration here are only "Rankin–Selberg convolutions" in the special case $n = m$. Nevertheless, these integrals are called Rankin–Selberg convolutions in the literature [41]. One may argue that the two different types of integral are actually not as different as one might think if a philosophical view is taken. To illustrate this point, one may cite the example of the integral representations of the exterior square L-functions on $\mathrm{GL}(n)$ that were exhibited by Jacquet, Piatetski–Shapiro, and Shalika [40]. If n is even, these integrals involve Eisenstein series. If n is odd, they do not. Nevertheless, this difference between the two cases is not so extreme that one would care to consider them fundamentally different.

As especially important property of these convolutions is that $L(s, \phi \times \psi)$ has a pole at $s = 1$ if and only if ϕ and ψ are contragredient forms. This observation has important applications—a typical example occurs in the proof of certain cases of Artin's conjecture by Langlands [51]. Another important consequence is the settling of some technical problems in representation theory by Jacquet and Shalika [44].

The theory of Jacquet, Piatetski–Shapiro, and Shalika will be the main topic of this section. Our object will be to give an elementary discussion of this theory, avoiding the use of representation theory at the cost of some generality. We feel that this will be of more use than simply repeating what is in the literature. It should be mentioned that another "classical" survey of this topic may be found in the survey article of Jacquet [38].

In general, I have avoided the problems caused by a finite number of "bad primes" here by considering only forms of level one. See the end of Section 3.4 for further remarks.

2.1. *Fourier Expansions and a Theorem of Shintani*

We shall restrict our attention to the analog of (even) Maass forms for $\Gamma = \mathrm{GL}(n, \mathbf{Z})$. Let $G = \mathrm{GL}(n, \mathbf{R})$, $K = O(n)$, and let Z be the center of G. Our automorphic functions will be functions on the homogeneous space $\mathscr{H} = G/KZ$. It will be convenient to introduce coordinates on $\mathscr{H}$

as follows. Each element of $\mathscr{H}$ has a unique coset representative of the form

$$g = \begin{pmatrix} 1 & x_{12} & x_{13} & \cdots & x_{1n} \\ & 1 & x_{23} & \cdots & x_{2n} \\ & & \ddots & & \vdots \\ & & & 1 & x_{n-1,n} \\ & & & & 1 \end{pmatrix} \begin{pmatrix} y_1 \cdots y_{n-1} & & & & \\ & y_2 \cdots y_{n-1} & & & \\ & & \ddots & & \\ & & & y_{n-1} & \\ & & & & 1 \end{pmatrix} \tag{2.1.1}$$

where the y_i are positive. We will find the x_{ij} and the y_i to be a convenient set of coordinates on $\mathscr{H}$.

Let $\Gamma = GL(n, \mathbf{Z})$, let G_∞ be the subgroup of unipotent upper triangular matrices, and let $\Gamma_\infty = G_\infty \cap \Gamma$. The algebra $\mathscr{D}$ of G-invariant differential operators on $\mathscr{H}$ is a polynomial ring over $\mathbf{C}$ in $n-1$ variables. Generators for this ring were found by Selberg [82]. We shall consider functions ϕ on $\mathscr{H}$ that satisfy:

1. We have $\phi(\gamma g) = \phi(g)$ for $\gamma \in \Gamma$.
2. There exists a character $\rho = \rho_\phi$ of $\mathscr{D}$ such that $D\phi = \rho(D)\phi$ for $D \in \mathscr{D}$.
3. In terms of the coordinates (2.1.1), the function $\phi(g)$ is of at most polynomial growth as $y_1, \ldots, y_{n-1} \to \infty$.

Such a function will be called an *automorphic form*. It will be further called a *cusp form* if, as usual, the integral over the group of horocycles attached to each maximal parabolic subgroup of G is zero.

We may define *Hecke operators* acting on the functions ϕ as follows (cf. Tamagawa [94] or Shimura [86] for the algebraic part of this theory). Let p be any prime. Let $\xi = \xi(p, i)$ be the matrix

$$\xi(p, i) = \begin{pmatrix} p & & & & & \\ & \ddots & & & & \\ & & p & & & \\ & & & 1 & & \\ & & & & \ddots & \\ & & & & & 1 \end{pmatrix}.$$

Here there are to be i diagonal entries equal to p, and $r - i$ equal to 1. The Hecke operator $T_{p,i}$ is defined as follows. Let us decompose the double coset $\Gamma\xi(p, i)\Gamma$ into a finite number of left cosets $\bigcup_\nu \Gamma\gamma_{p,i,\nu}$. Then by definition

$$(T_{p,i}\phi)(g) = \sum_\nu \phi(\gamma_{p,i,\nu}g).$$

The Hecke operators are normal with respect to the inner product on the space of cusp forms, and are commutative, so they may be simultaneously diagonalized. Thus there exist Eigenvalues $\lambda_{p,i}$ such that

$$T_{p,i}\phi = \lambda_{p,i}p^{i(r-i)/2}\phi.$$

There is also a contragredient form

$$\hat{\phi}(g) = \phi({}^\iota g),$$

where $\iota\colon G \to G$ is the involution

$${}^\iota g = w{}^tg^{-1}w, \qquad w = \begin{pmatrix} & & & 1 \\ & & -1 & \\ & \cdot^{\cdot^{\cdot}} & & \\ \pm 1 & & & \end{pmatrix}.$$

The eigenvalues of the Hecke operators on $\hat{\phi}$ are the eigenvalues of ϕ in reverse order:

$$\hat{\lambda}_{p,i} = \lambda_{p,r-i}.$$

Let

$$L(s, \phi) = \prod_p (1 - \lambda_{p,1}p^{-s} + \lambda_{p,2}p^{-2s} + \cdots \pm p^{-rs})^{-1}.$$

We may factor these Euler local factors product as follows. Let

$$1 - \lambda_{p,1}p^{-s} + \lambda_{p,2}p^{-2s} + \cdots \pm p^{-rs} = \prod_{k=1}^{n} (1 - \alpha_k(p)p^{-s}). \quad (2.1.2)$$

It is a consequence of the local Multiplicity One Theorem of Shalika [85] that there exists a unique function $W = W_\rho$ on $\mathscr{H}$ such that

1. We have

$$W\left(\begin{pmatrix} 1 & x_{12} & \cdots & x_{1n} \\ & 1 & \cdots & x_{2n} \\ & & \ddots & \vdots \\ & & & 1 \end{pmatrix} g\right) = e(x_{12} + \cdots + x_{n-1,n})W(g) \quad (2.1.3)$$

for real x_{ij}, $g \in \mathscr{H}$.

2. For the character ρ of $\mathscr{D}$, we have $DW = \rho(D)W$ for all $D \in \mathscr{D}$.
3. In terms of the coordinates (2.1.1), the function $W(g)$ is of at most polynomial growth as $y_1, \ldots, y_{n-1} \to \infty$.

In fact, this function W is of very rapid decay. For example, if $n = 2$,

$$W\begin{pmatrix} y & \\ & 1 \end{pmatrix} = \sqrt{y}K_{\nu - 1/2}(2\pi y),$$

where the parameter ν depends on ρ. For further information about characters ρ of $\mathscr{D}$ and about Whittaker functions, we refer to Sections 2.3 and 2.6 below.

Without condition (3), the space of solutions to (1) and (2) would have dimension $n!$, as was shown by Casselman and Zuckerman.

If ϕ is an automorphic form, and $m_1, \ldots, m_{n-1}$ are integers, let

$$\phi_{m_1,\ldots,m_{n-1}}(g) = \int_0^1 \cdots \int_0^1 \phi\left(\begin{pmatrix} 1 & x_{12} & \cdots & x_{1n} \\ & 1 & \cdots & x_{2n} \\ & & \ddots & \vdots \\ & & & 1 \end{pmatrix} g\right) \times e(-m_1 x_{12} - \cdots - m_{n-1} x_{n-1,n})\, dx_{12} \cdots dx_{n-1,n}. \tag{2.1.4}$$

If $m_1, \ldots, m_{n-1}$ are all nonzero, the local Multiplicity One Theorem implies that there exists a "Fourier coefficient" $a(m_1, \ldots, m_{n-1})$ such that

$$\phi_{m_1,\ldots,m_{n-1}}(g) = a(m_1, \ldots, m_{n-1}) \prod_{k=1}^{n-1} m_k^{-k(n-k)/2} \times W\left(\begin{pmatrix} m_1 \cdots m_{n-1} & & & \\ & m_2 \cdots m_{n-1} & & \\ & & \ddots & \\ & & & 1 \end{pmatrix} g\right). \tag{2.1.5}$$

If ϕ is cuspidal, then $\phi_{m_1,\ldots,m_{n-1}} = 0$ unless $m_1, \ldots, m_{n-1}$ are all nonzero. On the other hand, if ϕ is not cuspidal, there will be other types of Fourier coefficients, involving "degenerate" Whittaker functions, which are built up from Whittaker functions on lower rank groups.

If ϕ is cuspidal, there is a Fourier expansion, due to Piatetski–

Shapiro [68] and (independently), to Shalika [85]. Let Γ_0 be the subgroup of Γ whose elements have bottom row $(0, \ldots, 0, 1)$. Then

$$\phi(g) = \sum_{m_i} \sum_{\gamma \in \Gamma_\infty \backslash \Gamma_0} a(m_1, \ldots, m_{n-1}) \prod_{k=1}^{n-1} m_k^{-k(n-k)/2}$$

$$\times W\left(\begin{pmatrix} m_1 \cdots m_{n-1} & & & \\ & \ddots & & \\ & & m_{n-1} & \\ & & & 1 \end{pmatrix} \gamma g\right). \tag{2.1.6}$$

If ϕ is not cuspidal, there is still a Fourier expansion, but this is complicated by the presence of degenerate terms.

Suppose now that ϕ is cuspidal and a Hecke eigenform. Shintani [91] proved a theorem in representation theory that implies a useful expression for the Fourier coefficients. Firstly, let us point out that they are multiplicative: If $m_1 \cdots m_{n-1}$ and $m'_1 \cdots m'_{n-1}$ are coprime, then

$$a(m_1 m'_1, \ldots, m_{n-1} m'_{n-1}) = a(m_1, \ldots, m_{n-1}) a(m'_1, \ldots, m'_{n-1}). \tag{2.1.7}$$

Thus it is most essential to compute $a(p^{k_1}, \ldots, p^{k_{n-1}})$ where p is a prime. Shintani's Theorem shows that in terms of the parameters α_i defined by (2.1.2), we have

$$a(p^{k_1}, \ldots, p^{k_{n-1}}) = S_{k_1, \ldots, k_{n-1}}(\alpha_1, \ldots, \alpha_n), \tag{2.1.8}$$

where $S_{k_1, \ldots, kn-1}$ is a *Schur polynomial* whose definition we now recall.

The symmetric group S_n, and hence also its group algebra, act on the ring of polynomials $\mathbf{C}[x_1, \ldots, x_{n-1}]$ by permuting the subscripts of the variables. Let $\mathscr{A}$ be the alternator $\sum \operatorname{sgn}(\sigma) \cdot \sigma$ in the group algebra of S_n. Every alternating polynomial is divisible by the discriminant

$$\mathscr{A}\left(\prod_{r=1}^{n-1} x_{r+1}^r\right) = \prod_{i>j} (x_i - x_j).$$

Thus, if $k_1, \ldots, k_{n-1}$ are non-negative integers,

$$S_{k_1, \ldots, k_{n-1}}(x_1, \ldots, x_n) = \frac{\mathscr{A}\left(\prod_{r=1}^{n-1} x_{r+1}^{k_{n-r} + \cdots + k_{n-1} + r}\right)}{\mathscr{A}\left(\prod_{r=1}^{n-1} x_{r+1}^r\right)}$$

is a symmetric polynomial in $x_1, \ldots, x_n$. These are the polynomials that appear in (2.1.7).

If ϕ had a nontrivial character χ, we would have $\prod \alpha_i(p) = \chi(p)$. However, our assumptions imply that $\chi = 1$.

The Schur polynomials occur in the Weyl character formula for $\mathrm{GL}(n, \mathbf{C})$. Indeed, there exists a representation $\rho_{k_1,\ldots,k_{n-1}}$ of $\mathrm{GL}(n, \mathbf{C})$ such that if A is semisimple and has eigenvalues $\alpha_1, \ldots, \alpha_n$, then

$$\mathrm{tr}\, \rho_{k_1,\ldots,k_{n-1}}(A) = S_{k_1,\ldots,k_{n-1}}(\alpha_1, \ldots, \alpha_n).$$

In Shintani's formula, one applies this to the semisimple conjugacy class A that is associated with ϕ by the conjectures of Langlands [54]. This point of view suggests a generalization to reductive groups, and indeed such a generalization has been found by Kato [46] and by Casselman and Shalika [17].

2.2. A Classical Formula of D. E. Littlewood and MacMahon

The problem arises of generalizing the lemma of Section 1.1 to cover convolutions on $\mathrm{GL}(n)$ with $\mathrm{GL}(m)$. We find the precise formula that we require in Macdonald [56], formula I.4.3 (cf., also example 6 on p. 38). The same formula may be found in Littlewood [55], IV.6.4 Theorem V, who relates it to a formula of MacMahon. The discussion in Jaquet and Shalika [44], Section 2 essentially gives a new proof of this classical formula by giving it a concrete interpretation in terms of the representation theory of the symmetric group. This is a model for how things like this should be proved. In the discussion in the last two sections of Jacquet, Piatetski–Shapiro, and Shalika [41], it is shown that the local convolutions are compatible with parabolic induction, which implies the result at hand and much more. A result of this type is the most satisfactory that one could ask for.

Let $(\alpha) = (\alpha_1, \ldots, \alpha_n)$ and $(\beta) = (\beta_1, \ldots, \beta_m)$ be given. Let $A = \prod \alpha_i$, $B = \prod \beta_i$. Littlewood's formula is equivalent to

Theorem. *If $n = m$, we have*

$$\prod (1 - \alpha_i \beta_j x)^{-1}$$
$$= (1 - ABx^n)^{-1} \sum S_{k_1,\ldots,k_{n-1}}(\alpha) S_{k_1,\ldots,k_{n-1}}(\beta) x^{k_1 + 2k_2 + \cdots + (n-1)k_{n-1}}.$$

If $n < m$, we have

$$\prod (1 - \alpha_i \beta_j x)^{-1} = \sum S_{k_1,\ldots,k_{n-1}}(\alpha) S_{k_1,\ldots,k_n,1,\ldots,1}(\beta) A^{k_n} x^{k_1 + 2k_2 + \cdots + nk_n}.$$

We shall briefly indicate the connection of this with the formula as stated by Macdonald. Note that if λ is a partition, the Schur polynomial

$s_\lambda(\alpha)$ (in Macdonald's notation) vanishes if the length of λ is greater than n, while if

$$\lambda = (k_1 + \cdots + k_n, k_2 + \cdots + k_n, \ldots, k_n),$$

we have

$$s_\lambda(\alpha) = A^{k_n} S_{k_1,\ldots,k_{n-1}}(\alpha).$$

From this, one easily obtains our theorem from the formula as stated by Macdonald, and from the fact that $S_{k_1,\ldots,k_{n-1}}$ is homogeneous of degree $k_1 + 2k_2 + \cdots + (n-1)k_{n-1}$. In the case $n = m$, the extra factor $(1 - ABx^n)^{-1}$ comes from summing a geometric series.

If $n = m = 2$, this is equivalent to the lemma in Section 1.1.

Combining this formula (with $x = p^{-s}$) with formulas (2.1.6) and (2.1.7), we obtain a formula for the convolution. Let ϕ and ψ be as in the introduction to this section. The assumptions in Section 2.1 imply that the characters associated with ϕ and ψ are trivial, so that $\prod \alpha_i = \prod \beta_j = 1$. Let a and b be the Fourier coefficients of ϕ and ψ. We obtain in the case $n = m$,

$$L(s, \phi \times \psi) = \zeta(ns) \sum a(m_1, \ldots, m_{n-1}) b(m_1, \ldots, m_{n-1}) \times (m_1 m_2^2 \cdots m_{n-1}^{n-1})^{-s}, \tag{2.2.1}$$

and in the case $n < m$, we obtain

$$L(s, \phi \times \psi) = \sum a(m_1, \ldots, m_{n-1}) b(m_1, \ldots, m_n, 1, \ldots, 1) \times (m_1 m_2^2 \cdots m_n^n)^{-s}. \tag{2.2.2}$$

As a special case, we have the formula

$$L(s, \phi) = \sum a(n, 1, \ldots, 1) n^{-s},$$

where $L(s, \phi)$ is defined in the introduction to this section.

2.3. *Gamma Factors, Whittaker Functions and Functional Equations*

Let B be the subgroup of G of matrices of the form

$$b = \begin{pmatrix} y_1 \cdots y_n & * & * & \cdots & * \\ & y_2 \cdots y_n & * & \cdots & * \\ & & \ddots & & \vdots \\ & & & y_{n-1} y_n & * \\ & & & & y_n \end{pmatrix} \tag{2.3.1}$$

where the y_i are assumed positive. Also, let B_1 be the subgroup of B such that $y_n = 1$. By the Iwasawa decomposition, each coset in $\mathscr{H}$ has a unique representative that is an element of B_1 (i.e., the matrix (2.1.1)).

For some purposes, one prefers on B the coordinates

$$\eta_i = y_i y_{i+1} \cdots y_n.$$

Let δ be the *modulus* on B, i.e., the ratio between left and right Haar measures. Thus

$$\delta(b) = \prod_{i=1}^{n-1} y_i^{i(n-i)} = \prod_{i=1}^{n} \eta_i^{n-1-2i}. \tag{2.3.2}$$

The characters ρ of $\mathscr{D}$ may be classified as follows. Let $a = (a_1, \ldots, a_n)$ be given such that $a_1 + \cdots + a_n = 0$. We have a character χ_a of B given by

$$\chi_a(b) = \sqrt{\delta(b)} \prod_{i=1}^{n} \eta_i^{a_1}. \tag{2.3.3}$$

We extend χ_a to all of G by requiring it to be right invariant by K. This function is an eigenfunction of $\mathscr{D}$, and we thus obtain a character ρ_a defined by

$$D\chi_a = \rho_a(D)\chi_a \qquad \text{for } D \in \mathscr{D}. \tag{2.3.4}$$

It may be shown that the character ρ_a is invariant under permutations of $a_1, \ldots, a_n$ and that every character is obtained in this manner.

We shall find it convenient to use the following normalization for the Whittaker functions. If $\rho = \rho_a$ denote

$$K_a(y_1, \ldots, y_{n-1}) = \delta(g)^{-1/2} W_\rho(g), \tag{2.3.5}$$

where

$$g = \begin{pmatrix} y_1 \cdots y_{n-1} & & & \\ & y_2 \cdots y_{n-1} & & \\ & & \ddots & \\ & & & 1 \end{pmatrix}.$$

Thus if $n = 2$, $K_{\nu, -\nu}(y)$ coincides with the Bessel function $K_\nu(2\pi y)$. We may also use the notation K_ρ for this function K_a.

Our object is to state the functional equations for $L(s, \phi)$ and $L(s, \phi \times \psi)$. Let $\hat{\phi}$ be the contragredient form as before. If $\rho_\phi = \rho_a$, then

$\rho_{\hat{\phi}} = \rho_{-a}$, where $-a = (-a_1, \ldots, -a_n)$. The Fourier coefficients of $\hat{\phi}$ are the same as the Fourier coefficients of ϕ, but in reverse order:

$$\hat{a}(m_1, \ldots, m_{n-1}) = a(m_{n-1}, \ldots, m_1). \tag{2.3.6}$$

There is a gamma factor

$$\Phi(s, \phi) = \pi^{-ns/2} \prod_{i=1}^{n} \Gamma\left(\frac{s - a_i}{2}\right). \tag{2.3.7}$$

We have the functional equation

$$\Phi(s, \phi)L(s, \phi) = \Phi(1 - s, \hat{\phi})L(1 - s, \hat{\phi}). \tag{2.3.8}$$

This was proved by Godement and Jacquet [30]. A second proof, which is a special case of the technique used in the Rankin–Selberg convolutions, was obtained by Jacquet, Piatetski–Shapiro, and Shalika. The case $n = 3$ is explained in [42].

If also $\rho_{\psi} = \rho_b$, where $b = (b_1, \ldots, b_m)$, then we have the gamma factor

$$\Phi(s, \phi \times \psi) = \pi^{-nms/2} \prod_{i=1}^{n} \prod_{j=1}^{m} \Gamma\left(\frac{s - a_i - b_j}{2}\right). \tag{2.3.9}$$

Then we have

$$\Phi(s, \phi \times \psi)L(s, \phi \times \psi) = \Phi(1 - s, \hat{\phi} \times \hat{\psi})L(1 - s, \hat{\phi} \times \hat{\psi}). \tag{2.3.10}$$

Our object in this section is to give an indication of how these functional equations come about.

2.4. *The Convolution When $n = m$*

When $n = m$, the Rankin–Selberg convolution is an integral against an Eisenstein series. We shall consider this particular case in some detail. A still more detailed account of the case $n = m = 3$ may be found in Friedberg [21].

Let $I(g, s)$ be the function defined on $\mathscr{H}$ whose value on the matrix (2.1.1) is the power $(y_1 y_2^2 \cdots y_n^n)^s$ of the determinant. Let Γ_0 be the subgroup of Γ whose elements have bottom row $(0, \ldots, 0, 1)$. The Eisenstein series or Epstein zeta-function

$$E(g, s) = \pi^{-ns/2}\Gamma\left(\frac{ns}{2}\right)\zeta(ns) \sum_{\Gamma_0 \backslash \Gamma} I(\gamma g, s)$$

is convergent if $\mathrm{re}(s) > 1$. It has meromorphic continuation to all s with simple poles at $s = 0, 1$ and a functional equation

$$E(g, s) = E({}^{\iota}g, 1 - s). \tag{2.4.1}$$

We consider the convolution

$$\int_{\Gamma\backslash\mathscr{H}} \phi(g)\psi(g)E(g, s)\, dg, \tag{2.4.2}$$

where dg is the measure on $\mathscr{H}$ induced by Haar measure on G. The integral (2.4.2) is rapidly convergent if ϕ and ψ are cuspidal. If $\mathrm{re}(s) > 1$, this integral may be transformed as follows. Unfolding the definition of $E(g, s)$, (2.4.2) equals

$$\pi^{-ns/2}\Gamma\left(\frac{ns}{2}\right)\zeta(ns)\int_{\Gamma_0\backslash\mathscr{H}} \phi(g)\psi(g)I(g, s)\, dg.$$

We may now further unfold the summation over γ in the Fourier expansion (2.1.6) to obtain

$$\pi^{-ns/2}\Gamma\left(\frac{ns}{2}\right)\zeta(ns)\sum_{m_i} a(m_1, \dots, m_{n-1})\prod_{k=1}^{n-1} m_k^{-k(n-k)/2}$$
$$\times \int_{\Gamma_\infty\backslash\mathscr{H}} W_{\rho(\phi)}\left(\begin{pmatrix} m_1 \cdots m_{n-1} & & & \\ & \ddots & & \\ & & m_{n-1} & \\ & & & 1\end{pmatrix} g\right)\psi(g)I(g, s)\, dg,$$

where $\rho(\phi)$ is the character of $\mathscr{D}$ associated with ϕ.

If we choose the coset representative (2.1.1) for g, which is an element of B_1, so that $\delta(g)$ is defined, we may write

$$dg = \delta(g)^{-1} dx_{12} \cdots dx_{n-1,n}\frac{dy_1}{y_1}\cdots\frac{dy_{n-1}}{y_{n-1}}.$$

Using property (2.1.3) of the Whittaker function we obtain

$$\pi^{-ns/2}\Gamma\left(\frac{ns}{2}\right)\zeta(ns)\sum_{m_i} a(m_1, \dots, m_{n-1})\prod_{k=1}^{n-1} m_k^{-k(n-k)/2}$$
$$\times \int_0^\infty \cdots \int_0^\infty W_{\rho(\phi)}\left(\begin{pmatrix} m_1 \cdots m_{n-1} & & & \\ & \ddots & & \\ & & m_{n-1} & \\ & & & 1\end{pmatrix}\right.$$

$$\times \begin{pmatrix} y_1 \cdots y_{n-1} & & & \\ & \ddots & & \\ & & y_{n-1} & \\ & & & 1 \end{pmatrix}\Bigg)\Bigg)$$

$$\times \int_0^1 \cdots \int_0^1 \psi\left(\begin{pmatrix} 1 & x_{12} & \cdots & x_{1n} \\ & 1 & \cdots & x_{2n} \\ & & \ddots & \vdots \\ & & & 1 \end{pmatrix}\begin{pmatrix} y_1 \cdots y_{n-1} & & & \\ & \ddots & & \\ & & y_{n-1} & \\ & & & 1 \end{pmatrix}\right)$$

$$\times e(m_1 x_{12} + \cdots + m_{n-1} x_{n-1,n})\, dx_{12} \cdots dx_{n-1,n} (y_1 y_2^2 \cdots y_{n-1}^{n-1})^s$$

$$\times \prod_{k=1}^{n-1} y_k^{-k(n-k)} \frac{dy_1}{y_1} \cdots \frac{dy_{n-1}}{y_{n-1}}.$$

Now invoking (2.1.4), while remembering that $b(m_1, \ldots, m_{n-1}) = b(-m_1, \ldots, -m_{n-1})$, and applying (2.2.1), we see that (2.4.2) equals

$$\Psi(s, \phi \times \psi) L(s, \phi \times \psi),$$

where

$$\Psi(s, \phi \times \psi) = \pi^{-ns/2} \Gamma\left(\frac{ns}{2}\right) \int_0^\infty \cdots \int_0^\infty W_{\rho(\phi)} \begin{pmatrix} y_1 \cdots y_{n-1} & & & \\ & \ddots & & \\ & & y_{n-1} & \\ & & & 1 \end{pmatrix}$$

$$\times W_{\rho(\psi)} \begin{pmatrix} y_1 \cdots y_{n-1} & & & \\ & \ddots & & \\ & & y_{n-1} & \\ & & & 1 \end{pmatrix}$$

$$\times (y_1 y_2^2 \cdots y_{n-1}^{n-1})^s \prod_{k=1}^{n-1} y_k^{-k(n-k)} \frac{dy_1}{y_1} \cdots \frac{dy_{n-1}}{y_{n-1}}$$

$$= \pi^{-ns/2} \Gamma\left(\frac{ns}{2}\right) \int_0^\infty \cdots \int_0^\infty K_{\rho(\phi)}(y_1, \ldots, y_{n-1})$$

$$\times K_{\rho(\psi)}(y_1, \ldots, y_{n-1}) (y_1 y_2^2 \cdots y_{n-1}^{n-1})^s \frac{dy_1}{y_1} \cdots \frac{dy_{n-1}}{y_{n-1}}.$$

Now, in order to deduce the functional equation (2.3.10) from the functional equation (2.4.1) of the Eisenstein series, it is necessary to show that $\Psi(s, \phi)$ has meromorphic continuation and satisfies the "local functional equation"

$$\frac{\Psi(s, \phi \times \psi)}{\Phi(s, \phi \times \psi)} = \frac{\Psi(1 - s, \hat{\phi} \times \hat{\psi})}{\Phi(1 - s, \hat{\phi} \times \hat{\psi})}. \tag{2.4.3}$$

This problem will be considered in Section 2.6.

2.5. *The Convolution When n < m*

If $n < m$, the Rankin–Selberg convolution no longer involves an Eisenstein series. The case where $m = n + 1$ is the easiest, and we consider this case now. We embed GL(n) in GL($n + 1$) into the upper right-hand corner and integrate against the determinant thus:

$$\int_{\Gamma\backslash G} \phi(g)\psi\begin{pmatrix} g & \\ & 1 \end{pmatrix} |\det(g)|^{s - 1/2}\, dg, \tag{2.5.1}$$

where dg is Haar measure on $G = \mathrm{GL}(n, \mathbf{R})$. If re $(s) > 1$, we may unfold this convolution as follows.

Note that we may rewrite the Fourier expansion (2.1.7) for GL($n + 1$) as

$$\phi\begin{pmatrix} g & \\ & 1 \end{pmatrix} = \sum_{m_i} \sum_{\gamma \in \Gamma_\infty \backslash \Gamma} b(m_1, \ldots, m_n) \prod_{k=1}^{n} m_k^{-k(n+1-k)/2}$$
$$\times W\left(\begin{pmatrix} m_1 \cdots m_n & & & \\ & \ddots & & \\ & & m_n & \\ & & & 1 \end{pmatrix}\begin{pmatrix} \gamma g & \\ & 1 \end{pmatrix}\right).$$

Unfolding the summation over γ, (2.5.1) equals

$$\sum_{m_i} b(m_1, \ldots, m_n) \prod_{k=1}^{n} m_k^{-k(n+1-k)/2} \int_0^\infty \cdots \int_0^\infty$$
$$\times \int_0^1 \cdots \int_0^1 \phi\left(\begin{pmatrix} 1 & x_{12} & \cdots & x_{1n} \\ & 1 & \cdots & x_{2n} \\ & & \ddots & \vdots \\ & & & 1 \end{pmatrix}\begin{pmatrix} y_1 \cdots y_n & & & \\ & \ddots & & \\ & & y_{n-1}y_n & \\ & & & y_n \end{pmatrix}\right)$$
$$\times e(m_1 x_{12} + \cdots + x_{n-1,n})\, dx_{ij}$$

$$\times W_{\rho(\phi)}\left(\begin{pmatrix} m_1 \cdots m_{n-1} & & & \\ & \ddots & & \\ & & m_{n-1} & \\ & & & 1 \end{pmatrix}\begin{pmatrix} y_1 \cdots y_n & & & \\ & \ddots & & \\ & & y_{n-1}y_n & \\ & & & y_n \end{pmatrix}\right)$$

$$\times \prod_{k=1}^{n} y_k^{-k(n+1/2-k)}(y_1y_2^2\cdots y_n^n)^s \frac{dy_1}{y_1}\cdots\frac{dy_n}{y_n}.$$

Now invoking (2.1.4) and (2.2.2), we see that (2.5.1) equals

$$\Psi(s, \phi \times \psi)L(s, \phi \times \psi),$$

where this time

$$\Psi(s, \phi \times \psi) = \int_0^\infty \cdots \int_0^\infty K_{\rho(\phi)}(y_1, \ldots, y_{n-1}) \times K_{\rho(\psi)}(y_1, \ldots, y_n)(y_1y_2^2\cdots y_n^n)^s \frac{dy_1}{y_1}\cdots\frac{dy_n}{y_n}. \tag{2.5.2}$$

Now the functional equation will follow from the substitution $g \to {}^{\iota}g$, provided we prove the local functional equation

$$\frac{\Psi(s, \phi \times \psi)}{\Phi(s, \phi \times \psi)} = \frac{\Psi(1-s, \hat{\phi} \times \hat{\psi})}{\Phi(1-s, \hat{\phi} \times \hat{\psi})} \tag{2.5.3}$$

In the cases where $n < m - 1$, some new phenomena occurs. Firstly, there are different variants of the proof—one has some freedom of choice in the global integral that one uses to represent the convolution. We shall only present one of the possibilities. Secondly, the local factors Ψ and $\hat{\Psi}$ are no longer symmetrical. The integral that one starts with is

$$\int_{\Gamma\backslash G}\int_0^1\cdots\int_0^1 \phi\begin{pmatrix} g & \begin{matrix} 0 & x_{1,n+2} & \cdots & x_{1,m} \\ & & & \\ 0 & & & \end{matrix} \\ & \begin{matrix} 1 & x_{n+1,n+2} & & \vdots \\ & 1 & & \\ & & \ddots & \\ & & & x_{m-1,m} \\ & & & 1 \end{matrix} \end{pmatrix} \tag{2.5.4}$$

$$\times \psi(g)\det(g)^{s-(m-n)/2}e(-x_{n+1,n+2} - \cdots - x_{m-1,m})\,dx_{ij}\,dg.$$

As before, if re $(s) > 1$ one may unravel this convolution to obtain

$$\Psi(s, \phi \times \psi)L(s, \phi \times \psi), \tag{2.5.5}$$

where now

$$\Psi(s, \phi \times \psi) = \int_0^\infty \cdots \int_0^\infty K_{\rho(\phi)}(y_1, \ldots, y_{n-1}) \times K_{\rho(\psi)}(y_1, \ldots, y_n, 1, \ldots, 1)(y_1 y_2^2 \cdots y_n^n)^s \frac{dy_1}{y_1} \cdots \frac{dy_n}{y_n}. \tag{2.5.6}$$

This time however, the transformation of the integral to obtain the functional equation is more difficult, and the function $\Psi(1 - s, \hat{\phi})$ does not occur, but rather a modified integral. The special case $m = 3$, $n = 1$ is typical and is worked out in Jacquet, Piatetski–Shapiro, and Shalika [42] or in Bump [9], Chapter 8. Let us consider this case, referring to these references for details. Writing ϕ in place of ψ, we have, provided re $(s) > 1$,

$$\int_0^\infty \int_0^1 \int_0^1 \phi \begin{pmatrix} y & & x_{33} \\ & 1 & x_{23} \\ & & 1 \end{pmatrix} y^{s-1} e(-x_{23})\, dx_{13}\, dx_{23} \frac{dy}{y} = \Psi(s, \phi)L(s, \phi),$$

where

$$\Psi(s, \phi) = \int_0^\infty W_{\rho(\phi)} \begin{pmatrix} y & & \\ & 1 & \\ & & 1 \end{pmatrix} y^{s-1} \frac{dy}{y}. \tag{2.5.7}$$

On the other hand, if re $(1 - s) > 1$, one may transform the integral so as to obtain

$$\Psi_1(1 - s, \hat{\phi})L(1 - s, \hat{\phi}),$$

where

$$\Psi_1(s, \hat{\phi}) = \int_0^\infty \int_{-\infty}^\infty W_{\rho(\phi)} \begin{pmatrix} y & & \\ x & 1 & \\ & & 1 \end{pmatrix} y^{s-1}\, dx \frac{dy}{y}.$$

The local functional equation that one must prove is

$$\frac{\Psi(s, \phi)}{\Phi(s, \phi)} = \frac{\Psi_1(1 - s, \hat{\phi})}{\Phi(1 - s, \hat{\phi})}. \tag{2.5.8}$$

2.6. *Gamma Factors and Local Functional Equations*

The main difficulty now is to establish the local functional equations at the Archimedean place, i.e., (2.4.3), (2.5.3), and (2.5.8) in the above discussion. Apparently, these have been proved by Jacquet, Piatetski–Shapiro, and Shalika in the case where the principal series representations of $\mathrm{GL}(n, \mathbf{R})$ and $\mathrm{GL}(m, \mathbf{R})$ associated with the characters $\rho(\phi)$ and $\rho(\psi)$ are unitary. This hypothesis should not be necessary, however.

We will ask here a different question: When is it true that $\Psi(s, \phi \times \psi)$ *equals* (up to a constant multiple) the gamma factor $\Phi(s, \phi \times \psi)$? The answer is sometimes "yes," sometimes "no." If $n = m = 2$, these are the same—this is the Weber–Schafheitlin integral

$$\int_0^\infty K_\nu(y)K_\mu(y)y^s \frac{dy}{y} = 2^{s-3} \frac{\Gamma\left(\frac{s+\mu+\nu}{2}\right)\Gamma\left(\frac{s+\mu-\nu}{2}\right)\Gamma\left(\frac{s-\mu+\nu}{2}\right)\Gamma\left(\frac{s-\mu-\nu}{2}\right)}{\Gamma(s)}, \tag{2.6.1}$$

valid for sufficiently large s. If $n = 2$, $m = 3$, it was proved by Bump [10] that the convolution $\Psi(s, \phi \times \psi)$ *equals* (up to a constant multiple) the gamma factor $\Phi(s, \phi \times \psi)$. Specifically, one starts with the formula

$$\begin{aligned}\int_0^\infty \int_0^\infty K_a(y_1, y_2) y_1^{s_1-1} y_2^{s_2-1} \frac{dy_1}{y_1}\frac{dy_2}{y_2} &= \pi^{-s_1-s_2}\Gamma\left(\frac{s_1-a_1}{2}\right)\Gamma\left(\frac{s_1-a_2}{2}\right)\Gamma\left(\frac{s_1-a_3}{2}\right)\Gamma\left(\frac{s_2+a_1}{2}\right) \\ &\quad \times \Gamma\left(\frac{s_2+a_2}{2}\right)\Gamma\left(\frac{s_2+a_3}{2}\right)\Gamma\left(\frac{s_1+s_2}{2}\right)^{-1}.\end{aligned} \tag{2.6.2}$$

Here $a = (a_1, a_2, a_3)$, and we are using the notation introduced in 2.3, and we are assuming that $\mathrm{re}\,(s_1)$, $\mathrm{re}\,(s_2)$ are sufficiently large. This was proved by Bump [9], Chapters 2 or 10—a better proof has been found more recently and is the basis for the proof of the corresponding formula on $\mathrm{GL}(3, C)$ by Bump and Friedberg [12]. This formula gives the local calculation at infinity for the theory of the "exterior square" of Bump and Friedberg [11] (cf. Section 2.7) for the group GL(3). Now it was deduced from (2.6.2) that (remembering that in terms of the Bessel function K_ν, we have $K_{\nu,-\nu}(y) = K_\nu(2\pi y)$)

$$\int_0^\infty \int_0^\infty K_{a_1,a_2,a_3}(y_1, y_2) K_{\nu,-\nu}(y_1)(y_1 y_2^2)^s \frac{dy_1}{y_1} \frac{dy_2}{y_2}$$

$$= \pi^{-3s} \Gamma\left(\frac{s-a_1-\nu}{2}\right) \Gamma\left(\frac{s-a_2-\nu}{2}\right) \Gamma\left(\frac{s-a_3-\nu}{2}\right)$$

$$\times \Gamma\left(\frac{s-a_1+\nu}{2}\right) \Gamma\left(\frac{s-a_2+\nu}{2}\right) \Gamma\left(\frac{s-a_3+\nu}{2}\right). \tag{2.6.3}$$

Thus if $n = 2$, $m = 3$, the Ψ factor is indeed equal to the product of gamma functions.

On the other hand, if $n = 1$, $m = 3$, this is definitely not the case, although, as we shall see, there is still something positive to be said. If $n = m = 3$, we have no information, since we have not seriously tried to evaluate the integral (2.4.3) in this case. If one seeks the most optimistic interpretation of what is known, one may hope that an integral of Whittaker functions tends to be evaluable as a product or ratio of gamma functions provided that

1. the corresponding integral of Whittaker functions on the p-adic group may be evaluated as an "Euler factor" that is a product or ratio of terms of the form $(1 - ap^{-s})^{-1}$;
2. the integral is over the *entire* torus.

Thus (2.6.2) and (2.6.3) satisfy the second condition, while (2.5.7) does not. If this philosophy is valid, one may hope that $\Psi(s, \phi \times \psi)$ is a product of gamma functions in the cases where $n = m$ or $n = m - 1$, but not otherwise.

In the case $n = 1$, $m = 3$, Hoffstein and M. R. Murty [34] have recently given an ingenious and instructive proof of (2.5.8). For (2.6.2) and the Mellin inversion formula imply that (if $\rho(\phi) = \rho_a$)

$$\frac{\Psi(s, \phi)}{\Phi(s, \phi)} = \frac{1}{2\pi i} \int_{\sigma - i\infty}^{\sigma + i\infty} \Gamma\left(\frac{s_2 + a_1}{2}\right) \Gamma\left(\frac{s_2 + a_2}{2}\right)$$

$$\times \Gamma\left(\frac{s_2 + a_3}{2}\right) \Gamma\left(\frac{s + s_2}{2}\right)^{-1} \pi^{s/2 - s_2} \, ds_2, \tag{2.6.4}$$

for σ sufficiently large. This Mellin–Barnes integral defines an entire function.

Hoffstein and Murty now consider the integral

$$\int_0^\infty \int_0^\infty \int_{-\infty}^\infty W\begin{pmatrix} y_1y_2 & & \\ y_2x & y_2 & \\ & & 1 \end{pmatrix} y_1^{s_1-1} y_2^{s_2-1}\, dx \frac{dy_1}{y_1}\frac{dy_2}{y_2}. \tag{2.6.5}$$

It follows from the right invariance of W by orthogonal matrices and from (2.1.3) that

$$W\begin{pmatrix} y_1y_2 & & \\ y_2x & y_2 & \\ & & 1 \end{pmatrix} = e(\Delta^{-2}y_1x)\,W\begin{pmatrix} \Delta^{-1}y_1y_2 & & \\ & \Delta y_2 & \\ & & 1 \end{pmatrix},$$

where $\Delta = \sqrt{x^2+1}$. Replacing y_1, y_2 by $\Delta^2 y_1$, $\Delta^{-1}y_2$, respectively, in (2.6.5) and using the standard integral

$$\int_{-\infty}^\infty (x^2+1)^{-\nu} e(y_1x)\, dx = \frac{\pi^\nu}{\Gamma(\nu)} y_1^{\nu-1/2} K_{\nu-1/2}(2\pi y_1),$$

with $\nu = -s_1 + s_2/2 + \frac{1}{2}$, we see that (2.6.5) equals

$$\frac{\pi^\nu}{\Gamma(\nu)} \int_0^\infty \int_0^\infty K_a(y_1, y_2) K_{\nu-1/2}(2\pi y_1)(y_1y_2^2)^{s_2/2} \frac{dy_1}{y_1}\frac{dy_2}{y_2},$$

which by (2.6.3) equals

$$\frac{\Phi(s_1)\Phi(s_2-s_1)}{\pi^{s_1-s_2/2-1/2}\Gamma\left(-s_1+\dfrac{s_2}{2}+\dfrac{1}{2}\right)}.$$

Now by the Mellin inversion formula, we see that

$$\frac{\Psi_1(s,\phi)}{\Phi(s,\phi)} = \frac{1}{2\pi i}\int_{\sigma-i\infty}^{\sigma+i\infty} \Gamma\left(\frac{s_2-s-a_1}{2}\right)\Gamma\left(\frac{s_2-s-a_2}{2}\right) \times \Gamma\left(\frac{s_2-s-a_3}{2}\right)\Gamma\left(-s+\frac{s_2+1}{2}\right)^{-1} \pi^{-s_2+s/2+1/2}\, ds_2. \tag{2.6.6}$$

Shifting the line of integration, replacing $\hat{\phi}$ by ϕ and s by $1-s$ and comparing with (2.6.6) we obtain (2.5.8).

2.7. *Other Convolutions on* GL(*n*)

It should be emphasized that there may be a variety of ways of proving any given functional equation. The functional equation (2.3.8) of $L(s, \phi)$ was proved by Godement and Jacquet [30] by a different method. They integrate the form over the entire group against the determinant function. The local integrals obtained this way are integrals not of the Whittaker function but rather of the spherical function. Terras [95] in the case $n = 3$ gave a rather different proof in which one integrates the form against a theta function and an Eisenstein series. We have also been told that Selberg had a method of proving (2.3.8), but we have not seen this.

Another class of L-functions for which the Rankin–Selberg method yields functional equations are the exterior square L-functions associated with an automorphic form. In terms of the $\alpha_i(p)$ defined by (2.1.2), this is the Euler product

$$L(s, \phi, \wedge^2) = \prod_p \prod_{i<j} (1 - \alpha_i(p)\alpha_j(p)p^{-s})^{-1}.$$

Two distinct Rankin–Selberg convolutions have been given that prove the functional equation of this L-function: one by Jacquet, Piatetski–Shapiro and Shalika [40], and one by Bump and Friedberg [11]. This L-function will have no pole if n is odd. If n is even, there may or may not be a pole. In [40] it is shown that if $n = 4$, the exterior square L-function has a pole if and only if the form is in the image of a lifting from GSp(4).

One has also the symmetric square L-function

$$L(s, \phi, \vee^2) = \prod_p \prod_{i \le j} (1 - \alpha_i(p)\alpha_j(p)p^{-s})^{-1}.$$

If $n = 2$, this was discussed above in Section 1.8. If $n = 3$, a similar metaplectic integral has been found by Patterson and Piatetski–Shapiro [66] involving a *degenerate* theta function on the double cover of GL(3), i.e., one whose nondegenerate Fourier coefficients all vanish. This work is similar to the convolutions found earlier in the "Method C" of Gelbart and Piatetski–Shapiro [25].

Bump and Goldfeld [13] considered the integral of an Eisenstein series (of minimal parabolic type—a function of two variables) for GL(3, **Z**) over an anisotropic torus associated with a totally real cubic field. It was found that this integral could be identified with an integral arising in a different context. The second integral is a Rankin–Selberg convolution obtained by restricting an Eisenstein series for the Hilbert modular group associated with the cubic field (acting on a product of three upper half-planes) to the Poincaré upper half-plane, embedded on the diagonal, and integrating against the standard Eisenstein series for $SL(2, \mathbf{Z})$. The latter integral is divergent, but the theory of Zagier [100] makes sense of such integrals. It is probable that this identity can be generalized in different ways.

3. The Classical Groups

The conjectures of Langlands [54] assert, among other things, that a large class of Euler products should have functional equations. Specifically, if G is any reductive group, one associates with G a complex reductive group ${}^LG$. If G is split over its field of definition and if one is content to ignore that aspect of the theory having to do with the phenomenon of "base change," then one cares mainly about the connected component ${}^LG^0$. If G is a classical group, then ${}^LG^0$ is also a classical group. For split semisimple groups, the situation may be summarized by saying that if G has Cartan type A_n (resp. B_n, C_n, D_n), then ${}^LG^0$ has Cartan type A_n (resp. C_n, B_n, D_n), and the center of G is naturally isomorphic to the fundamental group of ${}^LG^0$. Suppose that G is defined over a global field F. Let ϕ be an automorphic form on $G_F \backslash G_A$, where A is the adele ring of F. For each finite dimensional representation r of degree d of ${}^LG^0$ one associates an Euler product $L(s, \phi, r)$ whose typical factor corresponding to a prime p of F is a polynomial of degree d in Np^{-s}. These L-functions are conjectured to have functional equations. Generally speaking, if the highest weight of r is not too large, these functional equations may often be proved by means of the Rankin–Selberg method. The groups of Cartan type A_n were discussed in the previous section, and it is the purpose of this section to survey what is known about the remaining types B_n, C_n, and D_n.

It should be pointed out that there exists an alternative method of proving these types of functional equations in many cases, namely, the method of Langlands and Shahidi—cf. Langlands [53], Shahidi [84], and the recent survey of Gelbart and Shahidi [28] for further information. We shall only discuss here what can be proved using direct integral representations of Euler products. Some authors prefer to distinguish between the integral representations that involve Eisenstein series and those that do not, and from this point of view, some of the integral representations which we consider here are *not* Rankin–Selberg convolutions.

It is highly likely that there are developments of which I am unaware. Moreover, I have not found it possible to be very detailed here for reasons of space. I have written out some details for $GSp(4)$ so as to give the reader some feeling for the nature of this work. Even this may be misleading, because Rankin–Selberg convolutions for the classical groups are so varied. I have chosen to frame the discussion of $GSp(4)$ adelically, but I hope that the reader will see that there is not really much difference between the point of view here and that of the previous section.

3.1. The State of the Art

The Hecke algebra for $Sp(n)$ was investigated by Satake [79] and by Shimura [87]. So far as I am aware, it was Andrianov who first found integrals representing Euler products for the group $GSp(4)$. If $G = GSp(4)$, then ${}^LG^0 \cong GSp(4)$. These are two fundamental weights for ${}^LG^0$. The irreducible representations with these fundamental weights as highest weight have degrees 4 and 5, and we shall denote the corresponding Euler products as $L(s, \phi, 4)$ and $L(s, \phi, 5)$. Owing to the "accidental" coincidence of the Cartan types B_2 and C_2 (i.e., the isomorphism $Sp(4) \cong \mathrm{spin}(5)$), we may regard G as either a symplectic group or an orthogonal group. Theorems about $L(s, \phi, 4)$ should generalize to automorphic forms on orthogonal groups, and theorems about $L(s, \phi, 5)$ should generalize to automorphic forms on symplectic groups.

In [1] and [2], Andrianov gave Euler product representations for $L(s, \phi, 4)$. The method is as follows. Let F be the ground field, which was $\mathbf{Q}$ in the case considered by Andrianov, who was concerned with Siegel modular forms. Let K be a quadratic extension such that ϕ has a nonzero Fourier coefficient in Siegel's sense, parametrized by a symmetric matrix having eigenvalues in K. Thus for Andrianov, K is an imaginary quadratic field. $Sp(4)$ contains a copy of the groundfield reduction $R_{K/F}\mathrm{SL}(2)$. Andrianov proved that restricting ϕ to this group and integrating against an Eisenstein series yields the Euler product $L(s, \phi, 4)$. This method is reinterpreted from an adelic point of view by Piatetski–Shapiro [70].

Another integral representation $L(s, \phi, 4)$ may be found in the paper of Novodvorsky [60]. This method of proving the functional equation is only valid when ϕ is generic. If ϕ fails to be generic, as in certain examples of Saito–Kurokawa [50] and Howe–Piatetski–Shapiro [35], then this integral representation fails. On the other hand, in the generic case, it has the advantage of proving the holomorphicity of $L(s, \phi, 4)$ and indeed if ϕ is nongeneric, the L-function may have poles.

In [3], Andrianov gave an Euler product representation for $L(s, \phi, 5)$. Here one forms the Rankin–Selberg convolution of ϕ with a theta function to obtain the Euler product.

Piatetski–Shapiro, in his paper "Euler subgroups" [68], began to develop a philosophy which explains some of the possible integral representations of Euler products, such as Andrianov's integral for $L(s, \phi, 4)$—but not others, such as Andrianov's integral for $L(s, \phi, 5)$.

Novodvorsky also considers another global integral in which one convolves the form on $\mathrm{G}Sp(4)$ with an automorphic form on GL(2), to obtain an Euler product of degree 8. Unfortunately, this paper somewhat sketchy. However, the Euler product of degree 8 is considered in detail by Soudry [93]. See also Novodvorsky and Piatetski–Shapiro [64] for background on $L(s, \phi, 4)$.

If G is a classical group, then ${}^LG^0$ is also a classical group, and so one may associate with an automorphic form on G an Euler product corresponding to the standard representation of ${}^LG^0$. These Euler products have all been proved to have functional equations by Piatetski–Shapiro and Rallis [73], [71]. The method involves embedding $G \times G$ in a classical group H of rank $2n$, where n is the rank of G. One restricts an Eisenstein series of maximal parabolic type on H to $G \times G$, and integrates against $\phi \times \phi$. The convolution unfolds into the integral of a spherical function or "matrix coefficient" associated with ϕ on G_A, similar to the integrals considered by Godement and Jacquet [30]. Incidentally, the integrals for $L(s, \phi, 4)$ and $L(s, \phi, 5)$ on $\mathrm{G}Sp(4)$ by this method are different from the ones discussed above.

The special case where $G = Sp(2n)$ deserves special mention. In this case, ${}^LG^0$ is isogenous to the (split) orthogonal group $\mathrm{SO}(2n + 1)$. Corresponding to the standard representation of ${}^LG^0$ is an Euler product of degree $2n + 1$. In very special cases, the functional equation for this Euler product was proved by Andrianov and Kalinin [4]. The method of Piatetski–Shapiro and Rallis discussed above provides a proof of the functional equation in general. For Siegel modular forms, the same method of restriction of Eisenstein series that was used by Piatetski–Shapiro and Rallis was discovered independently by Böcherer [7], who gives a more classical treatment.

Functional equations for standard L-functions on orthogonal groups were also proved by Novodvorsky and Piatetski–Shapiro [63] and by Novodvorsky [61].

A convolution for automorphic forms on the group $\mathrm{SU}(2,1)$ was discovered independently by Shintani [92] and (somewhat earlier) Piatetski–Shapiro [69]. $\mathrm{SU}(2,1)$ is a quasisplit form of $\mathrm{GL}(3)$, whose L-group is essentially the semidirect product of $\mathrm{GL}(3,\mathbf{C})$ by a cyclic group of order two. The irreducible representation of smallest degree for this group has degree six, and so there should be an Euler product of degree six with a functional equation associated with an automorphic form on $\mathrm{SU}(2,1)$. It was shown by Shintani and Piatetski–Shapiro that this Euler product could be represented as the Rankin–Selberg convolution of the form with a theta function. Shintani did not publish his paper, but Kudla [49] gives some further discussion. Gelbart and Piatetski–Shapiro [26] give another discussion of this same integral representation.

Gelbart and Piatetski–Shapiro [25] have generalized the Euler product of degree 8 introduced by Novodvorsky and studied by Soudry in a far-reaching way. If G is a classical group of rank n, ϕ an automorphic form on G and ψ an automorphic form on $\mathrm{GL}(n)$, then they are able to prove a functional equation for an Euler product of degree nd, where d is the degree of the standard representation of ${}^LG^0$, associated with ϕ and ψ. Let us review their methods.

If $G = \mathrm{SO}(2n+1)$ is a split orthogonal group of odd degree, the subgroup $\mathrm{SO}(2n)$ has a maximal parabolic subgroup whose Levi component is essentially $\mathrm{GL}(n)$. Thus one may use the form ψ to construct an Eisenstein series on $\mathrm{SO}(2n)$, which one integrates against the restriction of ϕ to this subgroup. Due to the accidental isomorphism $Sp(4) \cong \mathrm{spin}(5)$, this includes the convolution of $GSp(4)$ with $\mathrm{GL}(2)$ as a special case (cf. Section 3.4).

If $G = \mathrm{SO}(2n)$ is a split orthogonal group of even degree, Gelbart and Piatetski–Shapiro embed G in the larger group $\mathrm{SO}(2n+1)$, which has a maximal parabolic subgroup with Levi factor essentially $\mathrm{GL}(n)$. One constructs an Eisenstein series involving ψ on this larger group, restricts it to G and integrates against ϕ. This method is thus similar to the technique of restriction of Eisenstein series used by Garrett (cf. Section 1.7), and by Rallis and Piatetski–Shapiro (cf. above).

On the other hand, if G is a symplectic group, the convolution is rather different. In the latter case, G contains a maximal parabolic subgroup whose Levi component is $\mathrm{GL}(n)$, and so one may construct with ψ an Eisenstein $E(g, s, \psi)$ series on G. One considers the integral $\int_{G_F \backslash G_A} \phi(g)E(g, s, \psi)\theta(g)\, dg$, where $\theta(g)$ is a *degenerate theta function*, which possesses only singular terms in its Fourier expansion. The degeneracy of θ means that its presence does not disturb the convolution too badly. This type of convolution is rather similar to those considered (somewhat later) by Patterson and Piatetski–Shapiro (cf. Section 2.7 above).

Gelbart and Piatetski–Shapiro mention that the three methods that they use can be applied in different contexts. They point out, for example, that the method that they use for the orthogonal groups of even degree (restriction of Eisenstein series) may also be applied to the orthogonal groups of odd degree, yielding, for example, a convolution of an automorphic form on $GSp(4)$ with an automorphic form on GL(3) (cf. Section 3.5). Consult Section 6 of [25] for a number of variations on their themes.

An an introduction to this work, and an illustration of the surprising variety of the techniques that come into play, we shall present below three convolutions of a generic cusp form ϕ on $GSp(4)$ with cusp forms ψ on GL(n) in the three cases $n = 1, 2, 3$. These include the convolutions of Novodvorsky [60], as well as one suggested by Gelbart and Piatetski–Shapiro [25], and illustrate two of the three techniques used by Gelbart and Piatetski–Shapiro.

In addition to having given us many concrete examples, Piatetski–Shapiro deserves a great deal of credit for attempting to understand the general principles through which global integrals with functional equations may be related to Euler products. Let us conclude with some general remarks about these general principles.

There are two classes of special functions for which uniqueness theorems hold—Whittaker functions and spherical functions (or more generally, matrix coefficients of representations). Because of these uniqueness properties, a Whittaker function or spherical function on the adele group decomposes into a product of Whittaker or spherical functions on the local groups. Thus if a global integral possessing a functional equation can be unfolded into an integral of a Whittaker function or spherical function, it will represent an Euler product. Examples of convolutions unfolding to spherical functions are the integrals of Godement and Jacquet [30] and Piatetski–Shapiro and Rallis [71].

Convolutions that reduce to integrals of Whittaker functions seem more prevalant. For groups other than GL(n), there is an added complication that some automorphic forms, such as Siegel modular forms of genus ≥ 2 are not *generic*, i.e. are not associated with Whittaker functions. Thus one must also consider generalizations of the "Whittaker function philosophy", such as the generalized Bessel models of Novodvorsky and Piatetski–Shapiro [62]. See also [64], [70] and [68].

Still, there are convolutions with Euler products that at first do not seem to fit into these philosophies, such as Andrianov's integral for $L(s, \phi, 5)$ on $GSp(4)$. The integrals of Bump and Hoffstein [14], [15] (cf. Section 4.1 below) have been compared with this integral for this reason.

3.2. *Whittaker Functions for* GSp(4)

Let J be the matrix

$$\begin{pmatrix} & & & 1 \\ & & 1 & \\ & -1 & & \\ -1 & & & \end{pmatrix},$$

and let $G = GSp(4)$ be the group of $g \in GL(4)$ such that

$$^t g J g = s(g) J \tag{3.2.1}$$

for some scalar $s(g)$.

Let F be a local field, and let e be a nontrivial additive character of F. Let π be an irreducible admissible *generic* representation of G_F. "Generic" means that π has a Whittaker model, i.e., there exists a linear functional T on the space V of π such that (with $x_3' = x_3 - x_1 x_2$)

$$T\left(\pi\begin{pmatrix} 1 & x_2 & x_3 & x_4 \\ & 1 & x_1 & x_3' \\ & & 1 & -x_2 \\ & & & 1 \end{pmatrix} v\right) = e(x_1 + x_2) T(v).$$

Let K be the "standard" maximal compact subgroup of G_F. Thus if $F = \mathbf{R}$, $K = U(2)$, if $F = \mathbf{C}$, then K is the compact real form of $Sp(4)$, and if F is nonarchimedean, then $K = G_{\mathcal{O}}$, where $\mathcal{O}$ is the ring of integers in F.

Now suppose that π has a K-stable vector v_0. Let $W(g) = T(\pi(g)v_0)$. Then W is called a (class one) *Whittaker function*. Suppose now that F

is nonarchimedean, and that $\mathcal{O}$ is the largest fractional ideal on which e is trivial. If T is normalized so that $W(1) = 1$, then there exists a formula for $W(g)$, analogous to Shintani's formula [91], which is the basis for (2.1.8). This formula is a special case of the formula of Kato [46] and of Casselman and Shalika [17] for arbitrary reductive groups.

The connected L-group ${}^LG^0$ (cf. Langlands [54] and Borel [8]) is $GSp(4, \mathbf{C})$. Let ${}^LT$ be the maximal torus of elements of the form

$$t(\alpha_1, \alpha_2, \alpha_3, \alpha_4) = \begin{pmatrix} \alpha_1 & & & \\ & \alpha_2 & & \\ & & \alpha_3 & \\ & & & \alpha_4 \end{pmatrix},$$

where $\alpha_1\alpha_4 = \alpha_2\alpha_3$. The fundamental dominant weights of the torus are λ_1 and λ_2 where

$$\lambda_1 t(\alpha_1, \alpha_2, \alpha_3, \alpha_4) = \alpha_1, \qquad \lambda_2 t(\alpha_1, \alpha_2, \alpha_3, \alpha_4) = \alpha_1\alpha_3^{-1}.$$

If n_1, n_2 are integers, there exists an irreducible finite dimensional representation ρ_{k_1,k_2} of highest weight $k_1\lambda_1 + k_2\lambda_2$. The degree of this representation is $\frac{1}{6}(k_1 + 1)(k_2 + 1)(k_1 + k_2 + 2)(k_1 + 2k_2 + 3)$. Every finite-dimensional irreducible representation of ${}^LG^0$ is the tensor product of a ρ_{k_1,k_2} with a power of the character s defined by (3.2.1). Let us also define $u(\xi_1, \xi_2, \xi_3, \xi_4) = t(\xi_1\xi_2, \xi_1\xi_3, \xi_2\xi_4, \xi_3\xi_4)$. (This set of coordinates will have the advantage that the exponents in the Weyl character formula and the Kato–Casselman–Shalika formula will have integer values.)

Let us define polynomials that are analogous to Schur polynomials. These are polynomials that occur in the Weyl character formula for ${}^LG^0 = GSp(4, C)$ and in the Kato–Casselman–Shalika formula.

The Weyl group W is a dihedral group of order 8. It has generators

$$w_1 = \begin{pmatrix} 1 & & & \\ & & -1 & \\ & 1 & & \\ & & & 1 \end{pmatrix}, \qquad w_2 = \begin{pmatrix} & 1 & & \\ 1 & & & \\ & & & 1 \\ & & 1 & \end{pmatrix}.$$

It acts on the polynomial ring $\mathbf{C}[x_1, x_1^{-1}, x_2, x_2^{-1}, x_3, x_3^{-1}, x_4, x_4^{-1}]$ in such a way that for $w \in W$, denoting $x = (x_1, x_2, x_3, x_4)$, we have $wu(wx)w^{-1} = u(x)$. Thus $w_1: x_1 \to x_2,\ x_2 \to x_1,\ x_3 \to x_4,\ x_4 \to x_3$ and

w_2: $x_1 \to x_1$, $x_2 \to x_3$, $x_3 \to x_2$, $x_4 \to x_4$. Now let us define a group algebra element

$$\mathscr{B} = \sum_{w \in W} (-1)^{l(w)} w,$$

where $l(w)$ is the length function on the Weyl group. Thus $(-1)^{l(w)}$ is the character of W that equals -1 when $w = w_1$ or w_2. Let

$$T_{k_1, k_2}(x_1, x_2, x_3, x_4) = \frac{\mathscr{B}(x_1^{k_1+k_2+3} x_2^{k_1+2} x_3 x_4^{-k_2})}{\mathscr{B}(x_1^3 x_2^2 x_3)} \tag{3.2.2}$$

For example, we have

$$T_{1,0}(x) = x_1 x_2 + x_1 x_3 + x_2 x_4 + x_3 x_4,$$
$$T_{0,1}(x) = x_1 x_4^{-1} + x_4 x_1^{-1} + x_2 x_3^{-1} + x_3 x_2^{-1} + 1.$$

The *Weyl character formula* asserts then that

$$\operatorname{tr} \rho_{k_1, k_2}(u(x)) = T_{k_1, k_2}(x). \tag{3.2.3}$$

We may now give the formula of Kato and of Casselman and Shalika for the Whittaker function. Langlands associates with π a semisimple conjugacy class in ${}^L G^0$, which we represent as $X_\pi = u(\xi_1, \xi_2, \xi_3, \xi_4)$. Then if $\operatorname{ord}(y_i) = k_i$, for $i = 0, 1, 2$, we have

$$W\begin{pmatrix} y_0 y_1 y_2 & & & \\ & y_0 y_1 & & \\ & & y_0 & \\ & & & y_0 y_2^{-1} \end{pmatrix}$$
$$= N p^{-3/2 k_1 - 2k_2} (\xi_1 \xi_2 \xi_3 \xi_4)^{k_0} T_{k_1, k_2}(\xi_1, \xi_2, \xi_3, \xi_4) \tag{3.2.4}$$

if $k_1, k_2 \geq 0$, 0 otherwise.

Now let F be a global field, A its adele ring, $e(x) = \prod_v e_v(x_v)$ a character of the additive group A/F, and π an irreducible generic cuspidal automorphic representation of $GSp(4, A)$ associated with the automorphic form ϕ. "Generic" in this context means essentially that the global Whittaker function

$$W(g) = \int_{(A/F)^4} \phi\left(\begin{pmatrix} 1 & x_2 & & \\ & 1 & & \\ & & 1 & -x_2 \\ & & & 1 \end{pmatrix}\begin{pmatrix} 1 & & x_3 & x_4 \\ & 1 & x_1 & x_3 \\ & & 1 & \\ & & & 1 \end{pmatrix} g\right)$$
$$\times\, e(-x_1 - x_2)\, dx_1\, dx_2\, dx_3\, dx_4 \tag{3.2.5}$$

is nonzero. Generic forms are known to exist on every group, and are probably in a sense typical. However, except on GL(n), nongeneric forms also exist. Siegel modular forms of genus ≥ 2 are never generic. Many constructions, such as those we consider in this section, work only for generic forms.

If ϕ is chosen correctly, the Whittaker function may be assumed to decompose locally as $W(g) = \prod_v W_v(g_v)$, a product of local Whittaker functions. This is a consequence of a uniqueness result for the local Whittaker models, which is due to Rodier [77] for split reductive groups, and to Gelfand and Kazhdan [29] for GL(n). (Strictly speaking, an Archimedean uniqueness result must also be supplied. For GL(n), this is due to Shalika [85].) A generalization of this uniqueness theorem to the generalized Whittaker models of Novodvorsky and Piatetski–Shapiro allows one to obtain Euler products for nongeneric cuspidal representations, as in Piatetski–Shapiro [70].

3.3. *The Convolution* GSp*(4)* × GL*(1)*

The convolution considered here was considered by Novodvorsky [60]. The proper generalization of this convolution is to automorphic forms on orthogonal groups of odd degree. Let ϕ be a generic cusp form on GSp(4, A), with associated Whittaker function $W(g)$ defined by (3.2.5). Let ψ be a Hecke character of $A^\times/F^\times$, i.e., an automorphic form on GL(1). Consider the integral

$$\int_{A^\times/F^\times}\int_{(A/F)^3}\phi\left(\begin{pmatrix}1 & x_2 & & x_4\\ & 1 & & \\ & z & 1 & -x_2\\ & & & 1\end{pmatrix}\begin{pmatrix}y & & & \\ & y & & \\ & & 1 & \\ & & & 1\end{pmatrix}\right) \times e(-x_2)\psi(y)\|y\|^{s-1/2}\,dz\,dx_2\,dx_4\,d^\times y. \tag{3.3.1}$$

Owing to the fact that ϕ is invariant under the matrix

$$\begin{pmatrix} & & 1 & \\ & & & 1\\ -1 & & & \\ & -1 & & \end{pmatrix},$$

this integral has a functional equation with respect to $s \to 1 - s$.

We have the following formula:

$$\int_{(A/F)^2} \phi\left(\begin{pmatrix} 1 & x_2 & & x_4 \\ & 1 & & \\ & & 1 & -x_2 \\ & & & 1 \end{pmatrix} g\right) e(-x_2)\, dx_2\, dx_4$$

$$= \sum_{\substack{\alpha \in F^\times \\ \beta \in F}} W\left(\begin{pmatrix} \alpha & & & \\ & \alpha & & \\ & \alpha\beta & 1 & \\ & & & 1 \end{pmatrix} g\right). \qquad (3.3.2)$$

One proves this in two steps. Firstly, note that

$$F(x_1) = \int_{(A/F)^3} \phi\left(\begin{pmatrix} 1 & x_2 & & \\ & 1 & & \\ & & 1 & -x_2 \\ & & & 1 \end{pmatrix}\right.$$

$$\times \left.\begin{pmatrix} 1 & & x_3 & x_4 \\ & 1 & x_1 & x_3 \\ & & 1 & \\ & & & 1 \end{pmatrix} g\right) e(-x_2)\, dx_2\, dx_3\, dx_4$$

is a function on A/F and hence has a Fourier expansion, which we may evaluate at zero to obtain

$$F(0) = \sum_{\alpha \in A/F} \int_{A/F} F(x_1) e(-\alpha x_1)\, dx_1.$$

We may actually sum over $\alpha \in A^\times/F^\times$ since ϕ is cuspidal (so the term with $\alpha = 0$ vanishes). Each integral on the right may be evaluated in terms of the Whittaker function, so (replacing α by α^{-1}) we obtain

$$\int_{(A/F)^2} \phi\left(\begin{pmatrix} 1 & x_2 & & \\ & 1 & & \\ & & 1 & -x_2 \\ & & & 1 \end{pmatrix}\begin{pmatrix} 1 & & x_3 & x_4 \\ & 1 & & x_3 \\ & & 1 & \\ & & & 1 \end{pmatrix} g\right) e(-x_2)\, dx_3\, dx_4$$

$$= \sum_{\alpha \in F^\times} W\left(\begin{pmatrix} \alpha & & & \\ & \alpha & & \\ & & 1 & \\ & & & 1 \end{pmatrix} g\right). \qquad (3.3.3).$$

Another dose of the same medicine shows that summing over the β gets rid of the x_3, and so we obtain (3.3.3).

Now by (3.3.2), we see that (3.3.1) equals

$$\int_{A^\times}\int_A W\begin{pmatrix} y & & & \\ & y & & \\ & yx & 1 & \\ & & & 1\end{pmatrix}\psi(y)\|y\|^{s-1/2}\,dx\,d^\times y$$

$$=\int_{A^\times}\int_A W\begin{pmatrix} y & & & \\ & y & & \\ & x & 1 & \\ & & & 1\end{pmatrix}\psi(y)\|y\|^{s-3/2}\,dx\,d^\times y.$$

This decomposes into an Euler product, where the Euler factor corresponding to the place v of F is

$$\int_{F_v^\times}\int_{F_v} W_v\begin{pmatrix} y & & & \\ & y & & \\ & x & 1 & \\ & & & 1\end{pmatrix}\psi(y)\|y\|^{s-3/2}\,dx\,d^\times y.$$

Suppose that v is any nonarchimedean place of F such that W_v is right invariant by $GSp(4,\mathcal{O}_v)$ and such that the largest fractional ideal on which e_v is trivial is $\mathcal{O}$. Then the Kato–Casselman–Shalika formula permits us to evaluate the last integral. Note that

$$W\begin{pmatrix} y & & & \\ & y & & \\ & x & 1 & \\ & & & 1\end{pmatrix}=W\begin{pmatrix} y & & & \\ & y & & \\ & & 1 & \\ & & & 1\end{pmatrix}$$

if $x\in\mathcal{O}$, or zero if $x\notin\mathcal{O}$. Let p is the prime of $\mathcal{O}$ associated with the valuation v, and let us temporarily denote $X=\psi(p)Np^{-s}$. We see that if the semisimple conjugacy class of ${}^LG^0$ associated with ϕ at the place v is $u(\xi_1,\xi_2,\xi_3,\xi_4)$, then (3.2.4) shows that the local integral equals

$$\sum_{k=0}^{\infty}X^kT_{k,0}(\xi_1,\xi_2,\xi_3,\xi_4).$$

Applying (3.3.2) and summing the geometric series, this equals

$$\begin{aligned}&\mathcal{B}(\xi_1^3\xi_2^2\xi_3)^{-1}\mathcal{B}((1-\xi_1\xi_2X)^{-1}\xi_1^3\xi_2^2\xi_3)\\&\quad=L_v(s,\phi\times\psi,4\times 1)\mathcal{B}(\xi_1^3\xi_2^2\xi_3)^{-1}\\&\qquad\times\mathcal{B}((1-\xi_1\xi_3X)(1-\xi_2\xi_4X)(1-\xi_3\xi_4X)\xi_1^3\xi_2^2\xi_3).\end{aligned}$$

Here $L_v(s, \phi \times \psi, 4 \times 1)$ is the Euler factor

$$(1 - \xi_1\xi_2 X)^{-1}(1 - \xi_1\xi_3 X)^{-1}(1 - \xi_2\xi_4 X)^{-1}(1 - \xi_3\xi_4 X)^{-1}.$$

Now, expanding the product into a sum of eight monomials, all but one are annihilated by $\mathscr{B}$ (two of them end up cancelling), leaving us with the Euler factor $L_v(s, \phi \times \psi, 4 \times 1)$. This proves a functional equation for $L(s, \phi, 4)$ twisted by the Hecke character ψ.

3.4. *The Convolution* GSp(4) × GL(2)

The convolution of this section is considered by Novodvorsky [60] and by Soudry [93], and generalized by Gelbart and Piatetski–Shapiro [25].

Let ϕ be an automorphic form on G_A, where $G = GSp(4)$, and let ψ be an automorphic form on GL(2). For simplicity, we will assume that ϕ and ψ are both right invariant by standard maximal compact subgroups of their ambient groups and that both have trivial central character.

For the purposes of this section only let H be the subgroup of $GL(2) \times GL(2)$ consisting of elements (h_1, h_2) with $\det(h_1) = \det(h_2)$. Define an embedding $J\colon H \to G$ by

$$J(h_1, h_2) = \begin{pmatrix} a_2 & & & b_2 \\ & a_1 & b_1 & \\ & c_1 & d_1 & \\ c_2 & & & d_2 \end{pmatrix}, \qquad h_i = \begin{pmatrix} a_i & b_i \\ c_i & d_i \end{pmatrix}.$$

Let B be the subgroup of GL(2) consisting of upper triangular matrices. Define a function I_s on GL(2, A), which is to be right invariant by the standard maximal compact subgroup, and satisfies

$$I_s\begin{pmatrix} y_1 & x \\ & y_2 \end{pmatrix} = \left\|\frac{y_1}{y_2}\right\|^s.$$

Let $\zeta(s)$ be the Dedekind zeta-function of F, including gamma factors. Define the Eisenstein series $E(g, s)$ for $g \in GL(2, A)$ to be

$$\zeta(2s) \sum_{\gamma \in B_F \backslash GL(2, F)} I_s(\gamma g).$$

Let Z_A be the group of scalar matrices in GL(4). Consider

$$\int_{Z_A H_F \backslash H_A} \phi(J(h_1, h_2)) E(h_2, s) \psi(h_1)\, d(h_1 \times h_2). \tag{3.4.1}$$

Here the measure is Haar measure on H. This integral inherits a functional equation (with respect to $s \to 1 - s$) from the Eisenstein series.

Unfolding the Eisenstein series, we obtain

$$\zeta(2s) \int_{\mathrm{GL}(2,F)\backslash \mathrm{GL}(2,A)} \int_{A/F} \phi\left(J\left(h_1, \begin{pmatrix} \det(h_1) & x \\ & 1 \end{pmatrix} \right)\right) \times \psi(h_1) \|\det(h_1)\|^{s-1}\, dx\, dh_1.$$

Now we have the following lemma, whose proof we omit. Let N be the subgroup of upper triangular unipotent matrices in GL(2). Then it may be shown that

$$\int_{A/F} \phi\left(J\left(1, \begin{pmatrix} 1 & x \\ & 1 \end{pmatrix} \right) g \right) dx = \sum_{N_F \backslash \mathrm{GL}(2,F)} W_\phi(J(\gamma, 1)g). \tag{3.4.2}$$

Applying this to our previous integral, we obtain

$$\zeta(2s) \int_{N_F \backslash \mathrm{GL}(2,A)} W_\phi\left(J\left(h_1, \begin{pmatrix} \det(h_1) & \\ & 1 \end{pmatrix} \right)\right) \psi(h_1) \|\det(h_1)\|^{s-1}\, dh_1.$$

Under our assumptions that ϕ and ψ are both right invariant by standard maximal compact subgroups of their ambient groups, and have trivial central character, this is easily seen to equal

$$\zeta(2s) \int_{A^\times} \int_{A^\times} W_\phi \begin{pmatrix} y_1 y_2 & & & \\ & y_1 & & \\ & & 1 & \\ & & & y_2^{-1} \end{pmatrix} \times W_\psi \begin{pmatrix} y_1 & \\ & 1 \end{pmatrix} \|y_1 y_2^2\|^s \|y_1 y_2\|^{-2}\, d^\times y_1\, d^\times y_2,$$

where

$$W_\psi(g) = \int_{A/F} \psi\left(\begin{pmatrix} 1 & x \\ & 1 \end{pmatrix} g \right) e(x)\, dx.$$

Note that this Whittaker function is formed with the opposite additive character as W_ϕ, but this does not affect the values given by the Kato–Casselman–Shalika formula (i.e., Shintani's formula in this case) on the diagonal elements.

Since the Whittaker functions decompose locally, this has an Euler product, which may be evaluated by the same method as in Section 3.3, although this requires a great deal more calculation if one takes a naive approach. This same calculation is done very elegantly (in more generality) in the appendix [27] by Gelbart, Piatetski–Shapiro, and Rallis to [25]—still based on the Kato–Casselman–Shalika formula. The Euler factor at a nonramified place v turns out to be (in the obvious notation) $L_v(s, \phi \times \psi, 4 \times 2)$, an Euler product of degree 8.

We have not considered here the Euler factors at the ramified places. At the very least, one ought to show that the Euler factor at a nonarchimedean ramified place p is a rational function of Np^{-s}. At an Archimedean place, the Euler factor should be related to a product of gamma functions. These questions are closely connected with the (known) asymptotics of the Whittaker functions at zero. A different approach, at least to the nonarchimedean problem, has been suggested by Gelbart and Piatetski–Shapiro in Section 12 of [25], who prefer to use a theorem of Bernstein.

3.5. *The Convolution* GSp(4) × GL(3)

The convolution in this section does not appear in the literature, but if one takes into account the "accidental" isomorphisms spin(5) $\cong Sp(4)$, spin(6) $\cong$ SL(4), this is implicit in the remark of Gelbart and Piatetski–Shapiro [25], beginning of Section I.1, that one may use their "Method B" to obtain a convolution $G \times \mathrm{GL}(n+1)$ when G has Cartan type B_n (beginning of I.1.)

Let ϕ and ψ be automorphic forms on the adele groups of $G = \mathrm{G}Sp(4)$ and GL(3), respectively. Again, we make the simplifying assumptions that the forms are right invariant by the standard maximal compact subgroups of their respective ambient groups and that their central characters are trivial. Let $M = \mathrm{GL}(2)$, and let $j\colon M \to \mathrm{GL}(3)$, $i\colon \mathrm{GL}(3) \to \mathrm{GL}(4)$ be the maps

$$j(m) = \begin{pmatrix} \det(m) & \\ & m \end{pmatrix}, \qquad i(g) = \begin{pmatrix} g & \\ & 1 \end{pmatrix}.$$

Let P be the subgroup of elements of $H = \mathrm{GL}(4)$ whose bottom row is $(0, 0, 0, 1)$, and let Z be the center of H. Let us define an Eisenstein series on H as follows. We let $I_s(h)$ be the function that is right invariant by the standard maximal compact subgroup of H, and that satisfies

$$I_s(pz) = \hat{\psi}(h)\|\det(h)\|^s \qquad \text{when } p = \begin{pmatrix} h & * \\ 0 & 1 \end{pmatrix} \in P, \qquad h \in \mathrm{GL}(3).$$

Here $\hat{\psi}(g) = \psi({}^t g)$ is the contragredient form. Let

$$E(h, s) = L(4s - 1, \psi) \sum_{Z_F P_F \backslash H_F} I_s(\gamma h).$$

This Eisenstein series is convergent for re (s) sufficiently large and has analytic continuation and a functional equation with respect to $s \to 1 - s$. Here $L(s, \psi)$ is the L-function associated with the cusp form ψ, with gamma factors.

Let $Q = G \cap P = U(i \circ j)(M)$, where

$$U = \left\{ \begin{pmatrix} 1 & x_2 & x_3 & x_4 \\ & 1 & & x_3 \\ & & 1 & -x_2 \\ & & & 1 \end{pmatrix} \right\}.$$

Let $c\colon U \to \mathrm{GL}(3)$ be the map that sends

$$\begin{pmatrix} 1 & x_2 & x_3 & x_4 \\ & 1 & & x_3 \\ & & 1 & -x_2 \\ & & & 1 \end{pmatrix} \to \begin{pmatrix} 1 & x_2 & x_3 \\ & 1 & \\ & & 1 \end{pmatrix}.$$

The integral that we consider is

$$\int_{Z_A G_F \backslash G_A} E(g, s)\phi(g)\, dg, \tag{3.5.1}$$

where dg is Haar measure on the symplectic group G. This is a restriction of Eisenstein series, analogous to those considered by Garrett and by Piatetski–Shapiro and Rallis. The integral obviously has a functional equation—we shall show that it has an Euler product.

The key observation is that the coset space $P_F \backslash H_F$ is the same as $Q_F \backslash G_F$, since every matrix may be matched with a symplectic matrix with the same bottom row. Let us remark that in other restrictions of Eisenstein series the corresponding assertion would not be true, and one must prove a lemma at this point which shows that the cosets that do not restrict do not contribute to the integral—for example, condition (3) in Section 3 of Piatetski–Shapiro and Rallis [73], or (1.2.3) of Gelbart and Piatetski–Shapiro [25].

Thus we may unfold the Eisenstein series in (3.5.1) to obtain

$$L(4s - 1, \psi) \int_{Z_A Q_F \backslash G_A} I_s(g)\phi(g)\, dg,$$

and since we are assuming that the integrand is right invariant by the standard maximal compact subgroup, we may use the Iwasawa decomposition to rewrite this integral as

$$L(4s-1,\psi)\int_{Q_F\backslash Q_A} I_s(g)\phi(g)\,dg.$$

Note that if we represent $g \in Q_A$ as a product $u(i \circ j)(m)$ where $u \in U_A$, $m \in M_A$, the Haar measures are related by

$$dg = \|\det(m)\|^{-2}\,du\,dm,$$

so our last integral equals

$$L(4s-1,\psi)\int_{U_F\backslash U_A}\int_{M_F\backslash M_A} \|\det(m)\|^{2s-2}\psi(c(u)j(m))\phi(u(i\circ j)(m))\,dm\,du.$$

Now we use the Fourier expansion (which would require a minor modification if we were not assuming that the central character of ψ is trivial):

$$\hat{\psi}(g) = \sum_{N_F\backslash M_F} W_{\hat{\psi}}(j(\gamma)g).$$

This is essentially the same as the Fourier expansion (2.1.5) due to Piatetski–Shapiro and Shalika, except that we are using the "opposite" parabolic subgroup for our element γ. Here $W_{\hat{\psi}}$ is the Whittaker function associated with ψ in the usual way, and N is the subgroup of unipotent upper triangular matrices in $M = \mathrm{GL}(2)$. Thus we obtain

$$L(4s-1,\psi)\int_{U_F\backslash U_A}\int_{N_F\backslash M_A}$$
$$\times\ \|\det(m)\|^{2s-2}W_{\hat{\psi}}(c(u)j(m))\phi(u(i\circ j)(m))\,dm\,du$$

$$= L(4s-1,\psi)\int_{A^\times}\int_{A^\times} W_\phi\begin{pmatrix} y_1y_2 &&& \\ & y_1 && \\ && 1 & \\ &&& y_2^{-1}\end{pmatrix} W_{\hat{\psi}}\begin{pmatrix} y_1y_2 && \\ & y_1 & \\ && 1\end{pmatrix}$$
$$\times\ \|y_1y_2^2\|^{2s}\|y_1^3y_2^4\|^{-1}\,d^\times y_1\,d^\times y_2.$$

This evaluation shows that (3.5.1) does indeed have an Euler product. Although I have not actually carried out the calculation (which is likely to be difficult), it seems clear that the Euler factor at an unramified place v will turn out to equal $L(2s-\frac{1}{2}, \phi\times\psi, 4\times 3)$ in the obvious notation. (Note that when $s \to 1-s$, the variable $s' = 2s-\frac{1}{2} \to 1-s'$.)

4. The Metaplectic Group

Weil [98] defined a twofold cover of the symplectic group $Sp(2n, A)$, where A is the adele ring of a global field F, as the natural domain for theta functions. As we shall explain, there also exist covers of $\mathrm{GL}(r, A)$ that, following Weil, are referred to as "metaplectic groups," and that are also natural domains for theta functions. One of the most striking uses of theta functions is the phenomenon that the Rankin–Selberg convolution of an automorphic form with a theta function sometimes turns out to have an Euler product. The convolution in Section 1.8 is an example of this—other examples have been mentioned in Section 3.

In this section, we shall be concerned with Rankin–Selberg convolutions of forms on the metaplectic group itself with theta functions. In this context, the term "Rankin–Selberg convolution" refers to global integrals that are similar to those in Section 2. As is pointed out there, only in the "diagonal" case of two groups of equal rank do these integrals involve Eisenstein series. Thus, as is explained in the introduction to Section 2, some writers would not consider all these convolutions as belonging to the Rankin–Selberg method.

4.1. The Shimura Correspondence

Let $G = \mathrm{GL}(r)$. Let F be a global field whose group μ_n of n-th roots of unity has cardinality n, and let A be its adele ring. Weil [98], Kubota [48], and Matsumoto [57] have, in various degrees of generality, defined a group $\tilde{G}_A$ which is an n-fold cover of G_A; theta functions on this group have been defined as residues of Eisenstein series by Kazhdan and Patterson [47]. The paper of Patterson and Piatetski–Shapiro [47] shows that in some cases (conjecturally when $r|n$), there should also be cuspidal theta functions, which must be constructed in some other way. It seems probable that the construction of these will eventually emerge as an important problem. We will let $\theta_{n,r}$ denote a "standard" theta function on the n-fold cover of $\mathrm{GL}(r)$.

It is believed that there is a correspondence between automorphic forms on $\tilde{G}_A$ and automorphic forms on G_A. The first example of this is the work of Shimura [89], which deals with automorphic forms on the double cover of GL(2)—modular forms of half-integral weight. Shimura shows that the Rankin–Selberg convolution $L(s, f \times \bar{\theta})$ of a cusp form f of half-integral weight $k/2$ (with k odd) with the standard theta function considered in Section 1.8—or twist thereof—has an Euler product. Indeed, by means of a converse theorem, he is able to deduce from the functional equations he thus obtains that there exists a modular form ϕ of weight $k-1$ such that $L(s, f \times \bar{\theta}) = L(2s - \frac{1}{2}, \phi)$. This theory was restated in generality by Gelbart and Piatetski–Shapiro [24] in terms of the Weil representation.

Bump and Hoffstein [14], [15] obtained evidence that this is but a particular instance of a general phenomenon. Let us consider automorphic forms on the n-fold cover of GL(r, A). Given an automorphic form f on the $\tilde{G}_A$, let us consider the Rankin–Selberg convolution $L(s, f \times \bar{\theta}_{n,n})$ with a theta function on the n-fold cover of GL(n). By this, we mean that one considers a global integral similar to those of Jacquet, Piatetski–Shapiro, and Shalika, described in Section 2 above. It appears that this convolution should have an Euler product and that there should exist an automorphic form ϕ on G_A such that $L(s, f \times \bar{\theta}_{n,n}) = L(ns - \frac{1}{2}(n-1), f)$. We refer to the introduction to [15] for more details.

It should be pointed out that Kazhdan and Patterson, and Flicker and Kazhdan, have investigated (with some success) an alternative approach to the Shimura correspondence, based on the trace formula.

4.2. *Rankin–Selberg Convolutions with Other Theta Functions*

Let f be a cusp form on the n-fold cover of GL(r).

The result that $L(s, f \times \bar{\theta}_{n,n})$ has an Euler product is accessible given the current state of knowledge, because the nondegenerate Fourier coefficients of $\theta_{n,n}$ are known—this follows from the uniqueness of the exceptional Whittaker models obtained by Kazhdan and Patterson [47] for theta functions on the n-fold cover of GL(n). If $k < n$ one no longer has uniqueness of the exceptional Whittaker models associated with $\theta_{n,k}$ (although for $k = n - 1$ the Whittaker models are *almost* unique). Nevertheless, Bump and Hoffstein [16] have recently obtained some conjectures on the nature of $L(s, f \times \bar{\theta}_{n,k})$, which imply information about the Fourier coefficients of $\theta_{n,k}$ above and beyond what is implied by the theory of Kazhdan and Patterson. We will now summarize these conjectures.

Assuming that $k < n$, GL($r + n - k$) has a parabolic subgroup whose Levi component is GL(r) $\times$ GL($n - k$), and starting with the automorphic form $f \times \theta_{n-k}$ on the n-fold cover of this group, one may construct an Eisenstein series on the n-fold cover of GL($r + n - k$). Bump and Hoffstein conjectured that the Rankin–Selberg convolution $L(s, f \times \bar{\theta}_{n,k})$ is a Fourier coefficient of this Eisenstein series.

Patterson has made a conjecture on the nature of $\theta_{4,2}$, and these conjectures of Bump and Hoffstein are consistent with the conjecture of Patterson.

If we replace the cusp form f by the theta function $\theta_{n,r}$ then the Eisenstein series side of this equation is symmetrical, leading to a relation between $L(s, \theta_{n,r} \times \bar{\theta}_{n,k})$ and $L(s, \theta_{n,n-k} \times \bar{\theta}_{n,n-r})$. Thus, if $k + r \geq n$, this relation has the form (in terms of the Dedekind zeta-function)

$$L(s, \theta_{n,r} \times \bar{\theta}_{n,k}) = \zeta\left(ns + \frac{k+r}{2} - n\right)\zeta\left(ns + \frac{k+r}{2} - n - 1\right) \cdots \zeta\left(ns + \frac{k-r}{2} + 1\right)L(s, \theta_{n,n-k} \times \bar{\theta}_{n,n-r}).$$

We refer to [16] for some evidence for these conjectures. It is interesting to note that T. Suzuki (personal communication) has independently been led to consider Rankin–Selberg convolutions of theta functions and has introduced a new ingredient that may lead to progress on these conjectures. When $n = 4$, he considers the convolution $L(s, \theta_{4,2} \times \bar{\theta}_{4,4})$, but by introducing an extra parameter he gains some flexibility, which may lead him to a proof of Patterson's conjecture.

Acknowledgements

It is a great pleasure to dedicate this account of the Rankin–Selberg method to Professor Selberg. I hope it will be found useful. Another recent survey of the current state of the theory of Euler products, which I think will be found to be complementary to this paper, is Gelbart and Shahidi [28].

I would like to thank Siegfried Böcherer, Solomon Friedberg, Jeffrey Hoffstein, Jay Jorgenson, Ilya Piatetski–Shapiro, Peter Sarnak, and several participants in the Selberg symposium for helpful comments that have saved me from some errors; David Joyner with whom I first worked out some formulas in Section 3.2; and Solomon Friedberg, Dorian Goldfeld, and Jeffrey Hoffstein, with whom I have studied particular Rankin–Selberg convolutions. In particular, it is from Dorian Goldfeld that I came to understand the enormous extent to which our field is molded by Professor Selberg's ideas.

Preparation of this survey article was partially supported by NSF grant #DMS 8612896.

References

[1] Andrianov, A. "Zeta Functions and the Siegel Modular Forms." In: *Lie Groups and their Representations*, John Wiley and Sons (1975).

[2] Andrianov, A. "Dirichlet with Euler product in the theory of Siegel modular forms of genus 2," *Trudy Mat. Inst. Steklov.* **112** (1971), 73–94.

[3] Andrianov, A. "On zeta functions of Rankin type associated with Siegel Modular forms." In: *Springer Lecture Notes in Mathematics.* **627** (1977).

[4] Andrianov, A. and Kalinin, V. "On the analytic properties of standard zeta functions of Siegel modular forms," *Mat. Sb.* **35** (1979), 1–17.

[5] Asai, T. "On certain Dirichlet series associated with Hilbert modular forms and Rankin's method," *Math. Ann.* **226** (1977).

[6] Barnes, E. "A transformation of generalized hypergeometric series," *Quart. J. Math.* **41** (1910), 136–40.

[7] Böcherer, S. "Über die funktionalgleichung automorpher L-Funktionen zur Siegelschen modulgruppe," *Jour. Reine Ang. Math.* **362** (1985), 146–168.

[8] Borel, A. "Automorphic L-functions." In: *Automorphic Forms, Representations and L-functions*, AMS Proc. Symp. Pure Math. **33**(2) (1979).

[9] Bump, D. "Automorphic Forms on GL(3, **R**)," *Springer Lecture Notes in Mathematics.* **1083**, 1984.

[10] Bump, D. "Barnes' second lemma and its application to Rankin–Selberg convolutions," *Am. J. Math.*, to appear.

[11] Bump, D. and Friedberg, S. "The "exterior square" automorphic L-functions on GL(n)," to appear.

[12] Bump, D. and Friedberg, S. "On Mellin transforms of unramified Whittaker functions on GL(3, **C**)," to appear.

[13] Bump, D. and Goldfeld, D. "A Kronecker Limit formula for GL(3)." In: *Modular Forms*, Rankin, R.A. ed. (1984), 43–49.

[14] Bump, D. and Hoffstein, J. "Some Euler products associated with cubic metaplectic forms on GL(3)," *Duke Math. J.* **53** (1986), 1047–1072.

[15] Bump, D. and Hoffstein, J. "On Shimura's correspondence," *Duke Math. J.*, to appear.

[16] Bump, D. and Hoffstein, J. "Some conjectured relationships between theta functions and Eisenstein series on the metaplectic group," to appear.

[17] Casselman, W. and Shalika, J. "The unramified principal series of p-adic groups II: The Whittaker function," *Compositio Math.* **41** (1980), 207–231.

[18] Deligne, P. "Formes modulaires et représentations l-adiques." In: *Sém. Bourbaki Vol.* 1968/69 *exposés* 347–363, *Springer Lecture Notes in Mathematics.* **179** (1971).

[19] Doi, K. and Naganuma, H. "On the algebraic curves uniformized by arithmetical automorphic functions," *Ann. Math.* **86** (1967), 449–460.

[20] Doi, K. and Naganuma, H. "On the functional equation of certain Dirichlet series," *Invent. Math.* **9** (1969), 1–14.

[21] Friedberg, S. "A global approach to the Rankin–Selberg convolution for GL(3, **Z**), *Trans. AMS.* **300** (1987), 159–178.

[22] Garrett, P. "Decomposition of Eisenstein series: Rankin triple products," *Ann. Math.* **125** (1987), 209–237.

[23] Gelbart, S. and Jacquet, H. "A relation between automorphic representations of GL(2) and GL(3)," *Ann. Sci. Ecole Normale Sup.* 4e *série.* **11** (1978), 471–552.

[24] Gelbart, S. and Piatetski–Shapiro, I. "On Shimura's correspondence for modular forms of half-integral weight." In: *Automorphic Forms, Representation Theory and Arithmetic*, Springer Verlag and the Tata Institute (1981).

[25] Gelbart, S. and Piatetski–Shapiro, I. "L-functions for $G \times \mathrm{GL}(n)$." In: *Springer Lecture Notes in Mathematics.* **1254** (1985).

[26] Gelbart, S. and Piatetski–Shapiro, I. "Automorphic forms and L-functions for the unitary group." In: *Springer Lecture Notes in Mathematics.* **1041** (1983).

[27] Gelbart, S., Piatetski–Shapiro, I. and Rallis, S. appendix to *Springer Lectures Notes in Mathematics.* **1254** (1985).

[28] Gelbart, S. and Shahidi, F. "Analytic properties of automorphic L-functions," Academic Press (1988).

[29] Gelfand, I. and Kazhdan, D. "Representations of the group GL(n, K) where K is a local field." In: *Lie Groups and their Representations*, John Wiley and Sons (1975).

[30] Godement, R. and Jacquet, H. "Zeta functions of simple algebras," *Springer Lecture Notes in Mathematics.* **260** (1972).

[31] Goldfeld, D. "Analytic and arithmetic theory of Poincaré series," *Journées Arithmétiques de Luminy, Asterisque.* **61** (1979).

[32] Good, A. "Cusp forms and eigenfunctions of the Laplacian," *Math. Ann.* **255** (1981), 523–548.

[33] Hecke, E. "Über modulfunktionen und die Dirichletsche reihe mit Eulerscher produktentwicklungen," *Math. Ann.* **114** (1937), 1–28, 316–351.

[34] Hoffstein, J. and Murty, M. "Hyperkloosterman sums and twisted L-functions on GL(3)," to appear (tentative title).

[35] Howe, R. and Piatetski–Shapiro, I. "A counterexample to the "generalized Ramanujan conjecture" for (quasi-) split groups." In: *Automorphic Forms, Representations and L-functions*, AMS Proc. Symp. Pure Math. **33**(1) (1979).

[36] Ingham, A. "Note on Riemann's ζ-function and Dirichlet's L-functions," *J. London Math. Soc.* **5** (1930), 107–112.

[37] Jacquet, H. "Automorphic forms on GL(2) part II," *Springer Lecture Notes in Mathematics.* **278** (1972).

[38] Jacquet, H. "Dirichlet series for the group GL(n)." In: *Automorphic Forms, Representation Theory and Arithmetic*, Springer Verlag and the Tata Institute (1981).

[39] Jacquet, H. and Langlands, R. "Automorphic forms on GL(2)," *Springer Lecture Notes in Mathematics*. **114**, (1970).

[40] Jacquet, H., Piatetski–Shapiro, I. and Shalika, J. in preparation.

[41] Jacquet, H., Piatetski–Shapiro, I. and Shalika, J. "Rankin–Selberg Convolutions" *Am. J. Math.* **105** (1983), 367–464.

[42] Jacquet, H. Piatetski–Shapiro, I. and Shalika, J. "Automorphic forms on GL(3), I and II," *Ann. Math.* **109** (1979), 169–258.

[43] Jacquet, H., Piatetski–Shapiro, I. and Shalika, J. Relèvement cubique non normal," *C. R. Paris Sér. I Math.* **193** (1981), 13–18.

[44] Jacquet, H. and Shalika, J. "On Euler products and the classification of automorphic representations," *Am. J. Math.* **103** (1981), 499–588, 777–815.

[45] Jacquet, H. and Shalika, J. "A nonvanishing theorem for zeta functions of GL_2," *Invent. Math.* **38** (1976), 1–16.

[46] Kato, S. "On an explicit formula for class-1 Whittaker functions on split reductive groups over p-adic fields," preprint, University of Tokyo (1978).

[47] Kazhdan, D. and Patterson, S. "Metaplectic Forms," *Publ. Math. IHES* **59** (1984).

[48] Kubota, T. *Automorphic Forms and Reciprocity in a Number Field.* Kiyokuniya Book Store, (1969).

[49] Kudla, S. "On certain Euler products for SU(2, 1), *Compositio Math.* **42** (1981), 321–344.

[50] Kurokawa, N. "Examples of eigenvalues of Hecke operators on Siegel cusp form of degree two," *Invent. Math.* **49** (1978), 149–165.

[51] Langlands, R. "Base change for GL(2)," *Annals of Mathematics Studies.* **96** (1980).

[52] Langlands, R. "On the Functional Equations satisfied by Eisenstein Series," *Springer Lecture Notes in Mathematics.* **544** (1976).

[53] Langlands, R. "Euler Products," *Yale Mathematical Monographs.* **1** (1971).

[54] Langlands, R. "Problems in the theory of automorphic forms," In: *Springer Lecture Notes in Mathematics.* **170** (1970).

[55] Littlewood, D. *The Theory of Group Characters*, second edition. Oxford, 1950.

[56] Macdonald, I. *Symmetric Functions and Hall Polynomials.* Oxford University Press, Oxford, 1979.

[57] Matsumoto, H. "Sur les sous-groupes arithmetiques des groupes semi-simples déployés," *Ann. Scient. Ec. Norm. Sup.* (1969), 1–62.

[58] Mordell, L. "On Mr. Ramanujan's empirical expansions of modular functions," *Proc. Cambridge Phil. Soc.* **19** (1917), 117–124.

[59] Naganuma, H. "On the coincidence of two Dirichlet series associated with cusp forms of Hecke's Nebentypus and Hilbert modular forms over a real quadratic field," *J. Math. Soc. Jpn.* **25** (1973), 547–555.

[60] Novodvorsky, M. "Automorphic L-functions for the symplectic group GSp_4." In: *Automorphic Forms, Representations and L-functions, AMS Proc. Symp. Pure Math.* **33**(2) (1979).

[61] Novodvorsky, M. "Fonction J pour les groupes orthogonaux," *C. R. Paris Sér. A-B Math.* **280** (1981), 1421–1422.

[62] Novodvorsky, M. and Piatetski–Shapiro, I. "Generalized Bessel models for the symplectic group of rank two," *Mat. Sb.* **90**(132) (1973), 246–256.

[63] Novodvorsky, M. and Piatetski–Shapiro, I. "On zeta functions of infinite-dimensional representations," *Mat. Sb.* **92**(134) (1973), 499–510.

[64] Novodvorsky, M. and Piatetski–Shapiro, I. "Rankin–Selberg method in the theory of automorphic forms." In: *Several Complex Variables, AMS Proc. Symp. Pure Math.* **30**(2) (1977), 297–302.

[65] Patterson, S. and Piatetski–Shapiro, I. "A cubic analog of the cuspidal theta series," *J. Math. Pures et Appl.* **63** (1984).

[66] Patterson, S. and Piatetski–Shapiro, I. in preparation.

[67] Phillips, R. and Sarnak, P. "On cusp forms for co-finite subgroups of PSL(2, **R**)," *Invent. Math.* **80** (1985), 339–364.

[68] Piatetski–Shapiro, I. "Euler Subgroups." In: *Lie Groups and their Representations*, John Wiley and Sons (1975).

[69] Piatetski–Shapiro, I. "Tate theory for reductive groups and distinguished representations." In: *Procedings of the International Congress of Mathematicians, Helsinki* (1978), 585–590.

[70] Piatetski–Shapiro, I. L-functions for GSp(4), preprint (1979).

[71] Piatetski–Shapiro, I. and Rallis, S. "L-functions for the Classical Groups," notes by J. Cogdell, *Springer Lecture Notes in Mathematics.* **1254** (1984).

[72] Piatetski–Shapiro, I. and Rallis, S. "Rankin triple L-functions," preprint (1985).

[73] Piatetski–Shapiro, I. and Rallis, S. "L-functions for the classical groups." In: *Modular Forms*, R. A. Rankin, ed. (1984), 251–262.

[74] Ramanujan, S. "On certain arithmetical functions," *Trans. Cambridge Phil. Soc.* **22** (1916), 159–184.

[75] Ramanujan, "Some formulae in the analytic theory of numbers," *Messenger of Math.* **45** (1916), 81–84.

[76] Rankin, R. "Contributions to the theory of Ramanujan's function $\tau(n)$ and similar arithmetical functions, I and II," *Proc. Cambridge Phil. Soc.* **35** (1939), 351–356, 357–372.

[77] Rodier, F. "Modèle de Whittaker des représentations admissibles ds groupes réductifs p-adiques deployés," *C. R. Paris Sér. A Math.* **283** (1976), 429–431.

[78] Saito, H. "Automorphic forms and algebraic extensions of number fields," *Lectures in Math., Kyoto Univ.* (1975).

[79] Satake, I. "On spherical functions over p-adic fields," *Proc. Japan Acad.* **38** (1962), 422–425.

[80] Selberg, A. "Bemerkungen über eine Dirichletsche reihe, die mit der theorie der modulformer nahe verbunden ist," *Arch. Math. Naturvid.* **43** (1940), 47–50.

[81] Selberg, A. "On the estimation of Fourier coefficients of modular forms." In: *Number Theory, AMS Proc. Symp. Pure Math.* **8** (1965).

[82] Selberg, A. "Harmonic analysis and discontinuous groups in weakly symmetric Riemannian spaces with applications to Dirichlet series," *J. Ind. Math. Soc.* **20** (1956), 47–50.

[83] Serre, J.-P. *Abelian l-adic Representations and Elliptic Curves.* W. A. Benjamin, New York (1968).

[84] Shahidi, F. "On certain L-functions," *Amer. J. Math.* **103** (1981), 297–355.

[85] Shalika, J. "The multiplicity one theorem on GL(n)," *Ann. Math.* **100** (1974), 171–193.

[86] Shimura, G. *Introduction to the Arithmetic Theory of Automorphic Forms*, Iwanami Shoten and Princeton University Press (1971).

[87] Shimura, G. "On modular correspondences for $Sp(n, \mathbf{Z})$ and their congruence subgroups," *Proc. Nat. Acad. Sci.* **49** (1963), 824–828.

[88] Shimura, G. "On the holomorphy of certain Dirichlet series," *Proc. London Math. Soc.* **31**(3) (1975), 79–98.

[89] Shimura, G. "On modular forms of half-integral weight," *Ann. Math.* **97** (1973).

[90] Shintani, T. "On liftings of holomorphic cusp forms." In: *Automorphic Forms, Representations and L-functions, AMS Proc. Symp. Pure Math.* **33**(2) (1979).

[91] Shintani, T. "On an explicit formula for class-1 'Whittaker Functions' on GL(n) over P-adic fields," *Proc. Japan Acad.* **52** (1976).

[92] Shintani, T. "On automorphic forms on unitary groups of order 3," preprint (1979).

[93] Soudry, D. "The L and γ factors for generic representations of $GSp(4, k) \times GL(2, k)$ over a local nonarchimedean field," *Duke Math. J.* **51** (1984), 355–194.

[94] Tamagawa, T. "The zeta function of a division algebra," *Ann. Math.* **77** (1963), 387–405.

[95] Terras, A. "On automorphic forms for the general linear group," *Rocky Mt. J. Math.* **12** (1982), 123–143.

[96] Tunnell, J. "Artin's conjecture for representations of octahedral type," *Bull. Amer. Math. Soc.* **5** (1981), 173–175.

[97] Weil, A. Über die bestimmung dirichletsche reihen durch funktionalgleichungen," *Math. Ann.* **168**, 149–156.

[98] Weil, A. "Sur certaines groupes des opérateurs unitaires," *Acta Math.* **111**, 143–211.
[99] Wilton, J. "A note on Ramanujan's arithmetical function $\tau(n)$," *Proc. Cambridge Phil. Soc.* **25** (1923), 121–129.
[100] Zagier, D. "The Rankin–Selberg method for automorphic functions which are not of rapid decay," *J. Fac. Sci. Tokyo Univ.* **28** (1981), 415–437.

Added in Proof

In the year that has gone by since this paper was written, a number of new Rankin–Selberg convolutions representing L-functions have been discovered. Kohnen and Skoruppa have invented yet another integral representation for the Euler product of degree four associated with a Siegel modular form of genus two. Bump has found a Rankin–Selberg convolution representing the "spin" L-function (an Euler product of degree eight) associated with a generic form on GSp(6). D. Ginzburg has obtained convolutions representing standard L-functions associated with generic forms on $\mathrm{GL}(r) \times G$, where $G = \mathrm{SO}(2n)$ or $\mathrm{SO}(2n+1)$, and $r \leq n$, simultaneously generalizing work of Novodvorsky and of Gelbart and Piatetski–Shapiro. Most interestingly, Piatetski–Shapiro and Rallis have obtained a convolution associated with a generic form on $G_2 \times \mathrm{GL}(2)$. The example is the first known Rankin–Selberg representation of a Langlands L-function that is not accessible by the Langlands–Shahidi method.

5 On the Base Change Problem: After J. Arthur and L. Clozel

HERVÉ JACQUET

1. Automorphic *L*-functions

Automorphic L-functions are by definition L-functions attached to automorphic forms on $\mathrm{GL}(n)$. We first recall their definition. Let F be a number field and $F_{\mathbb{A}}$ its ring of adeles. Then $\mathrm{GL}(n, F_{\mathbb{A}})$ is a locally compact group containing $\mathrm{GL}(n, F)$ as a discrete subgroup. An automorphic form for $\mathrm{GL}(n, F)$ is a function on $\mathrm{GL}(n, F_{\mathbb{A}})$ invariant on the left under $\mathrm{GL}(n, F)$, satisfying additional conditions, some of which we now describe.

For each finite place v of F let R_v be the ring of integers in the local field F_v and K_v the group $\mathrm{GL}(n, R_v)$; we denote by H_v the convolution algebra of bi-K_v-invariant functions (or rather measures) on the group $G_v = \mathrm{GL}(n, F_v)$. It is called the Hecke algebra at the place v; there is an isomorphism of H_v onto the algebra of symmetric polynomials in n indeterminates $X_1, X_2, \ldots, X_n$ and their inverses. For instance, if $n = 2$ and the cardinality of the residual field of F_v is q_v and ω is an uniformizer for F_v, then this isomorphism takes the characteristic function of the set

$$K_v \begin{vmatrix} \omega & 0 \\ 0 & 1 \end{vmatrix} K_v \quad \text{to} \quad q_v^{-1/2}(X_1 + X_2)$$

NUMBER THEORY,
TRACE FORMULAS and
DISCRETE GROUPS

ISBN 0-12-067570-6

and the characteristic function of the set

$$K_v \begin{vmatrix} \omega & 0 \\ 0 & \omega \end{vmatrix} K_v \text{ to } q_v^{-1/2} X_1 X_2 .$$

We may also regard our polynomials as class functions on $\mathrm{GL}(n, C)$, the polynomial P corresponding to the function that on the diagonal matrix $\operatorname{diag}(x_1, x_2, \ldots, x_n)$ takes the value $P(x_1, x_2, \ldots, x_n)$. In this way we get an isomorphism $f \to \hat{f}$ of H_v onto an algebra of class functions on $\mathrm{GL}(n, C)$. In particular, every character of the algebra H_v has the form $f \to \hat{f}(A)$, where A is a semisimple n by n complex invertible matrix, or rather a conjugacy class of such matrices. An automorphic form is then, at almost all places v, an eigenfunction of the Hecke algebra H_v, acting on the right by convolution.

In fact, rather than individual automorphic forms, one considers "automorphic representations." Roughly speaking, an irreducible automorphic representation π is defined as follows: Its space V is a space of functions on $\mathrm{GL}(n, F_{\mathbb{A}})$ invariant on the left under $\mathrm{GL}(n, F)$; the space V is invariant and irreducible under right shifts by the elements of $\mathrm{GL}(n, F_{\mathbb{A}})$; the representation obtained in this way is π. If S is a finite set of places, containing all places at infinity, let us denote by K^S the product of the groups K_v for v not in S, by G_S the product of the groups G_v for v in S and by V^S, the space of vectors in V invariant under K^S. Then, if S is sufficiently large, the space V^S is not 0; it is then irreducible under G_s and each vector in V^S an eigenvector of H_v for all v not in S; we will denote by $\pi(v)$ the matrix corresponding to the eigenvalue of H_v on V^S. Then the partial L-function attached to π is defined by the Euler product:

$$L(s, \pi) = \prod_{v \text{ not in } S} \det(1 - \pi(v) q_v^{-s})^{-1}.$$

In fact the theory of representations allows us to define a complete L-function. Indeed, denote by π_S the irreducible representation of G_S on V^S. It is the tensor product, over all v in S, of irreducible representations π_v of the groups G_v. Since we may enlarge S, the irreducible representations π_v are defined for all v. To the representation π_v we can attach a L-factor $L(s, \pi_v)$, and the complete L-function is the product over all v of these L-factors. Of course, for almost all v, the L-factor is actually the one appearing above. The partial or complete L-functions defined in this way are the automorphic L-functions. They have analytic properties analogous to the properties of the classical

Dedekind zeta-functions or Dirichlet L-functions. As a matter of fact, the classical Dirichlet L-functions or the Dedekind zeta-functions are special cases of the automorphic L-functions.

The most interesting automorphic representations are the cuspidal ones. We will not define this notion precisely but will content ourselves with the following remark. Let $(n_1, n_2, \ldots, n_a)$ be a partition of n in a integers. Let $P = MU$ be the corresponding standard parabolic subgroup; thus U is the unipotent radical of P and M a Levi subgroup; in particular, M is isomorphic to the product of the groups $\mathrm{GL}(n_i)$ with $1 \leq i \leq a$. Let $(\pi_1, \pi_2, \ldots, \pi_a)$ be an a-tuple of irreducible cuspidal representations of $\mathrm{GL}(n_1)$, $\mathrm{GL}(n_2), \ldots, \mathrm{GL}(n_a)$, respectively. This tuple defines a representation of M, or rather a representation of P trivial on U that we can induce to G. The induced representation may not be irreducible, but its irreducible components are automorphic. If π is an irreducible component then, for almost all finite places v, the n by n matrix $\pi(v)$ is the direct sum of the matrices $\pi_i(v)$. In other words, the partial L-function attached to π is the product of the L-functions attached to the representations π_i. All automorphic representations can be obtained in this way in terms of cuspidal ones.

There is a slight extension of the theory of L-functions. Namely, if π_1 and π_2 are automorphic representations for $\mathrm{GL}(n_1)$ and $\mathrm{GL}(n_2)$, respectively, then we define a partial L-function by

$$L(s, \pi_1 \times \pi_2) = \prod_{v \text{ not in } S} \det\left(1 - \pi_1(v) \otimes \pi_2(v) q_v^{-s}\right).$$

This partial L-function is meromorphic. Assuming the representations cuspidal and unitary, it is nonzero at $s = 1$ and has a pole if and only if the representations are contragredient to one another. We will denote by $\tilde{\pi}$ the representation contragredient to π. The matrices $\tilde{\pi}(v)$ and $\pi(v)$ are inverse to one another. (cf. [J–H], [R], [S1], [S2]).

2. Induction and Base Change

Let E be an Abelian extension of F, cyclic of prime degree r. Suppose π is an automorphic representation for $\mathrm{GL}(n, E)$. The L-function attached to π (complete or partial) is an Euler product on E; by regrouping the local factors above a given place v of F, we may regard this L-function as an Euler product over F. As such, it has the analytic properties of the L-function attached to an automorphic representation for $\mathrm{GL}(nr, F)$. It is then natural to conjecture it is indeed the L-function attached to an automorphic representation for $\mathrm{GL}(nr, F)$. Denote by $I(\pi)$ ("induction from E to F") this conjectural automorphic representation. Let us describe, for almost all places v of F, the corresponding matrix $I(\pi)(v)$. If v decomposes (necessarily completely) in E then there are r places $v_1, v_2, \ldots, v_r$ above v, and $I(\pi)(v)$ is then the direct sum of the matrices $\pi(v_j)$ for $1 \leq j \leq r$. Assume, on the contrary, that v does not split and is unramified in E. Then there is exactly one place w of E above v. Let u be any primitive complex r-th root of 1. Then $I(\pi)(v)$ is the direct sum of the matrices $u^j\pi(w)$ for $1 \leq j \leq r$.

If $n = 1$ and $r = 2$ then π is an idele-class character for E, and the existence of π is in substance due to Hecke. If $n = 1$ and π is the trivial representation, then $L(s, \pi)$ is the Dedekind zeta-function; according to class-field theory, there is an idele-class character η of F such that

$$L(s, \pi) = \prod_{0 \leq j \leq r-1} L(s, \eta^j).$$

The L-function on the right is indeed attached to an automorphic representation σ for $\mathrm{GL}(r, F)$. It is induced by the r-tuple $(1, \eta, \eta^2, \ldots, \eta^{r-1})$ in the sense of Section 1. Let us describe its space in some detail, as this is an especially simple case. In a precise way, consider a function f on $\mathrm{GL}(r, F_{\mathbb{A}})$ such that

$$f\left[\begin{vmatrix} a_1 & & & \\ & a_2 & & * \\ & & \ddots & \\ & 0 & & a_r \end{vmatrix} g\right] = \prod \eta^j(a_j)|a_j|^{s_j - j + 1/2(r-1)} f(g).$$

Thus f transforms on the left under a one-dimensional character of the parabolic subgroup P of type $(1, 1, 1, \ldots, 1)$ in $G = \mathrm{GL}(r)$. Define a function E by the "Eisenstein series":

$$E(g) = \sum_{\gamma \in P(F)\backslash G(F)} f(\gamma g).$$

The series converges if the differences $s_j - s_{j+1}$ have a sufficiently large real part. The analytic continuations of these functions to the point $s_1 = s_2 = \cdots = s_r = 0$ span the space of σ, the representation being by right shifts.

Thus the existence of the representations $I(\pi)$ appears as a natural generalization of the basic identity of class-field theory.

There is another possible generalization. Indeed, the composition with the norm map gives a "base change" from the idele class characters of F to the idele class characters of E. If we denote by $B(\pi)$ this operation then we have the basic relation:

$$L(s, B(\pi)) = \prod_{0 \leq j \leq r-1} L(s, \eta^j \pi).$$

It is therefore reasonable to conjecture the existence of a similar operation $B(\pi)$ ("base change from E to F") taking the automorphic representations of $\mathrm{GL}(n, F)$ to those of $\mathrm{GL}(n, E)$; the representation $B(\pi)$ must be such that

$$L(s, B(\pi)) = \prod_{0 \leq j \leq r-1} L(s, \eta^j \pi).$$

In this formula $\eta\pi$ denotes the tensor product of π and η, that is, the representation that acts on the same space as π and is such that $\eta\pi(g) = \eta(\det g)\pi(g)$. Indeed, the right-hand side has all the analytic properties of the L-function attached to an automorphic form for $\mathrm{GL}(n, E)$. The relation between the matrices $\pi(v)$ and $B(\pi)(w)$ is as follows. Suppose that $\pi(v)$ is defined. If v splits in E and v_j, $1 \leq j \leq r$, denotes the corresponding places of E then $B(\pi)(v_j)$ is defined and equal to $\pi(v)$. If v does not split and is unramified in E then there is one place w of E above v and $B(\pi)(w)$ is defined and equal to

$$B(\pi)(w) = \pi(v)^r.$$

These operations have a number of formal properties. We list a few of them:

$$L(s, B(\pi) \times \sigma) = L(s, \pi \times I(\sigma)),$$

where π is an automorphic representation of $\mathrm{GL}(n, F)$ and σ an automorphic representation of $\mathrm{GL}(m, E)$. We have also

$$I(\sigma)\eta = I(\sigma).$$

Finally let τ be a generator of the Galois group of E over F. Since the Galois group operates on the group $\mathrm{GL}(n, E_{\mathbb{A}})$ it operates on its representations; a representation of the form $B(\pi)$ is invariant under τ.

We are going to sketch a proof of the existence of the maps B and I ([A-C]). The key to the proof will be the last invariance property.

3. The Distribution $I(f)$

It will be convenient to fix a subgroup of the ideles of E or F isomorphic to the group of positive real numbers and a corresponding subgroup

Z_+ of the center of $\mathrm{GL}(n, F_{\mathbb{A}})$ or $\mathrm{GL}(n, E_{\mathbb{A}})$. Consider the space L of functions ϕ on $\mathrm{GL}(n, F_{\mathbb{A}})$ invariant on the left under $Z_+ \mathrm{GL}(n, F)$ and square integrable (modulo that subgroup); the group $\mathrm{GL}(n, F_{\mathbb{A}})$ operates by right shifts in this Hilbert space. Let $L(G)$ be the subspace spanned by the irreducible invariant subspaces of L; this is the "discrete part" of the spectrum. If f is a smooth function of compact support on $\mathrm{GL}(n, F_{\mathbb{A}})$ then it defines a convolution operator on L and $L(G)$. We would like to compute its trace on the space $L(G)$; unfortunately, it is not clear that this operator has a trace. In this introduction, we will ignore this difficulty and will denote by $I(f, G)$ this "trace." We will need to consider other distributions as well. They are attached to certain parabolic subgroups. Suppose that $n = ab$. Let P be the standard parabolic subgroup of type $(a, a, \ldots, a)$ in $\mathrm{GL}(n)$. Thus $P = MU$ where U is the unipotent radical of P and M a Levi subgroup, say the group of matrices of the form

$$m = \begin{vmatrix} m_1 & & & * \\ & m_2 & & \\ & & \ddots & \\ & 0 & & m_b \end{vmatrix}.$$

The representation of M on the discrete part of its spectrum induces a representation of $\mathrm{GL}(n, F_{\mathbb{A}})$; call $I(P)$ its space. Let s be a permutation matrix that normalizes M and permutes circularly the blocks of a matrix in M. To s is attached a certain intertwining operator $M(s)$ of $I(P)$ into itself. At a formal level it is defined as follows: If ϕ is in the space $I(P)$ then

$$M(s)\phi(g) = \int_{U(F_{\mathbb{A}}) \cap s^{-1} U(F_{\mathbb{A}}) s \backslash U(F_{\mathbb{A}})} \phi(sug)\, du.$$

In fact the integral does not converge and must be defined by analytic continuation. Composing the operator defined by f on $I(P)$ and the operator $M(s)$, we obtain an operator whose "trace" we denote by $I(f, P)$. The sum of $I(f, G)$ and the $I(f, P)$ (suitably weighted) we denote by $I(f)$. It is the discrete part of the trace formula. The trace formula gives an explicit expression for this sum. The main ingredient in this expression is the orbital integrals, which we now define. Assume f is a product of local functions f_v; suppose that g is a regular semisimple element of $\mathrm{GL}(n, F_v) = G_v$; let G_g be its centralizer. The orbital integral corresponding to g is defined by

$$\Phi(g, f_v) = \int_{G_g \backslash G_v} f_v(h^{-1}gh)\, dh.$$

4. The Distribution $I(f', \tau)$

Consider now the group $\mathrm{GL}(n, E)$. If we restrict the scalars from E to F we may regard $\mathrm{GL}(n, E)$ as an algebraic group defined over F, which we will denote by G'. In particular, we have $G'(F) = \mathrm{GL}(n, E)$ and $G'(F_{\mathbb{A}}) = \mathrm{GL}(n, E_{\mathbb{A}})$. The Galois element τ operates on G', hence also on functions on $G'(F_{\mathbb{A}})$ by the rule

$$\tau f(g) = f(g^{\tau}).$$

As before, we have the discrete part of the spectrum $L(G')$; it is invariant under the action of τ. Let f' be a smooth function of compact support on $G'(F_{\mathbb{A}})$. It defines an operator on $L(G')$; composing this operator with the action of τ we obtain a new operator whose "trace" we denote by $I(f', \tau, G')$. As before we must add to this distribution a weighted sum of other distributions defined in terms of certain parabolic subgroups. We obtain, in this way, a distribution $I(f', \tau)$. It is invariant by twisted conjugation: Twisted conjugation is a new action of G' on itself defined by the formula

$$g \to \gamma^{-\tau} g \gamma.$$

The twisted trace formula gives then an explicit expression for the distribution $I(f', \tau)$. The main ingredients of this formula are the twisted orbital integrals. As usual let us assume f' is the product over the places of F of local functions f'_v. If g is a regular semisimple element (in the "twisted sense") in G'_v then the corresponding local orbital integral is given by

$$\Phi(f'_v, g, \tau) = \int_{G_{g\tau} \backslash G'_v} f'(h^{-\tau} gh)\, dh,$$

where $G_{g\tau}$ denotes the twisted centralizer of g, that is, the group of elements of G'_v that fix g in the twisted action.

5. Equality of the Distributions

There is a relation between the twisted conjugacy classes in $G'(F)$ and the ordinary conjugacy classes in $G(F) = \mathrm{GL}(n, F)$ as follows. Let γ be an element of $G'(F)$. Set

$$n(\gamma) = \gamma \cdot \gamma^{\tau} \cdot \gamma^{\tau^2} \cdots \gamma^{\tau^{r-1}}.$$

In general this is not an element of $G(F)$. Indeed

$$n(\gamma)^{\tau} = \gamma^{-1} n(\gamma)\gamma.$$

However, this relation shows that the conjugacy class of $n(\gamma)$ is defined over F, in the sense that $n(\gamma)$ is conjugate by an element of $G'(E)$ to an element μ of $G(F)$. The conjugacy class of μ in $G(F)$ is uniquely determined by the twisted conjugacy class of γ; we denote it by $N(\gamma)$. Then N determines an injection of the twisted conjugacy classes of $G'(F)$ into the conjugacy classes of $G(F)$; roughly speaking, the image consists of those elements whose determinant is a norm.

Similar considerations apply locally. In a precise way, if v is a place of E that does not split in E and w is the place of E above v, then E_w is a simple p-extension of F_v and τ induces a generator of the Galois group of E_w over F_v; furthermore, $G'_v = G(E_w)$. We get, therefore, a norm map N_v from the set of twisted conjugacy classes of G'_v to the set of conjugacy classes of G_v. If v splits in E, then G'_v is isomorphic to the product of r copies of G_v and the action of τ on G_v is then by circular permutation. The norm map is again defined; however, it is then a quite simple object. Namely, if g is identified with an r-tuple $(g_1, g_2, \ldots, g_r)$, then $N_v(g)$ is simply the conjugacy class of the product $g_1 g_2, \ldots, g_r$. The global and local norms are compatible. In particular, the Hasse principle is valid in this situation: A conjugacy class in $G(F)$ is a norm if and only if it is so locally at all places of F inert in E.

This relation suggests the following definition. Suppose v is a place of E, f'_v (resp. f_v) a smooth function of compact support on G'_v (resp. G_v). We shall say that they have matching orbital integrals if the following condition is satisfied: For each regular semisimple element g of $G(F)$

$$\begin{aligned}\Phi(g, f_v) &= \Phi(g', f'_v, \tau) \qquad \text{if } N_v(g') = \text{class}(g),\\ &= 0 \text{ if class}(g) \text{ is not a norm.}\end{aligned}$$

Suppose first that v splits in E; then G'_v is the product of r copies of G_v. Every conjugacy class in G_v is a norm so that the second condition is vacuous. Furthermore, it is easy to see that given f'_v there is an f_v with matching orbital integrals. Indeed, we may assume that f'_v is the tensor product of r functions $f_1, f_2, \ldots, f_r$ on G_v. Then we can take for f_v the convolution product of these functions. In particular, this shows that for any f_v there is a function f'_v with matching orbital integrals.

Now suppose that v does not split in E. Given f'_v it is still true that there is a function f_v with matching integrals. Conversely, given f_v satisfying the second condition above there is a function f'_v with matching orbital integrals. Essentially, the second condition is always satisfied if the support of f_v is contained in the set of elements whose determinant is a norm.

We need to specialize these notions to the case of the Hecke functions. Suppose that v is a place of F which splits in E. Then the Hecke algebra H'_v of G'_v is the tensor product of r copies of the Hecke algebra H_v of G_v. We define an algebra homomorphism from H'_v to H_v as follows: If f'_v is the tensor product of r elements of H_v then $b_v(f'_v)$ is their product. It is clear that if A is any n by n invertible complex matrix then

$$b_v(f'_v)^\wedge(A) = \prod_{1 \leq j \leq r} f_j^\wedge(A).$$

Suppose now v does not split in E but is unramified; let w be the place of E above v. Then H'_v, the Hecke algebra of G'_v, is actually the Hecke algebra of G_w. There is again an algebra homomorphism of H'_v to H_v. It is determined by the condition

$$b_v(f'_v)^\wedge(A) = f'^\wedge_v(A^r), \qquad \text{for all } A.$$

Now one can show that in both cases the function $b_v(f'_v)$ has matching orbital integrals with f'_v. In fact, it is the only element of H_v with this property.

After these preliminaries we can state the main result. Suppose f' and f are given. We naturally assume that f' and f are product of local functions. Suppose that for all places v the functions f'_v and f_v have matching orbital integrals. Then

$$rI(f', \tau) = I(f).$$

Both sides are defined in terms of traces, or at any rate, harmonic analysis, while functions f and f' are related by a geometric condition. This identity illustrates the "duality" between geometry and harmonic analysis that is at the core of the trace formula.

As for the proof, we will only remark that it uses Hasse principle in an essential way. To explain this point, let us remark that in the expression for the right-hand side enter the global orbital integral for the elliptic regular elements; if γ is such an element then its global orbital integral is the product of the local orbital integrals obtained by regarding γ as an element of G_v for every v. If the formula is to be true then the global orbital integral must vanish if γ is not a norm. However, if γ is a not a norm, it is not a norm at one place v inert in E, and then the local orbital integral for γ at the place v vanishes, by the very choice of the function f_v. Thus the global orbital integral must vanish.

6. Exploiting the Basic Identity

To exploit the basic identity, we will select a finite of places of F that we denote by S. We will assume that S contains all places at infinity and all places that ramify in E. We will denote by T the set of places of E above the places of S. It will be convenient to introduce the following notations: We will denote by H^S the tensor product of the Hecke algebras H_v for v not in S. The notations H'^S and H'^T will have a similar meaning; they of course denote the same object. We will frequently also denote by f'^S the tensor product of functions f'_v in H'_v for v not in S, and by f^S the tensor product of functions f_v in H_v for v not in S. There is an algebra homomorphism b from H'^S to H^S. It is defined by the formula $b(f'^S) = f^S$, if f'^S is the product of functions f'_v and f_v is the product of the functions $b_v(f'_v)$.

If f^S is an element of H^S that is a product of functions f_v and π is any automorphic representation for $\mathrm{GL}(n, F)$ with a nonzero K^S fixed vector, we will set

$$f^{S\wedge}(\pi) = \prod_{v \text{ not in } S} f_v^{\wedge}(\pi(v)).$$

By linearity this can be defined for any element of H^S. Similarly, we can define $f'^{S\wedge}(\sigma)$ for any element f'^S of H'^S and any automorphic representation σ for $G'(F) = G(E)$ with a K^T fixed vector. With this notation it is clear that if $f^S = b(f'^S)$ then the representation $B(\pi)$ we are looking for must satisfy

$$f'^{S\wedge}(B(\pi)) = f^{S\wedge}(\pi).$$

Now let us consider the distribution $I(f)$. First take an arbitrary function f of the form $f = f_S \cdot f^S$ where f_S denotes a product of functions f_v for v in S and f^S is an arbitrary element of H^S. Suppose π is an automorphic cuspidal representation for $G(F)$. Then it appears in the discrete spectrum with multiplicity one. Its contribution to $I(f)$ is the trace of the operator $\pi(f)$, which we may write as the product

$$\operatorname{tr} \pi_S(f_S) f^{S\wedge}(\pi).$$

If we think of f_S as being fixed while f^S varies in the Hecke algebra H^S, we see this is a scalar multiple of a character of H^S. Other representations in the discrete spectrum or in the induced representations associated to the various parabolic subgroups also contribute multiples of characters. So, altogether, $I(f)$ is an infinite linear combination of characters of the Hecke algebra. These characters determine the representations, at least those that are cuspidal; that is, if π_1 is another automorphic representation and

$$f^{S\wedge}(\pi) = f^{S\wedge}(\pi_1)$$

for all f^S in H^S, then in fact π_1 is cuspidal and isomorphic to π. Indeed, this follows at once from the properties of the L-functions described in Section 1. Furthermore, the principle of linear independence of characters applies to this kind of linear combinations.

Now restrict our f to the class of functions that have matching orbital integrals with some f'. Essentially, this means that for v in S the support of f_v is in the set of elements whose determinant is a norm. On the other hand, take f^S of the form $b(f'^S)$ where f'^S is in H'^S. Then $I(f)$ may now be regarded as an infinite linear combination of characters of the Hecke algebra H'^S. In particular, it may now happen that

$$f^{S\wedge}(\pi) = f^{S\wedge}(\pi_1)$$

for all f^S. Indeed this means that $\pi(v) = \pi_1(v)$ if v splits and

$$\pi(v)^r = \pi_1(v)^r$$

if v does not split. It is easy to see this implies that the partial L-functions attached to π and π_1 must then satisfy the equality

$$\prod_{0 \leq j \leq r-1} L(s, \pi \times \pi^{\sim} \eta^j) = \prod_{0 \leq j \leq r-1} L(s, \pi \times \pi_1{}^{\sim} \eta^j).$$

Now the factor on the left-hand side corresponding to $j = 0$ has a pole at $s = 1$ that is not cancelled by a zero of the other factors. Thus at least one factor on the right-hand side must have a pole; this means that π_1 must be the tensor product of π with some power of η. Assume that π is not isomorphic to its tensor product with η. Then there are exactly r representations π_1, and the sum of their contributions is simply

$$r \operatorname{tr}(\pi_S)(f_S) f^{S\wedge}(\pi).$$

We now fix an automorphic cuspidal representation π of $G(F)$. We assume first that π is not equivalent to its tensor product with η. We want to show that there is a base change $B(\pi)$ for π. To this end we choose the set S so large that π has a nonzero K^S fixed vector. We want to show there is at least one function f_S such that $\operatorname{tr} \pi_S(f_S)$ is nonzero. Fix v in S. We have to show there is f_v such that $\operatorname{tr} \pi_v(f_v)$ is nonzero. This is clear if v is split. Assume v is inert. Since the tensor products of π_v with the powers of η are either pairwise inequivalent or all equivalent, there is a function h on G_v such that

$$\sum_{0 \leq j \leq r-1} \operatorname{tr} \pi_v \eta_v^j(h) \neq 0.$$

If we let f_v be the function defined by

$$f_v(g) = \sum_{0 \leq j \leq r-1} h(g) \eta_v^j(\det g),$$

then $f_v(g)$ is zero unless $\det g$ is a norm and $\operatorname{tr} \pi_v(f_v)$ is nonzero. Thus we see that for at least one choice of f_S the contribution of π to $I(f)$ is a nonzero multiple of $f^{S\wedge}(\pi)$.

We now fix f_S as above, choose a corresponding f'_S, and take an arbitrary f'^S in the Hecke algebra H'^S. The distribution $I(f', \tau)$ is again an infinite linear combination of characters of the algebra H'^S, these characters having the form $f'^{S\wedge}(\sigma)$ where σ is an automorphic representation of $G'(F_{\mathbb{A}})$ invariant under τ. Denote by f^S the image of f'^S under b and by f the product of f_S and f^S. The equality $I(f) = I(f', \tau)$ is then the equality of two (infinite) linear combinations of characters of the Hecke algebra H'^S. It implies that there is at least one automorphic representation σ of $\Gamma'(F_{\mathbb{A}})$ such that

$$f'^{S\wedge}(\sigma) = f^{S\wedge}(\pi) \qquad \text{where } f^S = b(f'^S)$$

for all f'^S in the Hecke algebra H'^S. In view of our definitions this simply says that $\sigma = B(\pi)$. Considerations involving L-functions imply that, in fact, σ is cuspidal.

We have assumed that π is not equivalent to its product with η. Now suppose it is the case. Then again there is an automorphic representation σ, which is the base change of π. However, it is no longer cuspidal. What one can prove is that $\pi = I(\sigma_1)$ where σ_1 is an automorphic cuspidal representation of $\mathrm{GL}(n/r, E_{\mathbb{A}})$, which is not τ invariant; then σ is the representation induced by the r-tuple

$$(\sigma_1, \sigma_1^{\tau}, \sigma_1^{\tau^2}, \ldots, \sigma_1^{\tau^{r-1}}).$$

In the reverse direction if σ is an automorphic representation of $G(E_{\mathbb{A}})$, cuspidal say, and invariant under τ, it can be shown that it is the base change of some representation. Actually a detailed study of the situation gives the existence of the operations B and I completely.

7. Concluding Remarks

We hope to have made clear that the problem discussed is intimately connected with class-field theory. A generalization of the construction of I and B to arbitrary extensions would constitute a nonabelian class-field theory; in particular, it would imply that the Dedekind zeta-function of a given number field E of degree r over another field F is the L-function attached to an automorphic form for $\mathrm{GL}(r, F)$, or what amounts to the same, is a product of L-functions attached to automorphic cuspidal representations for various groups $\mathrm{GL}(m, F)$. It would also imply that every Artin–Hecke L-function is attached to an appropriate automorphic representation; in particular, it would imply Artin conjecture. As a matter of fact, using the base change for $n = 2$, Langlands obtained the identity of Artin L-functions with automorphic L-functions in certain cases.

As for the history of the present topic, it has been told many times, so we will limit ourselves to a few remarks. We have used L-functions to motivate the existence of I and B. As a matter of fact, for low values of r and n it is possible to establish the existence of I and B by simple considerations involving L-functions. In general, however, the trace formula in the form stated here must be used. It was first discovered by Saito and Shintani, then extended by Langlands to a more general situation and then by Clozel and Arthur to arbitrary n. Each extension introduces new ideas so that the word "extension" is somewhat misleading. Finally, the operation I can also be constructed in terms of the trace formula.

References

[A-C] Arthur, J. and Clozel, L. "Base change for GL(n)," preprint.

[F] Flicker, Y. "The trace formula and base change for GL(3)," *Lecture Notes in Mathematics.* **927** (1982).

[J-H] Jacquet, H. and Shalika, J. "On Euler products and the classification of automorphic representations, I and II," *Am. J. Math.* **103** (1981), 499–558; *Am. J. Math.* **103** (1981), 777–815.

[L] Langlands, R. Base Change for GL(2). Princeton Univ. Press, Princeton (1980).

[R] Rankin, R. "Contributions to the theory of Ramanujan's functions $\tau(n)$ and similar arithmetical functions, I and II," *Proc. Camb. Phil. Soc.* **35** (1939), 351–356, 357–372.

[Sa] Saito, H. "Automorphic forms and algebraic extensions of number fields," *Lectures in Math*, Kyoto Univ. (1975).

[S1] Selberg, A. "Bemerkungen uber eune Dirichletsche reihe, die mit der theorie der modulformen nahe verbunden ist," *Arch. Math. Naturvid.* **43** (1940), 47–50.

[S2] Selberg, A. "On the estimation of Fourier coefficients of modular forms." In: *Proc. Symposia Pure Math.* **8** (1965), American Math. Society, Providence.

[Sh] Shintani, T. "On liftings of holomorphic cusp forms." In: *Automorphic Forms, Representations and L-functions*, A. Borel and W. Casselman, eds., *Proc. of Symposia in Pure Math.* **33** (I-II) (1979), Amer. Math. Soc., Providence.

6 Eisenstein Series, the Trace Formula, and the Modern Theory of Automorphic Forms

R. P. LANGLANDS

1. Eisenstein Series and Automorphic *L*-functions

The modern theory of automorphic forms is a response to many different impulses and influences, above all the work of Hecke, but also class-field theory and the study of quadratic forms, the theory of representations of reductive groups, and of complex multiplication, but so far many of the most powerful techniques are the issue, direct or indirect, of the introduction by Maaß and then Selberg of spectral theory into the subject.

The spectral theory has two aspects: (i) the spectral decomposition of the spaces $L^2(\Gamma\backslash G)$ by means of Eisenstein series; (ii) the trace formula, which can be viewed as a striking extension of the Frobenius reciprocity law to pairs (Γ, G), G a continuous group and Γ a discrete subgroup.

The attempt to discover a class of Euler products attached to automorphic forms that would include the Dirichlet series with Größencharakter attached to real quadratic fields led Maaß in 1946,

NUMBER THEORY,
TRACE FORMULAS and
DISCRETE GROUPS

ISBN 0-12-067570-6

under difficult circumstances in the chaos of immediate postwar Germany, to the study of eigenfunctions of the Laplacian

$$\Delta = y^2\left[\frac{\partial^2}{\partial x^2} + \frac{\partial^2}{\partial y^2}\right]$$

on the upper half-plane, eigenfunctions that transform simply, or are even invariant, under discrete groups, Γ, especially subgroups of the modular group [27]. Apparently he was influenced to some extent by the work of Fueter on quaternionic function theory.

It is simplest to consider functions actually invariant under the discrete group. If it is Fuchsian and the fundamental domain is not compact, then Δ has a continuous spectrum, and the corresponding eigenfunctions are given by analytic continuation of functions defined by infinite series, the Eisenstein series. They are attached to cusps. If, for example, the cusp is at infinity so that the group

$$\Gamma_\infty = \left\{\gamma = \begin{bmatrix} 1 & x \\ 0 & 1 \end{bmatrix} | \gamma \in \Gamma\right\}$$

is infinite, then the attached series is

$$\sum_{\gamma \in \Gamma_\infty \backslash \Gamma} y_\gamma^{(s+1)/2} = \sum \left[\frac{y}{|cz+d|^2}\right]^{(s+1)/2}. \tag{1.1}$$

This series is easily seen to converge for Re $s > 1$ and to yield eigenfunctions of Δ, but not the ones needed for the spectral decomposition, for they correspond to parameters satisfying Re $s = 0$. Thus an analytic continuation is required. It is often carried out in two steps as in [22], the functions being first continued to the region Re $s > 0$, either with the help of the resolvent or the Green's function, the method used by Roelcke, or by a truncation process as employed by Selberg. The continuation across Re $s = 0$ was first effected by Selberg by means of a further truncation and the reflection principle.

To recapitulate briefly, Maaß's work on Euler products and automorphic forms drew his attention to a problem in spectral theory that, in turn, led to a problem in analytic continuation. The purpose of the first part of this lecture is to recall how a much larger class of Euler products, the *automorphic L-functions*, one of the central notions of the modern theory of automorphic forms, arose, a little by accident, from the solution of the problem in analytic continuation.

Recall first that in the fifties there was a tremendous surge of interest in automorphic forms on groups of higher dimension—due largely, I suppose, to the papers of Siegel on orthogonal groups and the symplectic groups; so it is hardly surprising that Selberg and others attempted to extend to them his techniques and ideas, the analytic continuation of the Eisenstein series and the trace formula. Decisive progress, however, had to await the introduction by Gelfand in 1962 [15] of the general notion of cusp forms that lies at the center of the spectral theory in higher dimensions.

Although it is not necessary, it is extremely convenient, if only to avoid elaborate notational complications, to work with adelic groups. In addition, the spectral decomposition is then made with respect to the largest possible family of operators, including the Hecke operators and the differential operators. It entails working on the group rather than the symmetric space, but then the obvious symmetries are recognized and the same problem not solved repeatedly.

For example, if $G = \mathrm{GL}(n)$ then an automorphic form is a function ϕ on $G(\mathbb{Q}\backslash G(\mathbb{A})$. It is a cusp form if, for every block decomposition of the $n \times n$ matrices and every $g \in G(\mathbb{A})$, we have

$$\int \phi\left(\begin{bmatrix} I & X \\ 0 & I \end{bmatrix} g\right) dx = 0.$$

Here X is an $n_1 \times n_2$ matrix, $n_1 + n_2 = n$, with adelic entries.

The notion of a cusp form clearly isolated, the analytic continuation of the general Eisenstein series is effected in three steps.

A. Series in One Variable Attached to Cusp Forms. If, for example, G is $\mathrm{SL}(n)$ and $n = n_1 + n_2$, then such a series is associated to a cusp form ϕ on

$$M(A) = \left\{ m = \begin{bmatrix} m_1 & 0 \\ 0 & m_2 \end{bmatrix} \in G(A) \right\}$$

and to a parameter s. Here m_i is an $n_i \times n_i$ matrix. For simplicity, I am confining myself to functions invariant on the right under a maximal compact subgroup K.

If

$$N = \left\{ \begin{bmatrix} I & X \\ 0 & I \end{bmatrix} \right\}$$

$$P = \left\{ \begin{bmatrix} A & X \\ 0 & B \end{bmatrix} \right\}$$

then

$$P = MN = NM.$$

Set

$$F_s(nmk) = \phi(m)\left(\frac{|\det m_1|^{n_2}}{|\det |m_2|^{n_1}}\right)^{s+1/2}. \tag{1.2}$$

Then the Eisenstein series is

$$E_s(g) = \sum_{\gamma \in P(\mathbb{Q})\backslash G(\mathbb{Q})} F_s(\gamma g). \tag{1.3}$$

Provided that ϕ is a cusp form, the methods used for the Eisenstein series on the upper half-plane will deal with those that like (1.3) are associated to maximal parabolic subgroups P and thus involve only a single parameter, although the extension should perhaps not be thought of as entirely routine.

B. Series in Several Variables Associated to Cusp Forms. They are attached to non maximal parabolic subgroups, thus for SL(n), to partitions $n = n_1 + n_2 + \cdots + n_r$ with more than two elements. The group M is given by

$$M = \{\begin{bmatrix} m_1 & & & \\ & m_2 & & \\ & & \ddots & \\ & & & m_r \end{bmatrix} | \prod \det m_i = 1\},$$

and P and N are defined accordingly. The functions F_s and the series E_s now depend on several parameters, $s = (s_1, \ldots, s_r)$, $\sum n_i s_i = 0$.

$$F_s(nmk) = \phi(m) \prod_{i=1}^{r} |\det m_i|^{s_i + \rho_i}, \tag{1.4}$$

where the ρ_i are real numbers chosen to simplify the formulas for the functional equations. To deal with these Eisenstein series, one combines the results from A with forms of Hartog's lemma. I observe that Hartog's lemma was introduced into the subject quite early, and by several mathematicians independently (cf. Appendix I of [22]).

C. The General Eisenstein Series. Apart from a more complicated dependence on $k \in K$, these are defined by functions like (1.4), with ϕ being any square-integrable automorphic form on $M(\mathbb{A})$. That ϕ is no longer necessarily a cusp form entails altogether new difficulties. Even for classical groups of low dimension, the analytic continuation of these series and the spectral decomposition involve quite different ideas than those that suffice for A or B [22].

One need not look far for fatuous and misleading comments on the techniques involved in the three steps. They are regrettable, but fortunately need not concern us here, for it is the series of step A, or rather the calculation of their constant term, that led to the introduction of the general *automorphic L-functions* and the *principle of functoriality*. Thus these ideas could have appeared before the general theory of Eisenstein series and independently of it, but the psychological inhibitions may have been too great.

A perhaps too simple formulation of the principle of functoriality is that certain natural operations on the L-series attached to automorphic forms reflect possible operations on the forms themselves. The usual Hecke theory, for example, associates to an automorphic form an Euler product

$$\prod_{p \notin S} \frac{1}{(1-\alpha_p p^{-s})(1-\beta_p p^{-s})}. \tag{1.5}$$

Since it is only the unordered pair $\{\alpha_p, \beta_p\}$ that matters, we may think of it as a conjugacy class $\{t_p\}$ of complex 2×2 matrices with these eigenvalues. A natural operation on elements of GL(2) is to let them act on symmetric tensors of a given degree n. This transforms conjugacy classes in GL(2) into conjugacy classes in GL$(n+1)$. So we pass from

$$\{t_p\} = \left\{ \begin{bmatrix} \alpha_p & 0 \\ 0 & \beta_p \end{bmatrix} \right\}$$

to

$$\left\{ \begin{bmatrix} \alpha_p^1 & & \\ & \alpha_p^{n-1} & \beta_p \\ & \ddots & \\ & & \beta_{pr}^n \end{bmatrix} \right\} \left\{ \begin{bmatrix} \alpha_p^1 & & & & \\ & \ddots & & & \\ & & \alpha_p^{n-1} & & \\ & & & \beta_p^1 & \\ & & & & \ddots \\ & & & & & \beta_p^{n-1} \end{bmatrix} \right\}$$

This leads us from the Euler product (1.5) to

$$\prod_{p \notin S} \frac{1}{(1 - \alpha_p^n p^{-s})(1 - \alpha_p^{n-1}\beta_p p^{-s})(1 - \beta_p^n p^{-s})},$$

and the principle of functoriality predicts that there is an automorphic form on GL($n + 1$) to which this series is attached.

Before indicating how the calculation of the constant term of Eisenstein series suggested the introduction of automorphic L-functions, I recall, for it is easy to forget, that 20 years ago it was by no means clear how, or even whether, the Hecke theory could be extended to groups other than GL(2). Ideas of varying quality were proposed, and it is surprising that this calculation, carried out more to pass the time than with any precise aim, should yield not only specific series whose analytic continuation and functional equation could be proved, but also a class of series with a natural completeness. Repeated efforts to rework the Hecke theory had led nowhere, except perhaps to a clearer understanding of how it functioned, which could only later be turned to profit.

There is a classical paradigm for the calculation, that giving the constant term of the series (1.1) or, more precisely, since we have passed to the adèle group, of

$$E_s(g) = \sum_{P(\mathbb{Q})\backslash G(\mathbb{Q})} F_s(\gamma g),$$

with

$$F_s(g) = \left|\frac{a}{b}\right|^{s+1/2}, \qquad g = \begin{bmatrix} 1 & x \\ 0 & 1 \end{bmatrix}\begin{bmatrix} a & 0 \\ 0 & b \end{bmatrix} k.$$

Here P is the group of upper-triangular matrices in $G = \mathrm{GL}(2)$, and K a maximal compact subgroup of $G(\mathbb{A})$.

If N is the subgroup

$$N = \left\{\begin{bmatrix} 1 & x \\ 0 & 1 \end{bmatrix}\right\}$$

of P, then the constant term of E_s is the function

$$\int_{N(\mathbb{Q})\backslash N(\mathbb{A})} E_s(ng)\, dn = \int_{\mathbb{Q}\backslash\mathbb{A}} E_s\left(\begin{bmatrix} 1 & x \\ 0 & 1 \end{bmatrix} g\right) dx.$$

To calculate it, we substitute the series expansion for E_s, and combine terms appropriately to obtain

$$\sum_{\gamma \in P(\mathbb{Q})\backslash G(\mathbb{Q})/N(\mathbb{Q})} \int_{P(\mathbb{Q}) \cap \gamma^{-1}N(\mathbb{Q})\gamma \backslash N(\mathbb{A})} F_s(\gamma ng)\, dn.$$

The double coset space $P(\mathbb{Q}\backslash G(\mathbb{Q})/n(\mathbb{Q})$ appearing here has a simple structure, for it consists of only two elements with representatives

$$\gamma = \begin{bmatrix} 1 & 0 \\ 0 & 1 \end{bmatrix} = 1, \qquad \gamma = \begin{bmatrix} 1 & 0 \\ 0 & 1 \end{bmatrix} = \omega.$$

This is special case of the Bruhat decomposition that will appear later. For $\gamma = 1$ the integral is simply $F_s(g)$, because

$$\int_{N(\mathbb{Q})\backslash N(\mathbb{A})} dn = 1.$$

For $\gamma = \omega$ the integral is a product

$$\prod_v \int_{N(\mathbb{Q}_v)} F_s(\omega n_v g_v)\, dn_v, \tag{1.6}$$

because

$$P(\mathbb{Q}) \cap \omega^{-1}N(\mathbb{Q})\omega = 1.$$

The integrals appearing in (1.6) can be calculated place by place. Take g to be 1, so that each g_v is 1. For a nonarchimedean place, we calculate $F_s(wn_v)$ easily. First of all,

$$\omega n_v = \begin{bmatrix} 0 & 1 \\ 1 & 0 \end{bmatrix} \begin{bmatrix} 1 & x \\ 0 & 1 \end{bmatrix} = \begin{bmatrix} 0 & 1 \\ 1 & x \end{bmatrix}.$$

If x is integral this matrix belongs to K and

$$F_s(\omega n_v) = 1.$$

Otherwise

$$\begin{bmatrix} 0 & 1 \\ 1 & x \end{bmatrix} = \begin{bmatrix} 1 & * \\ 0 & 1 \end{bmatrix} \begin{bmatrix} x^{-1} & 0 \\ 0 & x \end{bmatrix} \begin{bmatrix} * & * \\ x^{-1} & 1 \end{bmatrix}, \qquad |x| > 1.$$

Thus

$$\int_{N(\mathbb{Q}_v)} F(\omega n_v)\, dn_v = 1 + \left(1 - \frac{1}{p}\right) \sum_{n=1}^{\infty} \frac{1}{p^{ns}} = \frac{1 - \dfrac{1}{p^{s+1}}}{1 - \dfrac{1}{p^s}}.$$

Taking the product over the finite places, we obtain

$$\frac{\zeta(s)}{\zeta(s+1)}.$$

The infinite place yields the usual supplementary Γ-factor.

In general one carries out the calculation in a similar fashion and, at the end, a result obtained, one looks for a transparent way to express it. It is at this stage that the automorphic L-functions suggest themselves. What are the ingredients of the calculation?

i) The Eisenstein series associated to a cusp form ϕ on a Levi subgroup M of a maximal parabolic subgroup of G. Because ϕ is a cusp form the constant term will be expressed in terms of cusp forms for the same group M. (I observe, in passing, that square-integrable forms are not closed in the same way. This is one of the reasons that C is more difficult than A and B.) The constant term is itself a sum of one or two terms, depending upon the nature of the Bruhat decomposition. (I observe, again in passing, that for the classical groups the concepts of a parabolic subgroup or the Bruhat decomposition are elementary notions of linear algebra that could profitably be included in the education of all pure mathematicians.)

ii) Since we are working adelically, the calculation is ultimately local, and we are able to draw on our experience with local harmonic analysis, especially Harish–Chandra's theory of spherical functions on real groups, and the explicit formulas that Bhanu–Murty and Gindikin–Karpelevich contributed to it.

The calculation itself was carried out for a large number of illustrative cases in [21]. One begins with P and writes an arbitrary element g of $G(\mathbb{A})$ as $g = nmk$, $n \in N(\mathbb{A})$, $m \in M(\mathbb{A})$, $k \in K$. The function F_s has the form

$$F_s(g) = \chi_s(m)\phi(m)\psi(k),$$

where χ_s is a character of $M(\mathbb{A})$ that depends on the complex parameter s. The Eisenstein series is

$$E_s(g) = \sum_{P(\mathbb{Q})\backslash G(\mathbb{Q})} F_s(\gamma g),$$

with constant term

$$\int_{N'(\mathbb{Q})\backslash N'(\mathbb{A})} E_s(n'g)\,dn'. \tag{1.7}$$

The group N' belongs to a parabolic subgroup P', that may or may not be P itself.

Since the Eisenstein series can be continued to the whole complex plane as a meromorphic function of s, so can (1.7). The Bruhat decomposition allows us to write (1.7) as a sum of one or two terms. If there are two, one is simply $F_s(g)$, which is entire, so that the remaining term, the one in which we are interested, is as well behaved as (1.7) itself. It is

$$\int_{N'(\mathbb{A})} F_s(\omega n'g)\,dn', \tag{1.8}$$

with a suitable ω.

To calculate (1.8) as a product it is necessary to be somewhat careful in the choice of ϕ and ψ and, in addition, to recall that for each g the function $\phi_g\colon m \to \phi(mg)$ belongs to a space V of functions on $M(\mathbb{Q})\backslash M(\mathbb{A})$ transforming according to an irreducible representation σ of $M(\mathbb{A})$. Thus ϕ may be regarded as a function ϕ_g on $G(\mathbb{A})$ with values in V and, more precisely, as an element of the space of the induced representation

$$\mathrm{Ind}_P^G \sigma_s,$$

σ_s being the tensor product of σ with χ_s. It is extended to $P(\mathbb{A})$ by making it trivial on $N(\mathbb{A})$.

It turns out—as a result of formal considerations—that the operation (1.8) depends only on the class of σ and not on its realization on a subspace of $L^2(G\backslash\mathbb{Q})\backslash G/\mathbb{A}))$. The only information we need from the realization, but this is of course decisive, is that (1.8) can be analytically continued. To calculate (1.8) as a product, we work abstractly, realizing σ as a product $\otimes\,\sigma_v$, and reduce (1.8) to a product of local integrals. They depend only on the class of σ_v, and at the places where there is no ramification we can take a simple model for σ_v to perform the calculations.

The calculations reduce finally to simple summations like those for GL(2), but this requires some understanding of the unramified representations of the local groups $G(\mathbb{Q}_p)$ or, what amounts to the same thing, of the structure of the local Hecke algebras. This is an elementary but somewhat elaborate topic.

Almost everywhere the reductive group G with which we began is split, or at worst quasisplit, and split over an unramified extension. It is only such groups that have representations that we can call unramified. Their unramified representations can be parametrized, and an elegant form of the parametrization is forced upon us by the need to express the results of the calculation in transparent form. The parameter attached to an unramified representation is a semisimple conjugacy class $\{t(\pi_p)\}$ in a *complex* reductive group ${}^LG$. (For $G = \mathrm{GL}(n)$ the group ${}^LG$ is again $\mathrm{GL}(n)$, as was implicit in our earlier remarks; for other groups the relation between G and ${}^LG$ is less direct.)

At almost all places the local factors of the constant term can be expressed in terms of the classes $\{t(\pi_p)\}$. There are finite-dimensional complex analytic representations $r_1, \ldots, r_n$ of ${}^LM$ and constants $a_1, \ldots, a_n$ such that the local factor is

$$\prod_{i=1}^{n} \frac{\det\,[I - r_i(t(\pi_p))p^{-a_i(s+1)}]}{\det\,[I - r_i(t(\pi_p))p^{-a_i s}]}.$$

This suggests the introduction, for any complex analytic representation r of any ${}^LG$, of the Euler product

$$L_S(s, \pi, r) = \prod_{p \notin S} \frac{1}{\det\,(I - r(t(\pi_p))p^{-s})}. \qquad (1.9)$$

Here S is a finite set of places that a finer treatment would remove. Thus the constant term (1.8) is in essence a product

$$\prod_{i=1}^{n} \frac{L_S(a_i s, \pi, r_i)}{L_S(a_i s + 1, \pi, r_i)}.$$

The analytic continuation of the series (1.9) can then be obtained in sufficiently many cases to justify their further study by choosing some group M, which becomes the group of primary interest, and then searching for a group G that contains a parabolic subgroup of which it is a Levi factor. A large number of examples were given in [21].

2. The Structure of Trace Formulas and their Comparison

The L-functions attached by Hecke to modular forms in the upper half-plane are of great arithmetical importance, and so are the general automorphic L-functions. Of course, as usual for objects attached to reductive groups, the more familiar the groups the more important the object, so that the functions attached to the general linear group or the symplectic group will appear more frequently and in more critical circumstances than the others. That does not make them any easier to treat, and methods must be found to establish in general the basic analytic properties: analytic continuation to the whole complex plane and the functional equation.

There are methods available that, in addition, have led to substantial progress with outstanding arithmetical problems and have suggested new concepts and theorems in the study of harmonic analysis or theta series. They are only partially explored, and further development promises a deeper understanding of the theory of automorphic forms, and not a few surprises. However, they have limits of which we are becoming ever more keenly aware and that temper our pleasure at the success of those who have pursued them—but not our admiration. However, not every stone has been turned.

The methods fall into three classes: a more profound exploration of the expansion of Eisenstein series at the cusps, looking beyond the constant term; the multitude of zeta integrals described by Gelbart–Shahidi [14] that include, in particular, the Rankin–Selberg technique and that sometimes involve theta series; and the trace formula.

The general problem of analytic continuation of automorphic L-function leads (with a little imagination) quickly to the circle of questions referred to by the convenient catch phrase *principle of functoriality*, which implies the possibility of transporting automorphic representations from one group to another. The first two methods effect the analytic continuation directly, but the use of the trace formulas proceeds through the principle of functoriality.

In contrast to its initial purpose, which was apparently to analyze the spectrum of the Laplace–Beltrami operator on the quotient of the upper half-plane by a Fuchsian group, the arithmetical applications of the trace formula usually involve a comparison of two or more trace formulas or of a trace formula with a Lefschetz formula, and this leads to a multitude of questions in arithmetic and harmonic analysis, appealing in themselves, in whose solution the trace formula, in turn, can often be put to good effect.

One such problem, which belongs to the elements of the general theory of the trace formula and of Eisenstein series, but which viewed historically is a deep arithmetical statement, the result of successive efforts by several of the best mathematicians, is the conjecture of Weil that the Tamagawa number of a simply-connected semisimple group over $\mathbb{Q}$ is 1.

Recall that if G is a semisimple group over $\mathbb{Q}$, and $G(\mathbb{A})$ the group of adelic-valued points on G, so that $G(\mathbb{Q})$ is a discrete subgroup of $G(\mathbb{A})$, then the Tamagawa number is the volume of the quotient $G(\mathbb{Q})\backslash G(\mathbb{A})$ with respect to a canonically defined measure. An explicit volume for

$$\tau(G) = \text{vol}\,(G(\mathbb{Q})\backslash G(\mathbb{A}))$$

is of immediate appeal when $G(\mathbb{R})$ is compact (for example, if it is the orthogonal group of a definite form), for it yields directly a class-number formula. This is the case to which Siegel confined himself in his Oslo lecture of 50 years ago, when he discussed his extensions of classical formulas of Eisenstein and Minkowski [31]. He also dealt with groups for which $G(\mathbb{R})$ is not compact and was concerned, as should be stressed, although they are not pertinent here, with more general formulas than those for volumes of fundamental domains.

Tamagawa's realization much later that Siegel's formula for volumes had a strikingly simple adelic form led Weil to his general conjecture [32], which was verified in many cases, but by no means all. The problem was to find a general, uniform method. That we now have, and as part of the general trace formula, although for reasons that will be explained later factors of type E_8 that are not quasisplit will have to be excluded.

Connected reductive algebraic groups over $\mathbb{Q}$ are broken up into families defined cohomologically, a family consisting of all groups that can be obtained one from the other by an inner twist. For example, all special orthogonal groups attached to forms of the same dimension and with discriminants differing by a square lie in the same family, for if one of the groups, G, is defined by a symmetric matrix X and the other, G', by X' then there is a matrix U with coefficients from $\bar{\mathbb{Q}}$ and determinant from $\mathbb{Q}$ such that $X' = UX^tU$. Then

$$\psi: A \to UAU^{-1}$$

defines an isomorphism from G to G' over $\bar{\mathbb{Q}}$ and the isomorphisms $\psi_\sigma = \psi^{-1}\sigma(\psi)$, $\sigma \in Gal(\bar{\mathbb{Q}}/\mathbb{Q})$ are given by

$$A \to V_\sigma A V_\sigma^{-1}, \qquad V_\sigma = U^{-1}\sigma(U),$$

and

$$\det V_\sigma = (\det U)^{-1}\sigma(\det U) = 1,$$

so that $\sigma \to V_\sigma$ defines a Galois cocycle $\{\psi^{-1}\sigma(\psi)\}$ with values in the adjoint group of G. Thus ψ defines G' as an inner twisting of G.

Each inner family contains a distinguished element, determined up to isomorphism, and this is the quasisplit element of the group. It is the group for which $G(\mathbb{Q})\backslash G(\mathbb{A})$ is most noncompact and for a family containing orthogonal groups is characterized as the orthogonal group associated to a form with isotropic subspace of largest dimension. Abstractly it is characterized by the property that it contains a Borel subgroup defined over $\mathbb{Q}$. Since the Levi factors of a Borel subgroup are tori, the Eisenstein series associated with it have functional equations expressed in terms of L-functions attached to *Größencharakter* and are thus particularly easy to handle. On the other hand, a Borel subgroup over $\mathbb{Q}$ is necessarily a minimal parabolic over $\mathbb{Q}$, and the constant function is always a residue of Eisenstein series associated with the minimal parabolic. Combining these two facts with the elements of spectral theory for self-adjoint operators, one readily calculates $\tau(G) =$ vol $(G(Q))\backslash G(\mathbb{A})$ [20], [19]. It is found to be 1 for simply connected semisimple groups.

The second, more difficult step is to show that if G^* is quasisplit and G an inner form of it then $\tau(G) = \tau(G^*)$. It has been carried out by Kottwitz [17], who compares the trace formula for G, in which $\tau(G)$ appears, with that for G^*, in which $\tau(G^*)$ appears. The L-groups of G and G^* are the same, and one of the simpler manifestations of the principle of functoriality will be an assertion that the L-functions associated with G are all found among the L-functions associated with G^*. Thus for each automorphic π of G there will be an automorphic representation π^* of G^* such that

$$L(s, \pi, r) = L(S, \pi^*, r)$$

for all representation of ${}^LG$. One expects to prove this by a comparison of trace formulas for G and G^*, and methods are being developed for this purpose. What Kottwitz has in effect done is to anticipate their complete development by a judicious choice of the functions to be substituted in the formulas, so that a great many difficult terms are removed but not the volumes of the fundamental domains.

The proof of Weil's conjecture by Kottwitz, and of cyclic base-change by Arthur–Clozel [12], apart from any other applications, contemplated or already accomplished, are of sufficient importance to justify a description of the gross structural features of Arthur's general trace formula and of the scheme for comparison, although this entails some overlap with [11]. Moreover, a clear warning is necessary that the relative trace formulas of Jacquet have not yet been fitted into this scheme, that a great deal remains to be done, and that many of the techniques have been most highly developed for comparisons that involve disconnected groups, the so-called twisted trace formulas, and these have been omitted from this discussion.

In contrast to the trace formula of Selberg, which was, as its name implies, a formula for a trace, but which could only be proven for groups of rank one, the formula developed by Arthur begins with an equality between two functions that are then integrated separately over $G(\mathbb{Q})\backslash G^1(\mathbb{A})$ and the integrals calculated in completely different manner. The resultant identity is the first form of the trace formula [1], [2]. (For semisimple groups $G^1(\mathbb{A}) = G(\mathbb{A})$; for reductive groups it is the kernel of a family of homomorphisms from $G(\mathbb{A})$ to $\mathbb{R}^+$.)

Recall that the spectral analysis of the action of $G(\mathbb{A})$ by right translation on $\mathfrak{H} = L^2(G(\mathbb{Q})\backslash G(\mathbb{A}))$ is best accomplished with the operators

$$K(f)\colon \phi \to \int_{G(\mathbb{A})} \phi(gh)f(h)\,dh,$$

where f is a smooth function of compact support on $G(\mathbb{A})$, and $K(f)$ is an integral operator with kernel

$$K(g, h) = \sum_{G(\mathbb{Q})} f(g^{-1}\gamma h).$$

In addition, there is available a direct sum decomposition of $\mathfrak{H}$ defined by the theory of Eisenstein series. It is parametrized by pairs (M, ρ), M being a Levi factor over $\mathbb{Q}$ of a parabolic subgroup of G defined over $\mathbb{Q}$ and ρ a cuspidal representation of $M(\mathbb{A})$. The pairs are subject to a suitable equivalence relation. Denote the set of parameters by $\mathfrak{X}$. The summand $\mathfrak{H}_\chi$ labelled by $\chi \in \mathfrak{X}$ has itself a direct-integral structure, but for the moment only the coarse decomposition matters.

The basic equality, or identity, is between two truncations of the kernel of K. The first, a geometrically defined truncation of the restriction of K to the diagonal, is integrated over $G(\mathbb{Q})\backslash G^1(\mathbb{A})$ to yield the coarse $\mathfrak{o}$-expansion

$$J^T_{\mathrm{geom}}(f) = \sum_{\mathfrak{o}} J^T_{\mathfrak{o}}(f).$$

The sum runs over conjugacy classes of semisimple elements in $G(\mathbb{Q})$, and T is a multidimensional parameter determining the location of the truncation. The sum converges absolutely, and each term is a polynomial in T. The distributions $J_{\mathfrak{o}}^T$ have at least two disadvantages. They are not invariant under conjugation because the truncation demands a choice of a maximal compact subgroup of $G(\mathbb{A})$, and they are not expressible as integrals over a single conjugacy class, or even a single type of conjugacy class. For example, the term corresponding to $\mathfrak{o} = \{1\}$ involves all unipotent classes, and the term

$$\operatorname{vol}(G(\mathbb{Q})\backslash G^1(\mathbb{A}))f(1), \tag{2.1}$$

that must be isolated if the trace formula is to be applied to Weil's conjecture, does not appear explicitly.

The second truncation is defined by first composing K with a truncation operator ΛT on functions on $G(\mathbb{Q})\backslash G(\mathbb{A})$ that converts slowly increasing functions into rapidly decreasing functions and that is an idempotent, and then restricting the kernel of the composition $K \circ \Lambda^T$ to the diagonal. The operator $K \circ \Lambda^T$ is again an integral operator, and its kernel can be expressed as an integral of truncated Eisenstein series. Integrating over the diagonal we obtain a sum over $\mathfrak{X}$, referred to as the course χ-expansion,

$$J_{\text{spec}}^T(f) = \sum_{\chi} J_{\chi}^T(f).$$

Once again the distributions J_{χ}^T are not invariants. Moreover, although the J_{χ}^T themselves are polynomials in T, they are calculated as an integral over the parameters (partly continuous, partly discrete) of the direct integral decomposition of $\mathfrak{H}_{\chi}$ of inner products of truncated Eisenstein series, and these are not polynomials.

Thus the first form of the trace formula

$$\sum_{\mathfrak{o}} J_{\mathfrak{o}}^T(f) = \sum_{\chi} J_{\chi}^T(f)$$

is not too useful. What Arthur does next is to transform the left and right sides of the formula into forms that, although not final, at least appear more applicable. The appearance of the variable T on these formulas is of no significance since the differences

$$J_{\text{geom}}^{T'}(f) - J_{\text{geom}}^T(f)$$

and

$$J_{\mathrm{spec}}^{T'}(f) - J_{\mathrm{spec}}^{T}(f)$$

can be expressed in terms of the geometrical and spectral sides of the trace formula for the Levi factors of proper parabolic subgroups of G. For example, if G is anisotropic, the polynomials are in fact of degree 0 and the dependence on T is fictitious. Arthur himself prefers to work with a carefully chosen and fixed truncation parameter that he suppresses from the notation, and it is best to follow him so that the coarse trace formula becomes

$$\sum J_{\mathfrak{o}}(f) = \sum J_{\chi}(f).$$

Both transformations are difficult, but the geometric side is perhaps easier. Here the transformation introduces some disagreeable new features, but it appears that there is no choice but to accommodate ourselves to them. They are aesthetically displeasing but do not obstruct the arguments. We have to choose some compact neighborhood Δ of the identity in $G(\mathbb{A})$, to which are then attached finite sets S of places, and S appears explicitly in the new form of the geometric side, the fine $\mathfrak{o}$-expansion, which is then only valid for functions that depend only on the coordinates in S and have support in Δ, being outside of S the product of characteristic functions of the standard maximal compact subgroups.

The fine $\mathfrak{o}$-expansion [7] is a sum over the Levi subgroups containing a fixed Levi subgroup of some fixed parabolic subgroup over $\mathbb{Q}$.

$$J_{\mathrm{geom}}(f) = \sum_{M} \frac{|\Omega_0^M|}{|\Omega_0^G|} J_M(f),$$

and $J_M(f)$ is a sum over conjugacy classes in M of terms

$$a^M(s, \gamma) J_M(\gamma, f).$$

The two groups, Ω_0^M, Ω_0^G, whose order appears in the formula are Weyl groups, and $a^M(S, \gamma)$ is a constant that is not determined explicitly, and for many purposes need not be, except for certain γ. It implicitly involves a measure and, for some groups of low dimension and presumably in general, values of apparently unmanageable Dirichlet series. The notion of conjugacy employed is somewhat unusual. For the semisimple part of γ it is conjugacy in $M(\mathbb{Q})$; for the unipotent part it is conjugacy in $\prod_{v \in S} M(\mathbb{Q}_v) = M(\mathbb{Q}_S)$. The function f is a product of a function f_S on $\prod_{v \in S} G(\mathbb{Q}_v) = G(\mathbb{Q}_S)$ with a standard characteristic function outside of S; and

$$J_M(\gamma, f) = J_M(\gamma, f_S)$$

is a weighted orbital integral over

$$\{g^{-1}\gamma ng \mid g \in G(\mathbb{Q}_S),\ n \in N(\mathbb{Q}_S)\}$$

if $P = MN$ is a parabolic subgroup with Levi factor M. Thus it is a finite sum of weighted orbital integrals.

The distributions J_M are still not all invariant, but the larger M is, the more invariant they are. In particular, J_G is invariant, and a sum of invariant orbital integrals, because

$$J_G(\gamma, f_S) = \int_{G_\gamma(\mathbb{Q}_S)\backslash G(\mathbb{Q}_S)} f_S(g^{-1}\gamma g)\, dg,$$

where G_γ is the connected centralizer of γ in G. If $\gamma = 1$ this is $f_S(1) = f(1)$. Since

$$a^G(S, 1) = \tau(G),$$

the fine $\mathfrak{o}$-expansion contains a large invariant contribution in which the term (2.1) appears explicitly. I observe in passing that $a^G(S, \gamma)$ is 0 if the semisimple part of γ is not elliptic.

The distributions $J_M(\gamma)$, $M \neq G$, are not so simply expressed, and in advanced applications of the trace formula [12] they must be treated without the help of explicit formulas, to which in earlier, simpler applications [23] recourse could be had. It is best, nonetheless, to work them out in simple cases in order to develop a feel for them.

For $G = \mathrm{SL}(2)$ the only pertinent M is the group of diagonal matrices. If

$$g = \begin{bmatrix} \alpha & 0 \\ 0 & \beta \end{bmatrix} \begin{bmatrix} 1 & x \\ 0 & 1 \end{bmatrix} k,$$

with $k \in K_S$, $x = (x_v) \in \mathbb{Q}_S$, set

$$v_M(g) = -\sum_{v \in S} \lambda(x_v)$$

with

$$\lambda(x_v) = \begin{bmatrix} \frac{1}{2}\ln(1 + x_v^2) & v \text{ infinite} \\ \ln(\max\{1, |x|\}) & v \text{ finite.} \end{bmatrix}$$

If $\gamma \in M(F)$ has distinct eigenvalues a, b, then $J_M(\gamma, f_S)$ is equal to

$$\left[\prod_{v \in S} \left|\frac{a}{b}\right|^{1/2} \left|1 - \frac{b}{a}\right|_v\right] \int_{M(\mathbb{Q}_S)\backslash G(\mathbb{Q}_S)} f_S(g^{-1}\gamma g) v_M(g)\, dg.$$

A change of variables in the integral turns this into the product of

$$\prod_{v \in S} \left|\frac{a}{b}\right|^{1/2}$$

and

$$-\sum_{v \in S} \int_{K_S} \int_{Q_S} f_S(k^{-1}\gamma n(x)k)\lambda((1 - b/a)^{-1}x_v)\, dx\, dk, \tag{2.2}$$

where

$$n(x) = \begin{bmatrix} 1 & x \\ 0 & 1 \end{bmatrix}.$$

Since $a^M(S, \gamma) = 1$ for such γ, the singular behavior of this integral as γ approaches ± 1 is brought into the trace formula. If $h_S \colon \gamma \to J_m(\gamma, f_S)$ were smooth on $M(\mathbb{Q}_S)$, the contribution $J_M(f)$ to the fine o-expansion would, apart from whatever the terms associated to $\gamma = \pm 1$ yield, be the fine o-expansion for a function on M obtained by multiplying h_S with the product of the characteristic functions of the maximal compact subgroups outside of S. This would facilitate comparisons enormously, but h_S is not smooth. Fortunately, there is a way to circumvent the difficulties this causes, Arthur's principle of the *cancellation of singularities*, to which we shall come later.

The expression (2.2) is a sum of two other expressions,

$$\sum_{v \in S} \ln \left|1 - \frac{b}{a}\right|_v \int_{K_S} \int_{\mathbb{Q}_S} f_S(k^{-1}\gamma n(x)k)\, dx\, dk, \tag{2.3}$$

all of whose singular behavior at ± 1 arises from the factor in front of the integral, and a second expression, which although also singular at ± 1, does have limiting values at these points. They are

$$-\sum_{v \in S} \int_{K_S} \int_{\mathbb{Q}_S} f(k^{-1}\gamma n(x)k) \ln |x|_v\, dx\, dk, \qquad \gamma = \pm 1. \tag{2.4}$$

The integral appearing in (2.3) has the limit

$$\int_{K_S} \int_S f_S(k^{-1}\gamma n(x)h)\, dx\, dk, \qquad \gamma \pm 1, \tag{2.5}$$

which is the orbital integral of (2.4) without the weight, and is, in fact, the sum of the integrals over the unipotent conjugacy classes of $G(\mathbb{Q}_S)$. The terms, $J_M(\gamma, f_S)$, $\gamma \pm 1$, are linear combinations of (2.4) and (2.5). The coefficient of (2.5) can be chosen freely, for the coefficients $a^G(S, \gamma n(x))$, $x \in \mathbb{Q}$, can then be modified accordingly. Arthur introduces a procedure for arriving at a definite choice that seems to be as good as any other, but it is by no means sacred.

Some arbitrariness notwithstanding, the fine $\mathcal{o}$-expansion is a definite advance over the coarse, but is still neither invariant nor stable and thus not yet useful for comparisons. Before describing how Arthur achieves an invariant form, we must consider the fine χ-expansion. The conversion of the coarse χ-expansion to the fine requires a Paley–Wiener theorem for reductive groups [4] and a remarkably astute and skillful use of ordinary Fourier analysis [5], [6].

In the fine χ-expansion the inner products of truncated Eisenstein series disappear and the factors describing the functional equations appear. These are matrix-valued transcendental functions, but they factor as a product of a scalar-valued transcendental function, expressible in terms of automorphic L-functions as in Section I, and a matrix-valued rational function (in arguments s and p^s).

Taking residues of the Eisenstein series associated to the parameter χ can yield square-integrable eigenfunctions on $G(\mathbb{Q})\backslash G^1(\mathbb{A})$ or on intermediate Levi subgroups. This leads to a fine direct-integral decomposition of the space $\mathfrak{H}_\chi$.

$$\mathfrak{h}_\chi = \int_{\Pi(G,\chi)} d\pi,$$

where $\Pi(G, \chi)$ is a well determined set of unitary representations. There are analogous sets $\Pi(M, \chi)$ for all Levi subgroups, that are empty if M does not contain the Levi factor in any pair defining χ.

The fine χ-expression is a double sum [6]

$$J_{\text{spec}}(f) = \sum_\chi \left\{ \sum_M \frac{|\Omega_0^M|}{|\Omega_0^G|} \int_{\Pi(M,\chi)} a^M(\pi) J_M(\pi, f)\, d\pi \right\}$$

that is not yet known to be absolutely convergent in general. The functions $a^M(\pi)$ contain a local contribution that is defined by intertwining operators and is quite subtle and a global contribution that is expressible in terms of logarithmic derivatives of automorphic L-functions.

Such derivatives result from formal operations with collections of functions called (G, M)–families by Arthur [3], operations that pervade his work, appearing already in the proof of the fine $\mathfrak{o}$-expansion. They can be described briefly.

If M is a Levi factor of a parabolic, let $\mathscr{P}(M)$ be the finite set of parabolic subgroups containing M. (The case to think of is the group M of diagonal matrices in $G = \mathrm{GL}(n)$, then $\mathscr{P}(M)$ is parametrized by the permutations of $\{1, \ldots, n\}$.) Suppose that for each $P \in \mathscr{P}(M)$ we are given a smooth function $c_P(\lambda)$ on $\mathfrak{a}_M^*$ (the coordinate space $\mathbb{R}^n$ in the example). The collection $\{c_P\}$ is said to form a (G, M)–family if whenever P and P' are adjacent (the corresponding permutations σ, σ' differ by right multiplication by the interchange of adjacent integers $i, i+1$), then c_P and $c_{P'}$ agree on the hyperplane passing through the wall common to the chambers associated to P and P' (the hyperplane $\lambda_{\sigma(i)} = \lambda_{\sigma(i+1)}$). To each $P \in \mathscr{P}(M)$ there is an associated set of simple roots Δ_P and their coroots. A simple but important lemma states that the formula

$$c_M(\lambda) = \sum_{P \in \mathscr{P}(M)} c_P(\lambda)\theta_P(\lambda)^{-1}$$

with

$$\theta_P(\lambda) = \prod_{\alpha \in \Delta_P} \lambda(\check{\alpha})$$

defines a smooth function on $\mathfrak{a}_M^*$.

If, for example, we take $G = \mathrm{GL}(2)$ and M as above, then $\mathscr{P}(M)$ consists of two elements P and P', and if we take X_P, $X_{P'}$ in $\mathfrak{a}_M$ with

$$X_P - X_{P'} = (x, -x),$$

then

$$c_P(\lambda) = e^{\lambda(X_P)}, \qquad c'_P(\lambda) = e^{\lambda(X_{P'})}$$

is a (G, M)–family and

$$c_M(0) = x$$

is essentially the length of the integral joining X_P and $X_{P'}$. The analogous construction in higher dimensions leads to the volumes of convex sets that are ubiquitous in Arthur's papers.

A second construction of a (G, M)–family starts from a set, $\{r_\alpha | \alpha \text{ a root}\}$, of nonzero complex-valued functions on $\mathbb{R}$. If Σ_P is the set of roots in P, and $\bar{P}$ the parabolic in $\mathscr{P}(M)$ opposite to P so that $P \cap \bar{P} = M$, $\Sigma_P \cap \Sigma_{\bar{P}} = \varnothing$, set

$$r_{P|P'}(\lambda) = \prod_{\alpha \in \Sigma_{P'} \cap \Sigma_{P'}} r_\alpha(\lambda(\check{\alpha})). \tag{2.6}$$

Then for each $P' \in \mathscr{P}(M)$, $\nu \in \mathfrak{a}_M^*$

$$r_P(\lambda; \nu, P') = r_{P|P'}(\nu)^{-1} r_{P|P'}(\nu + \lambda)$$

defines a (G, M)–family, and $r_M(0; \nu, P')$ can be expressed in terms of logarithmic derivatives.

The product $d_P = c_P d_P$ of two (G, M)–families is again a (G, M)–family, and there are formulas for calculating d_M that can be regarded as variants either of Leibniz's rule or of partitions of convex sets.

It is principally these formulas that lead to the product $a^M(\pi) J_M(\pi, f)$ appearing in the fine χ-expansion. The factors that appear in the functional equation of the Eisenstein series are intertwining operators $J_{P'|P}(\pi_\lambda)$ for which formulas like (2.6) are available. They now involve operator-valued functions, but the formalism of (G, M)–families is still applicable and a factorization

$$J_{P'|P}(\pi_\lambda) = r_{P'|P}(\pi_\lambda) R_{P'|P}(\pi_\lambda)$$

in which $r_{P'|P}(\pi_\lambda)$ is a complex-valued function given by automorphic L-functions and $R_{P'|P}(\pi_\lambda)$ is an intertwining operator given by elementary functions, yields the factorization $a^M(\pi) J_M(\pi, f)$, in which all lack of invariance is in $J_M(\pi, f)$. (The notation is unfortunate. $a^M(\pi)$ is defined by $r_{P'|P}(\pi_\lambda)$ and $J_M(\pi, f)$ with the help of $R_{P'|P}(\pi_\lambda)$.)

Arthur's final step in his development of the trace formula is to render it invariant by a brutal transposition of terms from the spectral side to the geometric side, a process that has little to recommend it but its successes, but these are overwhelming, for with his invariant formula, cancellation of singularities becomes a very supple tool and it becomes possible to substitute functions for which many disagreeable terms vanish while the essential ones remain.

The source of all lack of invariance in the trace formula lies in the initial truncation; so the lack of invariance, the difference between the values of the distribution on f and on f^x, where $f^x(g) = f(x^{-1}gx)$, or between its value on the two convolutions

$$h * f : g \to \int h(gg_1^{-1}) f(g_1)\, dg_1$$

and $f * h$, can be expressed in a universal way for all the distributions $J_M(\gamma)$ or $J_M(\pi)$ occurring in the trace formula.

Denote any one of these distributions by J_M. Then the simplest form of the universal formula is

$$J_M(f^x) = \sum_{Q \supseteq M} J_M^{M_Q}(f_{Q,x}). \tag{2.7}$$

The sum is over all parabolic subgroups over $\mathbb{Q}$ containing M; $J_M^{M_Q}$ is the analogue of J_M appearing in the trace formula for the Levi factor M_Q of Q; and $f_{Q,x}$ is a function on $M_Q(\mathbb{A})$ given in terms of f by a simple integral formula of the form ([3])

$$f_{Q,x}(m) = \delta_Q(m)^{1/2} \int_K \int_{N_Q(\mathbb{A})} f(k^{-1}mnk)u'_Q(k, x)\, dk\, dn.$$

In particular,

$$f_{G,x} = f.$$

Strictly speaking, the trace formula may not apply to f^x, because it may not be an element of the Hecke algebra, so that (2.7) has to be rewritten in terms of convolutions to obtain a correct formula.

When working with the trace formula, one fixes a finite set of places containing the infinite place, and all freedom is in f_S so that the achievement of invariance is really a local problem on

$$\prod_{v \in S} G(\mathbb{Q}_v) = G(Q_S).$$

Keeping this in mind, we now simplify the notation, often replacing f_S by f.

To attain invariance, Arthur introduces a map $f \to \phi_M(f)$ from functions on $G(Q_S)$ to functions on $M(Q_S)$, or rather from functions on $G(Q_S)$ to a class of functions on $M(Q_S)$, for only the orbital integrals of $\phi_M(f)$ on regular semisimple elements are specified. (In view of theorems of Harish–Chandra and Kazhdan, this is equivalent to specifying the values on $\phi_M(f)$ of distributions supported on characters in the sense of [8], Section 1.) The function $\phi_M(f)$ is defined by the equation

$$\operatorname{tr} \pi(\phi_M(f)) = J_M(\pi, f),$$

π running over the irreducible tempered representations of $M(Q_S)$, thus exactly those necessary for the local harmonic analysis. Then all distributions J_M, both those from the spectral and those from the geometrical side of the trace formula, are converted into invariant distributions I_M by the inductive formula

$$J_M(f) = \sum_{L \supseteq M} I_M^L(\phi_L(f)),$$

the sum running over all Levi factors over $\mathbb{Q}$ containing M.

Observe that to carry out the inductive definition implicit in this formula it must be shown that the distributions I_M^L are supported on characters. Notice also that the construction simplifies distributions from the spectral side. For example, if π is tempered then

$$I_M(\pi, f) = \begin{bmatrix} 0 & M \neq G, \\ \operatorname{tr} \pi(f) & M = G. \end{bmatrix}$$

However, it will tend to complicate distributions from the geometric side. Of course, for $M = G$ the procedure effects no change, and

$$I_G(\gamma) = J_G(\gamma), \qquad I_G(\pi) = J_G(\pi).$$

Since the distributions $I_M(\gamma)$, $I_M(\pi)$ are obtained from $J_M(\gamma)$, $J_M(\pi)$ by a uniform procedure, they can be substituted for $J_M(\gamma)$, $J_M(\pi)$ in the trace formula to obtain an *invariant trace formula*. Moreover, the distributions appearing in the invariant trace formula being supported on characters, it makes sense to ask whether they simplify, or even vanish, on functions whose orbital integrals are subject to suitable conditions. This is indeed so and yields simpler terms of the trace formula that are very effective for specific purposes [13, 18].

It has been observed that arithmetical applications of the trace formula usually involve comparisons. In essence one shows that some linear combination of the geometric sides is zero, infers that the same linear combination of the spectral sides is zero, and then from this deduces relations between automorphic representations of the groups involved.

Since the very purpose and nature of the trace formula is to allow a term-by-term analysis of the geometric side, the comparison can proceed only if there is some correspondence between conjugacy classes in the groups involved. This is simplest for base change for $G(n)$ under cyclic extensions of the ground field, for then the norm map is a well-defined function from the conjugacy classes of one group to those of another. For this application the concomitant problems in local harmonic analysis (transfer of orbital integrals and fundamental lemmas) are consequently easier, and the methods for global comparison, especially cancellation of singularities, much more advanced [12].

For other problems the correspondence is not given by a function, and the notions of stabilization and endoscopy necessary to circumvent the attendant difficulties are perhaps the most startling that the trace formula has suggested to harmonic analysis, although not yet nearly so well understood as some others, such as Harish–Chandra's Selberg principle, Arthur's Paley–Wiener theorem, or formulas for characters as weighted orbital integrals.

Suppose G^* is quasisplit and $\psi: G \to G^*$ is an inner twist so that $\psi^{-1}\sigma(\psi)$ is an inner automorphism of G for all $\sigma \in Gal(\bar{\mathbb{Q}}|\mathbb{Q})$. Thus if γ is an element of $G(\mathbb{Q})$, $\psi(\gamma)$ is an element of $G^*(\bar{\mathbb{Q}})$. It is a theorem of Steinberg and Kottwitz that the conjugacy class of $\psi(\gamma)$ in $G^*(\bar{\mathbb{Q}})$ always meets $G^*(\mathbb{Q})$, and this allows us to define a correspondance between conjugacy classes in $G(\mathbb{Q})$ and $G^*(\mathbb{Q})$; but to obtain a function one has to introduce the notion of *stable* conjugacy classes in $G(\mathbb{Q})$, which is for simply-connected, semisimple groups just conjugacy in $G(\bar{\mathbb{Q}})$, but which is slightly more delicate for other groups [16]. The notion is also defined over local fields. Thus for a term-by-term comparison of the trace formulas of G and G^*, one needs a trace formula in which the geometric side is expressed as a sum over stable conjugacy classes and in which the distributions are stably invariant. Primitive forms of such a stable trace formula are available in general. In a very few special cases it is completely developed [25], [29].

It is the primitive form that leads to the proof of Weil's conjecture, but in order to illustrate cancellation of singularities we proceed with more general considerations. Stabilization and stably-invariant harmonic analysis lead to the introduction of endoscopic groups of G. These are quasisplit reductive groups H such that the L-group ${}^LH$ is imbedded in ${}^LG$. Among them is the group G^* with ${}^LG^* = {}^LG$. There is a map from regular semisimple conjugacy classes of $G(\mathbb{Q})$ to stable semisimple conjugacy classes of $H(\mathbb{Q})$ and a notion of transfer $f \to f^H$ from functions on $G(A)$ to functions on $H(A)$ [26]. The function f^H is not well defined; only its stably invariant orbital integrals are. In addition to each elliptic endoscopic group there is attached [17] a cohomological invariant $\iota(G, H)$.

If

$$I_M^{\mathrm{geom}}(f) = \sum_{\gamma} I_M(\gamma, f)$$

and if $I_M^{\text{spec}}(f)$ is defined in a similar fashion, then the invariant trace formula is an equality

$$\sum_M I_M^{\text{geom}}(f) = \sum_M I_M^{\text{spec}}(f).$$

Recall that the sum is over the Levi factors containing a fixed minimal one. The stably-invariant trace formula will be a similar identity, but it is best to take the sum over the Levi factors M^* of G^* containing a fixed minimal one, noting that to every M there is associated a unique M^*. Thus the identity will take the form

$$\sum_{M^*} SI_{M^*}^{\text{geom}}(f) = \sum_{M^*} SI_{M^*}^{\text{spec}}(f), \tag{2.8}$$

in which all distributions appearing are stably invariant so that their value on a function f is determined by the stable orbital integrals of f on semisimple elements.

One can associate to each Levi factor M_H of H a Levi factor M^* of G^*. We write $M_H \to M^*$. When f^H can be defined for all H, we define $SI_{M^*}^{\text{geom}}$

$$SI_{M^*}^{\text{geom}} = I_{M^*}^{\text{geom}}(f) - \sum_M{}' \iota(G, H) \sum_{M_H \to M^*} SI_{M_H}(f^H) \tag{2.9}$$

and, in a similar fashion, $SI_{M^*}^{\text{spec}}(f)$. The prime indicates that the sum is over all elliptic endoscopic groups, with the exception of G^*. Such a definition at least guarantees the validity of (2.8).

It is the distribution $SI_{G^*}^{\text{spec}}$ that carries the interesting information about automorphic forms, and one would like to show that

$$SI_{G^*}^{\text{spec}}(f) = SI_{G^*}^{\text{spec}}(f^*), \qquad f^* = f^{G^*}, \tag{2.10}$$

from which it follows, in particular, that $SI_{G^*}^{\text{spec}}$ is stable, for f^* is specified by the stable orbital integrals of f. Observe that SI_{G^*} denotes two distributions, one on G and one on G^*, distinguished only by their arguments.

As a first application of these ideas, one can, following Kottwitz, prove for G simply-connected and semisimple the existence of pairs f, f^* such that $f^* = f^{G^*}$ and such that both f^H and f^{*H} can be taken to be zero for $H \neq G^*$. In addition

$$I_{M^*}^{\text{geom}}(f) = I_{M^*}^{\text{spec}}(f) = 0, \qquad M^* \neq G^*,$$

$$I_{M^*}^{\text{geom}}(f^*) = I_{M^*}^{\text{spec}}(f^*) = 0, \qquad M^* \neq G^*.$$

For this pair, we thus obtain the identity

$$SI_{G^*}^{\mathrm{geom}}(f) - SI_{G^*}^{\mathrm{geom}}(f^*) = SI_{G^*}^{\mathrm{spec}}(f) - SI_{G^*}^{\mathrm{spec}}(f^*). \tag{2.11}$$

The right side is an infinite linear combination of irreducible traces. That f^H and f^{*H} can be taken to be zero for $H \neq G^*$ implies strong relations among the orbital integrals, and if Hasse's principle is valid for G as well as Weil's conjecture for groups of lower dimension, then the left side of (2.11) reduces to

$$\tau(G)f(1) - \tau(G^*)f^*(1) = (\tau(G) - \tau(G^*)).$$

This can be equal to a sum of traces only if it is zero, so Weil's conjecture follows inductively.

Hasse's principle intervenes because it is more generally necessary to the calculations that allow one to show that when f^H exists for all H then the contribution of the regular semisimple conjugacy classes to the right side of (2.9) is stable. One might hope that if the Hasse principle is so strongly enmeshed in these calculations it could be possible to use the trace formula to prove it. So far as I know, this has not been attempted.

Further general development of the stable trace formula awaits the proof of the existence, both locally and globally, of the transferred functions f^H. One can hope that once this is done the methods developed by Arthur will overcome the other obstacles.

The results of Arthur–Clozel suggest that not only will (2.10) be valid but, more generally,

$$SI_{M^*}^{\mathrm{geom}}(f) = SI_{M^*}^{\mathrm{geom}}(f^*) \tag{2.12}$$

$$SI_{M^*}^{\mathrm{spec}}(f) = SI_{M^*}^{\mathrm{spec}}(f^*) \tag{2.13}$$

and that, moreover, it will be possible to express the two sides of (2.12) and (2.13) as sums of more primitive stably invariant distributions. On the geometric side will appear, in particular, stably invariant orbital integrals and on the spectral side, among others, characters of tempered L-packets. The identities (2.12) and (2.13), of which (2.10) is the most interesting, realize a final form of the trace formula, the stably invariant trace formula, that can presumably be applied without further transformation.

A proof along the lines of [12] would of course involve an elaborate induction, upwards on the dimension of G and downwards on that of M.

The definition of f^* and the manipulation of [17] and [24] then show that the difference

$$SI_{G^*}^{\mathrm{geom}}(f) - SI_{G^*}^{\mathrm{geom}}(f^*) \tag{2.14}$$

is the sum of very few terms, those corresponding to unipotent elements. Thus if it could be shown that it was a discrete sum of characters, then a little harmonic analysis on one of the noncompact factors $G(\mathbb{Q}_v)$ should show that it is zero.

According to (2.8) the difference (2.14) can be expressed in terms of the differences

$$SI_{M^*}^{\mathrm{spec}}(f) - SI_{M^*}^{\mathrm{spec}}(f^*) \tag{2.15}$$

and

$$SI_{M^*}^{\mathrm{geom}}(f) - SI_{M^*}^{\mathrm{geom}}(f^*), \qquad M^* \neq G^*. \tag{2.16}$$

Now $I_G(\gamma, f)$ is equal to $J_G(\gamma, f)$ by its very definition. However, for $M \neq G$ there is an inexplicit element introduced in the passage from $J_M(\gamma, f)$ to $I_M(\gamma, f)$, namely, $\phi_L(f)$, so that the only handle on the distributions $I_M(\gamma)$ is their formal properties of splitting and descent.

In the case of base change, which one will take as a model for the development of the stable trace formula, one uses these properties and a progressively less restrictive choice of functions f and f^* to show that all differences (2.14), (2.15), and (2.16) are zero.

One begins with a class of functions for which (2.14) and all but one of (2.16), that corresponding to the M^* at which we have arrived by the downward induction, are zero. More precisely, it is analogues of these differences that appear, but rather than continuously qualifying my remarks with references to base change, I prefer to speak of the deduction of the stably invariant trace formula as being already achieved, with no intention of slighting the difficulties that are still to be overcome. Now $SI_{M^*}^{\mathrm{geom}}(f)$ or $SI_{M^*}^{\mathrm{geom}}(f^*)$ when expanded as a sum over stable conjugacy class of M^* presumably resemble very closely the expansion of the stable trace formula for some function on $M^*(\mathbb{A})$, but because of singularities like those we have seen in (2.2) cannot be equal to such an expansion. However, one can expect that the singularities cancel when the difference is taken, so that there is a function ε on $M^*(\mathbb{A})$ to which the trace formula can be applied such that

$$SI_{M^*}^{\mathrm{geom}}(f) - SI_{M^*}^{\mathrm{geom}}(f^*) = SI_{M^*}^{\mathrm{geom}}(\varepsilon).$$

Finally, one uses (2.8) for M^* rather than G and for ε rather than f to replace $SI_{M^*}^{\text{geom}}(\varepsilon)$ by a spectral-theoretic expansion, probably just $SI_{M^*}^{\text{spec}}(\varepsilon)$, the other terms very likely vanishing. This gives

$$\sum_{M^*} SI_{M^*}^{\text{spec}}(f) - \sum_{M^*} SI_{M^*}^{\text{spec}}(f^*) = SI_{M^*}^{\text{spec}}(\varepsilon).$$

This spectral equality involves measures of Lebesgue type of various dimensions, and they must vanish separately. In particular, the term corresponding to the measure of lowest dimension, the split rank of the center of G, is just

$$SI_{G^*}^{\text{geom}}(f) - SI_{G^*}^{\text{spec}}(f^*);$$

so this difference must be zero.

Once this difference is shown to be zero for the restricted class of functions, it can be shown that it is zero for an even larger class, for which one then deduces the vanishing of (2.14) and (2.16). To deal with completely general f and f^* it is necessary to work one's way down through the M^* to the minimal Levi factor, verifying at each stage that the vanishing of (2.16) for that M^* and the restricted class of functions implies its vanishing in general. The vanishing of (2.14), in general, is only obtained at the last stage, at which one also has the vanishing of (2.15) for $M^* = G^*$ and general f and f^*. For $M^* \neq G^*$, the vanishing of (2.15) is proved with the help of local results linking the distributions $I_{M^*}(\gamma)$ and $I_{M^*}(\pi)$ and a supplementary induction.

All of this looks to be elaborate and extremely difficult, as indeed it is; so it is very helpful to understand it for simple examples, where the convoluted inductive arguments and much of the technique are unnecessary and the constructions more explicit. The group $U(3)$ of [29] is perhaps the best choice, but even $G = \mathrm{SL}(2)$ yields considerable insight.

Here $G = G^*$, and the only Levi factor besides G itself that need be considered is the group M of diagonal matrices. It is Abelian so that $\Phi = \Phi_M(f)$ can be defined more directly than usual as the Fourier transform of

$$\pi \to J_M(\pi, f),$$

π a character of $M(\mathbb{Q}_S)$. However, this is not too direct, and a real understanding of Φ_M requires an examination of the normalized intertwining operators $R_{P'|P}(\pi_\lambda)$, but that would entail an elementary but extended digression.

Since

$$I_M(\gamma, f) = J_M(\gamma, f) - \Phi(\gamma),$$

the singularities of $I_M(f)$ are those of $J_M(\gamma, f)$. Since $G = G^*$ we may write M rather than M^* in the definitions of the stable trace formula. Moreover, the elliptic endoscopic groups other than G itself are all anisotropic tori, so that (2.9) reduces to

$$SI_M^{\text{geom}}(f) = I_M^{\text{geom}}(f)$$

and

$$SI_M^{\text{geom}}(f) = \sum_{\gamma \in M(\mathbb{Q})} I_M(\gamma, f).$$

Since

$$\gamma \to I_M(\gamma, f)$$

is not smooth, the right side is not the trace of a smooth function on $M(\mathbb{A})$, but the difference

$$\sum_{\gamma \in M(\mathbb{Q})} I_M(\gamma, f) - \sum_{\gamma \in M(\mathbb{Q})} I_M(\gamma, f^*)$$

is, because

$$\gamma \to I_M(\gamma, f) - I_M(\gamma, f^*)$$

is smooth. Of course this is trivial if we take $f^* = f$, but the point is that we need not do so.

The first sign that it is smooth is that the difference of (2.3) for f and (2.3) for f^* is 0 because the integrals appearing there are stable, and thus equal, so that the difference is zero, and the singular factor in front innocuous. The difference between (2.2) and (2.3) involves the factor $\prod_{v \in S} |a/b|_v^{1/2}$, which we may ignore, and a sum over v, each term of the sum being smooth away from ± 1 but not necessarily at ± 1. For example, for v nonarchimedian the corresponding term is the sum of

$$-\int_{K_S} \int_{\mathbb{Q}_S} f_S(k^{-1}\gamma n(x)k) \ln |x|_v \, dx \, dk$$

and

$$-\int_{K_S} \int_{|x|_v \le |1 - b/a|_v} f_S(k^{-1}\gamma n(x)k) \ln |x|_v \, dx \, dk.$$

The first expression is smooth, and the singularity of the second is that of the product of

$$-\int_{|x_v| \leq |1-b/a|_v} \left(\ln \left| \frac{1-b}{a} \right|_v - \ln |x_v| \right) dx_v$$

and

$$\int_{K_S} \int_{\mathbb{Q}_{S'}} f(k^{-1}\gamma n(x)k)\, dx\, dk \qquad S' = S - \{v\}. \tag{2.17}$$

Since (2.17) is a stable orbital integral, this singularity is cancelled by the corresponding one for f^*. The contributions to the singularity from the Archimedean place also cancel, but the proof is more elaborate.

References

[1] Arthur, J. "A trace formula for reductive groups I: Terms associated to classes in $G(\mathbb{Q})$," *Duke Math. Jour.* **45** (1978).

[2] Arthur, J. "A trace formula for reductive groups II: Applications of a truncation operator," *Comp. Math.* **40** (1980).

[3] Arthur, J. "The trace formula in invariant form," *Ann. of Math.* **114** (1981).

[4] Arthur, J. "A Paley Wiener theorem for real reductive groups," *Acta Math.* **150** (1983).

[5] Arthur, J. "On a family of distributions obtained from Eisenstein series I: Applications of the Paley-Wiener theorem," *Amer. Jour. Math.* **104** (1982).

[6] Arthur, J. "On a family of distributions obtained from Eisenstein series II: Explicit formulas," *Amer. Jour. Math.* **104** (1982).

[7] Arthur, J. "On a family of distributions obtained from orbits," *Can. Jour. Math.* **37** (1986).

[8] Arthur, J. "The invariant trace formula I: Local theory," *J. Amer. Math. Soc.*, to appear.

[9] Arthur, J. "The invariant trace formula II: Global theory," *J. Amer. Math. Soc.*, to appear.

[10] Arthur, J. "Intertwining operators and residues II: Invariant distributions," *Comp. Math.*, to appear.

[11] Arthur, J. "The trace formula and Hecke operators," this volume.

[12] Arthur, J. and Clozel, L. *Simple algebras, base change, and the advanced theory of the trace formula*, Ann. of Math. Studies, to appear.

[13] Bernstein, J-N. et al. "Représentations des groupes réductifs sur un corps local," *Hermann*, Paris (1984).

[14] Gelbart, S. and Shahidi, F. *Analytic properties of automorphic L-functions*, Academic Press (1988).

[15] Gelfand, I.M. "Automorphic functions and the theory of representations." *Proc. of the ICM, Stockholm* (1962).

[16] Kottwitz, R. "Rational conjugacy classes in reductive groups," *Duke Math. Jour.* **49** (1982).

[17] Kottwitz, R. "Stable trace formula: elliptic singular terms," *Math. Ann.* **275** (1986).

[18] Kottwitz, R. "Tamagawa numbers," *Ann. of Math.*, 127 (1988).

[19] Lai, K.F. "Tamagawa number of reductive algebraic groups," *Comp. Math.* **41** (1980).

[20] Langlands, R.P. "The volume of the fundamental domain for some arithmetical subgroups of Chevalley groups." In: *Algebraic Groups and Discontinuous Subgroups, Proc. Symp. Pure Math.* **9** (1966).

[21] Langlands, R.P. *Euler Products*. Yale Mathematical Monographs (1971).

[22] Langlands, R.P. *On the functional equations satisfied by Eisenstein series*, Springer Lecture Notes **544** (1976).

[23] Langlands, R.P. Base change for GL(2), Ann. of Math. Studies **96** (1980).

[24] Langlands, R.P. *Les débuts d'une formule des traces stable*, Publ. Math. de l'Univ. de Paris VII. **13** (1983).

[25] Langlands, R.P. and Labesse, J.-P. "*L*-indistinguishability for SL(2)," *Can. Jour. Math.* **31** (1979).

[26] Langlands, R.P. and Shelstad, D. "On the definition of transfer factors," *Math. Ann.*, 278 (1987).

[27] Maaß, H. "Über eine neue Art von nichtanalytischen automorphen Formen," *Math. Ann.* **121** (1949).

[28] Roelcke, W. "Analytische Fortsetzung der Eisensteinreihen zu den parabolischen Spitzen von Grenzkreisgruppen erster Art," *Math. Ann.* **132** (1956).

[29] Rogawski, J. *Automorphic representations of unitary groups in three variables*, Ann. of Math. Studies, to appear.

[30] Selberg, A. "Harmonic analysis and discontinuous groups in weakly symmetric Riemannian spaces with applications to Dirichlet series," *J. Indian Math. Soc.* (N.S.) **20** (1956).

[31] Siegel, C.L. "Analytische Theorie der quadratischen Formen," *Com Rend du Cong inter. des Math., Oslo* (1937).

[32] Weil, A. *Adeles and Algebraic Groups*. Birkhäuser (1982).

7 Selberg's Work on the Zeta-Function

HUGH L. MONTGOMERY

The bulk of Selberg's publications concerning the Riemann zeta-function are concentrated in the years 1942–1946. These papers appeared in Norwegian journals at a time when communications were disrupted by World War II. Thus it was that Siegel, when he first returned to Europe after the war, asked Bohr what had happened in the meantime, and Bohr responded, "Selberg." Since these papers received only limited circulation, some of their details are not well known, although, of course, the main results are very familiar. In our description below, some results are presented only in a simplified (and hence weakened) form, in order that the most essential elements can be best appreciated.

1. Zeros on or Near the Critical Line

In 1914, G.H. Hardy [5] announced that infinitely many of the zeros of the Riemann zeta-function lie on the critical line $\sigma = \frac{1}{2}$, and finally in 1921 he and J.E. Littlewood [6] proved that $N_0(T) \gg T$, where $N_0(T)$ denotes the number of zeros of $\zeta(\frac{1}{2} + it)$ in the interval $0 < t \leq T$.

NUMBER THEORY,
TRACE FORMULAS and
DISCRETE GROUPS

ISBN 0-12-067570-6

Letting $N(T)$ denote the total number of zeros of $\zeta(s)$ in the rectangle $0 \le \sigma \le 1$, $0 < t \le T$, we know that

$$N(T) \sim \frac{1}{2\pi} T \log T, \tag{1}$$

so there remained a considerable gap between the lower bound of $N_0(T)$ and the size of $N(T)$. We recall the strategy adopted by Hardy and Littlewood: If we put

$$\eta(s) = \left(\pi^{1/2 - s} \frac{\Gamma\left(\frac{s}{2}\right)}{\Gamma\left(\frac{(1-s)}{2}\right)} \right)^{1/2},$$

then $|\eta(\frac{1}{2} + it)| = 1$, and by the functional equation

$$Z(t) = \eta\left(\frac{1}{2} + it\right)\zeta\left(\frac{1}{2} + it\right)$$

is real valued. One observes that if $J > |I|$ where

$$I(t) = \int_t^{t+h} Z(u)\, du, \qquad J(t) = \int_t^{t+h} |Z(u)|\, du,$$

then $Z(u)$ has a change of sign in the interval $(t, t+h)$. By using the estimates

$$\int_T^{2T} |I(t)|^2\, dt \ll hT \tag{2}$$

and

$$\int_T^{2T} \left| \int_t^{t+h} \zeta(\tfrac{1}{2} + iu) - 1\, du \right|^2 dt \ll T \tag{3}$$

one can show that $J > |I|$ for most t, with the result that $N_0(T) \gg T/h$. The argument succeeds if h is larger than a certain constant, giving $N_0(T) \gg T$.

To this argument Selberg [12, 13, 14] added the idea of replacing $\zeta(s)$ by $\zeta(s)M(s)$ where $M(s)$ is a suitable "mollifier" that one hopes may dampen the large wobbles in $\zeta(\frac{1}{2} + it)$. The first use of a mollifier of this kind had been made by H. Bohr and E. Landau [1] in 1914 in order to show that $N(\sigma, T) = 0(T)$ for any fixed $\sigma > \frac{1}{2}$, where $N(\sigma, T)$ is the number of zeros $\rho = \beta + i\gamma$ of $\zeta(s)$ in the rectangle $\sigma \le \beta \le 1$, $0 < \gamma \le T$.

They took $M(s)$ to be a partial product of the Euler product for $1/\zeta(s)$.

$$M(s) = \prod_{p \leq X} (1 - p^{-s}).$$

Later F. Carlson [4] discovered that quantitatively sharper results can be derived by using instead a partial sum for $1/\zeta(s)$, namely,

$$M(s) = \sum_{n \leq X} \mu(n) n^{-s}.$$

After a succession of refinements Ingham [7] showed that

$$N(\sigma, T) \ll T^{3(1-\sigma)/(2-\sigma)} (\log T)^5. \tag{4}$$

Here we see that the effect of the mollifier is less when σ is near $\frac{1}{2}$ and that if σ is sufficiently near $\frac{1}{2}$, say $\sigma < \frac{1}{2} + (\log \log T)/(\log T)$, then the above estimate is weaker than the trivial estimate $N(\sigma, T) \leq N(T)$. Thus it is audacious to think that an improvement can be obtained by using a mollifier on the critical line. Moreover, the complications are compounded by the fact that $M(\frac{1}{2} + it)$ must be non-negative if one is to be able to conclude that $Z(t)$ changes sign. To this end, Selberg took $M(s)$ to be of the form $M(s) = \phi(s)\phi(1 - s)$ where ϕ is real on the real axis.

In his first realization of this program, Selberg [13] took

$$\phi(s) = \prod_{p \leq X} \left(1 - \frac{1}{2} p^{-s} - \frac{1}{8} p^{-2s}\right),$$

which is to say that $\phi(s)$ is comparable to a partial product for $\zeta(s)^{-1/2}$. Selberg showed that one could take $X = \sqrt{\log \log T}$, which gave

$$N_0(T) \gg T \log \log \log T.$$

He also mentioned that if he had used the Hardy-Littlewood approximate functional equation, one could then take $X = \sqrt{\log T}$, which would give

$$N_0(T) \gg T \log \log T.$$

It is evident that in order to obtain still better results one should try taking $\phi(s)$ to be a Dirichlet polynomial, say

$$\phi(s) = \sum_{\nu \leq X} \beta_\nu \nu^{-s} \tag{5}$$

with $\beta_1 = 1$. Then the remaining coefficients should be chosen so as to make the resulting integrals as small as possible. This, however, is a very formidable undertaking. In place of the integrals (2) and (3) considered by Hardy and Littlewood, one must now estimate the more complicated integrals

$$\int_T^{2T} \left| \int_t^{t+h} Z(u) \left| \phi\left(\frac{1}{2} + iu\right) \right|^2 du \right|^2 dt, \tag{6}$$

and

$$\int_T^{2T} \left| \int_t^{t+h} \zeta\left(\frac{1}{2} + iu\right) \phi\left(\frac{1}{2} + it\right)^2 - 1\, du \right|^2 dt. \tag{7}$$

To deal with (6), Selberg [14] multiplied out the square and inverted the order of integration, which led to the consideration of integrals of the form

$$\int_T^{2T} Z(t)Z(t+\delta) \left| \phi\left(\frac{1}{2} + it\right) \phi\left(\frac{1}{2} + it + i\delta\right) \right|^2 dt, \tag{8}$$

where δ is small, say $\delta \ll 1/\log T$. Ingham had discussed the mean value

$$\int_T^{2T} \zeta\left(\frac{1}{2} + it\right) \zeta\left(\frac{1}{2} - it - i\delta\right) dt,$$

but the introduction of the factors involving ϕ make the estimations substantially harder. Not only that, but the resulting main term is a complicated form of degree four in the coefficients β_ν. With remarkable insight, Selberg saw that it was appropriate to take

$$\beta_\nu = \alpha_\nu (1 - (\log \nu)/\log X)$$

where

$$\zeta(s)^{-1/2} = \sum_{\nu=1}^{\infty} \alpha_\nu \nu^{-s}$$

for $\sigma > 1$. By means of a long sequence of delicate manipulations, Selberg then showed that the integral (6) is $\ll T h^{3/2} (\log X)^{-1/2}$. He treated the integral (6) similarly,[1] and found that it satisfies the same estimate. These suffice to give the estimate

$$N_0(T) \gg T \log T$$

so that a positive proportion of the zeros lie on the critical line.

[1] In a subsequent exposition of this analysis, Titchmarsh [24; Sections 10.9–10.22] avoided the need to bound the integral (6), but for some purposes the original method is advantageous.

At the same time, Selberg [14] showed that

$$\sum_{0 \le \gamma \le T} \left|\beta - \frac{1}{2}\right| \ll T, \tag{9}$$

where $\rho = \beta + i\gamma$ is a zero of the zeta-function. From this we see that for any $\tau < 1$ there is a constant $c = c(\tau) > 0$ such that at least τ of the zeros ρ satisfy the inequality $|\beta - \frac{1}{2}| < c/\log \gamma$. The above is a natural consequence of the estimate

$$\int_T^{2T} \left|\zeta\left(\frac{1}{2} + it\right)M\left(\frac{1}{2} + it\right)\right|^2 dt \ll T, \tag{10}$$

where $M(s)$ is a certain Dirichlet polynomial

$$M(s) = \sum_{n \le X} a_n n^{-s}$$

with $a_1 = 1$, which we now discuss. The integral in (10) is simpler to treat than that in (8), partly because we now have $\delta = 0$, but also because the asymptotic main term now involves only a quadratic form in the a_n, which turns out to be

$$\sum_{m,n \le X} \frac{a_m a_n}{mn} (m, n) \log\left(\frac{c(m, n)\sqrt{T}}{\sqrt{mn}}\right), \tag{11}$$

where $c > 0$ is a certain absolute constant. Even this is too complicated to minimize exactly, but Selberg saw that the simpler form

$$\sum_{m,n \le X} \frac{a_m a_n}{mn} (m, n) \tag{12}$$

is minimized (subject to the linear constraint $a_1 = 1$) by taking

$$a_n = \mu(n) \frac{nL(X/n; n)}{\phi(n)L(X; 1)}, \tag{13}$$

where

$$L(y; n) = \sum_{\substack{m \le y \\ (m,n) = 1}} \frac{\mu^2(m)}{\phi(m)}. \tag{14}$$

Selberg showed that with this choice of the coefficients, the expression (11) is $\ll (\log T)/\log X$. Here X can be taken to be of the form T^a, $a > 0$. This gives (10) and hence (9).

We now are in a position to motivate Selberg's choice of the coefficients β_ν in (5). For fixed n it is not difficult to show that

$$L(y; n) \sim \frac{\phi(n)}{n} \log y,$$

and it may be expected that $L(y; n)$ is approximately this size for most n, even when n is large. This suggests that the a_n in (13) are approximately $\mu(n)(1 - \log n/\log X)$. Indeed, the bilinear forms above can be satisfactorily bounded with this choice of the a_n. With this knowledge, it is then reasonable to take $\phi(s)$ to be a similarly weighted partial sum of the Dirichlet series for $\zeta(s)^{-1/2}$.

The estimate (9) constitutes a refinement of Ingham's bound (4) when σ is near $\frac{1}{2}$. Selberg [17] further refined his own bound by considering the integral (10) with $\frac{1}{2}$ replaced by σ, $\frac{1}{2} < \sigma < 1$. In this way he showed that

$$N(\sigma, T) \ll T^{(9-2\sigma)/8} \log T. \tag{15}$$

Although there is no mention of sieve problems in these papers, one finds in them all the intellectual ingredients of the so-called Selberg λ^2 sieve method. In the first place, there is the simple observation that if one needs a function to be non-negative, one may try using a function that is expressed as a square of some other function. More importantly, the bilinear forms that arise in the λ^2 method are of the same type as those already considered above. Indeed, if one seeks to prove the Brun–Titchmarsh inequality by the λ^2 method, the bilinear form that arises is precisely the form (12) above. If the a_n are taken as in (13), then the value of this form is exactly $1/L(X; 1)$. Now by elementary methods (see Ward [25]) it is easy to show that

$$L(y; 1) = \log y + C_0 + \sum_p \frac{\log p}{p(p-1)} + o(1),$$

but to bound the form (12) it suffices to derive a lower bound for L. To this end it is enough to observe that

$$L(y; 1) = \sum_{n \le y} \frac{\mu(n)^2}{n} \prod_{p|n} (1 + p^{-1} + p^{-2} + \cdots) = \sum_{s(n) \le y} \frac{1}{n},$$

where $s(n)$ denotes the largest square-free number dividing n, and then that this latter sum is

$$\ge \sum_{n \le y} \frac{1}{n} > \log y.$$

Even this expository finesse is already found in Selberg's discussion [14, p. 57] of zeros of the zeta-function.

2. The Distribution of $\log \zeta(\frac{1}{2} + it)$

The estimate (1) for $N(T)$ can be expressed in a more precise form, namely

$$N(T) = \frac{T}{2\pi} \log \frac{T}{2\pi} - \frac{T}{2\pi} + \frac{7}{8} + S(T) + O\left(\frac{1}{T}\right),$$

where

$$S(T) = \frac{1}{\pi} \arg \zeta\left(\frac{1}{2} + it\right),$$

and the argument is calculated by continuous variation along the path from $+\infty + it$ to $\frac{1}{2} + it$. Since $S(T) \ll \log T$, we have rather precise information concerning the vertical distribution of the zeros of the zeta-function. Littlewood [8] had shown that

$$\int_0^T S(t)\, dt \ll \log T$$

and had further shown that if the Riemann Hypothesis is true then

$$S(T) \ll \frac{\log T}{\log \log T}.$$

Littlewood [9] also showed that if RH holds then

$$\int_0^T |S(t)|\, dt \ll T \log \log T.$$

and under the same hypothesis Titchmarsh [23] showed that

$$\int_0^T |S(t)|^2\, dt \gg T \log \log T,$$

but our knowledge concerning the usual size of $S(T)$ remained very sketchy until Selberg's work on this subject. Selberg [16, 17] derived estimates for the moments of $S(T)$, first assuming RH and then unconditionally. Assuming RH, he showed that if k is a positive integer then

$$\int_0^T S(t)^{2k}\, dt = \frac{(2k)!}{k!\,(2\pi)^{2k}}\, T(\log \log T)^k \left[1 + O\left(\frac{1}{\log \log T}\right)\right]. \tag{16}$$

Selberg made no mention of the odd moments, but it is clear that his method also gives

$$\int_0^T S(t)^{2k+1}\, dt \ll T(\log\log T)^k. \tag{17}$$

Unconditionally, Selberg [17] proved a slightly weakened form of (16) in which the error term is replaced by $O(1/\sqrt{\log\log T})$.[2] Again he made no mention of the odd moments, but it is clear that he could have also proved (17) unconditionally. From these estimates it follows that $S(t)$ is usually of the order of $\sqrt{\log\log t}$. Indeed, it follows that

$$\lim_{T\to\infty} \frac{1}{T} \operatorname{meas}\left\{t \in [5, T]: S(t) \le \frac{c}{\pi}\sqrt{\log\log t}\right\} = \frac{1}{\sqrt{\pi}} \int_{-\infty}^{c} e^{-u^2}\, du. \tag{18}$$

That is, the asymptotic distribution of $S(t)(\log\log t)^{-1/2}$ is normal with mean 0 and standard deviation $1/\pi\sqrt{2}$. It might seem surprising that in 1946 Selberg failed to draw this last conclusion from his estimates, but it must be remembered that the general results concerning the moment problem were not as widely known then as they are today.

Subsequent to this work, Selberg (unpublished) established similar results for $\log \zeta(\frac{1}{2} + it)$. In particular, he proved that if h and k are non-negative integers, then

$$\int_0^T \left[\log \zeta\left(\frac{1}{2} + it\right)\right]^h \left[\log \zeta\left(\frac{1}{2} - it\right)\right]^k dt$$
$$= \delta_{h,k} k!\, T(\log\log T)^k + O(T(\log\log T)^{(h+k-1)/2}),$$

where $\delta_{h,k} = 1$ if $h = k$, $\delta_{h,k} = 0$ otherwise. As in his earlier work, if RH is assumed then the exponent in the error term can be reduced to $(h + k - 2)/2$. As Selberg noted, these estimates imply that if Ω is a set in the complex plane that has Jordan content, then

$$\lim_{T\to\infty} \frac{1}{T} \operatorname{meas}\left\{t \in [5, T]: \frac{\log \zeta(\frac{1}{2} + it)}{\sqrt{\log\log t}} \in \Omega\right\}$$
$$= \frac{1}{\pi} \iint_{\Omega} e^{-(x^2+y^2)}\, dx\, dy.$$

[2] Titchmarsh [24] gave an account of the conditional estimate (16) in the case $k = 1$ and states the estimate for general k but incredibly failed to indicate that such estimates had been obtained unconditionally.

That is, Re log $\zeta(\frac{1}{2}+it)$ and Im log $\zeta(\frac{1}{2}+it)$ behave like independent normally distributed random variables with expectation $\mu = 0$ and variance $\sigma^2 = \frac{1}{2}\log\log t$. The relation (17) is a special case of this.

The method by which these results are obtained has a number of instructive elements. The first step is to approximate to ζ'/ζ by a suitable weighted partial sum of its Dirichlet series. This Selberg [15] had already done in an earlier paper concerning the mean-square distribution of primes in short intervals. We recall that if $D(s) = \sum a_n n^{-s}$ is a Dirichlet series with abscissa of convergence σ_c then Perron's formula asserts that

$$\sum_{n \le x} a_n = \frac{1}{2\pi i}\int_{c-i\infty}^{c+i\infty} D(s)\frac{x^s}{s}\,ds$$

for $c > \max(0, \sigma_c)$. The integrand has a pole at $s = 0$, which gives a residue $D(0)$. Thus if the path of integration is moved to the left then (depending on analyticity, the size of $D(s)$, etc.), one may find that the integral is approximately $D(0)$ and hence the sum on the left approximates $D(0)$. In practice an argument of this kind may fail because the kernel $1/s$ decreases only like an inverse first power. This difficulty may be overcome by introducing other kernels, with the result that certain weighting factors are introduced in the sum on the left. For example, we have

$$\sum_{n \le x} a_n\left(1 - \frac{\log n}{\log x}\right) = \frac{1}{2\pi i \log x}\int_{c-i\infty}^{c+i\infty} D(s)\frac{x^s}{s^2}\,ds$$

for $c > \max(0, \sigma_c)$. Here the kernel now decreases like an inverse second power, but now the residue at $s = 0$ is $D(0) + D'(0)/\log x$. This new term is a nuisance if the object is to approximate to $D(0)$. However, Selberg found that by taking two values of x in this formula and then forming a suitable linear combination, one may construct a kernel so that the residue at $s = 0$ is $D(0)$, and yet the kernel still decreases like an inverse second power: If $x > 1$ and $y > 1$ then

$$\sum_{n} a_n w(n) = \frac{1}{2\pi i}\int_{c-i\infty}^{c+i\infty} D(s)\frac{x^s}{s^2}\frac{y^s - 1}{\log y}\,ds$$

for $c > \max(0, \sigma_c)$, where

$$w(n) = w(n; x, y) = \begin{cases} 1 & \text{if } 1 \le n \le x, \\ \dfrac{\log xy}{\log y} - \dfrac{\log n}{\log y} & \text{if } x \le n \le xy, \\ 0 & \text{if } xy \le n. \end{cases}$$

This kernel (and others like it), introduced by Selberg, has many uses and has become a standard part of the general theory. It may be applied to ζ'/ζ, and the resulting approximation is then integrated to provide an approximate formula for $\log \zeta$. The error in this approximation depends on nearby zeros and, in particular, on nearby zeros which are to the right of the critical line by a significant amount. The sharp zero-density estimate (15) plays a crucial rôle in estimating the contribution of such error terms. Thus it is found that $\log \zeta(\frac{1}{2} + it)$ is approximately

$$\sum_n \frac{\Lambda(n)}{\log n} w(n) n^{-1/2-it}.$$

The higher powers of the primes do not make much contribution, on average, so one may think of this as being

$$\sum_p w(p) p^{-1/2-it}.$$

From the fundamental theorem of arithmetic we know that the numbers $\log p$ are linearly independent over $\mathbb{Q}$. Thus the terms p^{-it} behave like independent unimodular random variables. These terms are not exactly independent on the interval $[0, T]$, but they are sufficiently close to being independent that the moment has approximately the anticipated value.

3. Other Work

In the papers already mentioned, Selberg [16, 17] also established important results concerning sign changes of $S(t)$, and Ω-theorems for both $S(t)$ and $S_1(t)$. Moreover, he [18] extended his results to Dirichlet L-functions, and in doing so introduced additional important ideas. This latter work greatly surpassed work of R.E.A.C. Paley [10] and C.L. Siegel [22] in this area. The papers of Selberg and Chowla [20, 21] and of Brun, Jacobsthal, Selberg, and Siegel [3] address somewhat different questions. Bombieri [2; pp. 27, 40] gave an account of an interesting form of the large sieve that Selberg found and that he used to prove a zero-density theorem for L-functions that is sharp when σ is near 1. Although Selberg has published little concerning the zeta-function in recent years, he has generously offered suggestions and communicated his ideas to others, even to the point of supplying unpublished results. For example, the basic argument in the paper of Montgomery and Vaughan [9] is entirely Selberg's. He has had a great influence on many researchers, and evidence of this is found not only in some of my own papers but also in some of the publications of such people as A. Fujii, J. Mueller, K. Ramachandra, and J. Vaaler.

References

[1] Bohr, H. and Landau, E. "Sur les zéros de la fonction $\zeta(s)$ de Riemann," *C. R. Acad. Sci. Paris* **158** (1914), 158–162.

[2] Bombieri, E. "Le grand crible dans la théorie analytique des nombres," *Astérisque* **18**, (1974), Soc. Math. France, Paris. 87 pp.

[3] Brun, V., Jacobsthal, E., Selberg, A. and Siegel, C.L. "Correspondence about a polynomial which is related to Riemann's zeta function," *Norsk. Mat. Tidsskr.* **24** (1946), 65–71.

[4] Carlson, F. "Über die Nullstellen der Dirichletschen Reihen und der Riemannschen ζ-funktion," *Arkiv för Mat. Astr. och Fysik.* **15**(20) (1920).

[5] Hardy, G.H. "Sur les zéros de la fonction $\zeta(s)$ de Riemann," *C. R. Acad. Sci. Paris.* **158** (1914), 1012–1014.

[6] Hardy, G.H. and Littlewood, J.E. "The zeros of Riemann's zeta-function on the critical line," *Math. Zeit.* **10** (1921), 283–317.

[7] Ingham, A.E. "On the estimation of $N(\sigma, T)$," *Quart. J. Math.* (Oxford) **11** (1940), 291–292.

[8] Littlewood, J.E. "On the zeros of Riemann's zeta-function," *Proc. Cambridge Philos. Soc.* **22** (1924), 295–318.

[9] Littlewood, J.E. "On the Riemann zeta function," *Proc. London Math. Soc.* (2)**24** (1925), 175–201.

[10] Montgomery, H.L. and Vaughan, R.C. "Hilbert's inequality," *J. London Math. Soc.* (2)**8** (1974), 73–82.

[11] Paley, R.E. "On the k-analogues of some theorem in the theory of the Riemann ζ-function," *Proc. London Math. Soc.* (2)**32** (1932), 273–311.

[12] Selberg, A. "On the zeros of the zeta-function of Riemann," *Norske Vid. Selsk. Forh.* Trondheim **15**(16) (1942), 59–62.

[13] Selberg, A. "On the zeros of Riemann's zeta function on the critical line," *Arch. Math. Naturvid.* **45**(9) (1942), 101–114.

[14] Selberg, A. "On the zeros of Riemann's zeta function," *Skr. Norske Vid. Akad. Oslo.* I. No. 10, (1942), 59 pp.

[15] Selberg, A. "On the normal density of primes in small intervals, and the difference between consecutive primes," *Arch. Math. Naturvid.* **47**(6) (1943), 87–105.

[16] Selberg, A. "On the remainder in the formula for $N(T)$, the number of zeros of $\zeta(s)$ in the strip $0 < t < T$," *Avh. Norske Vid. Akad. Oslo.* I. No. 1, (1944), 27 pp.

[17] Selberg, A. "Contributions to the theory of the Riemann zeta-function," *Arch. Math. Naturvid.* **48**(5) (1946), 89–155.

[18] Selberg, A. "Contributions to the theory of Dirichlet's L-functions," *Skr. Norske Vid. Akad. Oslo.* I. No. 3, (1946), 62 pp.

[19] Selberg, A. "The zeta-function and the Riemann hypothesis," *C. R. Dixième Congrès Math. Scandinaves* (1946), 187–200; *Jul. Gjellerups Forlag*, Copenhagen (1947).

[20] Selberg, A. and Chowla, S. "On Epstein's zeta function, I," *Proc. Nat. Acad. Sci. U.S.A.* **35** (1949), 371–374.

[21] Selberg, A. and Chowla, S. "On Epstein's zeta-function," *J. Reine Angew. Math.* **227** (1967), 86–110.

[22] Siegel, C.L. "Contributions to the theory of the Dirichlet L-series and the Epstein zeta-functions," *Ann. of Math.* (2)**44** (1943), 143–172.

[23] Titchmarsh, E.C. "On the remainder in the formula for $N(T)$, the number of zeros in the strip $0 < t < T$," *Proc. London Math. Soc.* **27**(2) (1928), 449–458.

[24] Titchmarsh, E.C. *The Theory of the Riemann Zeta Function*, Oxford University Press, Oxford (1951), 346 pp.

[25] Ward, D.R. "Some series involving Euler's function," *J. London Math. Soc.* **2** (1927), 210–214.

8 Selberg's Work on the Arithmeticity of Lattices and its Ramifications

*G. D. MOSTOW**

1. Introduction

We use the word *lattice* Γ in a locally compact group G for a discrete subgroup Γ such that G/Γ has finite Haar measure. If in addition G/Γ is compact, we call Γ a *cocompact* lattice in G. A lattice in a connected semisimple group G is called *reducible* if $G = G_1G_2$ with G_1, G_2 connected normal subgroups, $G_1 \cap G_2$ discrete, and $\Gamma/(\Gamma \cap G_1)(\Gamma \cap G_2)$ is finite. A lattice Γ in G is *irreducible* if it is not reducible. Two subgroups Γ and Γ' of G are called *commensurable* if $\Gamma \cap \Gamma'$ has finite index in both Γ and Γ'.

Prior to 1960, it had been observed that the only general constructions of lattices applicable in all dimensions were arithmetic. We owe to Selberg the intuition that what cannot be constructed does not likely exist and the judgement to act on that intuition. In his 1960 paper "On discontinuous groups in higher dimensional symmetric spaces," Selberg proved the

Theorem. *Any cocompact lattice in* $\mathrm{SL}_n(\mathbb{R})$ *is conjugate to a lattice with algebraic numbers as matrix entries, provided* $n > 2$.

* Supported in part by NSF Grant DMS-8507130

ISBN 0-12-067570-6

Thereafter, he took aim at the question of whether a lattice Γ in a suitable G is arithmetic. Such a goal entails ultimately associating an integral structure with the lattice. He began by considering lattices in the direct product G of groups $SL_2(\mathbb{R})$ or $SL_2(\mathbb{C})$ and assumed that Γ was a *non*cocompact lattice. Let X denote the symmetric space associated to G. Then conjecturally, the cusps of the quotient $\Gamma\backslash X$ arose from unipotent elements in G. Selberg gave the proof of this conjecture for $SL_2(\mathbb{C})$ at the 1965 Conference on Analytic Functions in Erevan and later published the result for products of $SL_2(\mathbb{R})$ and $SL_2(\mathbb{C})$ in the 1968 paper "Recent developments in the theory of discontinuous groups of motions in symmetric spaces" (cf. [S]2). In that talk he was influenced by the 1959 results of Piatetski–Shapiro on cusps in a direct product $SL_2(\mathbb{R})^n$ (cf., [P]1). To a noncocompact lattice Γ, he could associate, via the cusps, a number field k. The question of whether an irreducible noncocompact lattice in $SL_2(\mathbb{R})^n$ is commensurable with a Hilbert modular group

$$SL_2(0) \hookrightarrow SL_2(\mathbb{R})^n, \qquad 0 = \text{integers of } k$$
$$n = [k:Q]$$

he reduced to the question: Are the denominators of Γ bounded? This question, in turn, he reduced to a question about the nonvanishing in a neighborhood of $s = 1$ depending only on the field k of

$$L(s, \chi) = \sum_{\mathbb{Q}} \frac{\chi(\mathbb{Q})}{N(\mathbb{Q})^s},$$

where the sum extends over all integral ideals $\mathbb{Q}$, and χ is not the principal character. $L(s, \chi)$ is then an integral function with functional equation relating $L(s, \chi)$ and $L(1 - s, \bar{\chi})$.

The foregoing two papers inspired two separate attacks on the problem of arithmeticity of lattices, one for the case of noncompact lattices and another strategy for the case of cocompact lattices.

The first post-Selberg breakthrough in the noncocompact case was the result by Kazhdan–Margulis: "A proof of Selberg's conjecture," Mat Skornik 1968, in which they proved his 1965 Erevan conjecture that a noncocompact lattice in a semisimple Lie group has a unipotent element other than the identity.

Building on this, Margulis in 1973 was able to prove the arithmeticity of nonarithmetic lattices. Independently, Raghunathan obtained impressive results in that direction (cf. [R]1).

It turned out, however, that the strategy inspired by Selberg's 1960 paper for cocompact lattices was applicable to noncocompact lattices as well and touched off more ramifications. For that reason I shall trace the course initiated by the 1960 paper.

2. Selberg's Rigidity Theorem

Let Γ be a lattice in a Lie group G. A *lattice deformation* of Γ is a one-parameter family of monomorphisms $\theta^{(t)}\Gamma \to G$, $0 \leq t \leq a$ such that $\theta^{(t)}(\Gamma)$ is a lattice for all t, $\theta^{(0)}(\gamma) = \gamma$, and $\theta^{(t)}(\gamma)$ is continuous in t for each $\gamma \in \Gamma$.

The main theorems in Selberg's 1960 paper was asserted for a cocompact lattice Γ in the group $G = \mathrm{SL}_n(\mathbb{R})$:

I. *Γ can be deformed into a lattice whose matrix elements are algebraic integers.*

II. *Any deformation of Γ is via inner automorphisms of G, provided $n > 2$.*

Selberg stated that his method generalizes to the other classical groups. We sketch Selberg's proof of II.

From the finite measure of G/Γ, Selberg deduced

S1. Given any element $g \in G$ and any neighborhood U of 1 in G, then $g^n U \cap U\Gamma \neq \varnothing$ for infinitely many n.

From this, he proved that

S2. The $\mathbb{R}$-linear span of Γ = the $\mathbb{R}$-linear span of G.

Denote this $\mathbb{R}$-linear span by Q; it is the enveloping associative algebra.

Selberg then showed that the deformation $\theta^{(t)}: \Gamma \to \Gamma^{(t)}$ extends to a deformation $\theta^{(t)}: Q \to Q$, if only

S3. $\theta^{(t)}$ preserves the traces of elements of Γ.

Since Q is a semisimple algebra, $\theta^{(t)}$ is given by inner automorphisms of Q, and these stabilize G.

Thus it remains to prove S3. This is done after several steps.

S4. For any $\nu \in \Gamma$, set

$$G_\nu = \{g \in G; g\nu = \nu g\}, \qquad \Gamma_\nu = \Gamma \cap G_\nu$$

then G_ν/Γ_ν is compact.

S5. Using S1, Selberg showed that Γ contains an element s whose eigenvalues are real positive and distinct. Let G_s^0 denote the connected component of the identity in G_s, and set $\Gamma_s^0 = \Gamma_s \cap G_s^0$. Then $G_s^0 \approx \mathbb{R}^{n-1}$ and $\Gamma_s^0 \approx \mathbb{Z}^{n-1}$, and the same is true for the deformed lattices $\Gamma^{(t)}$.

Let $\{s_1, \ldots, s_{n-1}\}$ be a base for Γ_s^0, and we can assume $s_1 = \operatorname{diag}(s_{i,1}, \ldots, s_{1,n})$. Then for any $(m_1, \ldots, m_{n-1}) \in \mathbb{Z}^{n-1}$ and $1 \le i \le n-1$, $1 \le j_1, j_2 \le n$ any inequalities or equalities

$$\prod_{i=1}^{n-1} s_{i,j_1}^{m_i} \gtreqless \prod_{i=1}^{n-1} s_{s_i,j_2}^{m_i} \qquad (*)$$

are preserved under the deformation $\Gamma^{(t)}$ of Γ. Consequently, for all t

$$s_{i,j}^{(t)} = (s_{i,j})^{\tau(t)} \qquad 1 \le i \le n-1, \qquad 1 \le j \le n$$

where the exponent $\tau(t)$ is a real number depending continuously on t, but not on i, j.

S6. Using I, Selberg proves that Γ_s^σ contains an element ν with exactly two eigenvalues equal.

From this he concluded that $\tau(t) \equiv 1$ for all t. From this, in turn, he showed that the trace of $\nu(t)$ is constant in t for all $\nu \in \Gamma$. This gives his Rigidity Theorem II.

Remark 1. A. Borel proved that Selberg's result S2 is valid for a lattice Γ in an arbitrary semisimple matrix group G. This assertion is equivalent to the assertion that Γ is Zariski-dense in G—this is the well-known Borel Density Theorem (cf. [B]1).

Remark 2. A Weil generalized Selberg's rigidity theorem to cocompact irreducible lattices in arbitrary semisimple real Lie groups other than those isomorphic to $SL_2(\mathbb{R})$ (cf. [W], 2).

3. Strong Rigidity

The conjecture of a stronger rigidity phenomenon arose from my attempt to fit Selberg's rigidity theorem, as generalized by Weil, into a topological transformation group setting. One is led then to the following question.

Let G and G' be semisimple Lie groups, which for convenience we may take to have no center $\neq(1)$ and no compact factor. Let Γ be an irreducible lattice in G and let $\theta: \Gamma \to G'$ be a monomorphism such that $\theta(\Gamma)$ is a lattice Γ' in G'. Does θ extend to an analytic isomorphism of G to G'?

Let K be a maximal compact subgroup of G and let $X = G/K$; X is a symmetric Riemannian space of nonpositive curvature and is homeomorphic to Euclidean space. Similarly, let $X' = G'/K'$ denote the symmetric Riemannian space associated to G.

By a result of Selberg, Γ has a torsion-free subgroup of finite index (cf. [S]1). It is easy to see that for our question no generality is lost in replacing Γ by a subgroup of finite index, so we can add the hypothesis that Γ is torsion-free. It is then an elementary topological fact that there exists Γ-equivariant continuous maps

$$\phi: X \to X' \qquad \text{and} \qquad \phi': X' \to X$$

so that $\phi' \circ \phi$ and $\phi' \circ \phi$ are homotopic to the identity map.

Upon analyzing Selberg's proof of his rigidity theorem, the key relation (*) shows its force as the elements $\prod_1^{n-1} s_i^{m_i}$ go to infinity in the sublattice of the abelian subgroup G_s^0. Since G is nonabelian, it seemed to me desirable to exploit relations at infinity not only on abelian subfamilies but among *all* elements of Γ *near infinity.*

Combining these two considerations led to the question:

Does the Γ-map $\phi: X \to X'$ extend to a map of the *boundary* of X to the boundary of X'?

The term *boundary* of X had been introduced in different ways by Satake and by Furstenberg. The natural choice was the unique compact G-orbit on the Satake ad-boundary; this is the same as the Furstenberg boundary G/P where P is a minimal parabolic subgroup of G. For example, if $G = \mathrm{SL}_n(\mathbb{R})$, then the boundary X_0 is the flag manifold. Choosing P as a maximal parabolic $\mathrm{SL}_n(\mathbb{R})$, one can get as G/P the Grassmannian manifolds $\mathrm{Gr}_k(n)$ of k-planes in n space.

The proof of strong rigidity was completed in several stages. The first stage was treated in [Mo]2, Publ. IHES, Vol. 34 (1968), which asserts in Theorem 12.1

Theorem. *Let G be the group of isometries of real hyperbolic n-space X_σ. Let Γ and Γ' be lattices in G and $\theta: \Gamma \to \Gamma'$ an isomorphism. Let $\phi: X \to X$ be a Γ-equivariant quasiconformal homeomorphism. Then θ extends to an inner automorphism of G provided $n > 2$.*

The proof consisted of applying the theory of quasiconformal mappings to get the existence of the boundary map

$$\phi_0: X_0 \to X_0.$$

Here the boundary is the $(n-1)$ sphere and the map ϕ_σ is a Γ-equivariant quasiconformal map. By quasiconformal theory, ϕ_σ is absolutely continuous on almost all lines of $n > 2$. Using the ergodicity of Γ in X_σ, I could prove that ϕ_σ is smooth—in fact, a Möbius transformation. By Γ-equivariance $\phi_\sigma(\nu x) = \theta(\nu)\phi_\sigma(x)$, i.e., $\phi_\sigma \nu = \theta(\nu)\phi_\sigma$. Since ϕ_σ may be regarded as an element of G, the extension of θ to G is nothing but

$$\theta(g) = \phi_0 g \phi_0^{-1}.$$

The second stage is described in the 1970 paper [Mo 3] and in the 1973 monograph [Mo 4], *Strong Rigidity in Locally Symmetric Spaces*. Here the existence of the boundary map ϕ_σ is proved under the hypothesis that the Γ-map $\phi: X \to X'$ is a pseudo-isometry, i.e., there exist positive constants b and c such that

$$c^{-1} d(x, y) \leq d(\phi(x), \phi(y)) \leq cd(x, y) \qquad \text{if } d(x, y) \geq b$$

(ibid *Theorem* 15.2). *Let G, G' be semisimple real Lie groups and X, X' be their associated symmetric spaces. Let Γ, Γ' be lattices in G, G', respectively, and let $\theta: \Gamma \to \Gamma'$ be an isomorphism. Assume* (*Pisom*) *there exist Γ-equivariant pseudoisometries*

$$\phi: X \to X' \qquad \text{and} \qquad \phi': X' \to X.$$

Then ϕ induces a homeomorphism $\phi_\sigma: X_\sigma \to X'_\sigma$.

It is easy to verify that the assumption (Pisom) holds if Γ and Γ' are cocompact lattices. In 1972, Gopal Prasad proved that the assumption (Pisom) holds for arbitrary lattices, using results of Garland–Raghunathan in $\mathbb{R}$-rank 1 (cf. *Inv. Math.* **21**, 255–286, 1973).

At this point, the proof of strong rigidity divides into two cases.

In case $\mathbb{R}$-rank $G > 1$, one can exploit the rich combinatorial structure on the boundary that is preserved by the boundary map. The combinatorial structure allows one to associate a spherical Tits building to the boundary that is preserved by ϕ_σ. The analogue of the fundamental theorem of projective geometry allows us to conclude that

ϕ_0 induces an isomorphism of G to G'. The extension of θ again comes from

$$\theta(g) = \phi_0 g \phi_0^{-1}.$$

In case $\mathbb{R}$-rank $G = 1$, one is dealing with the isometry group of one of the hyperbolic spaces $\mathbb{R}h^n$, $\mathbb{C}h^n$, $\mathbb{H}h^n$, or $\mathbb{O}h^2$ over the real, complex, quaternion, or Cayley numbers. Here we combine some ideas from the theory of quasiconformal mappings with properties of ergodic action of Γ on the boundary X_σ and conclude that the boundary ϕ_σ induces an isomorphism of G to G'. The proof of strong rigidity in [Mo 4] is valid for the case of irreducible cocompact lattices in semisimple groups not locally isomorphic to $SL_2(\mathbb{R})$, and Prasad's paper extends its validity to arbitrary lattices in such groups.

4. Margulis' Arithmeticity Theorem

In 1974, Margulis sent a remarkable paper to be read at the Vancouver Congress that contained the following theorem for which I proposed the name

Super-rigidity Theorem. *Let G be a connected semisimple algebraic group without center defined over $\mathbb{R}$, with $G^0_{\mathbb{R}}$ having no compact factors. Let Γ be an irreducible lattice in $G^0_{\mathbb{R}}$. Let K be a local field. Let $\rho: \Gamma \to GL(n, K)$ be a homeomorphism.*

1. *$\mathbb{R}$-rank $G > 1$.*
2. *The Zariski-closure of $\rho(\Gamma)$ is a connected simple group.*
3. *The K-topology closure of $\rho(\Gamma)$ is not compact.*
4. *G/Γ is compact.*

Then 1. *$K = \mathbb{R}$ or $\mathbb{C}$.*
2. *ρ extends to a rational representation of G.*

Margulis showed that the arithmeticity of the lattice Γ is an immediate consequence of the super-rigidity theorem: namely, using super-rigidity with $K = \mathbb{R}$ yields the fact that in suitable coordinates the matrix entries of Γ are rational. Applying super-rigidity with $K = \mathbb{Q}_p$, the p-adic field yields the conclusion that the $\rho(\Gamma)$ is bounded in the K-topology, i.e., $\rho(\Gamma)$ has bounded denominators.

The proof of the super-rigidity theorem reduces to proving.

Set $X_0 = G^0_{\mathbb{R}}/P^0_{\mathbb{R}}$ with P a minimal $\mathbb{R}$-parabolic subgroup of G. Let $H = \mathrm{SL}_n(K)$. Then there exists a Γ-equivariant measurable map

$$\phi_0\colon X_0 \to \mathrm{Gr}_k(n, K)$$

to the Grassmanian of k-dimensional subspaces of K^n.

The fact that ϕ_σ is rational almost everywhere comes from exploiting the rich combinatorial structure of X_0 and is reduced by Margulis to the well-known fact: A measurable homomorphism of $\mathbb{R}$ is analytic almost everywhere.

Margulis first established the existence of ϕ_0 with the help of Oceledets' Multiplicative Ergodic Theorem, which is a noncommutative generalization of the Birkhoff Ergodic Theorem. The Multiplicative Ergodic Theorem had its origins in results going back to R. Bellman and Furstenberg–Kesten. Following Margulis' Vancouver paper, Furstenberg showed how to obtain the existence of ϕ_0 for any lattice Γ, i.e., dropping hypothesis (4).

The 1977 appendix written by Margulis for the Russian translation of Raghunathan's "Discrete subgroups of Lie groups" incorporated some ideas of Furstenberg in proving super-rigidity and hence arithmeticity for lattices in groups of $\mathbb{R}$-rank greater than 1 (cf. *Inventiones Math.*, 76, 93–120, 1984). In the case of noncocompact lattices Γ, Margulis needs to employ the theorem of Kazhdan (cf. [K]) that Γ is finitely generated. Thereafter this super-rigidity theorem yields bounded denominators by the same argument used above (where we tacitly used the fact that a cocompact lattice is finitely generated).

It should be noted that the arithmeticity theorem proved by Margulis applies to the more general problem of irreducible lattices contained in a direct product of semisimple algebraic groups defined over $\mathbb{R}$, and p-adic fields $\mathbb{Q}_p$:

Let S be a finite set of prime numbers including the infinite prime ω. Let G_p be a semisimple simply connected algebraic group defined over $\mathbb{Q}_p$ such that $G_p(\mathbb{Q}_p)$ has no infinite compact quotient group. Set $\mathbb{Q}_\infty = \mathbb{R}$ and set

$$G_S = \prod_{p \in S} G_p(\mathbb{Q}_p)$$

$$\mathbb{Q}(s) = \{q \in \mathbb{Q};\, q \in \mathbb{Z}_p \qquad \text{for } p \notin S\}.$$

Define rank $G_S = \sum_{p \in S} \mathbb{Q}_p$-rank G_p.

Theorem. *Assume* rank $G_S > 1$. *Then any irreducible lattice* Γ *of* G_S *is S-arithmetic in* G_S; *that is, there exists an algebraic group H defined over* $\mathbb{Q}$ *and closed subgroup* H_1 *and* H_2 *of* $H_S = \prod_{p \in S} H(\mathbb{Q}_p)$ *an isomorphism* $\theta: G_S \cong H_1$ *and an open compact* $K \subset H_2$ *such that*

$$H = H_1 \times H_2$$

$$\theta(\Gamma) \propto \mathrm{Proj}_{H_1} H(\mathbb{Q}(S) \cap (H_2 \times K)$$

This result had been conjectured by Piatetski–Shapiro in [Pr]2.

5. Existence of Nonarithmetic Lattices in O(*n*, 1) for *n* > 1

Selberg's original conjecture on arithmeticity of lattices was motivated by the thought that lattices that cannot be constructed do not likely exist. The question whether any $\mathbb{R}$-rank 1 group can contain a nonarithmetic lattice is still open.

In 1966 Makorov exhibited nonarithmetic lattices in the isometry group of $\mathbb{R}h^3$, three-dimensional real hyperbolic space. Subsequently, Vinberg carried out a systematic investigation of groups generated by reflections in $\mathbb{R}h^n$ and found nonarithmetic lattices only for $n \leq 5$. In 1966, Piatetski–Shapiro made the arithmeticity conjecture precise (cf. [Pr]2) and as a consequence of Margulis' 1974 paper, nonarithmetic lattices can exist only in $\mathbb{R}$-rank 1 simple Lie groups.

At the 1983 Warsaw Congress, Vinberg's paper proved the nonexistence of cocompact lattices generated by reflections for $n \geq 30$. In 1986 Gromov and Piatetski–Shapiro, by a method analogous to the construction of the connected sum of two manifolds, constructed out of two suitably related arithmetically defined compact n-manifolds of constant negative curvature a "hybrid" of constant curvature manifold whose fundamental group is nonarithmetic. Thereby they proved the existence of nonarithmetic lattices in $O(n, 1)$ for every $n > 1$. This method yields noncocompact nonarithmetic lattices in $O(n, 1)$ as well as cocompact lattices.

6. Nonarithmetic Lattices in *U*(*n*, 1)

The first construction of nonarithmetic lattices in $U(2, 1)$ was announced at the Helsinki Congress in 1978 (cf. [Mo 5]). The lattices were defined as groups $\Gamma(p, t)$ depending on two parameters, with p an integer and t real, which are generated by three complex reflections with Coxeter diagram

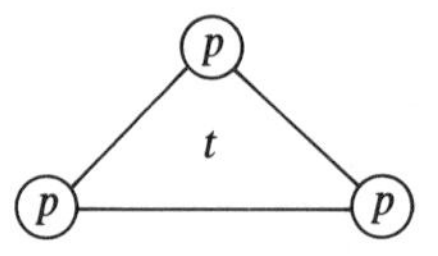

To each node one attaches a base vector e_i of a complex vector space $V = \oplus_i \mathbb{C}e_i$. On V, one defines the Hermitian form via

$$\langle e_i, e_i \rangle = 1$$

$$\langle e_1, e_2 \rangle = \langle e_2, e_3 \rangle = \langle e_3, e_1 \rangle = \frac{1}{2 \sin \frac{\pi}{p}} e^{\pi i t/3}.$$

One defines for each node the linear mapping

$$R_i \colon v \to v + (e^{2\pi i/p} - 1)\langle v, e_i \rangle, \qquad v \in V$$

and sets $\Gamma(p, t) = \langle R_1, R_2, R_3 \rangle$, the group generated by the R_i.

It is easy to see that each R_i is a complex reflection preserving the hermitian form, and the lattice has signature (one, $-$, two $+$) for $|t| < 3\pi(\frac{1}{2} - i/p)$. In [Mo 5] the values of p were restricted to 3, 4, 5; for these values the subgroups $\langle R_i, R_j \rangle$ were of finite order $24(p/6 - p)^2$. A sufficient condition was given for the group $\Gamma(p, t)$ to be discrete (and hence necessarily a lattice), and this sufficient condition was satisfied for 17 values of (p, t) with $3 \leq p \leq 5$, $o \leq t$. For seven of these values, $\Gamma(p, t)$ is a nonarithmetic lattice in $U(2, 1)$.

Subsequently, it turned out (cf. [Mo 7]) the lattices $\Gamma(p, t)$ are a special case of monodromy groups $\Gamma(\mu)$ of hypergeometric functions of n variables (cf. [DM])

$$F(x_2, \ldots, x_{n-1}) = \int_s^t z^{-\mu_\sigma}(z-1)^{-\mu_1}(z - x_2)^{-\nu_2} \cdots (z - x_{n+1})^{-\mu_{n+1}}\, dz,$$

where

$$\mu = (\mu_\sigma, \mu_1, \ldots, \mu_{n+2}), \qquad \sum_{\sigma}^{n+2} \mu_i = 2, \qquad 0 < \mu_i < 1,$$

$$= s, t \in \{0, 1, x_2, \ldots, x_{n+1}, \infty\}$$

For $n = 2$, such groups had been studied in 1885 by Picard, who claimed they yielded lattices in $U(2, 1)$ if the μ satisfies the *integrality condition*: if $i \neq j$, then $(1 - \mu_i - \mu_j) \in \mathbb{Z}$. Actually, Picard's proof was not complete (cf. [Mo 6]). In [DM] there is a correct proof of Picard's theorem: *If μ satisfies the integrality condition, then $\Gamma(\mu)$ is a lattice in* $PU(n, 1)$.

There exist μ satisfying the integrality condition only for $n \leq 5$. One finds seven cases (ignoring order) of μ with $\Gamma(\mu)$ nonarithmetic for $n = 2$, and one nonarithmetic Γ_μ for $n = 3$.

The lattices $\Gamma(p, t)$ coincide with lattices satisfying the weaker *half integral condition*: There is a subset $S_i \subset \{0, 1, \ldots, n + 2)$ such that for any $i \neq j$ with $\mu_i + \mu_j < 1$,

$$\begin{aligned} 1 - \mu_i - \mu_j &\in \tfrac{1}{2}\mathbb{Z} \qquad \text{if } i, j \in S_1 \\ &\in \mathbb{Z}, \qquad \text{otherwise.} \end{aligned}$$

In [Mo 7] it is proved: If μ satisfies the half-integral condition, then $\Gamma(\mu)$ is a lattice in $PU(n, 1)$. There exist μ satisfying the half-integral condition up to $n = 9$ but not greater. In [Mo 9], it is proved that *for $n > 2$, μ satisfies the half-integral condition if $\Gamma(\mu)$ is discrete, except for* $\mu = [\frac{1}{12}, \frac{3}{12}, \frac{5}{12}, \frac{5}{12}, \frac{5}{12}, \frac{5}{12}]$. Thus the only nonarithmetic lattices $\Gamma(\mu)$ occur for $n = 2$ and $n = 3$. At present, no other nonarithmetic lattices in $PU(n, 1)$ are known for $n > 1$.

7. Other Rigidity Phenomena

The strong rigidity theorem of Section 3 can be reformulated as:

Let M, M' be two complete locally symmetric Riemannian manifolds of nonpositive curvature and finite volume having no one- or two-dimensional geodesic factors. If M and M' have isomorphic fundamental groups, then for suitable choice of normalizing constants, M and M' are isometric.

The analogue of this result for homotopically equivalent compact Kahler manifolds M, M' of complex dimension ≥ 2 with "strongly negative sectional curvature" (as in the case of the complex ball) was proved by Y.T. Siu (cf. [Si]). He proved that such M and M' are biholomorphic or conjugate-biholomorphic.

In another direction, D. Sullivan extended the theme "rough isometry implies isometry" from finite volume hyperbolic manifolds to manifolds whose volume grows slower than that of hyperbolic space (cf. [Su]).

A third direction has been under investigation by R. Zimmer. According to Margulis' super-rigidity theorem, a lattice Γ of a simple group of ℝ-rank greater than 1 can be represented in a matrix group either faithfully as a lattice or degenerately in a compact group. Zimmer has found, to some extent, analogous results for the nonlinear action of lattices on manifolds (cf. [Z]).

The strong-rigidity of compact nonpositively curved locally symmetric spaces led to renewed interest in the geometry of finite volume nonpositively curved Riemannian spaces. Among the products of these investigations are a new geometric proof by Gromov of strong-rigidity for compact hyperbolic spaces, Pansu's special rigidity of quasiconformal mappings over the Cayley numbers and the theorems by Ballman, Brin, Eberlein, and Spatzier that nonpositively curved Riemannian spaces M of finite volume having rank ≥ 2 and no Euclidean factors are locally symmetric (cf. references). In the proof given by Burns and Spatzier, they associate a Tits building as in [Mo4] to the simply connected cover $\tilde{M}$ of M and exploit the structure of the topologized Tits building (cf. [BS]).

References

[Ba] Ballman, W. "Non-positively curved manifolds of higher rank," *Ann. of Math.* **222** (1985), 597–607.

[BBE] Ballman, W., Brin, M., and Eberlein, P. "Structure of manifolds of non-positive curvature I," *Ann. of Math.* **122** (1985), 171–203.

[BBS] Ballman, W., Brin, M., and Spatzier, R. "Structure of manifolds of non-positive curvature II," *Ann. of Math.* **122** (1985), 205–235.

[B] Borel, A.,
1. "Density properties of certain subgroups of semisimple groups," *Ann. of Math.* **72** (1960), 179–188.
2. *Introduction aux Groupes Arithmetiques.* Paris, Hermann (1969).

[BH] Borel, A. and Harish–Chandra, "Arithmetic subgroups of algebraic groups," *Ann. of Math.* **75** (1962), 485–535.

[BS] Burns, K. and Spatzier, R. "Manifolds of non-positive curvature and their buildings," *Publ. Math. I.H.E.S.*, to appear.

[DM] Deligne, P. and Mostow, G.D. "Monodromy of hypergeometric functions and non-lattice integral monodromy groups," *Publ. Math. I.H.E.S.* **63** (1986), 5–90.

[F] Furstenberg, H.
1. "A Poisson formula for semi-simple Lie groups," *Ann. of Math.* **77** (1963), 335–386.

2. "Boundary theory and stochastic processes on homogeneous spaces," *Proc. Sym. Pure Math.* **26**, AMS (1973), 193–229.

[GR] Garland, H. and Raghunathan, M.S. "Fundamental domains for lattices in rank one semi-simple groups," *Ann. of Math.* **92** (1970), 279–326.

[G] Gromov, M. "Asymptotic geometry of homogeneous spaces, *Proc. Conf. Torino* (1983); *Rend. Sem. Mat. Polyt. Torino, Fasc. Spe.* (1985), 59–60.

[GP] Gromov, M. and Piatetski–Shapiro, I. "Non-arithmetic groups in Lobachevsky spaces," to appear.

[K] Kazhdan, D. "Connection of the dual space of a group with the structure of its closed subgroups," *Funct. Anal. Appl.* **1** (1967), 63–65.

[KM] Kazhdan, D. and Margulis, G.A. "A proof of Selberg's conjecture," *Mat. S6.* **75** (117) (1968); *AMS Trans.* **4** (1968), 147–151.

[Mak] Markarov, V.S. "On a certain class of discrete groups of Lobachevsky space having an infinite fundamental region of finite measure," *Sov. Math. Dokl.* **7** (1966), 328–331.

[Ma] Margulis, G.A.

1. ,"On the arithmeticity of discrete subgroups," *Soviet Math. Dokl.*, **10** (1969), 900–902.
2. "Arithmetic properties of discrete groups," *Russian Mathematical Surveys*, **24** (1974), 107–156.
3. "Arithmeticity of non-uniform lattices in weakly non-compact groups," *Funct. Anal. Appl.* **8** (1975), 258–259.
4. "Discrete groups of motions of manifolds of non-positive curvature," *Proc. Int. Congress Math., Vancouver* 1974, **2**, 21–34 (translation in *AMS Transl.*, **109** (1977), 33–45.
5. "Arithmeticity of the irreducible lattices in the semi-simple groups of rank greater than 1," *Inv. Math.* **76** (1984), 93–120.

[Mo] Mostow, G.D.

1. "On the conjugacy of subgroups of semi-simple groups," *Algebraic Groups and Discrete Subgroups, Proc. Symposia in Pure Math.* **9** (1966), 413–420.
2. "Quasi-conformal mappings in n-space and the rigidity of hyperbolic space forms," *Publ. Math. I.H.E.S.* **34** (1967), 53–104.
3. "The rigidity of locally symmetric spaces," *Proc. Int. Congress Math.* **2** (1970), 187–198.
4. "Strong rigidity of locally symmetric spaces," *Ann. of Math. Studies.* **78** (1973).
5. "On a remarkable class of polyhedra in complex hyperbolic space," **86** (1980), 171–276.
6. "Existence of non-arithmetic monodromy groups," *Proc. NAS*, **78** (1981), 5948–5950.
7. "Generalized Picard lattices arising from half-integral conditions," *Publ. Math. I.H.E.S.* **63** (1986), 91–106.

8. "Braids, hypergeometric functions, and lattices," *Bull. AMS* (1987), 225–246.
9. "On discontinuous action of monodromy groups on the complex n-ball," to appear.

[Pa] Pansu, P.
1. "Dimension conforme et sphère á l'infini des varietés à coubure négative," *Ann. Acad. Sci. Fennicae*, to appear.
2. "Métriques de Carnot–Carethéodory et quasi-isométries des espaces symétrique de rang un," *Ecole Polytechnique Palaiseau.* (1984), preprint.

[P] Piatetsky–Shapiro, I.
1. "Discrete groups of analytic automorphisms of a polycylinder and automorphic forms," *Dokl. Akad. Nauk. USSR.* **124** (1959), 760–763.
2. "Discrete subgroups of Lie groups," *Trans. Moscow Math. Soc.* **18** (1968), 1–38.

[Pi] Picard, E. "Sur les fonctins hyperfuchsiennes provenant des sèries hypergeometriques de deux variables," *Ann. ENS, III.* **2** (1885), 357–384.

[Pr] Prasad, G. "Strong rigidity of $\mathbb{Q}$-rank 1 lattices," *Inv. Math.* **21** (1973), 255–286.

[R] Raghunathan, M.S.
1. "Discrete subgroups and $\mathbb{Q}$-structures on semi-simple groups." In: *Proc. Int. Coll. on Discrete Subgroups of Lie Groups and Applications to Moduli* (Bombay 1973), Oxford Univ. Press, Oxford (1975), 225–321.
2. *Discrete Subgroups of Lie Groups*, Springer (1972).

[S] Selberg, A.
1. "On discontinuous groups in higher dimensional symmetric spaces," *Contributions to Function Theory*, Tata Institute, Bombay (1960), 147–164.
2. "Recent developments in the theory of discontinuous groups of motions of symmetric spaces," *Proc.* 15*th Scand. Congress, Oslo*, 1968, *Lecture Notes in Math.* **118** (1970), 99–120.

[Si] Siu, Y.T. "Complex-Analyticity of Harmonic Maps and Strong Rigidity of Compact Kähler Manifolds," *Ann. of Math.* **112** (1980), 73–111.

[Su] Sullivan, D. "On the ergodic theory at infinity of an arbitrary discrete group of motions of hyperbolic motions."

[T] Thurston, W.P. "Three dimensional manifolds, Kleinian groups, and hyperbolic geometry," *Bull. AMS.* **6** (3) (1982), 357–381.

[V] Vinberg, E.B.
1. "Discrete groups generated by reflections in Lobachevsky spaces," *Mat. Sbornik* **72** (114) (1967), 417–478; correction ibid **73** (1967), 303.

2. "Discrete reflection groups in Lobachevsky space," *Proc. Int. Congress* (1983), 593–601.

[Wa] Wang, S.P. "The dual space of semi-simple Lie groups," *Am. J. Math.* **91** (1969), 921–937.

[W] Weil, A.

1. "Discrete subgroups of Lie groups I," *Ann. of Math.* **72** (1960), 369–384.
2. "Discrete subgroups of Lie groups II," *Ann. of Math.* **75** (1962), 578–602.

[Z] Zimmer, R.J. "*Ergodic Theory and Semi-simple Groups*, Birkhauser, Boston, Cambridge (1984).

RESEARCH ANNOUNCEMENTS

9 Mean Values of the Riemann Zeta-Function with Application to the Distribution of Zeros

J. B. CONREY, A. GHOSH, AND *S. M. GONEK*

1

The most precise results about the horizontal distribution of the zeros of the Riemann zeta-function are deduced from mean value theorems that involve the zeta-function multiplied by a Dirichlet polynomial. We are interested here in those results that give information about zeros on or near the critical line. The first result of this sort that required a detailed arithmetic argument involving the coefficients of the Dirichlet polynomial in order to accurately estimate the mean in question is due to Selberg [15] in his proof that a positive proportion of the zeros are on the critical line. His paper also contains a density result

$$N(\sigma, T) \ll \frac{T}{(\sigma - \frac{1}{2})}$$

uniformly in $\sigma > \frac{1}{2}$, where, as usual, $N(\sigma, T)$ is the number of zeros $\rho = \beta + i\gamma$ of $\zeta(s)$ with $0 < \gamma < T$ and $\beta \geq \alpha$; he later [16] strengthened this result to

$$N(\sigma, T) \ll T^{1 - 1/4(\sigma - 1/2)} \log T$$

uniformly for $\frac{1}{2} \leq \sigma \leq 1$.

NUMBER THEORY,
TRACE FORMULAS and
DISCRETE GROUPS

ISBN 0-12-067570-6

In 1973, Levinson [11], relying on the sort of mean value theorem mentioned above but using a different starting point, showed that at least $\frac{1}{3}$ of the zeros of the zeta-function are on the critical line. Levinson called the Dirichlet polynomial he used in his argument a "mollifier" because, as a rough approximation to $1/\zeta(s)$, it succeeded in smoothing the wild behavior of $\zeta(s)$ near the critical line. Improvements in the lower bound for the proportion of zeros on the critical line have depended in part on better choices for the mollifier, which have been found through the use of the calculus of variations.

Further developments in this method of mollifying have yielded lower bounds for the proportion of zeros of $\xi^{(m)}(s)$ on the critical line where $\xi(s) = \frac{1}{2}s(s-1)\pi^{-s/2}\Gamma(s/2)$; $\xi(s)$ is entire, real on the critical line, and has the same complex zeros as $\zeta(s)$ does. Most notably, this proportion tends to 1 as m tends to infinity. Lower bounds can also be obtained for the proportion of zeros of $\xi^{(m)}(s)$ that are simple and on the critical line (see Conrey [2] and [3]).

Jutila [10] has used the method of Selberg to improve his density result. He showed that for any $\delta > 0$,

$$N(\sigma, T) \ll_{\delta} T^{1-(1-\delta)(\sigma-1/2)} \log T.$$

2

Here we state a mean value theorem from which many of the above-mentioned results can be deduced. We also give two new corollaries. Then an analogous theorem about a discrete mean value is given, along with some of its consequences. Before stating the theorem, let us introduce some notation. Let $T > 0$ be large and let

$$B(s, P) = \sum_{n \le y} \frac{b(n, P)}{n^s},$$

where $y = T^{\theta}$, and

$$b(n, P) = \mu(n) P\left(\frac{\log y/n}{\log y}\right),$$

where P is entire with $P(0) = 0$ and μ is the usual Möbius function. Let $L = \log T$, $\alpha = a/L$, $\beta = b/L$ for complex numbers a and b. Let Q_1 and Q_2 be polynomials.

Theorem 1. *If $0 < \theta < \frac{1}{2}$ and α and β tend to 0 as $T \to \infty$ then for fixed c with $\frac{1}{2} \le c < \frac{3}{2} - \theta$,*

$$\frac{1}{i}\int_{c+i}^{c+iT} Q_1\left(\frac{-d}{da}\right)\zeta(s+\alpha)Q_2\left(\frac{-d}{db}\right)\zeta(1-s+\beta)B(s, P_1)B(1-s, P_2)\,ds$$

$$\sim T\left[Q_1(0)Q_2(0)P_1(1)P_2(1) + \frac{\partial}{\partial u}\frac{\partial}{\partial v}\frac{1}{\theta}\int T_a Q_1 T_b Q_2 \int P_1 P_2\Big|_{u=v=0}\right],$$

where the integrals are $\int_0^1 \ldots dx$, $P_1 = P_1(x+u)$, $P_2 = P_2(x+v)$,

$$T_a Q_1 = e^{-a(x+\theta u)}Q_1(x+\theta u), \qquad \textit{and} \qquad T_b Q_2 = e^{-b(x+\theta v)}Q_2(x+\theta v).$$

For example,

$$\int_1^T \left|\zeta\left(\frac{1}{2}+it\right)\right|^2 \left|B\left(\frac{1}{2}+it, P\right)\right|^2 dt \sim T\left(P(1)^2 + \frac{1}{\theta}\int_0^1 P'(x)^2\,dx\right).$$

We note, also, that Levinson's theorem follows with the choices $P_1(x) = P_2(x) = x$, $Q_1(x) = Q_2(x) = -1 - x$, $c = \frac{1}{2}$, $a = -b = -1.3$, and $\theta = \frac{1}{2} - \varepsilon$, $\varepsilon \to 0^+$. Also, Jutila's result follows with the choices $P_1(x) = P_2(x) = x$, $Q_1(x) = Q_2(x) = 1$, $a = b > 0$, $c = \frac{1}{2}$, and $\theta = \frac{1}{2} - \varepsilon$, $\varepsilon \to 0^+$.

Two other applications are as follows. First, let $N^d(T)$ denote the number of distinct zeros of the zeta-function in $0 < t < T$; then

$$N^d(T) \ge (0.628 + o(1))N(T),$$

where $N(T) \sim TL/(2\pi)$. This may be proved as follows. Let $N_r(T)$ denote the number of zeros of the zeta-function in $0 < t < T$ with multiplicity at most r, where zeros are counted according to their multiplicity. Then it is easy to show that

$$N^d(T) \ge \sum_{r=1}^{R} \frac{N_r(T)}{r(r+1)} + \frac{N_{R+1}(T)}{R+1}$$

for any $R \ge 1$. The above-mentioned result on N^d follows from the constants in Conrey [3] that may be deduced from Theorem 1.

As a second application we mention some results on the distribution of zeros of $\zeta^{(k)}(s)$. This topic is of interest because of its connection with the Riemann Hypothesis (see Levinson [11], Levinson and Montgomery [13], and Speiser [17]). In particular, the Riemann Hypothesis is equivalent to the assertion that all complex zeros of $\zeta'(s)$ have real part at least $\frac{1}{2}$. Levinson's method is based on a quantitative version of this. We mention that Levinson and Montgomery [13] have shown that

$$\sum_{\gamma_k < T} (\beta_k - \tfrac{1}{2}) \sim \frac{kT}{2\pi} \log \log T,$$

where $\rho_k = \beta_k + i\gamma_k$ denotes a zero of $\zeta^{(k)}(s)$; the number of terms in the sum is $\sim TL(2\pi)$. On RH at most finitely many terms in the sum are negative. On the other hand, it can be shown that

$$\sum_{0 < \gamma_k < T} T^{1/2 - \beta_k} \gg T$$

so that, e.g., there exist $R > 0$ and $c > 0$ such that at least cT zeros satisfy $0 < \gamma_k < T$ and $\beta_k < \frac{1}{2} + R/L$. Using Theorem 1 we can show that for any $R > 0$ there is a $c > 0$ such that

$$\sum_{\substack{0 < \gamma_k < T \\ \beta_k > (1/2) + R/L}} 1 > cTL$$

for all large T. The question of the precise horizontal distribution of the zeros of $\zeta'(s)$ remains open to conjecture.

3

Now we turn to discrete mean value theorems. We use the same notation mentioned before Theorem 1.

Theorem 2. *If $0 < \theta < \frac{1}{2}$ and α and β tend to 0 as $T \to \infty$, then*

$$\sum_{0 < \gamma < T} Q_1\left(-\frac{d}{da}\right)\zeta(\rho + \alpha)Q_2\left(-\frac{d}{db}\right)\zeta(1 - \rho + \beta)B(\rho, P_1)B(1 - \rho, P_2)$$

$$\sim \frac{TL}{2\pi}\frac{\partial}{\partial u}\frac{\partial}{\partial v}\left\{\left[\frac{1}{\theta}\int P_1P_2 + \int P_1 \int P_2\right]\right.$$

$$\times \left[\int T_aQ_1T_bQ_2 - \int T_aQ_1 \int T_bQ_2\right]$$

$$\left. + \int P_1 \int P_2\left(Q_1(0) - \int T_aQ_1\right)\left(Q_2(0) - \int T_bQ_2\right)\right\}\Bigg|_{u=v=0},$$

with the same notation conventions as in Theorem 1.

Theorem 3. *If* $0 < \theta < \frac{1}{2}$ *and* α *tends to* 0 *as* $T \to \infty$, *then*

$$\sum_{0<\gamma<T} Q\left(\frac{-d}{da}\right)\zeta(\rho+\alpha)B(\rho,P) \sim \frac{-TL}{2\pi}\frac{d}{du}\left(\left(Q(0)-\int T_a Q\right)\int P\right)\Bigg|_{u=0},$$

where

$$P = P(x+u) \qquad \text{and} \qquad T_a Q = e^{-a(x+\theta u)}Q(x+\theta u).$$

As a first application of these theorems, we mention the results of Conrey, Ghosh, and Gonek [6] on $N^d(t)$ and on the number of simple zeros $N_1(T)$ in $0 < t < T$; assuming RH, Montgomery [14] proved that

$$N_1(T) \geq \left(\frac{2}{3} + o(1)\right)N(T)$$

using his pair correlation method. In [6] we show that on RH

$$N_1(T) \geq \left(\frac{19}{27} + o(1)\right)N(T);\ N^d(T) \geq \left(\frac{5}{6} + \frac{1}{81} + o(1)\right)N(T).$$

The first inequality is obtained via the Cauchy-Schwarz inequality in the form

$$N_1(T) \geq \frac{\left|\sum_{0<\gamma<T} \zeta'(\rho)B(\rho,P)\right|^2}{\sum_{0<\gamma<T} |\zeta'(\rho)B(\rho,P)|^2}.$$

Clearly, the right side can be evaluated by our theorems: The choice $P(x) = -\theta x^2 + (1+\theta)x$ is optimal (with $\theta \to 1/2^-$). The second inequality may be deduced from the first using Montgomery's theorem [14]

$$\text{(on RH)} \qquad \sum_{\gamma<T} m(\rho) \leq \left(\frac{4}{3} + o(1)\right)N(T),$$

where $m(\rho)$ denotes the multiplicity of the zero ρ.

A second application is to bound $N_2(T)$, the number of simple and double zeros in $0 < t < T$, from below. Again by Cauchy's theorem

$$N_2(T) \geq \frac{\left|\sum_{\gamma<T} (\zeta'(\rho)B(\rho,P_1) + \zeta''(\rho)B(\rho,P_2))\right|^2}{\sum_{\gamma<T} |\zeta'(\rho)B(\rho,P_1) + \zeta''(\rho)B(\rho,P_2)|^2}.$$

The right side may be evaluated using Theorems 2 and 3 and RH. Then using

$$P_1(x) = 0.866x - 0.115x^2 - 0.082x^3$$

and

$$P_2(x) = 0.761x - 0.362x^2 - 0.024x^3,$$

we obtain the new result:

Theorem 4. *Assuming the Riemann Hypothesis,*

$$N_2(T) \geq (0.955 + o(1))N(T).$$

Thus, on RH, fewer than 4.5 % of the zeros have multiplicity three or greater.

We may also use Theorems 2 and 3 to obtain some information about the number $N(T, U)$ of pairs of zeros of the Riemann zeta-function with imaginary parts γ, γ' between 0 and T for which $0 < \gamma' - \gamma \leq U$. Montgomery [14] has conjectured that

$$N(T, U) \sim N(T) \int_0^{UL} \left\{1 - \left(\frac{\sin \pi\alpha}{\pi\alpha}\right)^2\right\} d\alpha$$

uniformly for $0 < \alpha_0 \leq UL \leq \alpha_1 < \infty$, and Gallagher [9] has shown, on RH, that

$$N(T, U) \leq \left(A + \frac{1}{2\pi^2 A} + O(A^{-2})\right)N(T),$$

where $A = UL$ is a positive integer or half-integer, and that if, in addition, almost all the zeros are simple then

$$N(T, U) \geq \left(A - 1 + \frac{1}{2\pi^2 A} + O(A^{-2})\right)N(T).$$

Gallagher's results allow for the possibility that for some $U \approx 1/L$,

$$N(T, U^+) - N(T, U^-) > \left(1 - O\left(\frac{1}{A^2}\right)\right)N(T).$$

We can apply Theorems 2 and 3 above to deduce:

Theorem 5. *Assuming the Riemann Hypothesis,*

$$N(T, U^+) - N(T, U^-) \le \left(\frac{2}{3} + O\left(\frac{1}{A}\right)\right)N(T)$$

uniformly for $0 < \alpha_0 \le UL = A \le \alpha_1 < \infty$.

To prove this theorem we first of all note that on RH it is equivalent to the following assertion:

$$\sum_{\substack{0<\gamma<T \\ \zeta(\rho+iU)\neq 0}} 1 \ge \left(\frac{1}{3} + O\left(\frac{1}{A}\right)\right)N(T).$$

By the Cauchy Schwarz inequality, the left side is

$$\ge \left|\sum_{\gamma<T} \zeta(\rho+iU)B(\rho+iU, P)\right|^2 \left(\sum_{\gamma<T} |\zeta(\rho+iU)B(\rho+iU, P)|^2\right)^{-1},$$

which can be evaluated asymptotically via Theorems 2 and 3. (The fact that the argument of B is shifted presents no problem as

$$B(s + a/L, P) = B(s, P_1)$$

where $P_1(x) = e^{-ia\theta(1-x)}P(x)$.) Then, we find that if $P(1) = 1$, then

$$\sum_{\gamma<T} \zeta\left(\rho + \frac{iA}{L}\right)B\left(\rho + \frac{iA}{L}, P\right) \sim -N(T)\left\{1 - J(iA) + iAJ(iA)\int_0^1 e^{-iA\theta(1-x)}P(x)\,dx\right\}$$

and

$$\sum_{\gamma<T} \left|\zeta\left(\frac{\rho+iA}{L}\right)B\left(\rho + \frac{iA}{L}, P\right)\right|^2$$

$$\sim N(T)\left\{1 + (1 - |J(iA)|^2)\frac{1}{\theta}\int_0^1 P'(x)^2\,dx + \left|1 - iA\theta\int_0^1 e^{-iA\theta(1-x)}P(x)\,dx\right|^2 - 2\,\mathrm{Re}\left\{J(iA)\left(1 - iA\theta\int_0^1 e^{-iA\theta(1-x)}P(x)\,dx\right)\right\},\right.$$

where $J(r) = \int_0^1 e^{-rx}\,dx$. Now, since $P(0) = 0$ and $P(1) = 1$, we find by an integration by parts that

$$\int_0^1 e^{-iA\theta(1-x)}P(x)\,dx = \frac{1}{iA\theta} + O\left(\frac{1}{A^2}\right)$$

for fixed P. Also $J(iA) \ll 1/A$ so that

$$\sum_{\substack{\gamma < T \\ \zeta(\rho + iU) \neq 0}} 1 \geq N(T)\left(1 + O\left(\frac{1}{A}\right)\right)\bigg/\left(1 + \frac{1}{\theta}\int_0^1 P'(x)^2\,dx + O\left(\frac{1}{A}\right)\right).$$

The result now follows from the choices $P(x) = x$, $\theta \to 1/2^-$.

Finally, we mention two results of Conrey, Ghosh, and Gonek ([5] and [7]) that do not follow directly from Theorems 2 and 3 but are proven using similar techniques. Firstly, on GRH, a positive proportion of the zeros of the zeta-function of a quadratic number field are simple. (This result does not seem to be accessible via Montgomery's pair correlation method.) Secondly, on GRH, the gaps between consecutive zeros of the zeta-function are infinitely often larger than 2.68 times the average spacing.

4

We now give a description of the main steps in the proofs of Theorems 1 and 2. Many of the details are similar to the work done in [6]. Let $\chi(s)$ denote the usual factor from the functional equation for the zeta-function; $\zeta(s) = \chi(s)\zeta(1-s)$. Let c be a fixed number satisfying $1 < c < \frac{3}{2} - \theta$, and let

$$I(\alpha, \beta, P_1, P_2) = \frac{1}{i}\int_{c+i}^{c+iT} \chi(1-s)\zeta(s+\alpha)\zeta(s+\beta)B(s, P_1)B(1-s, P_2)\,ds$$

and

$$I_1(\alpha, \beta, P_1, P_2) = \frac{1}{2\pi i}\int_{c+i}^{c+iT} \chi(1-s) \times \frac{\zeta'}{\zeta}(s)\zeta(s+\alpha)\zeta(1-s+\beta)B(s, P_1)B(1-s, P_2)\,ds.$$

Let M and M_1 denote the means in question in Theorems 1 and 2 with $Q_1 = Q_2 \equiv 1$.

As a first step, we use the approximations

$$\chi(1-s+\alpha) = \left(\frac{t}{2\pi}\right)^{-\alpha} \chi((1-s)(1+O(1/|t|)) \sim e^{-a}\chi(1-s)$$

and

$$\frac{\chi'}{\chi}(s) \sim -L$$

for $t \approx T$. This gives

$$M \sim e^{-b} I(\alpha, -\beta, P_1, P_2)$$

and

$$M_1 \sim e^{-b} I_1(\alpha, -\beta, P_1, P_2) + e^{-a}\overline{I_1(-\bar{\alpha}, \bar{\beta}, \bar{P}_2, \bar{P}_1)}$$
$$+ \frac{e^{-a}L}{2\pi} \overline{I(-\bar{\alpha}, \bar{\beta}, \bar{P}_2, \bar{P}_1)}.$$

Next, in view of the relationship

$$\frac{1}{2\pi i}\int_{c+i}^{c+iT} \chi(1-s) r^{-s}\, ds \sim e(-r),$$

for $0 < r < T/2\pi$ we can show that

$$I(\alpha, \beta, P_1, P_2) \sim \sum_{h,k \le y} \frac{b(h, P_1) b(k, P_2)}{k} \sum_{mn \le (Tk/2\pi h)} m^{-\alpha} n^{-\beta} e\left(-mn\frac{H}{K}\right),$$

where $H = h/(h, k)$ and $K = k/(h, k)$, and that

$$I_1(\alpha, \beta, P_1, P_2) \sim -\sum_{h,k \le y} \frac{b(h, P_1) b(k, P_2)}{k}$$
$$\sum_{\ell mn \le (Tk/2\pi h)} \Lambda(\ell) m^{-\alpha} n^{-\beta} e\left(\frac{-\ell mnH}{K}\right).$$

Now we estimate the inner sums here using Perron's formula; this requires knowledge of the generating functions. If $(H, K) = 1$, then

$$\sum_{m,n} m^{-s-\alpha} n^{-s-\beta} e\left(\frac{-mnH}{K}\right) - K^{1-\alpha-\beta-2s}\zeta(s+\alpha)\zeta(s+\beta)$$

is an entire function. Also,

$$\sum_{\ell, m, n} \frac{\Lambda(\ell)}{\ell^s} m^{-s-\alpha} n^{-s-\beta} e\left(\frac{-\ell m n H}{K}\right)$$

$$-\frac{\zeta(s+\alpha)\zeta(s+\beta)}{K}\left(\frac{\zeta'}{\zeta}(s) + \sum_{p \mid K} p^{s+\alpha+\beta-1} \log p\right)$$

$$\dot{\times}\ (\mu * T_{1-s-\alpha} 1 * T_{1-s-\beta} 1)(K)$$

has poles that are either not $\ll 1/L$ from 1 or are a distance $\ll 1/L$ from 1 but have residues that are small when averaged over h and k. In this formula $*$ denotes Dirichlet convolution, and

$$T_r 1(n) = n^r$$

for any r and n. Thus

$$I \sim \frac{1}{i} \int_{(c)} \sum_{h, k \leq y} \frac{b(h, P_1) b(k, P_2)}{h^s k^s} (h, k)^{1-\alpha-\beta-2s} \left(\frac{T}{2\pi}\right)^s \zeta(s+\alpha)\zeta(s+\beta) \frac{ds}{s}$$

and

$$I_1 \sim \frac{1}{2\pi i} \int_{(c)} \sum_{h, k \leq y} \frac{b(h, P_1) b(k, P_2)}{h^s k^{2-s}} (h, k) \zeta(s+\alpha)\zeta(s+\beta)$$

$$\times \left(\frac{-\zeta'}{\zeta}(s) + \sum_{p \mid K} p^{s+\alpha+\beta-1} \log p\right)$$

$$\times\ (\mu * T_{1-s-\alpha} 1 * T_{1-s-\beta} 1)(K) \left(\frac{T}{2\pi}\right)^s \frac{ds}{s},$$

where (c) denotes the straight line path from $c - i\infty$ to $c + i\infty$. The main terms arise from the poles of the integrand; using $\zeta(s) \sim 1/(s-1)$ we find

$$I \sim TL \frac{(e^{-a} S(-\alpha, \beta, P_1, P_2) - e^{-b} S(-\beta, \alpha, P_1, P_2))}{b - a}$$

and

$$I_1 \sim \frac{-TL^2}{2\pi}\left(\frac{S_2(\alpha, \beta, P_1, P_2)}{ab}\right.$$

$$+ \frac{e^{-a}}{b-a}\left(\frac{S(-\alpha, \beta, P_1, P_2)}{-a} + \frac{1}{L} S_1(-\alpha, \beta, P_1, P_2)\right)\Bigg)$$

$$- \frac{e^{-b}}{b-a}\left(\frac{S(-\beta, \alpha, P_1, P_2)}{-b} + \frac{1}{L} S_1(-\beta, \alpha, P_1, P_2)\right),$$

where

$$S(\alpha, \beta, P_1, P_2) = \sum_{h,k \le y} \frac{b(h, P_1)b(k, P_2)}{h^{1+\alpha}k^{1+\beta}} (h, k)^{1+\alpha+\beta},$$

$$S_1(\alpha, \beta, P_1, P_2) = \sum_{h,k \le y} \frac{b(h, P_1)b(k, P_2)}{h^{1+\alpha}k^{1+\beta}} (h, k)^{1+\alpha+\beta} \sum_{p|K} p^{\beta} \log p,$$

and

$$S_2(\alpha, \beta, P_1, P_2) = \sum_{h,k \le y} \frac{b(h, P_1)b(k, P_2)}{hk} (h, k)(\mu * T_{-\alpha}1 * T_{-\beta}1)(K).$$

The error terms are estimated using large sieve techniques and a Vaughan type identity.

Now we are to the arithmetic part of the argument. We can show that

$$S(\alpha, \beta, P_1, P_2) \sim \frac{1}{\theta L} \frac{\partial}{\partial u} \frac{\partial}{\partial v} e^{a\theta u + b\theta v} \int_0^1 P_1(x+u)P_2(x+v)\, dx \bigg|_{u=v=0}$$

$$= \frac{1}{\theta L} \int_0^1 (P_1'(x) + a\theta P_1(x))(P_2'(x) + b\theta P_2(x))\, dx,$$

and that

$$S_1(\alpha, \beta, P_1, P_2) \sim -\theta L S(\alpha, \beta, P_1, P_2^{(-1)}),$$

where

$$P_2^{(-1)}(x) = \int_0^x P_2(t)\, dt;$$

also,

$$S_2(\alpha, \beta, P_1, P_2) \sim S(-\alpha, \beta, P_1, P_2) + \frac{a}{L} P_1(1)P_2(1) + \frac{ab\theta}{L} P_1(1)P_2^{(-1)}(1).$$

Moreover,

$$S(\alpha, \beta, P_1, P_2) = S(\beta, \alpha, P_2, P_1)$$

and

$$S_2(\alpha, \beta, P_1, P_2) = S_2(\beta, \alpha, P_1, P_2)$$

so that in view of the above relationship between S and S_2 it follows that

$$S(-\beta, \alpha, P_1, P_2) \sim S(-\alpha, \beta, P_1, P_2) - \frac{b-a}{L} P_1(1)P_2(1).$$

Thus, we are led to

$$I(\alpha, \beta, P_1, P_2) \sim TL\left(\frac{e^{-a} - e^{-b}}{b-a} S(-\alpha, \beta, P_1, P_2) + \frac{e^{-b}}{L} P_1(1)P_2(1)\right)$$
$$\sim TL\left(\frac{e^{-a} - e^{-b}}{b-a} S(-\beta, \alpha, P_1, P_2) + \frac{e^{-a}}{L} P_1(1)P_2(1)\right)$$

and

$$I_1(\alpha, \beta, P_1, P_2) \sim \frac{-TL^2}{2\pi}\left(S(-\alpha, \beta, P_1, P_2)\left(\frac{1}{ab} - \frac{e^{-a}}{a(b-a)} + \frac{e^{-b}}{b(b-a)}\right)\right.$$
$$+ \theta S(-\alpha, \beta, P_1, P_2^{(-1)}) \frac{e^{-b} - e^{-a}}{b-a}$$
$$+ \frac{P_1(1)P_2(1)}{L}\left(\frac{1-e^{-b}}{b}\right)$$
$$\left. + \frac{\theta P_1(1)P_2^{(-1)}(1)}{L}(1-e^{-b})\right).$$

Then

$$M \sim T\left(P_1(1)P_2(1) + \frac{\partial}{\partial u}\frac{\partial}{\partial v} e^{-a\theta u - b\theta v}\frac{1}{\theta}\int_0^1 P_1(x+u)P_2(x+v)\,dx\Big|_{u=v=0}\right.$$

follows, as well as

$$M_1 \sim \frac{TL^2}{2\pi}\left\{S(-\alpha, -\beta, P_1, P_2)\left(\frac{1-e^{-a-b}}{a+b} + \frac{e^{-a}}{ab} - \frac{e^{-a-b}}{b(a+b)} - \frac{1}{a(b+a)}\right.\right.$$
$$\left. + \frac{e^{-b}}{ab} - \frac{e^{-a-b}}{a(b+a)} - \frac{1}{b(b+a)}\right)$$
$$+ \theta S(-\alpha, -\beta, P_1, P_2^{(-1)})\left(\frac{1-e^{-b-a}}{a+b}\right)$$
$$+ \theta S(-\alpha, -\beta, P_1^{(-1)}, P_2)\left(\frac{1-e^{-b-a}}{a+b}\right)$$
$$+ \frac{P_1(1)P_2(1)}{L}\left(1 - \frac{1-e^{-a}}{a} - \frac{1-e^{-b}}{b}\right)$$
$$\left. + \frac{\theta}{L}(P_1(1)P_2^{(-1)}(1)(1-e^{-b}) + P_1^{(-1)}(1)P_2(1)(1-e^{-a}))\right\}$$

$$= \frac{TL^2}{2\pi}\Bigg\{S(-\alpha, -\beta, P_1, P_2)(J(a+b) - J(a)J(b))$$

$$+ \theta J(a+b)(S(-\alpha, -\beta, P_1, P_2^{(-1)}) + S(-\alpha, \beta, P_1^{(-1)}, P_2))$$

$$+ \frac{P_1(1)P_2(1)}{L}(1 - J(a) - J(b)) + \frac{\theta}{L}(P_1(1)P_2^{(-1)}(1)bJ(b)$$

$$+ P_1^{(-1)}(1)P_2(1)aJ(a))\Bigg\},$$

where $J(r) = \int_0^1 e^{-rx}\,dx$. Now let

$$f(u, v) = \int_0^1 (P_1^{(-1)}(x+u)P_2(x+u) + P_1(x+u)P_2^{(-1)}(x+v))\,dx$$

$$= P_1^{(-1)}(x+u)P_2^{(-1)}(x+v)\Big|_0^1$$

$$= P_1^{(-1)}(1+u)P_2^{(-1)}(1+v) - P_1^{(-1)}(u)P_2^{(-1)}(v)$$

and let

$$g(u, v) = \int_0^1 P_1(x+u)\,dx \int_0^1 P_2(x+v)\,dx.$$

Then $f(0,0) = g(0,0)$, $f_u(0,0) = g_u(0,0)$, $f_v(0,0) = g_v(0,0)$ and $f_{uv}(0,0) = g_{uv}(0,0)$. We will use this to replace $f(u, v)$ by $g(u, v)$ in the formula for M_1. By our earlier formula for S we now have

$$M_1 \sim \frac{TL}{2\pi}\frac{\partial}{\partial u}\frac{\partial}{\partial v}\Bigg\{e^{-a\theta u - b\theta v}\left[\frac{1}{\theta}\int P_1P_2 + \int P_1 \int P_2\right][J(a+b) - J(a)J(b)]$$

$$+ \int P_1 \int P_2(1 - e^{-a\theta u}J(a))(1 - e^{-b\theta v}J(b))\Bigg\}\Bigg|_{u=v=0},$$

where the integrals are $\int_0^1 \dots dx$ and $P_1 = P_1(x+u)$, $P_2 = P_2(x+v)$.

Finally, our formulas are uniform in α and β and may be differentiated with respect to these variables (using Cauchy's formula for example). Since $Q(-d/da)e^{-ay} = Q(y)e^{-ay}$, Theorems 1 and 2 follow.

5. Concluding Remarks

In conclusion, we mention possible directions for further development that this work may suggest. The most obvious possibility regards the range of θ in Theorem 1. For the special choice $Q_1 = Q_2 \equiv 1$, $a = b = 0$, the range $(0, \frac{9}{17})$ is admissible for θ, as shown by the work of Balasubramanian, Conrey, and Heath-Brown [1]. There is no reason why this shouldn't work for arbitrary Q_1, Q_2, a, and b. More significant, however, is the work of Iwaniec and Deshouillers [8], which suggests that the range $(0, \frac{4}{7})$ should be admissible for θ.

Next is the question of optimal choices of functions P and Q for various applications. In general, if our application requires that $P_1 = P_2$ and $Q_1 = Q_2$ with P_1 (resp. Q_1) specified, then the optimal choice of Q_1 (resp. P_1) can be determined through the calculus of variations in a straightforward way. This is the situation with the result on simple zeros. However, in the case of the lower bound for N_r with $r \geq 2$ and in the case of the lower bound for the proportion of zeros of ζ on the critical line, the optimal choices have not been completely determined. (See Conrey [4] for a description of what choices of P and Q are admissible in the latter problem).

Finally, it may well be that there are direct applications of these theorems to the questions of how large and how small the gaps between consecutive zeros of the zeta-function can be.

References

[1] Balasubramanian, R., Conrey, B. and Heath-Brown, D.R. "Asymptotic mean square of the product of the Riemann zeta function and a Dirichlet polynomial," *J. Reine Angew. Math.* **357** (1985), 161–181.

[2] Conrey, J.B. "Zeros of derivatives of Riemann's xi-function on the critical line," *J. Number Theory*, **16** (1983), 49–74.

[3] Conrey, J.B. "Zeros of derivatives of Riemann's xi-function on the critical line II," *J. Number Theory*, **17** (1983), 71–75.

[4] Conrey, J.B. "On the distribution of the zeros of the Riemann zeta function." In: *Topics in Analytic Number Theory*, University of Texas Press, Austin (1985), 28–41.

[5] Conrey, J.B. Ghosh, A. and Gonek, S.M. "Large gaps between zeros of the zeta function," *Mathematika* **33** (1986), 212–238.

[6] Conrey, J.B. Ghosh, A. and Gonek, S.M. "Simple zeros of the Riemann zeta function," preprint.

[7] Conrey, J.B. Ghosh, A. and Gonek, S.M. "Simple zeros of the zeta function of a quadratic number field. II. "In Analytic Number Theory and Diophantine Problems. *Proceedings of a Conference at Oklahoma State University*, Birkhaüser, Basel, Stuttgart, (1987).

[8] Deshouillers, J.M. and Iwaniec, H. "Power mean-values for Dirichlet's polynomials and the Riemann zeta function. II." *Acta Arith.* **48** (1984), 305–312.

[9] Gallagher, P.X. "Pair correlation of zeros of the zeta function," *J. Reine Angew. Math.* **362** (1985), 72–86.

[10] Jutila, M. "Zeros of the zeta function near the critical line." In: *Studies in Pure Mathematics, to the Memory of Paul Turán*, Birkhaüser, Basel, Stuttgart, (1982), 385–394.

[11] Levinson, N. "More than one third of zeros of Riemann's zeta function are on $\sigma = \frac{1}{2}$," *Adv. Math.* **13** (1974), 383–436.

[12] Levinson, N. "Zeros of derivative of Riemann's ξ-function," *Bull. Amer. Math. Soc.* **80** (1974), 951–954.

[13] Levinson, N. and Montgomery, H.L. "Zeros of the derivatives of the Riemann zeta function," *Acta Math.* **133** (1974), 49–65.

[14] Montgomery, H.L. "The pair correlation of zeros of the zeta-function," *Analytic Number Theory, Proc. Symp. Pure Math.* **24** (1973), 181–193, Amer. Math. Soc., Providence, RI.

[15] Selberg, A. "On the zeros of Riemann's zeta-function," *Skr. Norske Vid. Akad. Oslo* No. 10, (1942).

[16] Selberg, A. "Contributions to the theory of the Riemann zeta-function," *Arch. for Math. og Naturv.* **B, 48** (5) (1946).

[17] Speiser, A. "Geometrisches zur Riemannschen zetafunktion," *Math. Ann.* **110** (1934), 514–521.

Note added in proof.

We are unable, at present, to estimate some of the error terms in Theorem 2 without assuming the Generalized Lindelöf Hypothesis. Consequently, the applications of Theorem 2 mentioned here depend on RH and GLH.

10 Geometric Ramanujan Conjecture and Drinfeld Reciprocity Law*

YUVAL Z. FLICKER AND
DAVID A. KAZHDAN

The purpose of this article is to describe and explain some of our recent work, which concerns, in particular, the following themes:

- The Ramanujan or purity conjecture for cuspidal automorphic forms π with a super cuspidal component of $\mathrm{GL}(r)$ over a global field F of characteristic $p > 0$
- The reciprocity law relating the above π with irreducible continuous r-dimensional $\ell(\neq p)$-adic representations ρ of the galois group $\mathrm{Gal}(\bar{F}/F)$ whose restriction to some decomposition group $\mathrm{Gal}(\bar{F}_v/F_v)$ is irreducible
- Drinfeld's explicit reciprocity law, which realizes the above conjectured correspondence $\pi \leftrightarrow \rho$ as the irreducible factors $\pi \otimes \rho$ in the composition series as a $G(\mathbf{A}) \times \mathrm{Gal}(\bar{F}/F)$-module of ℓ-adic cohomology with compact support and coefficients in a smooth sheaf of the geometric generic fiber of the Drinfeld Moduli scheme

* Based on a lecture at the Selberg Symposium, Oslo, June 1987. Partially supported by NSF grants, and also a Seed grant (first author).

NUMBER THEORY,
TRACE FORMULAS and
DISCRETE GROUPS

ISBN 0-12-067570-6

- Deligne's conjecture on the Lefschetz fixed point formula in étale topology for a finite flat correspondence, multiplied by a sufficiently high power of the Frobenius, on a separated scheme of finite type over a finite field
- Higher congruence relations

Our proof of the purity conjecture and our deduction of the Drinfeld explicit reciprocity law from Deligne's conjecture are based, in particular, on a new form of the Selberg trace formula for a test function with at least one supercusp component. This new trace formula is a representation theoretic analogue of Deligne's conjecture; some of its applications to lifting problems are studied in [FK], [Fi] ($i = 2, 3, 4$). It is our pleasure to dedicate this paper to Atle Selberg in appreciation of his work in general and his trace formula in particular.

Let F be a geometric global field, namely the field of rational functions on a smooth projective absolutely irreducible curve over a finite field of characteristic p. Let $\mathbf{A}$ be the ring of F-adèles, $G = \mathrm{GL}(r)$, and π an irreducible admissible $G(\mathbf{A})$-module over the field of complex numbers. Then π is the restricted direct product $\otimes_v \pi_v$ over all places v of F of irreducible admissible $G_v = G(F_v)$-modules π_v. Let R_v be the ring of integers in the completion F_v of F at v, and put $K_v = G(R_v)$. For almost all v the component π_v is unramified, namely, has a K_v-fixed nonzero vector, and consequently there are nonzero complex numbers $z_{1,v}, \ldots, z_{r,v}$, uniquely determined up to permutation by π_v, with the following property: π_v is the unique irreducible unramified constituent $\pi((z_{i,v}))$ of the unramified G_v-module $I(\mathbf{z}_v) = \mathrm{Ind}(\delta^{1/2}\mathbf{z}_v; B_v, G_v)$, which is normalizedly induced from the unramified character $\mathbf{z}_v\colon (b_{ij}) \to \prod_i z_{i,v}^{\mathrm{val}_v(b_{ii})}$ of the upper triangular subgroup B_v of G_v; val_v is the order valuation of F_v, normalized by $\mathrm{val}_v(\boldsymbol{\pi}_v) = 1$ where $\boldsymbol{\pi}_v$ is any generator of the maximal ideal in R_v.

Definition. The elements $z_{1,v}, \ldots, z_{r,v}$ are called the *Hecke eigenvalues* of π at v.

In this work we are concerned with cuspidal $G(\mathbf{A})$-modules π. The space $L_0(G, C)$ of complex-valued cusp forms on $G(\mathbf{A})$ consists of all functions ϕ on $G(F)\backslash G(\mathbf{A})$ which are compactly supported modulo the center $Z(\mathbf{A})$ of $G(\mathbf{A})$ and transform under $Z(\mathbf{A})$ according to a unitary character, with the property that for every proper F-parabolic subgroup P of G, whose unipotent radical is denoted by N, we have

$\int_{N(F)\backslash N(A)} \phi(nx)\, dn = 0$ for all x in $G(\mathbf{A})$. A *cuspidal* $G(\mathbf{A})$-module is an admissible irreducible $G(\mathbf{A})$-module π which occurs as a direct summand in the representation of $G(\mathbf{A})$ on $L_0(G, C)$ by right translation. These cuspidal π are representation theoretic analogues, for function fields, of the (spaces spanned by the translates of) holomorphic cusp forms on the upper-half complex plane.

Our proofs involve ℓ-adic techniques. We can obtain results concerning modules over the complex numbers, due to a rationality property of the cuspidal π, analogous to that which holds for holomorphic cusp forms. Let $\overline{Q}$ be the field of algebraic numbers.

Proposition. *Let π be a cuspidal $G(\mathbf{A})$-module with algebraically valued central character ω: $Z(\mathbf{A})/Z(F) = \mathbf{A}^\times/F^\times \to \overline{Q}^\times$. Then* (1) *$\pi$ can be realized in the space $L_0(G, \overline{Q})$ of $\overline{Q}$-valued cusp forms, and* (2) *all Hecke eigenvalues $z_{i,v}$ of π are algebraic.*

In fact we obtain more precise results. Namely, for every cuspidal π with central character of finite order, the galois closure $Q(\pi)$ of the field generated by all Hecke eigenvalues of π at the unramified places is a finite extension of Q. We call $Q(\pi)$ the *field of definition* of π. Moreover, this π can be realized in the space $L_0(G, Q(\pi))$ of $Q(\pi)$-valued cusp forms on G.

Let $L^2(G, \mathbf{C})$ be the space of complex valued functions ϕ on $G(F)\backslash G(\mathbf{A})$ which transform under $Z(\mathbf{A})$ by a unitary character, such that $|\phi|^2$ is integrable on $Z(\mathbf{A})G(F)\backslash G(\mathbf{A})$. An admissible irreducible $G(\mathbf{A})$-module π is called *discrete-series* if it occurs as a direct summand of the representation of $G(\mathbf{A})$ on $L^2(G, \mathbf{C})$ by right translation. Every cuspidal $G(\mathbf{A})$-module is discrete series. Let q_v be the cardinality of the residue field $R_v/(\pi_v)$, and $|\,.\,|_{\mathbf{C}}$ is the absolute value on $\mathbf{C}$, normalized by $|a|_{\mathbf{C}} = a$ for every positive real number a. Our first theme is the following

Integrality Conjecture. *The absolute value of each Hecke eigenvalue of any unramified component of a discrete series $G(\mathbf{A})$-module π is equal to an integral power $q_v^{i/2}$ of $q_v^{1/2}$ with $|i| < n$.*

This conjecture has an obvious extension to all components of π in terms of their central exponents. It can be made also for number fields, and other reductive groups.

Recall that an irreducible G_v-module π_v is *supercuspidal* if it has a nonzero coefficient ϕ_v which satisfies $\int_{N_v} \phi_v(xny)\, dn = 0$ for all x, y in

G_v, and every proper parabolic subgroup P_v of G_v, whose unipotent radical is denoted by N_v. Let ∞ be a place of F. We prove the Integrality Conjecture for almost all components of any cuspidal π which has a supercuspidal component π_∞. Now each component of a cuspidal π is unitary and nondegenerate. A well-known estimate asserts that if π_v is an irreducible nondegenerate unramified unitary G_v-module over $\mathbf{C}$, then $q_v^{-1/2} < |z_{i,v}|_{\mathbf{C}} < q_v^{1/2}$ for every Hecke eigenvalue $z_{i,v}$ of π_v. As noted by Laumon, combining these two estimates we obtain the following

Purity Theorem. *Let π be a cuspidal $G(\mathbf{A})$-module with a supercuspidal component π_∞. Then each conjugate of each Hecke eigenvalue $z_i(\pi_v)$, for almost all unramified components π_v of π, lies on the unit circle in* $\mathbf{C}$.

The proof of this result is algebro-geometric. It is patterned along lines suggested by the work of Langlands [Ls] and Drinfeld [D2]. It relies on a comparison of the Grothendieck form [G] of the Lefschetz fixed-point formula for powers of the Frobenius, with a new form of the Selberg trace formula. The Integrality Conjecture is analogous to Deligne's theorem [De 2] on the eigenvalues of the action of the Frobenius on ℓ-adic cohomology groups, which plays a key role in our proof. The Purity Theorem is a representation theoretic analogue for GL(r) over a function field of Ramanujan's well-known conjecture concerning the Hecke eigenvalues (or rather Fourier coefficients) of the cusp form $\Delta(z) = e^{2\pi i z}\prod_1^\infty(1 - e^{2\pi i z n})$ of weight 12 on the upper half-plane $\mathrm{Im}(z) > 0$ for the group $\mathrm{SL}(2, \mathbf{Z})$. Our methods are likely to extend and prove the Integrality Conjecture for any discrete-series π with an elliptic component π_∞ on using a stronger form of the trace formula, but this has not been done as yet.

The Purity Theorem is one of our two main absolute results. The other concerns a higher rank analogue of the Eichler-Shimura and Ihara congruence relations, and not the trace formula. These are relations between certain Frobenius eigenvalues and the Hecke eigenvalues. These relations imply strong purity results for these Frobenius eigenvalues. We delay stating this result until the proof of the Purity Theorem is explained, as it concerns objects which we anyway introduce below for the proof of the Purity Theorem. We next explain our main relative result, which concerns the reduction of the reciprocity law, and Drinfeld's explicit form of it, to Deligne's conjecture. The proof is similar to that of the Purity Theorem and relies on the trace formula.

The reciprocity law concerns continuous r-dimensional irreducible ℓ-adic ($\ell \neq p$) representations $\rho: W(\overline{F}/F) \to \mathrm{GL}(r, \overline{Q}_\ell)$ of the Weil group of F. We say that ρ is *constructible* if it is unramified for almost all v. Here $\overline{F}$ denotes a separable closure of F. Note that it is not yet known whether every continuous irreducible finite-dimensional representation of $W(\overline{F}/F)$ is necessarily constructible. From now on we deal only with constructible ρ. For constructible ρ, for almost all v the restriction ρ_v of ρ to the decomposition subgroup $W(\overline{F}_v/F_v)$ at v factorizes through the Weil group $W(\overline{\mathbf{F}}_v/\mathbf{F}_v) \simeq \mathbf{Z}$ of the field $\mathbf{F}_v$ of constants of F_v, and the isomorphism class of ρ_v is determined uniquely by the unordered r-tuple $\{u_{i,v} = u_i(\rho_v); 1 \leq i \leq r\}$ of eigenvalues of the Frobenius automorphism $\rho_v(\mathrm{Fr}_v)$. Here $\mathrm{Fr}_v: x \to x^{q_v}$, where $q_v = |\mathbf{F}_v|$, is a generator of the subgroup $W(\overline{F}_v/F_v)$ of $\mathrm{Gal}(\overline{\mathbf{F}}_v/\mathbf{F}_v) \simeq \hat{\mathbf{Z}} = \lim \mathbf{Z}/n\mathbf{Z}$. It is useful to note that in our case of function fields we have the following

Proposition. *Let ρ be a continuous irreducible finite dimensional ℓ-adic representation of $W(\overline{F}/F)$ whose determinant* det ρ *has finite order. Then ρ extends by continuity to a representation of the galois group* $\mathrm{Gal}(\overline{F}/F)$.

Definition. We say that a constructible ρ and the $G(\mathbf{A})$-module $\pi = \otimes \pi_v$ *correspond* if for almost all v the unordered r-tuples $\{u_i(\rho_v)\}$ and $\{z_i(\pi_v)\}$ are equal.

We state below Deligne's conjecture and indicate by a superscript * any statement which depends on it. The first is the following higher

Reciprocity Law*. *Fix a place ∞ of F and a rational prime $\ell \neq p$. The correspondence defines a bijection between the sets of equivalence classes of* (1) *cuspidal G-modules π whose component π_∞ at ∞ is supercuspidal, and* (2) *irreducible r-dimensional continuous ℓ-adic constructible representations ρ of $W(\overline{F}/F)$ whose restriction ρ_∞ to $W(\overline{F}_\infty/F_\infty)$ is irreducible. The determinant* det ρ *of ρ corresponds by class-field theory to the central character of π. In particular, the π of* (1) *whose central characters are of finite order correspond to the ρ of* (2) *whose determinants* det ρ *are of finite order.*

In particular, for each constructible ρ whose determinant has finite order there is a finite galois extension $Q(\rho)$ of Q which contains the Frobenius eigenvalues $u_i(\rho_v)(1 \leq i \leq r)$ for almost all v. Moreover, each constructible ρ belongs to a compatible system of ℓ-adic representations, in the following sense. For each prime $\ell' \neq \ell$, p there exists a continuous irreducible constructible r-dimensional ℓ'-adic representation ρ' of $W(\bar{F}/F)$ which is irreducible at ∞ such that the unordered r-tuples $\{u_i(\rho_v)\}$ and $\{u_i(\rho'_v)\}$ are equal for all v where ρ_v and ρ'_v are unramified.

When $r = 1$ the Reciprocity Law above reduces to global class-field theory for function fields. For $r = 2$ this is a theorem of Drinfeld [D1], [D2], and our work is merely a higher rank extension of Drinfeld's amazingly original work.

It is clear that at most one cuspidal π can correspond to a given ρ, by virtue of the rigidity theorem for cusp forms of $\mathrm{GL}(r)$, and at most one ℓ-adic ρ can correspond to a given π. Indeed, by the Chebotarev Density Theorem if K is a galois extension of F which is unramified outside a finite set, then the Frobenius elements of the unramified places of K are dense in $\mathrm{Gal}(K/F)$. The difficulty is in proving the *existence* of a π corresponding to ρ, and ρ to π.

The first existence assertion (given π there is ρ) is reduced to Deligne's conjecture using the trace formula by the same proof which establishes the Purity Theorem. It is the following

Existence Theorem 1*. *For every cuspidal G-module π whose component π_∞ is supercuspidal and its central character is of finite order, and for any rational prime $\ell \neq p$, there exists a corresponding irreducible r-dimensional ℓ-adic continuous representation ρ of* $\mathrm{Gal}(\bar{F}/F)$ *(which is necessarily constructible).*

Before explaining Deligne's conjecture and the proof* of the Existence Theorem, we proceed to discuss the reduction of the Reciprocity Law* to the Existence Theorem. This reduction is done by induction on r, using the theory of L-functions. Let $\rho\colon \mathrm{Gal}(\bar{F}/F) \to \mathrm{Aut}_{\bar{Q}_\ell} V$ be an r-dimensional constructible ℓ-adic representation, and $L(t, \rho) = \prod_v L(t, \rho_v)$ the Euler product attached to ρ by Grothendieck. Here

$$L(t, \rho_v) = \det[(1 - \rho(\mathrm{Fr}_v)t^{\log_p q_v})|\, V]^{-1} \in Q_\ell[[t]],$$

for almost all v. Then $L(t, \rho)$ is a rational function in t which satisfies the functional equation

$$L(p/t, \check{\rho}) = \varepsilon(t, \rho)L(t, \rho),$$

where $\check{\rho}$ is the dual of ρ, and $\varepsilon(t, \rho)$ is a monomial in t depending on ρ. We use a result of Laumon [L], conjectured by Deligne, that when ρ is constructible and has virtual degree zero, then $\varepsilon(t, \rho)$ is equal to the product $\Pi_v\, \varepsilon(t, \rho_v, \psi_v)$ over all v of the local constants $\varepsilon(t, \rho_v, \psi_v)$ of [De1].

Using this product formula for $\varepsilon(t, \rho)$ we apply a variant of the Piatetski–Shapiro converse theorem [PS]; assuming the validity of the Reciprocity Law by induction for $r - 1$, we prove* the following converse direction of the Existence Theorem 1* (given ρ there is π):

Existence Theorem 2*. *Given an irreducible continuous ℓ-adic r-dimensional constructible representation ρ of $W(\bar{F}/F)$, there exists an automorphic G-module π which corresponds to ρ.*

The variant of the Converse Theorem which we use is the following. Put $G'' = \mathrm{GL}(r - 1)$.

Converse Theorem. *Fix a place ∞ of F and a supercuspidal G''_∞-module π^0_∞. Suppose that $\pi = \otimes \pi_v$ is an admissible nondegenerate (i.e., a constituent of the space of Whittaker functions) $G(\mathbf{A})$-module such that the standard Euler product $L(s, \pi, \tau)$ attached to the pair (π, τ) is entire (namely, it is a polynomial in p^s and p^{-s}) and satisfies the usual functional equation for all cuspidal G''-modules τ whose component τ_∞ at ∞ is the fixed supercuspidal τ^0_∞. Then there is a constituent of τ which is automorphic.*

The proof of this form of the Converse Theorem, which assumes the functional equation of $L(s, \pi, \tau)$ only for τ with the component τ^0_∞, is similar to that of Piatetski–Shapiro [PS]. We need this form since our induction assumption can be made only for the τ provided by the Existence Theorem 1*, namely, for those with a supercuspidal component τ^0_∞.

Arguing along lines suggested by [De1], we use the Existence Theorems*, the functional equations of the L-functions, and properties of variation of local L and ε factors under twists by highly ramified characters to establish* the existence of a local correspondence $\pi_v \leftrightarrow \rho_v$ with the property that the global π and ρ correspond if and only if π_v and ρ_v correspond for all places v of F. More precisely, we prove the

Local Reciprocity Law*. *For every local field F_v of positive characteristic and $r \geq 1$ there is a unique bijection $\pi_v \leftrightarrow \rho_v$ between the sets of*

equivalence classes of (1) *irreducible supercuspidal G_v-modules π_v, and* (2) *continuous ℓ-adic r-dimensional irreducible representations ρ_v of $W(\bar{F}_v/F_v)$, which reduces to local class-field theory for $r = 1$, with the following properties*: (*A*) *If π_v corresponds to ρ_v then* (1) *$\pi_v \otimes \chi_v$ corresponds to $\rho_v \otimes \chi_v$ for every character χ_v of $F_v^\times \simeq \overline{W}(F_v/F_v)^{ab}$*; (2) *the central character of π_v corresponds to* det *ρ_v by local class-field theory*; (3) *the contragredient of π_v corresponds to the contragredient of ρ_v.* (*B*) *If the* GL(*n*, F_v)*-module π_n corresponds to ρ_n, and the* GL(*m*, F_v)*-module π_m corresponds to ρ_m, then*

$$L(s; \pi_m, \pi_n) = L(p^{-s}; \rho_m \otimes \rho_n), \qquad \varepsilon(s; \pi_m, \pi_n) = \varepsilon(p^{-s}; \rho_m \otimes \rho_n).$$

where $L(s; \pi_m, \pi_n)$ is the L-function of [JPS].

By [GK] or the converse theorem, the supercuspidal π_n is uniquely determined by the family $L(s; \pi_n, \pi_{n-1})$ for all π_{n-1}. By virtue of the work of Bernstein and Zelevinsky [Z], there is a unique natural extension of this correspondence to relate the sets of equivalence classes of all (1) irreducible G_v-modules π_v, and (2) continuous ℓ-adic r-dimensional representations ρ_v of $W(\bar{F}_v/F_v)$, which satisfies (A), commutes with induction, and bijects square integrable π_v with indecomposable ρ_v. Moreover, π corresponds to ρ if and only if π_v corresponds to ρ_v for almost all v.

The Local Reciprocity Law* is stated for every $\ell \neq p$, and for every local field F_v of characteristic p. However, it is possible that the translation principle of [K2] and [De3] can be used to deduce the validity of the Local Reciprocity Law* also for F_v of characteristic zero and residual characteristic p, for any rational prime $\ell \neq p$ (cf. Henniart (in preparation)).

Our study of the Purity Theorem and the Existence Theorem 1* is based on Drinfeld's ([D1], [D2]) explicit construction of a $\mathrm{Gal}(\bar{F}/F) \times G(\mathbf{A}_f)$-module H. Here $\mathbf{A}_f$ is the ring of finite adèles, namely the adèles without component at ∞. Drinfeld's Explicit Reciprocity Law would conjecture that the irreducible constituents $\tilde{\rho} \times \tilde{\pi}_f$ of H realize the Reciprocity Law*. In [D1], [D2] Drinfeld introduces the notion of elliptic modules and their level structures, and constructs a moduli scheme M of isomorphism classes of such pairs (of rank r). When $r = 2$ the moduli scheme is a curve, and Drinfeld proves the Purity Theorem and Reciprocity Law, in particular, for cuspidal GL(2)-modules π with a supercuspidal component π_∞ on studying $H = H_c^1(\bar{M})$. Here $H_c^1(\bar{M})$ is the first ℓ-adic cohomology group with compact support and coefficients in a smooth sheaf defined by π_∞, of the geometric generic fiber $\overline{M} = M \times_A \overline{F}$ of the moduli curve M. In the higher dimensional case we work instead with the virtual representation $H^+ - H^-$, where $H^+ = \oplus_i H_c^i(\overline{M})(r-1-i$ is even) and $H^- = \oplus_i H_c^i(\overline{M})(r-1-i$ is odd), with coefficients in a sheaf determined by the supercuspidal π_∞. We conjecture (see below) that $H_c^i(M) = 0$ for $i \neq r-1$. Had this been proven we would take $H = H_c^{r-1}(\overline{M})$.

We use below the following definitions and notations. Let A be the ring of functions in F which are regular outside ∞, namely, the elements of F which are integral outside ∞. Let I be a nonzero ideal in A. Let $\hat{A}$ be the profinite completion of A, and $U_I = \mathrm{GL}(r, \hat{A}) \cap [1 + M(r, I\hat{A})]$ the congruence subgroup mod I of $\mathrm{GL}(r, \hat{A})$. Fix the Haar measure on $G(\mathbf{A}_f)$ which assigns U_I the volume one. Let $\mathbf{H}_I$ be the convolution algebra of compactly supported U_I-biinvariant functions on $G(\mathbf{A}_f)$. It is naturally isomorphic to the algebra (under product in $G(\mathbf{A}_f)$) spanned by the double cosets $U_I g U_I$, g in $G(\mathbf{A}_f)$. If π_f is an admissible $G(\mathbf{A}_f)$-module, let π_f^I denote the space of U_I-fixed vectors in π_f. Then π_f^I is an $\mathbf{H}_I$-module, and $\pi_f \to \pi_f^I$ is a bijection from the set of equivalence class of irreducible $G(\mathbf{A}_f)$-modules generated by their U_I-fixed vectors to the set of equivalence classes of irreducible $\mathbf{H}_I$-modules.

In [D1] it is shown that there exists an affine scheme $M_{r,I}$ of finite type over A, parametrizing the set of isomorphism classes of elliptic modules of rank r with structure of level I. It is affine, smooth but not proper. The adèle group $G(\mathbf{A}_f)$ acts on the scheme $M_r = \varprojlim M_{r,I}$. We have that $U_I \backslash M_r$ is equal to $M_{r,I}$.

The scheme which plays a key role in the work is a covering scheme of $M_{r,I}$, introduced in [D2]. Let D_∞ be a division algebra of dimension r^2 central over F_∞, with invariant $1/r$. Let π be a uniformizer in F_∞. Let U_∞ be a congruence subgroup in the multiplicative group $D_\infty^\times$. Then U_∞ is normal, compact, and open in $D_\infty^\times$, and has finite index in $D_\infty^\times/\langle\pi\rangle$. There exists a finite étale galois covering $\tilde{M}_{r,U_\infty}$ of M_r, and $\tilde{M}_{r,I,U_\infty}$ of $M_{r,I}$, with galois group $D_\infty^\times/U_\infty\langle\pi\rangle$, and the direct product $[(U_\infty\backslash D_\infty^\times) \times \mathrm{GL}(r, \mathbf{A}_f)]/F^\times$ acts on $\tilde{M}_{r,U_\infty}$.

Let $\bar{\rho}$ be an irreducible representation of $D_\infty^\times/U_\infty\langle\pi\rangle$, and $\ell \neq p$ a rational prime. Then there is a smooth $\overline{Q}_\ell$-adic sheaf $\mathbf{L} = \mathbf{L}(\overline{\rho})$ on $X = M_r$ associated with $\bar{\rho}$, and one defines the ℓ-adic cohomology groups $H_c^i(\overline{X}, \mathbf{L})$ of the geometric generic fiber $\overline{X} = X \times_A \overline{F}$ with compact support and coefficients in the $\overline{Q}_\ell$-sheaf $\mathbf{L}(\bar{\rho})$. Given a finite correspondence $(f,\ h\colon X' \to X)$ with flat f and a sheaf morphism

$\alpha\colon h^*\mathbf{L} \to f^!\mathbf{L}$, one defines an endomorphism $H_c^i(f, \alpha, h)$ of the $\bar{Q}_\ell$-module $H_c^i(X, \mathbf{L})$ as the composition of the following natural maps:

$$H_c^i(\bar{X}, \mathbf{L}) \overset{h^*}{\to} H_c^i(\bar{X}', h^*\mathbf{L}) \overset{\alpha}{\to} H_c^i(\bar{X}', f^!\mathbf{L}) = H_c^i(\bar{X}, f_* f^!\mathbf{L}) \overset{t}{\to} H_c^i(\bar{X}, \mathbf{L}).$$

Here $t\colon f_* f^!\mathbf{L} \to \mathbf{L}$ is the morphism adjoint to the identity morphism $f^!\mathbf{L} \to f^!\mathbf{L}$. In our case, for any double coset $U_I g U_I$ in $U_I \backslash G(\mathbf{A}_f)/U_I$ the action of $G(\mathbf{A}_f)$ on M_r defines a finite flat correspondence

$$(f_g, h_g\colon X_g \to X)$$

on X such that the natural map $j\colon f_g^*\mathbf{L}(\bar{\rho}) \to f_g^!\mathbf{L}(\bar{\rho})$ is an isomorphism, and a sheaf morphism $\alpha = \alpha(g)\colon h_g^*\mathbf{L}(\rho) \to f_g^!\mathbf{L}(\rho)$. Then $g \mapsto H_c^i(g) = H_c^i(f_g, \alpha(g), h_g)$ defines an action of the algebra $\mathbf{H}_I$ on $H_c^i(\bar{X}, \mathbf{L}(\bar{\rho}))$. Taking the direct limit over I one obtains an action of $G(\mathbf{A}_f)$ on $H_c^i(M_r \times_A \bar{F}, \mathbf{L}(\bar{\rho}))$.

Now the galois group $\mathrm{Gal}(\bar{F}/F)$ acts on $H^i = H_c^i(\bar{X}, \mathbf{L}(\bar{\rho}))$; so does the Hecke algebra $\mathbf{H}_I$. Denote the irreducible composition factors of $H = \sum_i (-1)^i H^i (0 \le i \le 2(r-1))$ as a virtual $\mathbf{H}_I \times \mathrm{Gal}(\bar{F}/F)$-module by $\tilde{\pi}_f \times \tilde{\rho}$. Let $\pi_\infty = \pi_\infty(\bar{\rho})$ be the square-integrable G_∞-module which corresponds to the $D_\infty^\times$-module $\bar{\rho}$ (thus π_∞ and $\bar{\rho}$ satisfy the character relation $\chi_{\pi_\infty}(\gamma) = (-1)^{r-1}\chi_{\bar{\rho}}(\gamma')$ for all elliptic regular γ in G_∞ and γ' in $D_\infty^\times$ with equal characteristic polynomials). Suppose that π_∞ is supercuspidal and its central character has finite order. Then we prove* the following

Drinfeld Reciprocity Law*. (1) *For every constituent $\tilde{\pi}_f \times \tilde{\rho}$ of H there is a cuspidal G-module $\pi_f \otimes \pi_\infty$ such that $\tilde{\pi}_f \simeq \pi_f^I$ as $\mathbf{H}_I$-modules.* (2) *For every cuspidal G-module $\pi_f \otimes \pi_\infty$ with $\pi_v^I \neq 0$ there is a unique factor $\tilde{\pi}_f \otimes \tilde{\rho}$ in H such that $\pi_f^I \simeq \tilde{\pi}_f$.* (3) *The dimension of every irreducible $\tilde{\rho}$ in H is r. The restriction $\tilde{\rho}_\infty$ of $\tilde{\rho}$ is irreducible and corresponds to π_∞ by the Local Reciprocity Law.* (4) *For every constructible irreducible ℓ-adic representation ρ of dimension r such that ρ_∞ is irreducible, there exists $I \neq 0$ and a supercuspidal π_∞, such that $\rho \simeq \tilde{\rho}$ for some constituent $\tilde{\pi}_f \times \tilde{\rho}$ of H.* (5) *If $\tilde{\pi}_f \times \tilde{\rho}$ is a constituent of H then $\tilde{\rho}$ and $(\tilde{\pi}_f \times \pi_\infty)v^{-(r-1)/2}$ correspond (also in the strong sense of all places), where v is the volume character on $\mathbf{A}^\times/F^\times$.* (6) *The multiplicity of $\tilde{\pi}_f \otimes \pi_\infty$ in H is one.* (7) *H is the direct sum of the $\tilde{\pi}_f \times \tilde{\rho}$ if $H^i = 0$ for all i such that $r - 1 - i$ is odd.*

This we reduce to Deligne's conjecture. We conjecture that

$$H_c^i(\bar{X}, \mathbf{L}(\bar{\rho})) = H^i(\bar{X}, \mathbf{L}(\bar{\rho}))$$

for all i, and, in particular, $H^i = 0$ for $i \neq r-1$. Using congruence relations we derive below some evidence for the conjecture that H^i vanishes for $i \neq r-1$. In our case of positive characteristic the $\tilde{\pi}_f$ are *a priori* only $G(\mathbf{A}_f)$-modules; the $\tilde{\pi}_f \otimes \pi_\infty$ are shown* to be cuspidal, and the sum is shown* to be direct, using a comparison of Deligne's conjecture and the Selberg Formula. This is in contrast with the analogous theory in characteristic zero, where one uses the de Rham cohomology to have the *a priori* statement that the sum is direct and the $\tilde{\pi}_f$ that occur in H are automorphic. The Explicit Reciprocity Law is likely to follow* from our methods also when π_∞ is square-integrable but not supercuspidal. However, then the $\tilde{\pi}_f$ are conjecturally (parts of) discrete series automorphic representations which are not necessarily cuspidal (for example, $\tilde{\pi}_f$ may be one-dimensional if π_∞ is the Steinberg representation, namely, dim $\bar{\rho} = 1$). As the reduction of this statement to Deligne's conjecture requires a stronger form of the Trace Formula then we use, we do not discuss this here. Of course, in this case we do not expect H^i to vanish for $i \neq r-1$.

The proof of the Purity Theorem and the Existence Theorem 1* depends on certain fixed-point formulae. These formulae apply to schemes over finite fields and in our case to the special fiber $X_v = X \times_A \mathbf{F}_v$ of $X = M_{r,I}$ at the place v of F; here $\mathbf{F}_v$ is the residue field A/v, and $\overline{\mathbf{F}}_v (= \overline{F}/\bar{v}$, where $\bar{v}$ is an extension of v to $\overline{F}$) is its algebraic closure. Put $\overline{X}_v = X_v \times_{F_v} \overline{\mathbf{F}}_v$. By constructibility of our smooth sheaf $\mathbf{L} = \mathbf{L}(\bar{\rho})$ on X, for almost all primes v of A not in I we have that $H^i = H^i_c(\overline{X}, \mathbf{L}(\bar{\rho}))$ is isomorphic to $H^i_v = H^i_c(\overline{X}_v, \mathbf{L}(\bar{\rho}))$ as $\mathbf{H}_I$-modules. Moreover, H^i and H^i_v are isomorphic as $\mathbf{H}_I \times \mathrm{Gal}(\overline{F}_v/F_v)$-modules, for almost all v.

The fixed point formula used in the proof of the Purity Theorem is the following. Let X be a separated scheme of finite type over the finite field $\mathbf{F}_q$ of $q = p^d$ elements. Let $\mathbf{L}$ be a smooth $\overline{Q}_\ell$-adic sheaf. Put $\overline{X} = X \times_{\mathbf{F}_q} \overline{\mathbf{F}}_q$. The geometric $\mathrm{Fr}_q \times 1$ and arithmetic $1 \times \mathrm{Fr}_q$ Frobenii act on X, and on the sheaf $\mathbf{L}$, and their product as an endomorphism of $H^i_c(\overline{X}, \mathbf{L})$ acts trivially. Let m be an integer. At each point x in the set $X(\mathbf{F}_{q^{|m|}})$, $(\mathrm{Fr}_q \times 1)^m$ fixes x, and it acts on the stalk $\mathbf{L}_x \simeq \overline{Q}^t_\ell$. Put $\mathrm{tr}((\mathrm{Fr}_q \times 1)^m | \mathbf{L}_x)$ for the trace of $(\mathrm{Fr}_q \times 1)^m$ on the stalk $\mathbf{L}_x$.

Grothendieck Fixed Point Formula. *For every separated scheme X of finite type over $\mathbf{F}_q$, and an $\overline{Q}_\ell$-adic sheaf $\mathbf{L}$ on X, for every $m \neq 0$ we have*

$$\sum_{x \in X(\mathbf{F}_{q^{|m|}})} \mathrm{tr}((\mathrm{Fr}_q \times 1)^m | \mathbf{L}_x) = \sum_i (-1)^i \, \mathrm{tr}((\mathrm{Fr}_q \times 1)^m | H^i_c(\overline{X}, \mathbf{L})).$$

This is due to Grothendieck [G]; see also [SGA5], Exp. III, (6.13.3), p. 134 and [SGA4$\frac{1}{2}$], p. 86. Here X is not required to be (smooth and) proper. In particular, it applies with our noncompact scheme $M_{r,I,v}$. Underlying the proof is the observation that in characteristic $p > 0$ one has $\frac{d}{dx}(x^p) = 0$, hence the graph of the Frobenius is transverse to the diagonal. In particular the fixed points of the Frobenius are isolated.

If $\overline{X}$ is proper and smooth over an algebraically closed field k, and $\mathbf{L}$ is a smooth $\overline{Q}_\ell$- sheaf on $\overline{X}$, then a stronger variant of the Fixed-Point Formula (which we do not use) is known; see [SGA4$\frac{1}{2}$], p. 151, for the case of a constant $\mathbf{L}$, and [SGA5; Exp. III], Theorem 4.4 (p. 102) and (4.12) (p. 111) for the case of any smooth $\overline{Q}_\ell$-sheaf $\mathbf{L}$. To state this, let $i: \overline{X}' \hookrightarrow \overline{X} \times_k \overline{X}$ be a closed subscheme which is transverse to the diagonal morphism $\overline{\Delta}: \overline{X} \hookrightarrow \overline{X} \times_k \overline{X}$. Suppose that $f = \mathrm{pr}_1 \circ i$ is finite and flat. Put $h = pr_2 \circ i$. Let $\alpha: h^*\mathbf{L} \to f^!\mathbf{L}$ be a sheaf morphism. Suppose that α factorizes through the morphism $j: f^*\mathbf{L} \to f^!\mathbf{L}$, which is obtained by adjunction from the trace map tr: $f_!f^*\mathbf{L} \to \mathbf{L}(f_! = f_*$ for our proper f) of [SGA4; Exp. XVIII], Theorem 2.9 (p. 553)(SLN 305 (1973)). Then $\alpha = j \circ \beta$ for some $\beta: h^*\mathbf{L} \to f^*\mathbf{L}$.

For each point x' of $\overline{X}'$ we have $(h^*\mathbf{L})_{x'} = \mathbf{L}_{h(x')}$ by definition. If $h(x') = x$ and $f(x') = x$, then the sheaf morphism $\beta: h^*\mathbf{L} \to f^*\mathbf{L}$ induces a morphism $\beta_{x'}: (h^*\mathbf{L})_{x'} \to (f^*\mathbf{L})_{x'}$ on the stalks, namely $\beta_{x'}$ is an endomorphism of the finite dimensional vector space $\mathbf{L}_x$ over $\overline{Q}_\ell$. Then we have

Lefschetz Fixed-Point Formula. *If $\overline{X}$ is proper and smooth over an algebraically closed field, and $\mathbf{L}$ is a smooth $\overline{Q}_\ell$-sheaf on $\overline{X}$, then*

$$\sum_{\substack{x' \in \overline{X}' \\ h(x') = f(x') = x}} \mathrm{tr}[\beta_{x'} | \mathbf{L}_x] = \sum_i (-1)^i \, \mathrm{tr}[H_c^i(f, \alpha, h) | H_c^i(\overline{X}, \mathbf{L})].$$

However, the Drinfeld moduli scheme $M_{r,I,v}$ is not proper. In the late 1970's Deligne suggested that the following variant could be used to imply Drinfeld's reciprocity Law.

Deligne's Conjecture. *Suppose that X is a separated scheme of finite type over $\mathbf{F}_q$; $(f, h: X' \to X)$ is a correspondence where f is finite and flat; $\mathbf{L}$ a smooth $\overline{Q}_\ell$-adic sheaf on X, and $\alpha: h^*\mathbf{L} \to f^!\mathbf{L}$ a sheaf morphism which factorizes as the composition of the natural morphism $j: f^*\mathbf{L} \to f^!\mathbf{L}$*

and a morphism $\beta: h^*\mathbf{L} \to f^*\mathbf{L}$. Then there exists an integer m_0 such that for every integer m with $|m| \geq m_0$ we have

$$\sum_{x'} \mathrm{tr}[(\beta \circ (\mathrm{Fr}_q \times 1)^m)_{x'} | \mathbf{L}_x]$$

$$= \sum_i (-1)^i \,\mathrm{tr}[H_c^i(f, \alpha \circ (\mathrm{Fr}_q \times 1)^m, h \circ (\mathrm{Fr}_q \times 1)^m) | H_c^i(\overline{X}, \mathbf{L})].$$

On the left the sum ranges over all x' in $\overline{X}'$ with $(h \circ (\mathrm{Fr}_q \times 1)^m)(x') = f(x')$, and we put $x = f(x')$.

Underlying this conjecture is the hope that multiplying the correspondence $(f, h: X' \to X)$ by a sufficiently high power of the Frobenius one obtains a correspondence transverse to the diagonal $\Delta: X \hookrightarrow X \times_{\mathbf{F}_q} X$. In [D2] Drinfeld already worked only with high powers of the Frobenius. In [SGA 5; Exp. III], Theorem 4.4, Illusie expresses the alternating sum on the right in terms of local data for a quasifinite, flat correspondence (see (4.12), p. 111), and a complex $\mathbf{L}$ of sheaves in $D_c^b(\overline{X}, \overline{Q}_\ell)$. When X is one-dimensional, for a correspondence multiplied by a high power of the Frobenius, this local data is known to be the trace on the stalk $\mathbf{L}_x$ as in the equality above, and so Deligne's conjecture follows. In addition Deligne's conjecture holds in the cases when X is proper and smooth, and when $f = h = id$, as mentioned above. Deligne-Lusztig [DL], p. 119, noted that Deligne's conjecture holds for an automorphism of finite order of the scheme X; they multiplied the automorphism by a Frobenius and considered the result as a Frobenius (with respect to another structure on the scheme) for which the Grothendieck Fixed-Point Formula is valid. Our form [FK] of the Simple Trace Formula is a representation theoretic analogue of Deligne's conjecture.

A brief sketch of our study of the Purity Theorem (and Drinfeld's Law*) by means of the Grothendieck Formula (and Deligne's conjecture, respectively) will now follow. As noted above, the Fixed-Point Formula expresses the alternating sum of traces of the action of the Frobenius on the modules H^i by means of the set of points in $M_{r,I,v}(\overline{\mathbf{F}}_v)$ fixed by the action of the Frobenius, and the traces of the resulting morphisms on the stalks of the sheaf $\mathbf{L}(\bar{\rho})$ at the fixed points. Drinfeld describes in [D2] the set $M_{r,I,v}(\overline{\mathbf{F}}_v)$ as a disjoint union of isogeny classes of elliptic modules over $\overline{\mathbf{F}}_v$. Their types are studied in analogy with the Honda-Tate Theory. A type is described in group theoretic terms as an elliptic torus in $G(F)$, and the cardinality of the set

$M_{r,I,v}(\mathbf{F}_{v,m})([\mathbf{F}_{v,m}:\mathbf{F}_v] = m)$ is expressed in terms of orbital integrals of conjugacy classes γ in $G(F)$ which are elliptic in $G(F_\infty)$. Drinfeld showed (unpublished; we give a simpler proof, based on [K1]) that these are the orbital integrals of a test function whose component at v is the spherical function $f_m = f_m^{(r)}$ on G_v defined by the relation $\mathrm{tr}(\pi_v(z))(f_m) = q_v^{m(r-1)/2} \sum_{i=1}^r z_i^m$, where $\mathbf{z} = (z_i)$. This explains the shift $q_v^{(r-1)/2}$ in the relation between the Frobenius and Hecke eigenvalues and the fact that when $\tilde{\rho}_v$ is unramified, so is $\tilde{\pi}_v$.

Our form of the Trace Formula is the following

Trace Formula. *Let $f = \otimes f_w$ be a test function on $G(\mathbf{A})$ whose component f_∞ is a supercusp form. Suppose that its component f_v at v is the above spherical $f_m^{(r)}$, where m is sufficiently large, depending on the support of the other components $f_w (w \neq v)$. Then*

$$\sum_\pi \mathrm{tr}\, \pi(f) = \sum_\gamma c(\gamma)\Phi(\gamma, f).$$

On the left the sum ranges over all cuspidal G-modules. On the right the sum is finite and ranges over the elliptic conjugacy classes γ in G. $\Phi(\gamma, f)$ is the orbital integral of f at γ, and $c(\gamma)$ are standard volume factors.

The computations mentioned prior to the statement of the Trace Formula imply that the right (geometric) side of the Selberg Trace Formula is equal to the left (stalk) side of the Fixed Point Formula, namely, we have

$$\sum_\gamma c(\gamma)\Phi(\gamma, f) = \sum_{x \in M_{r,I,v}(\mathbf{F}_{q_v^{|m|}})} \mathrm{tr}[(\mathrm{Fr}_v \times 1)^m | \mathbf{L}(\bar{\rho})_x] \qquad (*)$$

for a function $f = f_\infty f^{v,\infty} f_{m,v}^{(r)}$ specified below. By virtue of the Fixed-Point Formula the right side of (*) is equal to the following *cohomological side* (of the Fixed-Point Formula):

$$\sum_{i=0}^{2(r-1)} (-1)^i \sum \mathrm{tr}\, \tilde{\pi}_f(f^\infty)\, \mathrm{tr}\, \tilde{\rho}_v((\mathrm{Fr}_v \times 1)^m).$$

The inner sum ranges over all irreducible constituents $\tilde{\pi}_f \times \tilde{\rho}$ in H^i, and f^∞ is the product of the characteristic function f_v^0 of K_v in G_v, with a locally constant compactly supported function $f^{v,\infty}$ on $G(\mathbf{A}^{v,\infty})$, which is specified below, where $\mathbf{A}^{v,\infty}$ is the ring of adèles without components at v and ∞. Note that

$$\mathrm{tr}\, \tilde{\rho}_v((\mathrm{Fr}_v \times 1)^m) = \sum_{j=1}^r u_j(\tilde{\rho}_v)^m,$$

where $u_j(\tilde{\rho}_v)$ are the Frobenius eigenvalues. By virtue of the Trace Formula, the left side of $(*)$ is equal to the following *automorphic side* (of the Selberg Trace Formula):

$$\sum_{\pi} \operatorname{tr} \pi(f) = (-1)^{r-1} \sum_{\pi} \operatorname{tr} \pi^{v,\infty}(f^{v,\infty}) \operatorname{tr} \pi_v(f_{m,v}^{(r)}).$$

The sums range over all cuspidal G-modules $\pi = \otimes \pi_w$ whose component at ∞ is the supercuspidal π_∞ which corresponds to the $D_\infty^\times$-module $\bar{\rho}$. The component f_∞ of $f = f^{v,\infty} f_\infty f_{m,v}^{(r)}$ satisfies $\operatorname{tr} \pi_\infty(f_\infty) = (-1)^{r-1}$ and $\operatorname{tr} \pi'_\infty(f_\infty) = 0$ for every irreducible G_∞-module π'_∞ inequivalent to π_∞. Recall that

$$\operatorname{tr} \pi_v(f_{m,v}^{(r)}) = q_v^{m(r-1)/2} \sum_{i=1}^{r} z_i(\pi_v)^m,$$

where $z_i(\pi_v)$ are the Hecke eigenvalues.

All in all, we obtain the following *fundamental identity*:

$$\sum_{i=0}^{2(r-1)} (-1)^i \sum_{\tilde{\pi}_f} \operatorname{tr} \tilde{\pi}_f(f^\infty) \sum_{j=1}^{r} u_j(\tilde{\rho})^m$$

$$= (-1)^{r-1} \sum_{\pi} \operatorname{tr} \pi^{v,\infty}(f^{v,\infty}) \sum_{j=1}^{r} (z_j(\pi_v) q_v^{(r-1)/2})^m.$$

Now the scheme $M_{r,I,v}$ is not proper, and the Grothendieck Formula is available only for powers of the Frobenius. Hence the components $f_w(w \neq v, \infty)$ of the test function f are taken to be the characteristic function of $U_I \cap G_w$. Then $\operatorname{tr} \pi^{v,\infty}(f^{v,\infty})$ is a non-negative integer, namely, the dimension of the space of U_I-fixed vectors in $\pi^\infty = \otimes_{w \neq \infty} \pi_w$. Since both sides of the fundamental identity consist of finite sums (for a fixed I and π_∞), m ranges over the infinite set of $m \geq m_0$, and the coefficients of the Hecke eigenvalues are all positive (if r is odd) or negative (if r is even), we conclude from linear independence of the characters $m \mapsto u_j^m$ and $m \mapsto (q_v^{(r-1)/2} z_i)^m$ the following.

For every cuspidal G-module π with a supercuspidal component π_∞ and a nonzero U_I-fixed vector, there is a finite set of places of F such that for every v outside this set, for every Hecke eigenvalue $z_i(\pi_v)$ there is a Frobenius eigenvalue $u_j(\tilde{\rho}_v)$ of some constituent $\tilde{\pi}_f \times \tilde{\rho}$ of H, such that $u_j(\tilde{\rho}_v) = q_v^{(r-1)/2} z_i(\pi_v)$. Deligne's purity result [De 2] asserts that the complex absolute values of the algebraic numbers $u_j(\tilde{\rho}_v)$ are integral powers of $q_v^{1/2}$, while the unitarity of the nondegenerate component π_v of the cusp form π implies the bound $q_v^{-1/2} < |z_i(\pi_v)| < q_v^{1/2}$. Consequently all conjugates of the Hecke eigenvalues lie on the unit circle in the complex plane. This completes our sketch of the proof of the Purity Theorem for cusp forms.

The proof does not imply any relation between the cuspidal π and the $G(\mathbf{A}_f)$-module $\tilde{\pi}_f$ attached to $\tilde{\rho}$. It does not show that the $\tilde{\pi}_f$ are automorphic. It does not show that every Frobenius eigenvalue is related to a Hecke eigenvalue and hence has complex absolute values equal to $q_v^{(r-1)/2}$; indeed, there might be cancellations among the coefficients in the cohomological side. However, assuming Deligne's conjecture, we may take any correspondence associated to a U_I-biinvariant function $f^\infty = f^{v,\infty} f_v^0$, in the fundamental identity. Using linear independence of characters of the Hecke algebra of U_I-biinvariant functions, we conclude that the sum in the cohomological side is taken over the same set as in the automorphic side, namely, each $\tilde{\pi}_f$ is the component outside ∞ of a cuspidal π, the multiplicity of $\tilde{\pi}_f \times \tilde{\rho}$ is one, and $\tilde{\pi} = \tilde{\pi}_f \otimes \pi_\infty$ corresponds to $\tilde{\rho}$ (multiplied by $\nu^{(r-1)/2}$). This completes our sketch of the reduction of the Existence Theorem 1* (for every cuspidal π there is an ℓ-adic ρ) to Deligne's conjecture.

Our last theme in this work is a higher rank generalization of the classical theory of congruence relations. This method is entirely different than the previous one. It relies on a study of the geometry of certain correspondences on $M_{r,I,v}$, and not on the Trace Formula. It applies to $H_c^i(M_{r,I} \times_A \overline{F}, \mathbf{L}(\bar{\rho}))$ with any $D_\infty^\times$-module $\bar{\rho}$, not necessarily one which corresponds to a supercuspidal G_∞-module π_∞, and also to cohomology without compact supports. The result is the following; it does not depend on Deligne's conjecture.

Congruence Relations. *For any $\bar{\rho}$, I and i as above, and for almost all v (depending on I), for every irreducible constituent $\tilde{\pi}_f \times \tilde{\rho}$ of $H_c^i(M_{r,I} \times_A \overline{F}, \mathbf{L}(\bar{\rho}))$ we have the following. For every eigenvalue u of the endomorphism $\tilde{\rho}(\mathrm{Fr}_v \times 1)$ there is a Hecke eigenvalue $z(\tilde{\pi}_v)$ such that $u = q_v^{(r-1)/2} z(\tilde{\pi}_v)$.*

In particular, $\tilde{\rho}(\mathrm{Fr}_v \times 1)$ has at most r distinct eigenvalues.

This is the only result which we prove which relates the Frobenius and Hecke eigenvalues of $\tilde{\rho}$ and $\tilde{\pi}_f$ which occur together as an irreducible constituent $\tilde{\pi}_f \times \tilde{\rho}$ in the composition series of H^i as an $\mathbf{H}_I \times \mathrm{Gal}(\overline{F}/F)$-module. The proof is based on the study of the correspondence T_j defined by the coset $K_v g_j K_v$ in the Hecke algebra of compact K_v-double cosets in G_v; here $g_j = \mathrm{diag}(\pi_v, \ldots, \pi_v, 1, \ldots, 1)$ with $\det g_j = \pi_v^j$. We show that Fr_v satisfies the relation

$$\sum_{j\,\mathrm{odd}} q_v^{j(j-1)/2}\,\mathrm{Fr}_v^{r-j} \circ T_j = \sum_{j\,\mathrm{even}} q_v^{j(j-1)/2}\,\mathrm{Fr}_v^{r-j} \circ T_j \qquad (0 \le j \le r).$$

In particular we have

$$p_{\tilde{\pi}_v}(q_v^{(r-1)/2}\tilde{\rho}(\mathrm{Fr}_v \times 1)) = 0, \qquad \text{where} \qquad p_{\tilde{\pi}_v}(t) = \det(tI - Z(\tilde{\pi}_v))$$

is the characteristic polynomial of $\tilde{\pi}_v$, namely $Z(\tilde{\pi}_v)$ is the matrix whose r eigenvalues are the Hecke eigenvalues $z_i(\tilde{\pi}_v)$ of the unramified component $\tilde{\pi}_v$ of $\tilde{\pi}_f$.

In the classical theory of Eichler-Shimura and Ihara which concerns $\mathrm{GL}(2, Q)$, one shows on studying the de Rham cohomology that *a priori* the $\tilde{\pi}_f$ are automorphic (after tensoring with π_∞) and $\dim \tilde{\rho} = 2$. The relations of eigenvalues for almost all v, obtained from the "congruence relations" study of the correspondence attached to $K_v\begin{pmatrix} \pi_v & 0 \\ 0 & 1 \end{pmatrix}K_v$, suffices to show that $\tilde{\pi} \to \tilde{\rho}$ so defined realizes the Reciprocity Law. Similar study is carried out in Drinfeld [D1] for GL(2) over a function field. It will be interesting to obtain such complete results also in the higher rank case on developing Drinfeld's de Rham Theory.

When π_∞ is supercuspidal, combining the Congruence Relations with the Purity Theorem, we obtain

Corollary. *If π_∞ is supercuspidal, for almost all v, each conjugate of each Frobenius eigenvalue of a constituent $\tilde{\rho}$ of $H_c^i(\overline{M}_{r,I}, \mathbf{L}(\tilde{\rho}))$ has complex absolute value $q_v^{(r-1)/2}$.*

This is used in the reduction of the Reciprocity Law* to Deligne's Conjecture.

References

[De1] Deligne, P. "Les constantes des equations fonctionelles des fonction L," *SLN* **349** (1973), 501–597.

[De2] Deligne, P. "La conjecture de Weil. II," *Publ. Math IHES* **52** (1980), 313–428.

[De3] Deligne, P. "Les corps locaux de caractéristique p, limites de corps lacaux de caractéristique 0." Dans: *Représentations des Groupes Réductifs sur un corps local.* Hermann, Paris (1984).

[DL] Deligne, P. and Lusztig, G. "Representations of reductive groups over finite fields," *Ann. of Math.* **103** (1976), 103–161.

[D1] Drinfeld, V. "Elliptic modules," *Mat. Sbornik* **94** (136) (1974) (4) = *Math. USSR Sbornik* **23** (1974), 561–592.

[D2] Drinfeld, V. "Elliptic modules. II," *Mat. Sbornik* **102** (144) (1977) (2) = *Math. USSR Sbornik* **31** (1977), 159–170.

[F1] Flicker, Y. "Rigidity for automorphic forms," *J. d'Analyse Math.* **49** (1987), 135–202.

[F2] Flicker, Y. "Regular trace formula and base change lifting," *Amer. J. Math.* 110 (1988), 739–764.

[F3] Flicker, Y. "Base change trace identity for $U(3)$," *J. d'Analyse Math.* **52** (1989).

[F4] Flicker, Y. "Regular trace formula and base change for $\mathrm{GL}(n)$," preprint.

[FK] Flicker, Y., Kazhdan, D. "A simple trace formula," *J. d'Analyse Math.* **50** (1987), 189–200.

[GK] Gelfand, I., Kazhdan, D. "On representations of the group $\mathrm{GL}(n, K)$, where K is a local field," In: Lie groups and their representations, 95–118. John Wiley and Sons (1975).

[G] Grothendieck, A. "Formule de Lefschetz et rationalité des fonctions L," Sém. Bourbaki 1964/5, No. 279. Dans: *Dix exposés sur la cohomologie des schémas*, Advanced Studies in Pure Math., North Holland (1968).

[JPS] Jacquet, H., Piatetski-Shapiro, I., Shalika, J. "Rankin–Selberg convolutions," *Amer. J. Math.* **105** (1983), 367–464.

[K1] Kazhdan, D. "Cuspidal geometry of p-adic groups," *J. d'Analyse Math.* **47** (1986), 1–36.

[K2] Kazhdan, D. "Representations of groups over close local fields," *J. d'Analyse Math.* **47** (1986), 175–179.

[L] Laumon, G. "Transformation de Fourier, constantes d'équations fonctionnelles et conjecture de Weil, " *Publ. Math. IHES.* **65** (1987), 131–210.

[Ls] Langlands, R. "Modular forms and ℓ-adic representations," *SLN* **349** (1973), 361–500.

[PS] Piatetski-Shapiro, I. "Zeta functions of $\mathrm{GL}(n)$," mimeographed notes, Maryland (1976).

[Z] Zelevinsky, A. "Induced representations of reductive p-adic groups II. On irreducible representations of $\mathrm{GL}(n)$," *Ann. scient. Ec. Norm. Sup.* **13** (1980), 165–210.

[SGA4$\frac{1}{2}$] Deligne, P. "Cohomologie étale," *SLN* **569** (1977).

[SGA5] Grothendieck, A. "Cohomologie ℓ-adic et fonctions L," *SLN* **589** (1977).

11 On the Brun–Titchmarsh Theorem*

JOHN B. FRIEDLANDER

Despite the failure of the sieve to fulfill its most obvious goal, the detection of primes in interesting integer sequences, it has retained its value as a useful tool in many arithmetical investigations. A large number of these applications stem from the sieve's ability to give good upper bounds and, indeed, as demonstrated already by Brun, to often give upper bounds of the expected order of magnitude.

In the case of primes in arithmetic progressions, a result of this type is the bound

$$\pi(X; q, a) \leq \frac{cX}{\phi(q) \log\left(\frac{X}{q}\right)}, \tag{1}$$

where c is constant and X is assumed to be sufficiently large, which for $1 \leq q < X^{1-\varepsilon}$ gives the expected order of magnitude and (when $q > X^{1/2}$) contains information apparently not even provided by the Generalized Riemann Hypothesis.

The bound (1) was given by Titchmarsh [24], who used it to study the sum $\sum_{p \leq X} d(p-1)$. Results of type (1) have since come to be known as

* Supported in part by N.S.E.R.C. grant A5123.

ISBN 0-12-067570-6

Brun–Titchmarsh theorems. The name is sometimes applied more generally, in particular, to the short interval analogue

$$\pi(X) - \pi(X - Y) \leq \frac{cY}{\log Y}, \tag{2}$$

which for $Y > X^{\varepsilon}$ again gives the expected order of magnitude.

The correctness of the order of magnitude in these results and their wide range of applicability has lent great importance to the determination of the admissible values for the constant c. It is this problem that concerns us here. The history falls naturally into two stages.

We recall that in estimating the number of primes in a finite sequence A of integers we consider sums

$$\sum_{n \in A} \sum_{d \mid n} \lambda_d = \sum_{d} \lambda_d \sum_{\substack{n \in A \\ d \mid n}} 1,$$

where the λ_d may be thought of as approximating the Möbius function.

For most sequences of interest we may write

$$\sum_{\substack{n \in A \\ d \mid n}} 1 = \frac{X}{f(d)} + r_d,$$

where the function f is multiplicative and r_d may be thought of as an error term. Inserting this in the above we get

$$X \sum_{d} \frac{\lambda_d}{f(d)} + \sum_{d} \lambda_d r_d. \tag{3}$$

Stage 1. The first stage is associated with the development of the Selberg sieve [21, 22, 12, 8] and the Buchstab–Rosser sieve [23, 13].

Here one starts with a rather crude bound for the error of the type

$$\left| \sum_{d} \lambda_d r_d \right| \leq \sum_{d} |\lambda_d r_d|$$

and assuming the latter is small, for example, because the support of the $\{\lambda_d\}$ is small, one attempts to choose the $\{\lambda_d\}$ so as to minimize (or at least make fairly small) the first sum in (3) subject to the upper bound requirement

$$\lambda_1 \geq 1, \sum_{d|n} \lambda_d \geq 0 \qquad \text{for all } n.$$

In this fashion one obtained the results (1) and (2), with $c = 2 + \varepsilon$ as special cases of a much more general result dealing with a large class of sequences. This general result was optimal, as observed by Selberg [22], in the sense that $2 + \varepsilon$ could not be replaced by $2 - \varepsilon$ for certain sequences in the class. This did not preclude the possibility of further improvements of (1) and (2) but made it clear that, apart from the ε (both results with $c = 2$ were proved by Montgomery and Vaughan [17]; it is my understanding that this was done independently by Selberg), further developments would require a more careful treatment of the error term.

Stage 2. During the 1970s, in a sequence of important papers, the main ideas for the nontrivial treatment of the error term were developed by Hooley [10, 11], Motohashi [18, 19], and Iwaniec [14, 15]. The history of these developments is well documented in the introduction of the paper [15]. Here one chooses an optimal set $\{\lambda_d\}$ as in stage 1, then, using special properties of the sequence A, estimates $|\sum_d \lambda_d r_d|$ more efficiently than as $\sum_d |\lambda_d r_d|$.

This was first done by Motohashi in connection with the Selberg sieve (the earlier work of Hooley belongs more properly to the next section), who took advantage of the representation of that error term as a double sum

$$\sum_d \lambda_d r_d = \sum_{d_1 < D} \sum_{d_2 < D} \rho_{d_1} \rho_{d_2} r_{[d_1, d_2]},$$

where $[d_1, d_2]$ denotes the least common multiple. Subsequently, Iwaniec [14] discovered a modification of the Buchstab–Rosser weights that left essentially unchanged the first main sum in (3), while simultaneously replacing the error term by a number of double sums of the form

$$S = \sum_{m < M} \sum_{n < N} a_m b_n r_{mn},$$

with coefficients satisfying

$$|a_m| \leq 1, \qquad |b_n| \leq 1.$$

In comparison with Motohashi's work, one has the extra flexibility of not requiring $M = N$. The advantage of the double sum may be seen, for example, from the fact that after Cauchy's inequality

$$|S|^2 \leq \sum_m |a_m|^2 \sum_m \left| \sum_n b_n r_{mn} \right|^2 \leq M \sum_{n_1} \sum_{n_2} \left| \sum_m r_{mn_1} r_{mn_2} \right|,$$

the "unknown" coefficients have disappeared, and it may be possible in the last sum over m to take advantage of some cancellation amongst the r_{mn}.

As a result of these works there are now available a number of results with $c = 2 - \delta(\alpha)$, for some $\delta(\alpha) > 0$ where, X^α is the trivial level of distribution, that is, $q = X^{1-\alpha}$ in case (1) and $Y = X^\alpha$ in case (2). In case (2) such results are known for every fixed α with $0 < \alpha \leq 1$. In case (1) they are known unconditionally for $\frac{1}{3} < \alpha \leq 1$ and, subject to certain natural assumptions about incomplete Kloosterman sums (Hooley's R^* conjecture), they are known for the remaining range $0 < \alpha \leq \frac{1}{3}$.

We take this opportunity to mention some (five-year-old unpublished) results of this type for the case (2).

Theorem 1. *For $X > X_0(\alpha, \varepsilon)$ we have*

$$\pi(X) - \pi(X - X^\alpha) < \frac{(2 + \varepsilon - \delta(\alpha))X^\alpha}{\alpha \log X}$$

in any of the following cases.

(A) *If $0 < \alpha \leq 1$ we can take $\delta = K\alpha^2$ for some positive absolute K.*

(B) *If $(\varepsilon, \frac{1}{2} + \varepsilon)$ is an exponent pair for every $\varepsilon > 0$ (as has been conjectured) and $0 < \alpha < \frac{2}{3}$, then we can take $\delta = \frac{2}{5}$.*

(C) *If $\frac{1}{3} < \alpha < \frac{3}{5}$ then we can take $\delta = (6\alpha - 2)/(7\alpha - 1)$.*

(D) *If $\frac{1}{4} < \alpha < \frac{7}{10}$ then we can take $\delta = (8\alpha - 2)/(16\alpha - 1)$.*

The proofs of these use exponential sum ideas similar to those described in [15]. Parts (C) and (D) can be improved by using deep unpublished work on exponential sums due to Iwaniec (circa 1983) and by contour integration in case $\alpha > \frac{1}{2}$.

Results on Average

For many applications it is sufficient to have bounds that hold for most sequences of a given class. Moreover, in this case we may hope to improve the estimates of the sieve error $\sum \lambda_d r_d$ by taking advantage of the extra variable, which labels the sequences within the class. In the case of short intervals we have the following theorem [7]. In view of the fact that Selberg helped me with its proof it seems particularly appropriate to mention it here.

Theorem 2. *Let $h(t)$ be a real valued function with bounded derivative (this can be eased considerably) and satisfying $\lim_{t\to\infty} h(t) = \infty$. Then for all t in $[1, X]$, apart from a set of measure $o(X)$ and for a certain positive constant c, we have*

$$\pi(t) - \pi(t - h(t)\log t) < ch(t).$$

This result is essentially best possible with respect to the length of the intervals. The known value for c is not very good. In case $h(t) = \log^A t$ with $A > 5$ we can take $c = 4 + \varepsilon$. This corresponds to the $2 + \varepsilon$ in (1) and (2) and may be true more generally.

We now consider the case of bounds on average for primes in arithmetic progressions. It was here that the pioneering work of Hooley took place. Here there are two obvious ways to average, fix a and average over q or fix q and average over a. Hooley [10] dealt with both these situations. Subsequent authors concentrated almost exclusively (as shall we) on the case of fixed a and variable q, which is the case of greater interest for applications. In this last area a great deal of recent development has taken place by Iwaniec [15], Deshouillers and Iwaniec [4], and by Fouvry [5].

In the case where one bounds $\pi(X; q, a)$ for individual progressions as in (1) any range $q > X^{\varepsilon}$ is of interest, as our knowledge of the expected asymptotic formula is limited to the Siegel–Walfisz range $q < (\log X)^N$. By contrast, when one averages over q, say with $q \sim Q$ (that is, $Q < q \leq 2Q$), $Q = X^{\theta}$, then by virtue of the Bombieri–Vinogradov theorem the Asymptotic Formula holds for $\theta < \frac{1}{2}$, and by recent joint work [2] of the author together with Bombieri and Iwaniec holds also for $\theta = \frac{1}{2}$. It is thus only the range $\frac{1}{2} < \theta < 1$ that remains of interest. Typical results of this type are the following special cases of a result due to Fouvry [5].

Let $\varepsilon > 0$, $a \neq 0$, $A > 0$. Then, for $X > X_0(\varepsilon, a, A)$ and $Q = X^\theta$ we have

$$\pi(X; q, a) \leq (C + \varepsilon)\frac{X}{\phi(q)\log X}$$

for almost all q with $(q, a) = 1$, $Q < q \leq 2Q$, the number of exceptional q being $< Q(\log X)^{-A}$, where $C = C(\theta)$ is given by

$$C(0.6) = \frac{10}{3}, \qquad C(0.7) = \frac{80}{23}, \qquad C(0.8) = \frac{60}{17}, \qquad C(0.9) = \frac{40}{11}.$$

For θ slightly greater than $\frac{1}{2}$, say $\theta = \frac{1}{2} + \delta$ we have

$$\begin{aligned} C(\theta) &= 4 \qquad \text{(by Stage 1)} \\ &= \frac{8}{3} \qquad \text{(by Deshouillers–Iwaniec [4])} \\ &< 2 \qquad \text{(by Fouvry [5])} \\ &= 1.48 \qquad \text{(by Rousselet [20]).} \end{aligned}$$

The methods that led to the joint work [2] have led us also to the following result.

Theorem 3 (Bombieri–Friedlander–Iwaniec). *Let $a \neq 0$, $\theta > \frac{1}{2}$, $Q = X^\theta$. We then have for some absolute positive constant K, and $X > X_0(\theta, a)$*

$$\sum_{\substack{q \sim Q \\ (q,a)=1}} \left| \pi(X; q, a) - \frac{1}{\phi(q)}\, liX \right| < K(\theta - \tfrac{1}{2})^2\, liX.$$

Thus we get on average upper (and lower) bounds of Brun–Titchmarsh type where we may take the constant as close to one as we wish by choosing θ sufficiently close to $\frac{1}{2}$. Unfortunately we are not able to bound the exceptional set of q as in the earlier results.

An application of the above results is the following.

Corollary. *Let $a \neq 0$ and $\alpha > 1/2\sqrt{e}$. Then there exist infinitely many primes p such that all prime factors of $p + a$ are less than p^α.*

This result, which improves an estimate of Fouvry and Grupp [6], is proved by following closely the argument of Balog [1]. In fact, the corollary as stated follows already from the theorem of [2], the advantage of the current result being that it yields the required primes in greater number.

We conclude by giving an extremely brief sketch of the proof of Theorem 3 assuming some familiarity with the arguments of [2].

We begin with a combinatorial identity that replaces sums over primes by sums over multilinear forms. The identity of Heath–Brown [9] is often used for this purpose. We shall use here the older identity of Linnik [16].

Let $\xi_z(s) = \prod_{p \geq z}(1 - p^{-s})^{-1} = 1 + T(s)$, say, for $s > 1$ (real). For sufficiently large s we have $|T(s)| < 1$ so

$$\log \xi_z(s) = \sum_{k=1}^{\infty} \frac{(-1)^{k-1}}{k}(T(s))^k.$$

But also

$$\log \xi_z(s) = \sum_{n \geq z} \frac{\Lambda_1(n)}{n^s},$$

where Λ_1 is $\Lambda(n)/\log n$, but restricted to the powers of those primes $\geq z$. Thus, for $n \geq z$,

$$\Lambda_1(n) = \sum_{k=1}^{\infty} \frac{(-1)^{k-1}}{k} \sum_{\substack{n_1 \ldots n_k = n \\ n_j > 1 \forall j}}^{*} 1,$$

where $\sum^*$ restricts the summation to $n_1, \ldots, n_k$, each having no prime factor $<z$. Since each $n_j > 1$ the sum over k actually terminates. Inserting this formula into the definition of $\pi(X; q, a)$ we are led to the study of sums of the type

$$\sum_{\substack{q \sim Q \\ (q,a)=1}} \left| \sum^* \cdots \sum_{\substack{n_1 \ldots n_k \equiv a(q) \\ n_1 \ldots n_k \leq X}}^{*} 1 - \frac{1}{\phi(q)} \sum^* \cdots \sum_{\substack{(n_1 \ldots n_k, q)=1 \\ n_1 \ldots n_k \leq X}}^{*} 1 \right|.$$

After a splitting-up argument this sum may be replaced, with admissible error term, by the sum

$$\sum_{\substack{q \sim Q \\ (q,a)=1}} \left| \sum_{\eta} \sum^* \cdots \sum_{\substack{n_1 \ldots n_k a(q) \\ n \in \eta}}^{*} 1 - \frac{1}{\phi(q)} \sum^* \cdots \sum_{\substack{(n_1 \ldots n_k, q)=1 \\ n \in \eta}}^{*} 1 \right| = \sum_q \left| \sum_{\eta} (\quad) \right|,$$

where η runs through a family of boxes of the type

$$\begin{aligned}\eta &= \{n = (n_1, \dots, n_k);\ N_j \le n_j < N_j(1+\delta) \text{ for } 1 \le j \le k\}\\ &= (N_1, \dots, N_k)\end{aligned}$$

say, and δ is a small parameter.

In the paper [2] the above sum is majorized by $\sum_\eta \sum_q |(\quad)|$, and the inner sum over q is treated for each box η. Nontrivial estimates for sums of this type may be obtained by means of the dispersion method [16], the Weil and Deligne estimates for exponential sums, and the work of Deshouillers and Iwaniec [3] on sums of Kloosterman sums. A number of attacks are available leading to successful estimates for various configurations $(N_1, N_2, \dots, N_k)$. One such method, for instance, (Theorem 1 of [2], based on a method of Fouvry), roughly speaking, allows the treatment (provided θ is not much greater than $\frac{1}{2}$) of any box where any of the N_j satisfies $X^\varepsilon < N_j < X^{1/6-\varepsilon}$ and, hence (provided $z > X^\varepsilon$), allows us to reduce to the cases $1 \le k \le 6$.

For the most part, these methods yield estimates for more general sums where $\sum_{n_1}^* \cdots \sum_{n_k}^* 1$ is replaced by a convolution $\alpha_1 * \alpha_2 * \cdots * \alpha_k$, subject to fairly mild restrictions on the α_j. In some cases, however, a Fourier method is applied that requires one of the components, say α_1, to be essentially a characteristic function and, indeed, also requires the corresponding support δN_1 to be rather large. Moreover, since in the application, the variable n_1 is restricted to integers without prime factors $<z$, a modified estimate is required, which is deduced from the characteristic function estimate by means of the fundamental lemma of sieve theory.

In the case of the theorem in [2], Q is taken to be $X^{1/2}g(X)$ where $\log(g(X)) = o(\log X)$ (actually slightly more is required), and, under these circumstances, it turns out to be possible to satisfactorily estimate the contribution from all boxes save for the cases $k = 4, 5, 6$, and here the number of boxes that must be trivially estimated are sufficiently few in number (we call these the boundary) that their contribution is of lower order of magnitude.

In the case of our current Theorem 3, we wish to choose $\theta > \frac{1}{2}$ (fixed), and this causes the boundary to be somewhat larger. The trivial estimate of this boundary no longer gives a lower order of magnitude but does give a constant multiple of the main term (small when $\theta - \frac{1}{2}$ is small), provided that the parameter z is chosen to be a fixed power of X. The constant involved may be improved by noting that the contribution to the boundary from those integers having a prime factor p in the range $z \leq p < X^{1/6-\varepsilon}$ may, after another application of Theorem 1 of [2], be essentially neglected.

The above choice of the parameter z (in [2] the choice was slightly smaller) leads to some additional problems in the application of the fundamental lemma to deduce the modified Fourier estimate referred to earlier. In particular, it is found to be better to lump together the contribution of the various boxes (with fixed k), which are to be estimated by this method and, here, as with the boundary, the estimate is not of lower order of magnitude but rather a constant multiple of the main term. This constant may, however, be made small in comparison with the previous one simply by choosing z to be a sufficiently small power of X. Any payment for such a choice is exacted only in the value of X_0 and this leads to Theorem 3. The determination of a reasonable value for the constant K is in principle possible but has not yet been carried out.

References

[1] Balog, A. "$p + a$ without large prime factors," *Sem. Th. des Nombres, Bordeaux* 1983–84, exp. 31, 5 pp.

[2] Bombieri, E. Friedlander, J.B. and Iwaniec, H. "Primes in arithmetic progressions to large moduli II," *Math. Ann.*, **277**, (1987), 361–393.

[3] Deshouillers, J.-M. and Iwaniec, H. "Kloosterman sums and Fourier coefficients of cusp forms," *Invent. Math.* **70** (1982), 219–288.

[4] Deshouillers, J.-M. and Iwaniec, H. "On the Brun–Titchmarsh theorem on average," *Proc. Janos Bolyai Conf. Budapest* (1981).

[5] Fouvry, E. "Théorème de Brun–Titchmarsh; application au théorème de Fermat," *Invent. Math.* **79** (1985), 383–407.

[6] Fouvry, E. and Grupp, F. "On the switching principle in sieve theory," *J. Reine Angew.* **370** (1986), 101–126.

[7] Friedlander, J.B. "Sifting short intervals (I)," *Math. Proc. Cambridge Phil. Soc.* **91** (1982), 9–15; (II) ibid. **93** (1982), 381–384.

[8] Halberstam, H. and Richert, H.E. *Sieve Methods.* Academic Press, London (1974).

[9] Heath-Brown, D.R. "Prime numbers in short intervals and a generalized Vaughan identity," *Can. J. Math.* **34** (1982), 1365–1377.

[10] Hooley, C. "On the Brun–Titchmarsh theorem," *J. Reine Angew.* **255** (1972), 60–79.

[11] Hooley, C. "On the Brun–Titchmarsh theorem II," *Proc. London Math. Soc.* **30** (3) (1975), 114–128.

[12] Jurkat, W.B. and Richert, H.E. "An improvement of Selberg's sieve method I," *Acta Arith.* **11** (1965), 217–240.

[13] Iwaniec, H. "Rosser's sieve," *Acta Arith.* **36** (1980), 171–202.

[14] Iwaniec, H. "A new form of the error term in the linear sieve," *Acta Arith.* **37** (1980), 307–320.

[15] Iwaniec, H. "On the Brun–Titchmarsh theorem," *J. Math. Soc. Jpn.* **34** (1982), 95–123.

[16] Linnik, Yu. V. "The Dispersion Method in Binary Additive Problems," *Translations of Math. Monographs* **4**, A.M.S. (Providence) (1963).

[17] Montgomery, H.L. and Vaughan, R.C. "On the large sieve," *Mathematika.* **20** (1973), 119–134.

[18] Motohashi, Y. "On some improvements of the Brun–Titchmarsh theorem," *J. Math. Soc. Jpn.* **26** (1974), 306–323.

[19] Motohashi, Y. "On some improvements of the Brun–Titchmarsh theorem III," *J. Math. Soc. Jpn.* **27** (1975), 444–453.

[20] Rousselet, B. "Inégalités de type Brun–Titchmarsh en moyenne." In: *Groupe de Travail en Théorie Analytique des Nombres*, in press.

[21] Selberg, A. "On an elementary method in the theory of primes," *Norsk. Vid. Selsk. Forh.* Trondheim. **19** (1947), 64–67.

[22] Selberg, A. "The general sieve method and its place in prime number theory," *Proc. International Congress of Math.* (Harvard) **I** (1950), 286–292.

[23] Selberg, A. "Sieve methods," *Proc. Sympos. Pure Math.*, A.M.S. Providence, **20**, (1971), 311–351.

[24] Titchmarsh, E.C. "A divisor problem," *Rend. Circ. Mat. Palermo.* **54** (1930), 414–429.

Note Added in Proof.

In the meanwhile, the author with Bombieri and Iwaniec has improved Theorem 3 with respect to the uniformity in θ near $\frac{1}{2}$.

12 A Double Sum Over Primes and Zeros of the Zeta-Function*

P. X. GALLAGHER

The object of this paper is the estimation and asymptotic evaluation, for certain ranges of the parameters, of

$$B(X, T) = \sum_{0<\gamma\leq T} \sum_{n\leq X} \frac{\Lambda(n)}{n^{1/2+i\gamma}},$$

where Λ is von Mangoldt's function and $\frac{1}{2} + i\gamma$ ranges over the zeros of the Riemann zeta-function on the critical line. We assume the Riemann Hypothesis (RH) throughout, and write $U = \log X$, $L = \log T$, and $\beta = U/L$.

Theorem 1. *On* RH, *for X and* $T \to \infty$,

$$B(X, T) \sim -\frac{1}{2\pi} TL^2 \cdot \frac{1}{2}\beta^2 \qquad (X = O(T)); \tag{1}$$

$$B(X, T) \ll TL^2 \qquad (X = O(T^2)). \tag{2}$$

NUMBER THEORY,
TRACE FORMULAS and
DISCRETE GROUPS

ISBN 0-12-067570-6

We can improve (2) if we assume, in addition, Montgomery's pair correlation conjecture (MC), which states that the number of pairs γ, γ' with $0 < \gamma, \gamma' \le T$ and $0 < \gamma' - \gamma \le \varepsilon$ is asymptotic, for $T \to \infty$, to

$$TL \frac{1}{2\pi} \int_0^\alpha \left(1 - \left(\frac{\sin \pi a}{\pi a}\right)^2\right) da \qquad \left(\alpha = \varepsilon \frac{L}{2\pi} \simeq 1\right).$$

Theorem 2. *On* RH *and* MC, *for* X *and* $T \to \infty$,

$$B(X, T) \sim -\frac{1}{2\pi} TL^2\left(\beta - \frac{1}{2}\right) \qquad (T \ll X = o(T^2)). \tag{3}$$

For more on MC, see [4–8], [11–14].

In Section 1, formula (1) is derived from the Landau–Gonek Formula [9], [10]

$$\sum_{0<\gamma\le T} n^{-i\gamma} = -\frac{T}{2\pi}\frac{\Lambda(n)}{n^{1/2}} + O(n^{1/2}L \log L) \qquad (1 \ll \log n \ll L) \tag{4}$$

and standard results of Mertens and Tchebychev. Then, using the Riemann–von Mangoldt zero counting formula and an analogous prime counting formula, we get

$$B(X, T) = B_{R,S} + O(TL^2) \qquad (X = O(T^2)) \tag{5_0}$$

and a formula (5_o) in which the two O's are replaced by o's, where

$$B_{R,S} = \int_0^T \int_0^U e^{-itu}\, dR(u)\, dS(t)$$

and R and S are the remainder terms defined in Section 1 in the prime and zero counting formulas.

In Section 2, we first prove a general inequality for integrals of the shape of $B_{R,S}$ ("type II" integrals in Vaughan's terminology [18]). In what follows $\|\cdot\|$ is the $L^2(\mathbb{R})$ norm, $\Delta_h f(x) = f(x+h) - f(x)$, and f_X is the function on $\mathbb{R}$ that agrees with f on $[-X, X]$ and is constant on $(-\infty, -X]$ and on $[X, \infty)$.

Lemma 1. *For complex valued functions* F *and* G *of finite variation on* $[-U, U]$ *and* $[-V, V]$ *we have, for* $UV \ge 1$,

$$\int_{-V}^{V} \int_{-U}^{U} e(uv)\, dF(u)\, dG(v) \ll UV \max_{h \le (2V)^{-1}} \|\Delta_h F_U\| \, \|\Delta_{(2U)^{-1}} G_V\|. \tag{6}$$

The proof combines arguments of Bombieri and Iwaniec [1, Lemma 2.4] with [3, Lemma 1]. Inequality (6) reduces the estimation of $B_{R,S}$ to that of $\|\Delta_\varepsilon S_T\|$ for $\varepsilon \ll U^{-1}$ and $\|\Delta_\delta R_U\|$ for $\delta \ll T^{-1}$. The bounds

$$\int_0^U (R(u+\delta) - R(u))^2 \, du \ll \delta\Lambda^2 \qquad (U \ll \Lambda = \log \delta^{-1}) \tag{7}$$

$$\int_0^T (S(t+\varepsilon) - S(t))^2 \, dt \ll T \qquad (\varepsilon L \ll 1) \tag{8}$$

originating in work of Selberg [15], [16] are then applied. For a stronger form of (7), see Goldston and Montgomery, [8, Theorem 1]. For (8) and related asymptotic formulas, see Fujii [2] and Tsang [17]. The application is limited to $X \ll T^2$ by a difficulty near U in the transition between $\|\Delta_\delta R_U\|^2$ and the integral in (7).

In Section 3, a particular case of the Guinand explicit formula is used to express a smoothed version of $B_{R,S}$ as a double integral involving $dS(t')\,dS(t)$. This integral is then asymptotically evaluated, for each β, using MC in a way analogous to that by which Mueller and the author showed [6, Theorem 4] that MC implies that for $U \to \infty$

$$\int_0^U (R(u+\delta) - R(u))^2 \, du \sim \delta\Lambda^2\left(B - \frac{1}{2}\right) \qquad \left(1 \le B = \frac{U}{\Lambda} \ll 1\right). \tag{$\widehat{\mathrm{MC}}$}$$

Recently Goldston has shown [7] that $\widehat{\mathrm{MC}}$ is in fact the arithmetic equivalent of MC. Finally, a second application of Lemma 1 is used in the unsmoothing to finish the proof of Theorem 2. In this last step, $\widehat{\mathrm{MC}}$ and the Brun–Titchmarsh inequality are used in place of (7) to get from $O(TL^2)$ to $o(TL^2)$.

1. Application of the Landau–Gonek Formula and Reduction to $B_{R,S}$

Proof of (1). By (4),

$$B(X, T) = -\frac{T}{2\pi} \sum_{n \le X} \frac{\Lambda^2(n)}{n} + O\left(L \log L \sum_{n \le X} \Lambda(n)\right).$$

The first sum is $\sim \frac{1}{2}U^2$ for $U \to \infty$, and the second is $\lesssim X$, so $B(X, T) \sim -T/4\pi\, U^2$ for X and $T \to \infty$ provided $X = o(TL/\log L)$.

Proof of (5₀) and (5ₒ). Following the notation of [5], apart from a sign change in the definitions of P, Q, and R, we define R and S to be the odd functions for which $P = Q + R$ for $u \geq 0$ and $N = M + S$ for $t \geq 0$, where

$$P(u) = \sum_{n \leq e^u}' \frac{\Lambda(n)}{n^{1/2}}, \qquad Q(u) = 4 \sinh \tfrac{1}{2}u - 2u,$$

$$N(t) = \sum_{0 < \gamma \leq t}' 1, \qquad M(t) = \frac{1}{\pi} \int_0^t \arg G\left(\frac{1}{2} + it_1\right) dt_1,$$

and $G(s) = \pi^{-s/2}\Gamma(\frac{1}{2}s)$; the dashes indicate that terms with $n = e^u$ or $\gamma = T$ are to be halved.

In an obvious notation, $B(X, T)$ (modulo terms with $n = X$ or $\gamma = T$, which contribute $O(TL^2)$), is

$$B_{P,N} = -B_{Q,M} + B_{Q,N} + B_{P,M} + B_{R,S}.$$

It suffices to show that the first three terms on the right are $O(TL^2)$ for $X = O(T^2)$ and $o(TL^2)$ for $X = o(T^2)$.

From the fact that Q' is monotonic and is $\ll X^{1/2}$ on $[0, U]$ it follows that the u integral in $B_{Q,N}$ is $\ll X^{1/2}/t$, from which

$$B_{Q,N} \ll X^{1/2} \sum_{0 < \gamma \leq T} \frac{1}{\gamma} \ll X^{1/2}L^2.$$

Analogously, since M' is ultimately monotonic and is $\gg L$ on $[0, T]$ it follows that the t integral in $B_{P,M}$ is $\ll L/u$, from which

$$B_{P,M} \ll L \sum_{n \leq X} \frac{\Lambda(n)}{n^{1/2} \log n} \ll X^{1/2} \frac{L}{U};$$

quite similarly,

$$B_{Q,M} \ll X^{1/2} \frac{L}{U}.$$

2. Application of the Second Moment Bounds

Proof of Lemma 1. As in [1], starting with the identity

$$e(uv) = 2U\phi\left(\frac{u}{2U}\right) \int_{v-(2U)^{-1}}^{v} e(uw)\, dw,$$

with $\phi(\alpha) = 2\pi i\alpha/(1 - e(-\alpha))$, the double integral on the left of (6) is

$$\int_{-\infty}^{\infty}\int_{-\infty}^{\infty} e(uv)\, dF_U(u)\, dG_V(v)$$

$$= 2U\int_{-W}^{W}\int_{-\infty}^{\infty} - e(uw)\, dH(u)\Delta_{(2U)^{-1}}G_V(w)\, dw,$$

where $W = V + (2U)^{-1}$ and

$$H(u) = \int_0^u \phi\left(\frac{x}{2U}\right) dF_U(x).$$

It suffices by Cauchy's inequality in view of $UW \ll UV$ to show that

$$\int_{-W}^{W}\left|\int_{-\infty}^{\infty} e(uw)\, dH(u)\right|^2 dw \ll W^2 \max_{h \le (2V)^{-1}} \|\Delta_h F_U\|^2. \tag{9}$$

Since H is constant on $(-\infty, -U]$ and on $[U, \infty)$, the function $2W\Delta_{(2W)^{-1}}H$ belongs to L^2; its Fourier transform is

$$-\int_{-\infty}^{\infty} \frac{e(uw)}{2\pi iw/2W}\, d\Delta_{(2W)^{-1}}H(u) = \frac{1 - e(-w/2W)}{2\pi iw/2W}\int_{-\infty}^{\infty} e(uw)\, dH(u).$$

Since the first factor on the right is $\gg 1$ on $[-W, W]$, Plancherel's theorem gives

$$\int_{-W}^{W}\left|\int_{-\infty}^{\infty} e(uw)\, dH(u)\right|^2 dw \ll W^2\|\Delta_{(2W)^{-1}}H\|^2. \tag{10}$$

We have

$$\Delta_{(2W)^{-1}}H(u) = \int_u^{u+(2W)^{-1}} \phi\left(\frac{x}{2U}\right) d(F_U(x) - F_U(u)).$$

Since ϕ and ϕ' are bounded on $[-\frac{3}{4}, \frac{3}{4}]$, integration by parts gives

$$|\Delta_{(2W)^{-1}}H| \ll |\Delta_{(2W)^{-1}}F| + U^{-1}\int_0^{(2W)^{-1}} |\Delta_h F_U|\, dh.$$

It follows by Cauchy's inequality that

$$\|\Delta_{(2W)^{-1}}H\| \ll \max_{h \le (2W)^{-1}} \|\Delta_h F_U\|. \tag{11}$$

Combining (10) and (11) gives (9) since $V \le W \ll V$.

Proof of (2). By Lemma 1, for large T and U

$$B_{R,S} \ll TU \max_{\delta \ll T^{-1}} \|\Delta_\delta R_U\| \, \|\Delta_{(2U)^{-1}} S_T\|,$$

where from now on R_U agrees with R on $[0, U]$, is constant on $[U, \infty)$, and is zero on $(-\infty, 0]$, and S_T is defined similarly.

It follows from (8) that the second norm is $\ll T^{1/2}$ for $L \ll U$. In fact, the difference between $\|\Delta_\varepsilon S_T\|$ and the square root of the integral in (8) is $\ll \varepsilon^{1/2} L$ since $S \ll L$ on $[-\varepsilon, T + \varepsilon]$. To conclude via (7) that the first norm is $\ll T^{-1/2} L$ by a similar argument using the von Koch bound $R \ll U^2$ would require $U^2 \ll L$. Another argument allows $X \ll T^2$: It suffices to show that with $\delta \ll T^{-1}$

$$\int_{U-\delta}^{U} (R(u+\delta) - R(U))^2 \ll \delta L^2 \qquad (X \ll T^2).$$

To get this, we treat P and Q separately. On $[U - \delta, U]$, for $X \ll T^2$

$$Q(u+\delta) - Q(U) \ll \delta X^{1/2} \ll 1 = o(L),$$
$$P(u+\delta) - P(U) \ll (\delta X + 1) X^{-1/2} L = O(L),$$

estimating crudely.

It is important for the unsmoothing at the end of Section 3 to observe that with $\delta = (2T)^{-1}$

$$\int_{U-\delta}^{U} (R(u+\delta) - R(U))^2 \, du = o(\delta L^2) \qquad \left(X \ll T^2, \quad \frac{X}{T} \to \infty\right). \tag{12}$$

For this it suffices to use in the last estimation the Brun–Titchmarsh inequality in the weakened form

$$\psi(X+h) - \psi(X) = o(h \log X) \qquad (h \to \infty, h \ll X)$$

to get that for $\delta X \to \infty$, $\delta X^{1/2} \ll 1$,

$$P(U+\delta) - P(U) = o(\delta X) X^{-1/2} U = o(U).$$

3. Application of Guinand's Formula and MC

We write

$$B_{R,S} = B_{R,S}(f) + B_{R,S}(\phi - f),$$
$$B_{R,S} = \int_0^T \int_0^U f(u) e^{-itu} \, dR(u) \, dS(t),$$

with f a smooth approximation to the characteristic function ϕ of $[0, U]$.

We will need bounds for

$$g(\varepsilon) = \int_{-\infty}^{\infty} e^{i\varepsilon u} f(u)\, du$$

and for its first two derivatives, which are $O(\varepsilon^{-3})$ for large ε. For this reason we choose $f = H^{-2}\phi_1 * \phi_2 * \phi_2$ where $*$ is convolution and ϕ_1 and ϕ_2 are the characteristic functions of $(H, U - H)$ and $(-\frac{1}{2}H, \frac{1}{2}H)$, respectively; here H is a function of U satisfying $H = o(U)$ that will be specified more precisely below. We have

$$g^{(j)}(\varepsilon) \ll \frac{L^{j+1}}{(1 + LH^2|\varepsilon|^3)} \qquad (j = 0, 1, 2). \tag{13}$$

In fact, the Fourier transform ψ of the characteristic function of any subinterval of $[-\lambda, \lambda]$ satisfies $\psi^{(j)}(\varepsilon) \ll_j \lambda^{j+1}/(1 + \lambda|\varepsilon|)$, from which (13) follows, using $U \simeq L$.

By Guinand's Formula [4, Theorem 1] (with a sign change since R here is $-R$ in [4])

$$\int_0^U f(u) e^{-itu}\, dR(u) = -\int_{-\infty}^{\infty} g(t' - t)\, dS(t'). \tag{14}$$

From this, the bounds (13), and the fact that $S(t) \ll \log(2 + |t|)$, we get

$$B_{R,S}(f) = -\int_0^T \int_0^T g(t' - t)\, dS(t')\, dS(t) + o(TL^2), \tag{15}$$

provided $H/L \to 0$ sufficiently slowly. In fact, the contribution of $(-\infty, 0]$ and $[T, \infty)$ to the integral on the right of (14) is

$$g(-t)S(0) - \int_{-\infty}^{0} g'(t' - t)S(t')\, dt' - g(T - t)S(T) - \int_T^{\infty} g'(t' - t)S(t')\, dt',$$

so the contribution to $B_{R,S}(f)$ is a sum of 12 terms; a "worst" one is

$$\int_0^T \int_T^{\infty} g''(t' - t)S(t')S(t)\, dt'\, dt \ll L^5 \int_0^T \int_T^{\infty} \frac{dt'\, dt}{1 + LH^2(t' - t)^3},$$

provided $H/U \to 0$ sufficiently slowly, and the integral on the right is

$$\ll \int_0^{\infty} \frac{v\, dv}{1 + LH^2 v^3} \ll (LH^2)^{-2/3}.$$

The other terms may be dealt with similarly, and we get (15), with a remainder not much bigger than L^3.

The inner integral in (15) is

$$g(0)(S(t^+) - S(t^-)) + \int_0^T g(t' - t)\, d_{t'}(S(t') - S(t^\pm)),$$

where $\pm$ or neither is taken according as $t' - t \gtrless 0$ or $=0$. It follows that

$$\int_0^T \int_0^T g(t' - t)\, dS(t')\, dS(t) = g(0)S_T^* + \int_{-T}^T g(\varepsilon)\, dS_T^{[2]}(\varepsilon), \tag{16}$$

where generally for a function F of finite variation on $\mathbb{R}$,

$$F^* = \sum_t |F(t^+) - F(t^-)|^2,$$

and

$$F^{[2]}(\varepsilon) = \int_{-\infty}^{\infty} (F(t + \varepsilon) - F(t^\pm))\, dF(t),$$

where $\pm$ is taken according as $\varepsilon \gtrless 0$ and $F^{[2]}(0) = 0$.

In this notation, MC in a generalized formulation states that for $T \to \infty$

$$N_T^{[2]}(\varepsilon) = T\frac{L}{2\pi}\int_0^\alpha (1 - \mu(a))\, da + o(TL) \qquad \left(\alpha = \varepsilon\frac{L}{2\pi} \simeq 1\right) \tag{17_{N}}$$

for some real even continuous function μ in $L^1(\mathbb{R})$. It follows from (17_{N}) that

$$N_T^{[2]}(\varepsilon) = o(TL) \qquad (\alpha \to 0); \tag{18_{N}}$$

RH alone implies [6, (6)] that

$$N_T^{[2]}(\varepsilon) \sim T\frac{L}{2\pi}\alpha \qquad (|\alpha| \to \infty, |\varepsilon| \leq T). \tag{19_{N}}$$

It also follows [6, (8)] from (17_{N}) on the assumption that $\int \mu = 1$ that

$$N_T^* \sim T\frac{L}{2\pi}. \tag{20_{N}}$$

For use in (16), we next reformulate the last four equations as statements about S_T:

Lemma 2. *On* RH *and* MC, *as* $T \to \infty$

$$S_T^{[2]}(\varepsilon) = -T\frac{L}{2\pi}\int_0^\alpha \mu(a)\,da + o(TL) \qquad (\alpha \simeq 1) \tag{17$_S$}$$

$$S_T^{[2]}(\varepsilon) = o(TL) \qquad (\alpha \to 0) \tag{18$_S$}$$

$$S_T^{[2]}(\varepsilon) = o(TL) \qquad (|\alpha| \to \infty, |\varepsilon| \le T) \tag{19$_S$}$$

$$S_T^* \sim T\frac{L}{2\pi}. \tag{20$_S$}$$

Proof. From $N = M + S$ it follows that

$$N_T^{[2]} = M_T^{[2]} + 2[M_T, S_T] + S_T^{[2]}, \tag{21}$$

where generally

$$2[F, G] = \int_{-\infty}^{\infty} \Delta_\varepsilon F(t)\,dG(t) + \int_{-\infty}^{\infty} \Delta_\varepsilon G(t)\,dF(t).$$

The second integral here is $-\int \Delta_{-\varepsilon} F\,dG$, so

$$2[F, G] = \int_{-\infty}^{\infty} (F(t+\varepsilon) - F(t-\varepsilon))\,dG(t).$$

Since M'_T is monotonic and $\ll L$ and $S_T \ll L$,

$$2[M_T, S_T] = o(TL) \qquad (\varepsilon \le T). \tag{22}$$

Since $M'(t) \sim 1/2\pi \log t$,

$$M_T^{[2]}(\varepsilon) \sim T\frac{L}{2\pi}\alpha \qquad (\varepsilon \le T). \tag{23}$$

The $S^{[2]}$ statements follow from the corresponding $N^{[2]}$ statements with the help of (21), (22), and (23). The S^* and N^* statements are the same since S and N have the same jumps.

We next show that for $T \to \infty$, $\beta = U/L \simeq 1$, and $H/U \to 0$ sufficiently slowly,

$$B_{R,S}(f) = -\frac{1}{2\pi}TL^2\int_0^\beta (1 - \hat{\mu}(b))\,db + o(TL^2). \tag{24}$$

In view of (15), (16), and (20$_S$), it suffices to show that

$$\int_{-T}^{T} g(\varepsilon)\,dS_T^{[2]}(\varepsilon) = -\frac{1}{2\pi}TL^2\int_0^\beta \hat{\mu}(b)\,db + o(TL^2). \tag{25}$$

It follows from (17_S) that there are positive functions α_1 and α_2 of T so that as $T \to \infty$, we have $\alpha_1 \to 0$, $\alpha_2 \to \infty$, and

$$S_T^{[2]}(\varepsilon) = -T\frac{L}{2\pi}\int_0^{\alpha}\mu(a)\,da + o(TL) \qquad (\alpha_1 \le |\alpha| \le \alpha_2).$$

We denote by s, m, l the ranges $|a| \le \alpha_1$, $\alpha_1 \le |a| \le \alpha_2$, $|a| \ge \alpha_2$, where $a = \varepsilon L/2\pi$.

The contribution of m to the integral on the left in (25) is

$$T\frac{L}{2\pi}\left(\int_m g\left(\frac{2\pi a}{L}\right)\mu(a)\,da + \int_m g\left(\frac{2\pi a}{L}\right)do(1)\right).$$

Using the bound in (13) on g, the da integral is

$$\int_{\mathbb{R}} g\left(\frac{2\pi a}{L}\right)\mu(a)\,da + O(L\alpha_1 + L^3H^{-2}\alpha_2^{-3}).$$

Provided $H/U \to 0$ sufficiently slowly, the O-term here is $o(L)$. The integral over $\mathbb{R}$ is

$$L\int_{\mathbb{R}} f(bL)\hat{\mu}(b)\,db \to L\int_0^{\beta}\hat{\mu}(b)\,db.$$

Using the bounds in (13) on g and g', the $do(1)$ integral is

$$o(L + L^3H^{-2}\alpha_1^{-2}) = o(L),$$

provided $H/U \to 0$ and $\alpha_1 \to 0$ sufficiently slowly.

Using the bounds on g and g' with (18_S) in a similar way, the contribution of s to the integral on the left of (25) is

$$\ll L\cdot o(TL) + L^2\cdot o(TL)\cdot\frac{\alpha_1}{L} = o(TL^2).$$

Finally, using the bounds on g and g' with (19_S) in a similar way, the contribution of l to the integral on the left of (25) is

$$H^{-2}\left(\frac{\alpha_2}{L}\right)^{-3}\cdot o(TL\alpha_2) + \int_{\alpha_2}^{\infty} LH^{-2}\left(\frac{\alpha}{L}\right)^{-3}\cdot o(TL\alpha)\cdot L^{-1}\,d\alpha$$
$$= o(TH^{-2}L^4\alpha_2^{-1}) = o(TL^2),$$

provided $H/U \to 0$ sufficiently slowly.

It remains to show that for $T \to \infty$, $X \ll T^2$, and $H/U \to 0$,

$$B_{R,S}(\phi - f) = o(TL^2). \qquad (26)$$

For this we observe that $B_{R,S}(\phi - f)$ is a weighted average, over the interval $(0, 2H)$, of

$$B_{R,S,h} = \int_0^T \int_0^h e^{-itu}\, dR(u)\, dS(t) + \int_0^T \int_{U-h}^U e^{-itu}\, dR(u)\, dS(t),$$

so it suffices to show that $B_{R,S,h} = o(TL^2)$ uniformly for such h. For this we use Lemma 1, giving

$$B_{R,S,h} \ll TU(\|\Delta_{(2T)^{-1}} R_h\| + \|\Delta_{(2T)^{-1}}(R_U - R_{U-h})\|) \max_{\varepsilon \ll U^{-1}} \|\Delta_\varepsilon S_T\|.$$

By comparison with the proof of (2) and the remark at the end of Section 2, it suffices, for $X \ll T^2$, $X/T \to \infty$ to have

$$\int_0^h (\Delta_{(2T)^{-1}} R)^2\, du + \int_{U-h}^U (\Delta_{(2T)^{-1}} R)^2\, du = o(T^{-1}L^2),$$

and this follows from $\widehat{\text{MC}}$ at the end of the introduction.

For $T \le X = o(TL/\log L)$, formula (3) follows from the formula in (1), which is proved for this range in Section 1.

References

[1] Bombieri, E. and Iwaniec, H. "On the order of $\zeta(\frac{1}{2} + it)$," *Ann. Scuola Normale Sup. Pisa Cl. Sci. IV* **13** (1986), 449–472.

[2] Fujii, A. "On the zeros of Dirichlet L-functions I," *T.A.M.S.* **196** (1977), 249–276.

[3] Gallagher, P.X. "A large sieve density estimate near $\sigma = 1$," *Invent. Math.* **11** (1970), 329–339.

[4] Gallagher, P.X. "Pair correlation of the zeros of the zeta function," *J. Reine Angew. Math.* **362** (1985), 72–86.

[5] Gallagher, P.X. "Applications of Guinand's formula," *Analytic Number Theory and Diophantine Problems, Proceedings of a Conference at Oklahoma State University*, Birkhauser, Basel-Stuttgart (1987), 135–157.

[6] Gallagher, P.X. and Mueller, J.H. "Primes and zeros in short intervals," *J. Reine Angew. Math.* **303/304** (1978), 205–220.

[7] Goldston, D.A. "On the pair correlation conjecture for the zeros of the Riemann zeta function," *J. Reine Angew. Math.* **385** (1988), 24–40.

[8] Goldston, D.A. and Montgomery, H.L. "Pair correlation of zeros and primes in short intervals," *Analytic Number Theory and Diophantine Problems, Proceedings of a Conference at Oklahoma State University* Birkhauser, Basel-Stuttgart (1987), 187–201.

[9] Gonek, S.M. "An explicit formula of Landau and its applications," preprint.

[10] Gonek, S.M. "A formula of Landau and mean values of $\zeta(s)$," *Topics in Analytic Number Theory*, S.W. Graham and J.D. Vaaler, eds., University of Texas Press, Austin (1985), 92–97.

[11] Joyner, D. *Distribution Theorems of L-Functions*. Longman Scientific and Technical, New York (1986).

[12] Montgomery, H.L. "The pair correlation of zeros of the zeta function," *Proc. Sympos. Pure Math.* **24** (1973), 181–193.

[13] Mueller, J.H. "Arithmetic equivalent of essential simplicity of zeta zeros," *T.A.M.S.* **275** (1983), 175–183.

[14] Odlyzko, A.M. "On the distribution of spacings between zeros of the zeta function," *Math. Comp.* **48** (1987), 273–308.

[15] Selberg, A. "On the normal density of primes in short intervals and the difference between consecutive primes," *Archiv f. Math. og Naturvid.* **B47** (6) (1943), 87–105.

[16] Selberg, A. "On the remainder term for $N(T)$, the number of zeros of $\zeta(s)$ in the strip $0 < t < T$," *Avhandlinger Norske Vid. Akad. Oslo*. No. 1. (1944).

[17] Tsang, K.-M. "Some Ω-theorems for the Riemann zeta function," *Acta Arith.* **46** (1986), 369–395.

[18] Vaughan, R.C. "Adventures in arithmetick, or: How to make good use of a Fourier transform," *Math. Intelligencer* **9** (2) (1987), 53–60.

13 Integral Representations of Eisenstein series and L-functions

PAUL B. GARRETT

The idea of obtaining integral formulas by integrating against some sort of Eisenstein series (or restriction of Eisenstein series) is not new. The "Rankin–Selberg method" is one idea that has interesting analytic and arithmetic consequences. Here we consider a different use of Eisenstein series, obtaining not only integral representations of some L-functions, but also integral representations of more complicated Eisenstein series in terms of simpler ones. We find that this method naturally yields expressions of the general form

(Integral of cuspform against restriction of Siegel-type Eisenstein series)
= (L-function attached to cuspform) × (Eisenstein series attached to cuspform).

A slightly more precise statement of the main result (2.3) is as follows. We will consider families $\{G_n: n = 1, 2, \ldots\}$ of reductive linear algebraic groups G_n over $\mathbb{Q}$ so that there are morphisms

$$\iota_{mn}: G_m \times G_n \to G_{m+n}$$

NUMBER THEORY,
TRACE FORMULAS and
DISCRETE GROUPS

ISBN 0-12-067570-6

defined by

$$g \times g' \to \begin{pmatrix} a & 0 & b & 0 \\ 0 & a' & 0 & b' \\ c & 0 & d & 0 \\ 0 & c' & 0 & d' \end{pmatrix},$$

where

$$g = \begin{pmatrix} a & b \\ c & d \end{pmatrix}, \qquad g' = \begin{pmatrix} a' & b' \\ c' & d' \end{pmatrix}.$$

These groups G_n include certain rational forms of Sp_n, $U(n, n)$, $SO(n, n)$, and quaternion unitary groups (see (1.1)). Let f be an eigencuspform on $G_m(\mathbb{R})$ of level $\mathfrak{f}$, weight ρ, central character ω, and eigenvalues λ for the invariant operators (see (1.1)). In (1.1) we define a simple $\mathbb{C}$-antilinear isometry $f \to f^\natural$ from the space of such eigencuspforms to those of level f, weight ρ, central character ω, and eigenvalues λ. Let $E(g) = E_{m+n}(g; \mathfrak{n}, \rho, \chi, m)$ be a "Siegel-type" Eisenstein series on $G_{m+n}(\mathbb{A})$ as in (1.3.1), attached to the parabolic subgroup

$$P = \left\{ \begin{pmatrix} * & * \\ 0 & * \end{pmatrix} \in G_{m+n} \right\},$$

where $\mathfrak{n} \subset f$, and where $\rho, \omega, \rho_1, \chi$ have a suitable compatibility property (see (2.2)). Then, for $g \in G_n(\mathbb{A})$, assuming absolute convergence of the integral, we have what we call the *Main Formula* (2.3):

$$\int_{Z_m(\mathbb{A})G_m(\mathbb{Q})\backslash G_m(\mathbb{A})} (E \circ \imath_{mn}(g_1, g)\bar{f}(g_1)\, dg_1 = L_2(f, \chi; \mathfrak{n})(E_n(g; \mathfrak{n}, \rho_2, \bar{\chi}; f^\natural),$$

where $E_n(g; \mathfrak{n}, \rho_2, \bar{\chi}; f^\natural)$ is an Eisenstein series attached to the cuspform $f^\natural$, and where $L_2(f, \chi; \mathfrak{n})$ is essentially the "symmetric square" L-function attached to f (see (2.2)). In particular, for $m = n$, this is

$$\int_{Z_m(\mathbb{A})G_m(\mathbb{Q})\backslash G_m(\mathbb{R})} (E \circ \imath_{mn})(g_1, g)\bar{f}(g_1)\, dg_1 = L_2(f, \chi; \mathfrak{n}) f^\natural(g).$$

Further, for $\mathfrak{n}$ sufficiently small (contained in $\mathfrak{f}$), assuming absolute convergence of the integral for $m = n$, we have an important non-vanishing result

$$L_2(f, \chi; \mathfrak{n}) = \prod_{p \text{ finite}, p \text{ not dividing } \mathfrak{n}} L_2(f, \chi)_p \neq 0,$$

where $L_2(f, \chi)_p$ is the Euler p-factor of the L-function. This nonvanishing result is obtained without any consideration of the precise nature of the L-functions.

This yields several corollaries regarding the L-function and also regarding a comparison of possible poles of the L-function and the various Eisenstein series that appear.

The crux of the matter is the relatively elementary linear algebra computation in (3.1). From this, it is essentially formal that one obtains (in (3.2))

$$\int_{Z_m(\mathbb{A})G_m(\mathbb{Q})\backslash G_m(\mathbb{A})} (E \circ \iota_{mn})(g_1, g)\bar{f}(g_1)\, dg_1$$
$$= (\text{product of local integrals}) \times E_n(g; \mathfrak{n}, \rho_2, \bar{\chi}; f^{\natural}),$$

where at finite primes p the local integrals are power series in p^{-s}. Though we do not compute these integrals here, they are easily verified to be rational functions in p^{-s} and are the local factors of an automorphic L-function attached to the cuspform f (with some extra factors that are Abelian L-functions of number fields). At least for the case of *holomorphic* cuspforms, the local integrals at infinite primes are indeed expressible in terms of gamma functions. It seems to be an open problem to determine the precise nature of these Archimedean integrals in general.

The brief history of this method is as follows. In [G1], this idea (combined with arithmetic properties of Siegel's Eisenstein series) was applied to groups Sp_n to study the arithmetic of the holomorphic Eisenstein series that arise, and the special values of the L-functions. In [PSR], in effect a somewhat more general version of the case $m = n$ was considered (as a "compactification" of G_n) and the double integral

$$\int_{Z_m(\mathbb{A})G_m(\mathbb{Q})\backslash G_m(\mathbb{A})} \int_{Z_m(\mathbb{A})G_m(\mathbb{Q})\backslash G_m(\mathbb{A})} (E \circ \iota_{mn})(g_1, g)\bar{f}(g_1)\bar{f}^{\natural}(g_2)\, dg_1\, dg_2$$
$$= \langle f, f \rangle L_2(f, \chi; \mathfrak{n})$$

yields an integral representation of the L-function. In [G2] the special case $Sp_1 \times Sp_2 \to Sp_3$ was used as the starting point for a further integration against the restriction along $Sp_1 \times Sp_1 \to Sp_2$, obtaining an integral representation for Rankin triple product L-functions.

Such decomposition results (which are sharper than just integral representation of L-functions) also can be made to yield direct proofs of arithmetic properties of holomorphic automorphic forms (see a forthcoming paper of this author), depending only upon the linear algebra mentioned above and upon the arithmetic of Siegel's Eisenstein series. Of course, the special values of L-functions over totally real number fields are the main ingredient here.

In this paper we have made some unnecessary simplifying assumptions on the "type" of the cuspforms, for the sake of notational and conceptual clarity, but these assumptions may easily be dropped. The present situation does include holomorphic automorphic forms and "waveforms."

1. Automorphic Forms on Some Classical Groups

1.1. Notation and Terminology

Let $M(m, n, R)$ be the space of m-by-n matrices over a ring R, and let $M(n, R) = M(n, n, R)$. Let $S \to S^\tau$ denote the transpose of a matrix $S \in M(m, n, R)$, and denote by S_{ij} the (i,j)-th entry of S. Generally, if $\alpha \to \alpha^\sigma$ is an involution on a ring R, define

$$*: M(m, n, R) \to M(n, m, R)$$

by

$$(A^*)_{ij} = (A^\tau)^\sigma_{ij} = A^\sigma_{ji}.$$

For a ring R, define an element J_n of $M(2n, R)$ by

$$J_n = \begin{pmatrix} 0_n & -\eta 1_n \\ 1_n & 0_n \end{pmatrix}.$$

where $\eta = \pm 1$ is fixed throughout the sequel.

Let F be a number field. Let D be a division algebra of finite dimension N^2 over $\mathbb{Q}$, with an involution σ. Let $\delta = \delta_1 : D \to M(N^2, \mathbb{Q})$ be a fixed rational representation, and let

$$\delta = \delta_n : \mathrm{GL}(n, D) \to \mathrm{GL}(n, M(N^2, \mathbb{Q})) \approx \mathrm{GL}(nN^2, \mathbb{Q})$$

be the natural representation. We will also use $\delta = \delta_n$ to denote the natural extension

$$\delta = \delta_n : \mathrm{GL}(n, D \otimes_{\mathbb{Q}} A) \to \mathrm{GL}(N, A)$$

for any commutative $\mathbb{Q}$-algebra A. Let $\mathfrak{o}$ be an order in D so that $\mathfrak{o}$ contains $\mathbb{Z}$ and so that $\mathfrak{o}$ is stable under σ. For a commutative $\mathbb{Z}$-algebra A, define

$$G_n(A) = \{g \in \mathrm{GL}(2n, \mathfrak{o} \otimes_{\mathbb{Z}} A): g^* J_n g = J_n \quad \text{and} \quad \det \delta(g) = 1\},$$

where we extend σ to $\mathfrak{o} \otimes_{\mathbb{Q}} A$. This gives G_n the structure of a linear algebraic group over $\mathbb{Z}$. For an element of these groups, use an n-by-n block decomposition

$$g = \begin{pmatrix} a & b \\ c & d \end{pmatrix}.$$

Let $\mathbb{A}$ and $\mathbb{A}_0$ be the adeles and finite adeles of $\mathbb{Q}$, respectively. For a finite or infinite prime p of $\mathbb{Q}$, let $\mathbb{Q}_p$ be the p-adic completion of $\mathbb{Q}$, with integers $\mathbb{Z}_p$ if p is finite. Put

$$K_n = \{g \in G_n(\mathbb{R}): \delta(g) \in O(N)\},$$

where $O(N)$ is the usual orthogonal group in $\mathrm{GL}(N, \mathbb{R})$. One can check that K_n is a maximal compact subgroup of $G_n(\mathbb{R})$; further, it is clear that K_n is the group of real points of an algebraic subgroup of G_n defined over $\mathbb{Q}$. Let Z_n be the center of G_n.

Definition. Let $\mathfrak{f}$ be a nonzero ideal in $\mathbb{Z}$. Define the *principal congruence subgroup of $G_n(\mathbb{A}_0)$ of level $\mathfrak{f}$* by

$$\mathbb{K}_n(\mathfrak{f}) = \{g \in G_n(\mathbb{A}_0): g \equiv 1 \text{ modulo } \mathfrak{f}\}.$$

At primes p not dividing $\mathfrak{f}$, g is merely required to be in $G_n(\mathbb{Z}_p)$.

Definition. Let ρ be a $\mathbb{C}^x$-valued representation of K_n. Let ω be a $\mathbb{C}^x$-valued representation of $Z_n(\mathbb{A})/Z_n(\mathbb{Q})$. Fix a nonzero ideal $\mathfrak{f}$ of $\mathbb{Z}$. Let f be a continuous $\mathbb{C}$-valued function on $G_n(\mathbb{A})$ so that

(i) f is left $G_n(\mathbb{Q})$-invariant: $f(\gamma g) = f(g)$ for $\gamma \in G_n(\mathbb{Q})$, $g \in G_n(\mathbb{A})$;

(ii) f is right $\mathbb{K}_n(f)$-invariant for some f: $f(gk) = f(g)$ for $k \in \mathbb{K}(\mathfrak{f})$, $\mathrm{g} \in \mathrm{G_n}(\mathbb{A})$;

(iii) f is right (K_n, ρ)-equivalent: $f(gk) = \rho(k) f(g)$ for $k \in K_n$, $g \in G_n(\mathbb{A})$;

(iv) f is has central character ω: $f(zg) = \omega(z) f(g)$ for $z \in Z_n(\mathbb{A})$, $g \in G_n(\mathbb{A})$;

We say that f is a *weak automorphic form of level $\mathfrak{f}$, weight ρ, and central character ω*. Let $WMfm_n(\mathfrak{f}, \rho, \omega)$ denote the collection of such.

For a finite or infinite prime p, put a Haar measure on $G_n(\mathbb{Q}_p)$ so that, for each finite prime p, $G_n(\mathbb{Z}_p)$ has measure 1. Let $G_n(\mathbb{A})$ have the product measure.

Definition. The associated Petersson inner product $\langle\,,\rangle$ on $WMfm_n(f, \rho, \omega)$ is the integral (if convergent)

$$\langle f_1, f_2\rangle = \int_{Z_n(\mathbb{A})G_n(\mathbb{Q})\backslash G_n(\mathbb{A})} f_1(g)f_2(g)\,dg.$$

Then write $L^2(Z_n(\mathbb{A})G_n(\mathbb{Q})\backslash G_n(\mathbb{A}))$ for the space of measurable $\mathbb{C}$-valued functions f so that $\langle f, f\rangle$ exists. (Modulo the equivalence relation of equality almost everywhere) this is a Hilbert space with inner product $\langle\,,\rangle$.

Definition. Let ρ be a $\mathbb{C}^x$-valued representation of K_n. Let $\mathbb{H}(\infty, \rho)$ be the convolution algebra of infinitely differentiable functions ϕ on $G_n(\mathbb{Q}_\infty)$ so that for $k_1, k_2 \in K_n$ and $g \in G_n(\mathbb{Q}_\infty)$ we have

$$f(k_1gk_2) = \rho(k_1)\rho(k_2)f(g),$$

and which are compactly supported modulo $Z_n(\mathbb{R})$. Convolution is taken on $Z_n(\mathbb{R})\backslash G_n(\mathbb{R})$.

Definition. For a finite prime p of $\mathbb{Z}$, and for a $\mathbb{C}^x$-valued representation ω of $Z(\mathbb{Q}_p)$, let $\mathbb{H}_n(p, \omega)$ denote the convolution algebra of left and right $G_n(\mathbb{Z})$-invariant locally-constant $\mathbb{C}$-valued functions ϕ on $G_n(\mathbb{Q}_p)$ with compact support modulo $Z_n(\mathbb{Q}_p)$, so that $\phi(zg) = \omega(z)f(g)$ for $z \in Z(\mathbb{Q}_p)$ and $g \in G_n(\mathbb{Q}_p)$. We identify $\phi \in \mathbb{H}_n(p, \omega)$ with the convolution operator $f \to f * \phi$ for any (locally integrable) right $G_n(\mathbb{Z}_p)$-invariant function on $G_n(\mathbb{A})$. Convolution is taken on $Z_n(\mathbb{Q}_p)\backslash G_n(\mathbb{Q}_p)$.

Proposition. *Let p be a finite prime of $\mathbb{Z}$ so that p splits the division algebra D (i.e., so that $D \otimes \mathbb{Q}_p$ is isomorphic to a matrix algebra over F for some finite field extension F of $\mathbb{Q}_p$). Then the convolution algebra $\mathbb{H}(p, \omega)$ is commutative. Let f be an ideal of v not divisible by p. Then, for each $\phi \in \mathbb{H}(p, \omega)$ the operator*

$$f \to f * \phi$$

stabilizes $WMfm_n(f, \rho, \omega)$. Further, $\mathbb{H}(p, \omega)$ is closed under taking adjoints with respect to the Petersson inner product on $L^2(Z(\mathbb{A})G_n(\mathbb{Q})\backslash G_n(\mathbb{A}))$.

Proof. The commutativity property will follow from the existence of the anti-involution $g \to g^\tau$ on $G_n(\mathbb{Q}_p)$, which respects a Cartan decomposition of $G_n(\mathbb{Q}_p)$.

Let $\mathfrak{O}$ be the integral closure of $\mathbb{Z}_p$ in F. By elementary linear algebra, since p splits D, we have a decomposition

$$G_n(\mathbb{Q}_p) = G_n(\mathbb{Z}_p)\Delta G_n(\mathbb{Z}_p),$$

where $\Delta \subset G_n(\mathbb{Q}_p)$ consists of diagonal matrices with entries in F. (This is an elementary divisor form over $\mathfrak{O}$). Also, it is elementary that $G_n(\mathbb{Q}_p)$ is stable under $g \to g^\tau$. Note, further, that $Z_n(\mathbb{Q}_p) \subset \Delta$.

We claim first that, for all $g \in G_n(\mathbb{Q}_p)$, and for all $\phi \in \mathbb{H}(p, \omega)$,

$$\phi(g^\tau) = \phi(g).$$

Let $g = k_1\delta k_2$ with each $k_j \in G_n(\mathbb{Q}_p)$, and $\delta \in \Delta$. Note that elements of Δ are fixed by $g \to g^\tau$, and that $G_n(\mathbb{Z}_p)$ is stable under $g \to g^\tau$. Therefore,

$$\phi(g^\tau) = \phi((k_1\delta k_2)^\tau) = \phi(k_2^\tau \delta k_1^\tau) = \phi(\delta) = \phi(k_1\delta k_2) = \phi(g).$$

This proves the claim.

Now let ϕ and ψ be in $\mathbb{H}(p, \omega)$. We have

$$\begin{aligned}
(\phi * \psi)(g)(\phi * \psi)(g^\tau) &= \int_{Z(n,F_p)\backslash G(n,F_p)} \phi(g^\tau h^{-1})\psi(h)\, dh \\
&= \int_{Z_n(\mathbb{Q}_p)\backslash G_n(\mathbb{Q}_p)} \phi(h^{-1})\psi(hg^\tau)\, dh \\
&= \int_{Z_n(\mathbb{Q}_p)\backslash G_n(\mathbb{Q}_p)} \psi(gh^\tau)\phi(h^{-1})\, dh \\
&= \int_{Z_n(\mathbb{Q}_p)\backslash G_n(\mathbb{Q}_p)} \psi(gh^{-1})\phi(h^\tau)\, dh \\
&= \int_{Z_n(\mathbb{Q}_p)\backslash G_n(\mathbb{Q}_p)} \psi(gh^{-1})\phi(h)\, dh \\
&= (\psi * \phi)(g),
\end{aligned}$$

by repeated use of the left and right invariance and use of the previous claim. This proves the commutativity of the convolution algebra.

The remaining assertions are immediate.

Proposition. *The convolution algebra $\mathbb{H}(\infty, \rho)$ is commutative. For each $\phi \in \mathbb{H}(p, \omega)$ the operator*

$$f \to f * \phi$$

stabilizes $WMfm_n(f, \rho, \omega)$. *Further,* $\mathbb{H}(\infty, \rho)$ *is closed under taking adjoints with respect to the Petersson inner product on* $L^2(Z(\mathbb{A})G_n(\mathbb{Q})\backslash G_n(\mathbb{A}))$.

Proof. Similar to the previous.

Definition. Let $WCfm_n(\mathfrak{f}, \rho, \omega)$ be the collection of functions f in $WMfm_n(\mathfrak{f}, \rho, \omega)$ that are also in $L^2(Z_n(\mathbb{A})G_n(\mathbb{Q})\backslash G_n(\mathbb{A}))$ so that for every proper $\mathbb{Q}$-parabolic subgroup P of G_n with unipotent radical U,

$$\int_{U(\mathbb{Q})\backslash U(\mathbb{A})} f(ug)\, du = 0$$

for g in $G_n(\mathbb{A})$ outside a set of measure 0.

Definition. For every finite prime p of $\mathbb{Z}$ splitting D, let $\lambda_p\colon \mathbb{H}(p, \omega) \to \mathbb{C}$ be an algebra homomorphism. Let $\lambda_\infty\colon \mathbb{H}(\infty, \rho) \to \mathbb{C}$ be an algebra homomorphism, and let $\lambda = \{\lambda_p\colon p \le \infty\}$. Define the space of cuspforms on $G_n(\mathbb{A})$ of level $\mathfrak{f}$, weight ρ, central character ω, and type λ to be

$$\begin{aligned} Cfm_n(\mathfrak{f}, \rho, \omega, \lambda) = \{&f \in WCfm_n(\mathfrak{f}, \rho, \omega)\colon f * \phi = \lambda_p(\phi) f \\ &\text{for all finite } p \text{ splitting } D \text{ and } \phi \in \mathbb{H}(p, \omega), \\ &\text{and } f * \phi = \lambda_\infty(\phi) f \qquad \text{for all } \phi \in \mathbb{H}(\infty, \rho)\}. \end{aligned}$$

Corollary. *For distinct triples* (ρ, ω, λ) *and* $(\rho', \omega', \lambda')$ *of data as above,* $Cfm_n(\mathfrak{f}, \rho, \omega, \lambda)$ *is orthogonal to* $Cfm_n(\mathfrak{f}, \rho', \omega', \lambda')$ *in* $L^2(Z_n(\mathbb{A})G_n(\mathbb{Q})\backslash G_n(\mathbb{A}))$.

Definition. Define an involution $f \to f^\natural$ on $\mathbb{C}$-valued functions on $G_n(\mathbb{A})$ by

$$f^\natural(g) = \overline{f(g^\natural)},$$

where $g \to g^\natural$ is the involution on $G_n(\mathbb{A})$ defined by

$$g^\natural = \varepsilon g \varepsilon^{-1},$$

with

$$\varepsilon = \begin{pmatrix} 0_n & \eta 1_n \\ 1_n & 0_n \end{pmatrix},$$

Note that $g \to g^\natural$ stabilizes K_n, and stabilizes $G_n(\mathbb{Z}_p)$.

Lemma. *The ($\mathbb{C}$-antilinear) involution $f \to f^{\natural}$ maps $Cfm_n(\mathfrak{f}, \rho, \omega, \lambda)$ to $Cfm_n(\mathfrak{f}, \rho, \bar{\omega}, \bar{\lambda})$ isometrically. Further, for a finite prime p splitting D, and for $\phi \in \mathbb{H}(p, \omega)$, we have*

$$(f * \phi)^{\natural} = f^{\natural} * \bar{\phi}.$$

Proof. First, it is clear that $f \to f^{\natural}$ is an isometry of $L^2(Z_n(\mathbb{A})G_n(\mathbb{Q})\backslash G_n(\mathbb{A}))$ to itself. The first assertion is straightforward from the fact that $\mathbb{K}_n(f)$ is stabilized by the involution $g \to g^{\natural}$ on G_n and from the fact that $\zeta \to \zeta^{\natural}$ acts trivially on the center of $G_n(\mathbb{A})$ so that the character is inverted by application of $^{\natural}$. Further, $g \to g^{\natural}$ induces a $\mathbb{C}$-antilinear involution on $\mathbb{H}(\infty, \rho)$, since one can check (case by case ...) that $\bar{\rho}(g^{\natural}) = \rho(g)$.

We prove just the last assertion in detail. Once again, we use the fact that there is an elementary divisor decomposition

$$G_n(\mathbb{Q}_p) = G_n(\mathbb{Z}_p)\Delta G_n(\mathbb{Z}_p),$$

where Δ consists of diagonal matrices. Now $g \to g^{\natural}$ stabilizes $G_n(\mathbb{Z}_p)$, and the effect of this involution on Δ is just the same as the effect of conjugation by $J = J_n \in G_n(\mathbb{Z}_p)$, whether $\eta = +1$ or -1. Thus, for

$$g = A\delta B$$

with $\delta \in \Delta$, $A \in G_n(\mathbb{Z}_p)$, $B \in G_n(\mathbb{Z}_p)$, *and for* $\phi \in \mathbb{H}(p, \omega)$, we have

$$\begin{aligned} \phi(g^{\natural}) &= \phi(\varepsilon A\delta B\varepsilon^{-1}) = \phi(\varepsilon A\varepsilon^{-1}\varepsilon\delta\varepsilon^{-1}\varepsilon B\varepsilon^{-1}) \\ &= \phi(\varepsilon\delta\varepsilon^{-1}) = \phi(J\varepsilon J^{-1}) \\ &= \phi(\delta) = \phi(A\delta B) = \phi(g). \end{aligned}$$

Then the assertion of the lemma is immediate from the integral definition of the operator.

1.2. Some Eisenstein Series

For each $0 \le j \le n$, we have $\mathbb{Z}$-subgroups $P_n(j)$ of G_n as follows. Let

$$e_1 = (1 \quad 0 \quad \dots \quad 0)$$
$$e_2 = (0 \quad 1 \quad 0 \quad \dots \quad 0)$$
$$\dots$$

be the standard $\mathfrak{o}$-basis for $\mathfrak{o}^n$ (as row vectors). Let $\mathbb{O}$ be the 0-vector in $\mathfrak{o}^n$. Let

$$\tilde{e}_i = (\mathbb{O} \quad e_i) \in \mathfrak{o}^{2n}.$$

Let F_i be the $\mathfrak{O}$-subspace of $\mathfrak{O}^{2n}$ spanned by $e_1, \ldots, e_i$. Let $P_n(j)$ be the $\mathbb{Z}$-subgroup of G_n consisting of elements g of so that $F_i g = F_i$ for $1 \le i \le j$ (with right multiplication). Let $U_n(j)$ be the unipotent radical of $P_n(j)$. We will explicitly describe some Levi components of these parabolic subgroups.

For a commutative $\mathbb{Z}$-algebra A, define

$$t_j\colon \mathrm{GL}(n-j, \mathfrak{O} \otimes A) \to G_n(A),$$
$$\lambda_j\colon G_j(A) \to G_n(A),$$

by

$$t_j(\alpha) = \begin{pmatrix} \alpha & 0 & 0 & 0 \\ 0 & 1_j & 0 & 0 \\ 0 & 0 & \alpha^{*-1} & 0 \\ 0 & 0 & 0 & 1_j \end{pmatrix}$$

$$\lambda_j(g) = \begin{pmatrix} 1_{n-j} & 0 & 0 & 0 \\ 0 & a & 0 & b \\ 0 & 0 & 1_{n-j} & 0 \\ 0 & c & 0 & d \end{pmatrix},$$

for

$$g = \begin{pmatrix} a & b \\ c & d \end{pmatrix} \in G(j, A).$$

Let

$$L_n(j, A) = \{t_j(\alpha)\colon \alpha \in \mathrm{GL}(n-j, \mathfrak{O} \otimes A)\},$$
$$G_n(j, A) = \{\lambda_j(g)\colon g \in G_j(A)\},$$
$$H_n(j, A) = L_n(j, A)U_n(j, A).$$

This gives each of these subgroups the structure of a linear algebraic group over $\mathbb{Z}$, and they are $\mathbb{Z}$-subgroups of G_n. The product $G_n(j)L_n(j)$ is a Levi component for $P_n(j)$, and

$$U_n(j) = \left\{ g \in G_n\colon g = \begin{pmatrix} 1_{n-j} & * & * & * \\ 0 & 1_j & * & 0_j \\ 0 & 0 & 1_{n-j} & 0 \\ 0 & 0 & * & 1_j \end{pmatrix} \right\}.$$

We will only consider certain types of Eisenstein series. Fix $0 < n \in \mathbb{Z}$, $0 \le j < n$; let $G = G_n$, $P = P_n$, $U = U_n$. Fix an ideal $\mathfrak{m}$ of $\mathbb{Z}$, a

$\mathbb{C}^x$-valued representation ρ of $K = K_n$, a character ω on $Z(\mathbb{A})/Z(\mathbb{Q}) = Z_n(\mathbb{A})/Z_n(\mathbb{Q})$, and a $\mathbb{C}^x$-valued representation χ on $P(\mathbb{A})$ that is trivial on $P(\mathbb{Q})$. Let ϕ be a function on $G(\mathbb{A})$ so that

(i) ϕ is left $(P(\mathbb{A}), \chi)$-equivariant.
(ii) ϕ is left $U(\mathbb{A})$-invariant
(iii) ϕ is right $\mathbb{K}_n(\mathfrak{m})$-invariant
(iv) ϕ is $(Z(\mathbb{A}), \omega)$-equivariant
(v) ϕ is right (K, ρ)-equivariant.

Definition. With ϕ as just above, for $g \in G(\mathbb{A})$, define the Eisenstein series on $G(\mathbb{A})$ attached to ϕ by

$$E_n(g; \phi) = \sum_{\gamma \in P(\mathbb{Q}) \backslash G(\mathbb{Q})} \phi(\gamma g)$$

(if this sum converges). Clearly there are some compatibility conditions among the ϕ, χ, ω which are necessary to avoid trivial vanishing of this Eisenstein series.

First we will define some Eisenstein series of a very simple sort. Fix positive integers $m \leq n$ and put $m + n = N$. Define

$$Q_m = (0 \quad 1_m) \in M(m, n, D)$$

$$\xi_m = \begin{pmatrix} 1_m & 0 & 0 & 0 \\ 0 & 1_n & 0 & 0 \\ 0_m & \eta Q_m & 1_m & 0 \\ Q_m^* & 0_n & 0 & 1_n \end{pmatrix} \in G(\mathbb{Q}).$$

Let ξ_{m0} be the projection of ξ_m to $G_N(\mathbb{A}_0)$. Let $\mathfrak{n}$ be a nonzero ideal in $\mathbb{Z}$. Let χ be a $\mathbb{C}^x$-valued representation of $\mathbb{J}/\mathbb{Q}^x$, where $\mathbb{J}$ is the idele group of $\mathbb{Q}$. Let ρ be a $\mathbb{C}^x$-valued representation of K_N. Define a function ψ on $P_N(0, \mathbb{A})$ by

$$\psi(t_0(\alpha)u) = \chi(\det \delta(\alpha)),$$

where $\alpha \in \mathrm{GL}(N, \mathfrak{o} \otimes \mathbb{A})$, and t_0: $\mathrm{GL}(N, \mathfrak{o} \otimes \mathbb{A}) \to L_N(0, \mathbb{A}) \subset P_N(0, \mathbb{A}) \subset G_N(\mathbb{A})$ is the map defined above. For $z = \zeta 1_{2n} \in Z_N(\mathbb{A})$, with $\zeta \in \mathbb{J}$, put $\omega(z) = \chi(\det \zeta 1_N)$. Then define

$$\phi(zpk_\infty k_0 \xi_{m0}) = \omega(z)\psi(p)\rho(k_\infty)$$

for $z \in Z_N(\mathbb{A})$, $p \in P_N(0, \mathbb{A})$, $k_\infty \in K_N$, $k_0 \in \mathbb{K}_N(\mathfrak{n})$, and $\phi \equiv 0$ off the set

$$Z_N(\mathbb{A})P_N(0, \mathbb{A})K_N\mathbb{K}_N(\mathfrak{n})\xi_{m0}.$$

It is immediate that (without concern for convergence) this function ϕ satisfies the general hypotheses above to allow formation of an Eisenstein series. It is $(Z_N(\mathbb{A}), \tilde{\chi})$-equivariant, where for $\zeta \in \mathbb{J}$

$$\tilde{\chi}(\zeta 1_{2N}) = \chi(\det \delta(\zeta 1_N)).$$

Definition (1.3.1). With ϕ as just above, we define an Eisenstein series by

$$E_N(g; \mathfrak{n}, \rho, \chi, \mathfrak{m}) = E_N(g; \phi).$$

We have a second important class of examples of Eisenstein series: those attached to cuspforms on smaller groups in the same family. Let m, $\mathfrak{n}$, N, ξ_{m0}, ρ, n be as above. Let χ be a character on $\mathbb{J}/\mathbb{Q}^x$. Let $f \in Cfm_m(\mathfrak{f}, \rho_1, \omega, \lambda)$, where ρ restricted to K_m is ρ_1 and $\mathfrak{n} \subset \mathfrak{f}$. Define a function ψ on $P_N(m, \mathbb{A})$ by

$$\psi(t_m(\alpha)\lambda_m(g)u) = \chi(\det \delta(\alpha))f(g),$$

where $\alpha \in \mathrm{GL}(n, \mathfrak{o} \otimes \mathbb{A})$, $g \in G_n(\mathbb{A})$, $u \in U_N(m, \mathbb{A})$, and t_m and λ_m are the maps defined previously. Then define

$$\phi(\xi_m zpk_\infty \xi_{m0}) = \omega(z)\psi(p)\rho(k_\infty)$$

for $z \in Z_N(\mathbb{A})$, $p \in P_N(m, \mathbb{A})$, $k_\infty \in K_N$, $k_0 \in \mathbb{K}_N(\mathfrak{n})$, and $\phi = 0$ off the set

$$Z_N(\mathbb{A})P_N(m, \mathbb{A})K_N\mathbb{K}_N(n)\xi_{m0}.$$

Then this function satisfies the hypotheses necessary to form an Eisenstein series. It is $(Z_N(\mathbb{A}), \tilde{\chi})$-equivariant, where for $\zeta \in \mathbb{J}$

$$\tilde{\chi}(\zeta 1_{2N}) = \chi(\det \delta(\zeta 1_n))\omega(\zeta 1_{2m}).$$

Definition (1.3.2). With ϕ as just above, we define an Eisenstein series by

$$E_N(g; \mathfrak{n}, \rho, \chi; \mathfrak{f}) = E_N(g; \phi).$$

2. Statement of the Theorems

2.1. Imbeddings of Groups

For all positive integers m and n put $N = m + n$ and define a morphism

$$\iota_{mn}: G_m \times G_n \to G_N$$

of linear algebraic groups over $\mathbb{Z}$ by

$$g \times g' \to \begin{pmatrix} a & 0 & b & 0 \\ 0 & a' & 0 & b' \\ c & 0 & d & 0 \\ 0 & c' & 0 & d' \end{pmatrix},$$

where

$$g = \begin{pmatrix} a & b \\ c & d \end{pmatrix}, \quad g' = \begin{pmatrix} a' & b' \\ c' & d' \end{pmatrix}.$$

Remarks. This situation (i.e., with such groups and morphisms) appears to be the most general in which the present formalism works out, though in [PSR] and some other works of those authors integral representations of the L-functions alone are obtained for somewhat larger families of classical groups.

2.2. Some L-Functions Attached to Cuspforms

Let ϕ be as in (1.3.1) with $m = n$, associated to data $\mathfrak{n}$, ρ, χ, $\mathfrak{m}$, used to define the Eisenstein series $E_{2n}(g; \mathfrak{n}, \rho, \chi, \mathfrak{m})$ on $G_{2n}(\mathbb{A})$.

(i) Suppose that $\mathfrak{n}$ is small enough so that if D is not split at a finite prime p of $\mathbb{Z}$, then p divides $\mathfrak{n}$;

(ii) Suppose that $\mathfrak{n}$ is small enough so that, if p does not divide $\mathfrak{n}$, then χ is trivial on $\mathbb{Z}_p^x$.

(iii) Suppose that $\rho \circ \iota_{mm} \approx \rho_1 \otimes \rho_1$ as a representation of $K_m \times K_m$.

Let p be a finite or infinite prime of $\mathbb{Z}$. Define a function $\Phi_p = \Phi_{p,\mathfrak{n},\rho,\chi}$ on $g \in G_m(\mathbb{Q}_p)$ by

$$\Phi_p(g) = \phi(\xi_m \iota_{mm}(g, 1_{2m})).$$

Lemma. *For a finite prime p, the function $\Phi_{p,\mathfrak{n},\rho,\chi}$ is left and right $G_m(\mathbb{Z}_p)$-invariant if p does not divide n.*

Proof. Let k, k_1, $k_2 \in G_m(\mathbb{Z}_p)$. It is a direct linear algebra computation that

$$\xi_m \iota_{mm}(kg, 1_{2m}) = t_0(k^{\natural\,-1})\xi_m \iota_{mm}(g, k^{\natural\,-1}).$$

Further, $\iota_{mm}(G_m(\mathbb{Z}_p) \times G_m(\mathbb{Z}_p)) \subset G_{2m}(\mathbb{Z}_p)$. Therefore,

$$\begin{aligned}\Phi_p(k_1 g k_2) &= \phi(\xi_m \iota_{mm}(k_1 g k_2, 1_{2m})) = \phi(t_0(k_1^{\natural\,-1})\xi_m \iota_{mm}(g, k_1^{\natural\,-1})) \\ &= \chi(\det \delta(t_0(k_1^{\natural\,-1}))\phi(\xi_m \iota_{mm}(g^{-1}, 1_{2m})),\end{aligned}$$

by right $G_{2m}(\mathbb{Z}_p)$-invariance and left $(P_{2m}(0, \mathbb{A}), \chi)$-equivariance. Since χ is trivial on $\mathbb{Z}_p^x$, $\chi(\det \delta(t_0(k_1^{\natural -1})) = 1$. This gives the result.

Corollary. *Let ω be a character on $Z_m(\mathbb{A})/Z_m(\mathbb{Q})$ so that, for $\zeta \in \mathbb{J}$ so that $\zeta 1_{2m} \in Z_m(\mathbb{A})$, we have*

$$\omega(\zeta 1_{2m}) = \chi(\zeta 1_m).$$

Let $\mathfrak{f}$ be a nonzero ideal in $\mathbb{Z}$. Let $\mathfrak{n} \subset \mathfrak{f}$ be a nonzero ideal in $\mathbb{Z}$ contained in all primes $p\mathbb{Z}$ not splitting D. Let f be in $Cfm_m(\mathfrak{f}, \rho_1, \omega, \lambda)$. Assuming convergence of the integral, for a finite prime p not dividing $\mathfrak{n}$ we have

$$f * \Phi_{p,\mathfrak{n},\rho,\chi} = L_2(f, \chi)_p \times f,$$

for some $L_2(f, \chi)_p \in \mathbb{C}$ *independent of* $\mathfrak{n}$, *where the convolution is on* $G_m(\mathbb{Q}_p)$.

Proof. First, it is easy to check that $\Phi_{p,\mathfrak{n},\rho,\chi}$ is $(Z_m(\mathbb{A}), \omega)$-equivariant. Since $\Phi_{p,\mathfrak{n},\rho,\chi}$ is left and right $G_m(\mathbb{Z}_p)$-invariant, it is in some closure of $\mathbb{H}(p, \omega)$.

Lemma. *For $p = \infty$, the function $\Phi_{\infty,\mathfrak{n},\rho,\chi}$ is left and right (K_m, ρ_1)-equivariant.*

Proof. Let $k, k_1, k_2 \in G_m(\mathbb{Z}_p)$. It is a direct linear algebra computation that

$$\xi_m \iota_{mm}(kg, 1_{2m}) = t_0(k^{\natural -1})\xi_m \iota_{mm}(g^{-1}, k^{\natural -1}).$$

Further, $\iota_{mm}(K_m \times K_m) \subset K_{2m}$. Therefore,

$$\begin{aligned}\Phi_p(k_1 g k_2) &= \phi(\xi_m \iota_{mm}(k_1 g k_2, 1_{2m})) = \phi(t_0(k_1^{\natural -1})\xi_m \iota_{mm}(g, k_1^{\natural -1}))\rho_1(k_2) \\ &= \chi(\det \delta(t_0(k_2^{\natural -1}))\rho_1(k_1^{\natural -1})\rho_1(k_2)\phi(\xi_m \iota_{mm}(g, 1_{2m})),\end{aligned}$$

by right K_{2m}-invariance and left $(P_{2m}(0, \mathbb{A}), \chi)$-equivariance. Case-by-case one finds that

$$\det \delta(t_0(K_m) = \{1\}.$$

Further, (case-by-case) one has $\rho_1(k_1^{\natural -1}) = \rho_1(k_1)$. This gives the result.

Corollary. *Let ω be a character on $Z_m(\mathbb{A})/Z_m(\mathbb{Q})$ so that for $\zeta \in \mathbb{J}$, $\zeta 1_{2m} \in Z_m(\mathbb{A})$, we have*

$$\omega(\zeta 1_{2m}) = \chi(\zeta 1_m).$$

Let f be in $Cfm_m(\mathfrak{f}, \rho_1, \omega, \lambda)$. Assuming convergence of the integral, for $p = \infty$ we have

$$f * \Phi_{p,\mathfrak{n},\rho,\chi} = L_2(f, \chi,)_\infty \times f,$$

Proof. Again, $\Phi_{p,\mathfrak{n},\rho,\chi}$ is $(Z_m(\mathbb{A}), \omega)$-equivariant. Since $\Phi_{p,\mathfrak{n},\rho,\chi}$ is left and right K_m-invariant, it is in a weak closure of $\mathbb{H}(\infty, p)$.

Definition. For a nonzero ideal $\mathfrak{n} \subset \mathfrak{f}$ of $\mathbb{Z}$ containing all finite primes not splitting D, define

$$L_2(f, \chi; \mathfrak{n}) = L_2(f, \chi)_\infty \times \prod_{p\,\text{not dividing}\,\mathfrak{n}} L_2(f, \chi)_p.$$

2.3. The Main Formula

(i) Take $m \leq n$.

(ii) Let ω be a character on $Z_m(\mathbb{A})/Z_m(\mathbb{Q})$. Suppose that there is a character χ on $\mathbb{J}/\mathbb{Q}^x$ so that for $z = \zeta 1_{2m} \in Z_m(\mathbb{A})$ (with $\zeta \in \mathbb{J}$), $\omega(z) = \chi(\det \delta(\zeta 1_m))$.

(iii) Let n be a nonzero ideal of $\mathbb{Z}$ contained in an ideal f of $\mathbb{Z}$.

(iv) Let ρ be a $\mathbb{C}^x$-valued representation of K_{m+n}. Put $\rho \circ \iota_{mn} \approx \rho_1 \otimes \rho_2$ and suppose that $\rho_2 \circ \iota_{n-m,m} \approx \rho_3 \otimes \rho_1$ for some representation ρ_3 of K_{n-m}.

Remarks. The (iv) is the only hypothesis here that is really not sharp. This is because we are restricting our attention to one-dimensional representations of the maximal compact subgroups of the real points of the groups.

Theorem. *Let $f \in Cfm_m(\mathfrak{f}, \rho_1, \omega, \lambda)$. Let $E(g) = E_{m+n}(g; \mathfrak{n}, \rho, \chi, \mathfrak{m})$ be the Eisenstein series on $G_{m+n}(\mathbb{A})$ as in (1.3.1). Then for $g \in G_n(\mathbb{A})$, assuming absolute convergence of the integral, we have*

$$\int_{Z_m(\mathbb{A})G_m(\mathbb{Q})\backslash G_m(\mathbb{A})} (E \circ \iota_{mn})(g_1, g)\bar{f}(g_1)\, dg_1 = L_2(f, \chi; \mathfrak{n})E_n(g; \mathfrak{n}, \rho_2, \bar{\chi}; f^\natural),$$

where $E_n(g; \mathfrak{n}, \rho_2, \bar{\chi}; f^\natural)$ is the Eisenstein series attached to the cuspform

$$f^\natural \in Cfm_m(f, \rho_1, \bar{\omega}, \bar{\lambda})$$

as in (1.3.2), *and where* $L_2(f, \chi; \mathfrak{n})$ *is as in the previous section. In particular, for* $m = n$, *this is*

$$\int_{Z_m(\mathbb{A})G_m(\mathbb{Q})\backslash G_m(\mathbb{A})} (E \circ \iota_{mn})(g_1, g)\bar{f}(g_1)\, dg_1 = L_2(f, \chi; \mathfrak{n}) f^{\natural}(g).$$

Further, for $\mathfrak{n}$ *sufficiently small (contained in* $\mathfrak{f}$), *assuming absolute convergence of the integral for* $m = n$, *we have the nonvanishing*:

$$\prod_{p\,\text{finite}, p\,\text{not dividing}\,\mathfrak{n}} L_2(f, \chi)_p \neq 0.$$

2.4. Corollaries

Corollary. *With hypotheses and notation as above, we have*

$$\int_{Z_m(\mathbb{A})G_m(\mathbb{Q})\backslash G_m(\mathbb{A})} \int_{Z_m(\mathbb{A})G_m(\mathbb{Q})\backslash G_m(\mathbb{A})} (E \circ \iota_{mm})(g_1, g_2)\bar{f}(g_1) f^{\natural}(g_2)\, dg_1\, dg_2$$
$$= L_2(f, \chi; \mathfrak{n})\langle f^{\natural}, f^{\natural}\rangle = L_2(f, \chi; \mathfrak{n})\langle f, f\rangle.$$

Proof. From the case $m = n$ of the theorem, upon integrating in g_2 against $\bar{f}^{\natural}(g_2)$ we obviously obtain the result.

It is well known that every character χ on $\mathbb{J}/\mathbb{Q}^{x}$ is of the form

$$\chi(\alpha) = \|\alpha\|^{s}\chi_1(\alpha),$$

where χ_1 is unitary, and $s \in \mathbb{C}$. Write $s = s(\chi)$.

Corollary. *The L-function* $L_2(f, \chi; \mathfrak{n})$ *has a meromorphic analytic continuation in* $s(\chi)$ *to a meromorphic function on* $s \in \mathbb{C}$.

Proof. From [L], for example, the Eisenstein series $E_{m+n}(g; \mathfrak{n}, \rho, \chi, \mathfrak{m})$ has a meromorphic analytic continuation in $s(\chi)$ and is of "moderate growth at infinity." It is also well known that cuspforms are of "rapid decay at infinity." Therefore, the double integral in the previous corollary is absolutely convergent for $s(\chi)$ away from the poles of $E_{m+n}(g; \mathfrak{n}, \rho, \chi, \mathfrak{m})$.

Corollary. *Take* $\mathfrak{n}$ *sufficiently small so that* $L_2(f, \chi; \mathfrak{n})$ *is not identically* 0. *The Eisenstein series* $E_n(g; \mathfrak{n}, \rho_2, \bar{\chi}; f^{\natural})$ *has a meromorphic analytic continuation in* $s(\chi)$ *and has poles at worst at the poles of*

$$L_2(f, \chi; \mathfrak{n})^{-1} E_{m+n}(g; \mathfrak{n}, \rho, \chi, \mathfrak{m}).$$

Proof. The analytic continuation is already known, from [L]. Then the same argument as in the previous proof, applied to the integral representation of the theorem, gives the result.

3. Proof of the Main Formula

3.1. Double Coset Spaces

Let $\iota = \iota_{mn}$, and put $N = m + n$.

Proposition. *Take $m \leq n$. The double coset space*

$$P_N(0, \mathbb{Q})\backslash G_N(\mathbb{Q})/G_m(\mathbb{Q}) \times G_n(\mathbb{Q})$$

has irredundant representatives

$$\xi_k = \begin{pmatrix} 1_n & 0 & 0 & 0 \\ 0 & 1_{n'} & 0 & 0 \\ 0 & \eta Q_k & 1_n & 0 \\ Q_k^* & 0 & 0 & 1_{n'} \end{pmatrix}_k,$$

with $k = 0, 1, \ldots, m$, where

$$Q_k = \begin{pmatrix} 0 & 0 & 0 \\ 0 & 0 & 1_k \end{pmatrix}.$$

Proof. It is elementary that

$$P_N(0, \mathbb{Q})\backslash G_N(\mathbb{Q}) \approx \mathrm{GL}(N, D)\backslash Y,$$

where

$$Y = \{(c \quad d) \in M(N, 2N, D) \colon cd^* = \eta\, dc^* \quad \text{and} \quad (c \quad d) \text{ is of rank } N\} \subset M(N, 2N, D).$$

The isomorphism is given by

$$P_N(0, F)\begin{pmatrix} a & b \\ c & d \end{pmatrix} \to \mathrm{GL}(N, D)(c \quad d).$$

We will determine irredundant representatives for

$$\mathrm{GL}(N, D)\backslash Y/G_m(\mathbb{Q}) \times G_n(\mathbb{Q}).$$

By a right multiplication by a suitable element of

$$U_m(0, F)U_n(0, F) \subset G_m(\mathbb{Q}) \times G_n(\mathbb{Q})$$

we may suppose without loss of generality that $\operatorname{rank}(d) = N$. Then by left multiplication by $d^{-1} \in GL(N, D)$ we may suppose that $d = 1_N$. For $(c \quad 1) \in Y$, write

$$(c \quad 1) = \begin{pmatrix} c_{11} & c_{12} & 1_m & 0 \\ c_{21} & c_{22} & 0 & 1_n \end{pmatrix},$$

where c_{11} is m-by-m, etc. Since $(c \quad 1) \in Y$, $c^* = \eta c$. By a right multiplication by the obvious element of

$$J_m U_m(0, F) J_m^{-1} \times J_n U_n(0, F) J_n^{-1},$$

we reduce $(c \quad 1)$ to the form

$$(c \quad 1) = \begin{pmatrix} 0 & c_{12} & 1_m & 0 \\ c_{21} & 0 & 0 & 1_n \end{pmatrix},$$

with $c_{21} = \eta c_{12}^*$. There exist $A \in GL(m, D)$, $B \in GL(n, D)$ such that

$$A c_{12} B = \eta Q_k,$$

for $k \in \{0, 1, \ldots, n\}$; this is merely an elementary divisor form over D. Then left multiplication by

$$\begin{pmatrix} A & 0 \\ 0 & B^* \end{pmatrix} \in GL(N, F),$$

and right multiplication by

$$t_k(B) t_k(A^*) \in L_m(0, \mathbb{Q}) L_n(0, \mathbb{Q}) \subset G_m(\mathbb{Q}) \times G_n(\mathbb{Q})$$

puts $(c \quad d)$ into the form

$$\begin{pmatrix} 0 & \eta Q_k & 1_m & 0 \\ Q_k^* & 0 & 0 & 1_n \end{pmatrix}.$$

Clearly this is the image in X of ξ_k. This shows that $\{\xi_k : k = 0, \ldots, m\}$ is a set of representatives for the double coset space.

On the other hand, the action of $GL(N, D) \times G_m(\mathbb{Q}) \times G_n(\mathbb{Q})$ on

$$\begin{pmatrix} c_{11} & c_{12} & d_{11} & d_{12} \\ c_{21} & c_{22} & d_{21} & d_{22} \end{pmatrix} \in Y$$

does not alter the rank of either of

$$\begin{pmatrix} c_{11} & d_{11} \\ c_{21} & d_{21} \end{pmatrix} \qquad \begin{pmatrix} c_{12} & d_{12} \\ c_{22} & d_{22} \end{pmatrix}.$$

The rank of the corresponding matrices

$$\begin{pmatrix} 0 & 1 \\ Q_k^* & 0 \end{pmatrix} \qquad \begin{pmatrix} \eta Q_k & 0 \\ 0 & 1 \end{pmatrix}$$

is uniquely determined by k. This proves the irredundancy of these representatives.

Definition. Define (as before) an involution $G_k(\mathbb{Q}) \to G_k(\mathbb{Q})$ by

$$g^{\natural} = \varepsilon g \varepsilon^{-1},$$

where

$$\varepsilon = \begin{pmatrix} 0 & \eta 1_k \\ 1_k & 0 \end{pmatrix}.$$

Proposition. *Using notation of (1.2), the isotropy subgroup in $G_m(\mathbb{Q}) \times G_n(\mathbb{Q})$ of*

$$P_N(0, \mathbb{Q})\xi_k \in P_N(0,\mathbb{Q})\backslash G_N(\mathbb{Q})$$

is

$$\Theta_k(\mathbb{Q}) = \{(g_1, g_2) \in G_m(\mathbb{Q}) \times G_n(\mathbb{Q})\colon g_1 = \zeta_1 \lambda_k(h),\, g_2 = \zeta_2 \lambda_k(h^{\natural}),$$
$$\textit{with } \zeta_1 \in H_m(k, \mathbb{Q}) \subset G_m(\mathbb{Q}),\, \zeta_2 \in H_n(k, \mathbb{Q}) \subset G_n(\mathbb{Q}), \textit{ and } h \in G_k(\mathbb{Q})\};$$

we have

$$\Theta_k(\mathbb{Q}) \subset P_m(k, \mathbb{Q}) \times P_n(k, \mathbb{Q}).$$

Proof. Let $(g_1, g_2) \in G_m \times G_n$. The condition

$$P_N(0, \mathbb{Q})\xi_k \iota(g_1, g_2) = P_N(0, \mathbb{Q})\xi_k$$

is equivalent to the requirement that, in an N-by-N block decomposition, $\xi_k \iota(g_1, g_2)\xi_k^{-1}$ has (2, 1)-block 0. Put

$$g_1 = \begin{pmatrix} a & b \\ c & d \end{pmatrix}, \qquad g_2 = \begin{pmatrix} a' & b' \\ c' & d' \end{pmatrix}.$$

The (2, 1) block of $\xi_k \iota(g_1, g_2)\xi_k^{-1}$ is

$$\begin{pmatrix} 0 & \eta Q_k & 1 & 0 \\ Q_k^* & 0 & 0 & 1 \end{pmatrix} \begin{pmatrix} a & 0 & b & 0 \\ 0 & a' & 0 & b' \\ c & 0 & d & 0 \\ 0 & c' & 0 & d' \end{pmatrix} \begin{pmatrix} 1 & 0 \\ 0 & 1 \\ 0 & -\eta Q_k \\ -Q_k^* & 0 \end{pmatrix}$$
$$= \begin{pmatrix} b - \eta Q_k b' Q_k^* & \eta Q_k a' - \eta\, d Q_k \\ Q_k^* a - d' Q_k^* & c' - \eta Q_k^* c Q_k \end{pmatrix}.$$

We must determine necessary and sufficient conditions on g_1 and g' so that this is 0. Decompose each of a, b, c, d, a', b', c', d' into blocks a_{ij}, b_{ij},

c_{ij}, d_{ij}, a'_{ij}, b'_{ij}, c'_{ij}, d'_{ij} where a_{11}, b_{11}, c_{11}, d_{11}, a'_{11}, b'_{11}, c'_{11}, d'_{11} are k-by-k, etc. Then the latter expression is

$$\begin{pmatrix} b_{11} & b_{12} & 0 & -\eta d_{12} \\ b_{21} & b_{22}-\eta b'_{22} & \eta a'_{21} & \eta a'_{22}-\eta d_{22} \\ 0 & -d'_{12} & c'_{11} & c'_{12} \\ a_{21} & a_{22}-d'_{22} & c'_{21} & c'_{22}-\eta c_{22} \end{pmatrix}$$

Therefore, the condition that the (2, 1) block be 0 is that

$$g_1 = \begin{pmatrix} a_{11} & a_{12} & b_{11} & b_{12} \\ 0 & a_{22} & b_{21} & b_{22} \\ 0 & 0 & d_{11} & 0 \\ 0 & c_{22} & d_{21} & d_{22} \end{pmatrix} = \begin{pmatrix} * & * & * & * \\ 0 & 1 & * & * \\ 0 & 0 & * & 0 \\ 0 & 0 & * & 1 \end{pmatrix} \begin{pmatrix} 1 & 0 & 0 & 0 \\ 0 & a_{22} & 0 & b_{22} \\ 0 & 0 & 1 & 0 \\ 0 & c_{22} & 0 & d_{22} \end{pmatrix}$$

$$g_2 = \begin{pmatrix} a'_{11} & a'_{12} & b'_{11} & b'_{12} \\ 0 & d_{22} & b'_{21} & \eta c_{22} \\ 0 & 0 & d'_{11} & 0 \\ 0 & \eta b_{22} & d'_{21} & a_{22} \end{pmatrix} = \begin{pmatrix} * & * & * & * \\ 0 & 1 & * & * \\ 0 & 0 & * & 0 \\ 0 & 0 & * & 1 \end{pmatrix} \begin{pmatrix} 1 & 0 & 0 & 0 \\ 0 & d_{22} & 0 & \eta c \\ 0 & 0 & 1 & 0 \\ 0 & \eta b_{22} & 0 & a_{22} \end{pmatrix}.$$

This gives the indicated decomposition of g_1 and g_2.

Proposition. *Take* $m \leq n$, $\iota = \iota_{mn}$. *For* g, $g_1 \in G_m(\mathbb{A}) \approx G_n(m, \mathbb{A})$, $h \in H_n(m, \mathbb{A})$, $g_2 \in G_n(\mathbb{A})$,

$$\xi_n \iota(g^\natural g_1, ghg_2) = h_1 h_2 \xi_n \iota(g_1, g_2),$$

with

$$g = \begin{pmatrix} a & b \\ c & d \end{pmatrix},$$

$$h = \begin{pmatrix} A & u & S & v \\ 0 & 1 & w & 0 \\ 0 & 0 & A^{*-1} & 0 \\ 0 & 0 & u' & 1 \end{pmatrix}$$

$$h_1 = t_N \begin{pmatrix} a & 0 & b \\ 0 & 1 & 0 \\ c & 0 & d \end{pmatrix}$$

$$h_2 = t_N \begin{pmatrix} 1 & w & 0 \\ 0 & A^{*-1} & 0 \\ 0 & u' & 1 \end{pmatrix}.$$

Proof. Direct calculation.

3.2. Local Integrals

After the linear algebra of the previous section, the proof is quite formal. Let notation be as in previous sections and hypotheses as in Section 2.

Proposition. *Let $f \in Cfm_m(\mathfrak{f}, \rho_1, \omega, \lambda)$. For $g \in G_N(\mathbb{A})$, put*

$$\Psi(g) = \sum_{\zeta \in P_n(m, \mathbb{Q}) \backslash G_n(\mathbb{Q})} \phi(\xi_m \iota_{mn}(1_{2m}, \zeta) g).$$

Then with $g_2 \in G_n(\mathbb{A})$ we have

$$\int_{Z_m(\mathbb{A}) G_m(\mathbb{Q}) \backslash G_m(\mathbb{A})} E_N(\iota_{mn}(g_1, g_2); \mathfrak{f}, \rho, \chi, \mathfrak{m}) \bar{f}(g_1)\, dg_1$$
$$= \int_{Z_m(\mathbb{A}) \backslash G_m(\mathbb{A})} \Psi(\iota_{mn}(g_1, g_2)) \bar{f}(g_1)\, dg_1.$$

Proof. We use the coset calculations of (3.1) to write

$$E_N(g; \mathfrak{f}, \rho, \chi, \mathfrak{m}) = \sum_{k=0,\ldots,m} \sum_{\gamma \in \Theta_k(\mathbb{Q}) \backslash \iota_{mn}(G_m(\mathbb{Q}) \times G_n(\mathbb{Q}))} \phi(\xi_k \gamma g).$$

As noted in (3.1), we can take representatives

$$\iota_{mn}(H_m(k, \mathbb{Q}) \backslash G_m(\mathbb{Q}) \times P_n(k, \mathbb{Q}) \backslash G_n(\mathbb{Q}))$$

for the coset space

$$\Theta_k(\mathbb{Q}) \backslash \iota_{mn}(G_m(\mathbb{Q}) \times G_n(\mathbb{Q})).$$

Therefore,

$$\int_{Z_m(\mathbb{A}) G_m(\mathbb{Q}) \backslash G_m(\mathbb{A})} E_N(\iota_{mn}(g_1, g_2); \mathfrak{f}, \rho, \chi, \mathfrak{m}) f(g_1)\, dg_1$$
$$= \sum_{k=0,\ldots,m} \int_{Z_m(\mathbb{A}) G_m(\mathbb{Q}) \backslash G_m(\mathbb{A})} \sum_{\gamma, \zeta} \phi(\xi_k \iota_{mn}(\gamma g_1, \zeta g_2)) \bar{f}(g_1)\, dg_1$$

where γ is summed over $H_m(k, \mathbb{Q}) \backslash G_m(\mathbb{Q})$ and ζ is summed over $P_n(k, \mathbb{Q}) \backslash G_n(\mathbb{Q})$. For each index k, we unwind the integral in g_1:

$$\int_{Z_m(\mathbb{A}) G_m(\mathbb{Q}) \backslash G_m(\mathbb{A})} \sum_{\gamma, \zeta} \phi(\xi_k \iota_{mn}(\gamma g_1, \zeta g_2)) \bar{f}(g_1)\, dg_1$$
$$= \int_{Z_m(\mathbb{A}) H_m(k, \mathbb{Q}) \backslash G_m(\mathbb{A})} \sum_{\zeta} \phi(\xi_k \iota_{mn}(g_1, \zeta g_2)) \bar{f}(g_1)\, dg_1.$$

By the last proposition of (3.1), for $k < m$, the function $q(g_1) = \phi(\xi_k \iota_{mn}(g_1, \zeta g_2))$ of g_1 is left invariant by the normal subgroup $U_m(k, \mathbb{A})$ of $H_m(k, \mathbb{A})$. Write

$$Z_m(\mathbb{A})H_m(k, \mathbb{Q})\backslash G_m(\mathbb{A}) \approx H_m(k, \mathbb{Q})\backslash H_m(k, \mathbb{A}) \times Z_m(\mathbb{A})H_m(k, \mathbb{A})\backslash G_m(\mathbb{A}).$$

With $h \in L_m(k, \mathbb{Q})\backslash L_m(k, \mathbb{A})$, $u \in U_m(k, \mathbb{Q})\backslash U_m(k, \mathbb{A})$, $g \in Z_m(\mathbb{A})H_m(k, \mathbb{A})\backslash G_m(\mathbb{A})$

$$\int_{Z_m(\mathbb{A})H_m(k,\mathbb{Q})\backslash G_m(\mathbb{A})} \sum_\zeta \phi(\xi_k \iota_{mn}(g_1, \zeta g_2))\bar{f}(g_1)\, dg_1$$

$$= \iiint \sum_\zeta \Phi(\xi_k \iota_{mn}(uhg, \zeta g_2))\bar{f}(uhg)\, du\, dh\, dg.$$

Now use the fact that f is a cuspform, together with the $U_m(k, \mathbb{A})$-invariance of ϕ:

$$\int_{U_m(k,\mathbb{Q})\backslash U_m(k,\mathbb{A})} \sum_\zeta \phi(\xi_k \iota_{mn}(uhg, \zeta g_2))\bar{f}(uhg)\, du$$

$$= \sum_\zeta \phi(\xi_k \iota_{mn}(hg, \zeta g_2)) \int_{U_m(k,\mathbb{Q})\backslash U_m(k,\mathbb{A})} \bar{f}(uhg)\, du$$

$$= \sum_\zeta \phi(\xi_k \iota_{mn}(hg, \zeta g_2)) \times 0 = 0.$$

Therefore, the parts of the Eisenstein series with $k < m$ make no contribution to the integral against a cuspform. This gives the result.

Proposition. *Take* g_0 *and* $g_1 \in G_m(\mathbb{A})$, $h = t_m(A) \in L_n(m, \mathbb{A})$, $u \in U_n(m, \mathbb{A})$, $k_0 \in \prod_p G_n(\mathbb{Z}_p)$, $k \in K_n$. *Then*

$$\phi(\xi_m \iota_{mn}(g_1, \lambda_m(g_0)hukk_0)) = \rho_2(k)\chi(\det \delta(A)) \prod_{p \le \infty \text{ not dividing } \mathfrak{n}} \Phi_p(g_0^{\natural\,-1} g_1)$$

$$= \rho_2(k)\chi(\det \delta(A)) \prod_{p \le \infty \text{ not dividing } \mathfrak{n}} \Phi_p(g_1^{\natural\,-1} g_0)$$

if $\lambda_m(g_1^{\natural\,-1} g_0)hukk_0 \in H_n(m, \mathbb{A})$ *modulo* $\mathfrak{n}$. *Otherwise, the left-hand side is* 0.

Proof. From (3.1),

$$\phi(\xi_m \iota_{mn}(g_1, \lambda_m(g_0)hukk_0)) = \phi(\xi_m \iota_{mn}(1_{2m}, \lambda_m(g_1^{\natural\,-1} g_0)hukk_0)).$$

By the definition of ϕ, $\phi(g) = 0$ unless $g\xi_m^{-1} \in P_{m+n}(0, \mathbb{A})K_{m+n}$ modulo $\mathfrak{n}$. That is, $\phi(\xi_m \iota_{mn}(1_{2m}, \lambda_m(g_1^{\natural -1}g_0)hukk_0)) = 0$ unless

$$\xi_m \iota_{mn}(1_{2m}, \lambda_m(g_1^{\natural -1}g_0)hukk_0)\xi_m^{-1} \in P_{m+n}(0, \mathbb{A})K_{m+n} \text{ modulo } \mathfrak{n}.$$

The calculations of (3.1) apply modulo $\mathfrak{n}$ as well: for $g_2 \in G_n(\mathbb{A})$,

$$\xi_m \iota_{mn}(1_{2m}, g_2)\xi_m^{-1} \in P_{m+n}(0, \mathbb{A})K_{m+n} \text{ modulo } \mathfrak{n}$$

if and only if $g_2 \in H_n(m, \mathbb{A})K_{m+n}$ modulo $\mathfrak{n}$. This gives the last assertion of the proposition.

Now take $\lambda_m(g_1^{\natural -1}g_0)hukk_0 \in H_n(m, \mathbb{A})$ modulo $\mathfrak{n}$, with g_0, g_1, h, u, k, k_0 as in the statement of the proposition. Then the last proposition of (3.1) gives

$$\begin{aligned}\phi(\xi_m \iota_{mn}(1_{2m}, \lambda_m(g_1^{\natural -1}g_0)hukk_0)) &= \phi(\xi_m \iota_{mn}(g_0^{\natural -1}g_1, hukk_0)) \\ &= \phi(h'u'\xi_m \iota_{mn}(g_0^{\natural -1}g_1, 1_{2n}))\rho_2(k) \\ &= \chi(\det \delta(A'))\phi(\xi_m \iota_{mn}(g_0^{\natural -1}g_1, 1_{2n}))\rho_2(k)\end{aligned}$$

for some $h' = t_0(A') \in P_{m+n}(0, \mathbb{A}')$, $u' \in U_{m+n}(0, \mathbb{A})$ with $A' \in \mathrm{GL}(m+n, \mathfrak{o} \otimes \mathbb{A})$ so that

$$\chi(\det \delta(A')) = \chi(\det \delta(A)).$$

This gives the first assertion of the proposition.

From the previous proposition, for g_1 and $\lambda_m(g_0)hukk_0$ as in the statement of that proposition, we have

$$\begin{aligned}&\int_{Z_m(\mathbb{A})G_m(\mathbb{Q})\backslash G_m(\mathbb{A})} \phi(\iota_{mn}(g_1, \lambda_m(g_0)hukk_0); f, \rho, \chi, m)\bar{f}(g_1)\, dg_1 \\ &\quad = \chi(\det \delta(A))\rho_2(k) \int_{Z_m(\mathbb{A})\backslash G_m(\mathbb{A})} \left\{ \prod_{p \le \infty \text{ not dividing } \mathfrak{n}} \Phi_p(g_1^{\natural -1}g_0)\bar{f}(g_1)\, dg_1 \right\}.\end{aligned}$$

Now

$$\begin{aligned}&\int_{Z_m(\mathbb{A})\backslash G_m(\mathbb{A})} \left\{ \prod_{p \le \infty \text{ not dividing } \mathfrak{n}} \Phi_p(g_1^{\natural -1}g_0)\bar{f}(g_1)\, dg_1 \right\} \\ &\quad = \int_{Z_m(\mathbb{A})\backslash G_m(\mathbb{A})} \left\{ \prod_{p \le \infty \text{ not dividing } \mathfrak{n}} \Phi_p(g_1^{\natural -1})\bar{f}(g_0^{\natural}g_1)\, dg_1 \right\} \\ &\quad = \int_{Z_m(\mathbb{A})\backslash Gm(\mathbb{A})} \left\{ \prod_{p \le \infty \text{ not dividing } \mathfrak{n}} \Phi_p(g_1)f^{\natural}(g_0 g_1^{-1})\, dg_1 \right\} \\ &\quad = L_2(f, \chi; \mathfrak{n})f^{\natural}(g_0),\end{aligned}$$

by definition of $L_2(f, \chi; \mathfrak{n})$. Therefore, for $g_2 \in G_n(\mathbb{A})$, using the previous proposition,

$$\int_{Z_m(\mathbb{A})\backslash Gm(\mathbb{A})} \phi(\iota_{mn}(g_1, g_2))\bar{f}(g_1)\, dg_1 = L_2(f, \chi; \mathfrak{n})\phi(\rho_2),$$

where we use ϕ to denote the function used in (1.3.2) to define an Eisenstein series attached to a cuspform f, etc. (with notation as immediately preceding (1.3.2).

From this discussion, for $g \in G_n(\mathbb{A})$,

$$\int_{Z_m(\mathbb{A})G_m(\mathbb{Q})\backslash G_m(\mathbb{A})} E(\iota_{mn}(g_1, g); \mathfrak{f}, \rho, \chi, \mathfrak{m})f(g_1)\, dg_1$$

$$= \sum_{\zeta \in P_n(m,\mathbb{Q})\backslash G_n(\mathbb{Q})} \phi(\zeta g) = L_2(f, \chi; \mathfrak{n})E(g; \mathfrak{n}, \rho, \chi; f),$$

as claimed in the theorem.

Regarding nonvanishing: for $\alpha \in \mathbb{J}$, we can write $\chi(\alpha) = \|\alpha\|^s\chi_1(\alpha)$, with $s \in \mathbb{C}$ and χ_1 unitary. For Re(s) sufficiently large the Eisenstein series is absolutely convergent. For such s, it is almost immediate that $L_2(f, \chi)_p$ is a convergent power series in p^{-s}, and

$$L_2(f, \chi; \mathfrak{n})_0 = \prod_{p \text{ finite not dividing } \mathfrak{n}} L_2(f, \chi)_p$$

is a convergent infinite product. In particular, this product can vanish only if some factor vanishes, and only finitely many factors $L_2(f, \chi)_p$ can vanish. We can choose $\mathfrak{n}$ small enough (yet nonzero) so that all p so that $L_2(f, \chi)_p = 0$ divide $\mathfrak{n}$.

This completes the proof of the results of (2.3).

References

[G1] Garrett, P.B. "Decomposition of Eisenstein series; Applications." In: *Automorphic Forms in Several Variables*, Birkhauser, Boston (1984).

[G2] Garrett, P.B. "Decomposition of Eisenstein series: Rankin triple products," *Annals of Math.* **125** (1987), 209–235.

[L] Langlands, R.P. "On the functional equations satisfied by Eisenstein series," *Springer Lecture Notes* **544**, (1976).

[PSR] Piatetskii–Shapiro, I.I. and Rallis, S. "Rankin Integral Representations of L-functions for Classical Groups." In: *Modular Forms*, R. Rankin, ed., Ellis-Horwood, Chichester (1985).

14 Recent Results on Automorphic L-Functions*

S. GELBART

Introduction

We shall describe some of the latest results concerning the analytic continuation of Langlands' automorphic L-functions. A lengthy discussion of these kinds of results, together with some historical background, is the subject matter of [GeSh], written jointly with F. Shahidi. The present paper may be viewed as a complement to [GeSh], and the reader is referred to that work for any unexplained notation or definitions.

Given a connected reductive group G defined over F, an automorphic cuspidal representation $\pi = \otimes_v \pi_v$ of $G_{\mathbb{A}_F}$, and a finite dimensional representation r of the L-group ${}^LG = {}^LG^0 \times \Gamma_F$, the problem is to analytically continue the (partial) Langlands L-function

$$L_S(s, \pi, r) = \prod_{v \notin S} L(s, \pi_v, r_v),$$

where S contains the primes ramified for π or G, and

$$L(s, \pi_v, r_v) = [\det(I - r_v(t_{\pi_v}) g_v^{-s})]^{-1}$$

for $v \notin S$.

* Talk delivered at the Conference on Number Theory, Trace Formulas, and Discrete Groups, in honor of Atle Selberg (Oslo, Norway, June 1987).

NUMBER THEORY,
TRACE FORMULAS and
DISCRETE GROUPS

ISBN 0-12-067570-6

We begin by recalling the two main methods used to establish the functional equation and meromorphic properties of these L-functions. In order to underscore the advantages and drawbacks of each approach, we focus first (in Section 3) on the example of $G = Sp(n)$. We also describe (in Section 4) how these methods can be combined to yield new results beyond the reach of either method applied alone (cf. [GePS] and [Sha1]). A discussion of further works in progress on exceptional groups and Poincaré series is left for the concluding section.

We note that our talk in Oslo managed to get through only the first two sections of this paper.

1. The Method of Zeta-Integrals

The idea here is to find an explicit integral expression for $L_S(s, \pi, r)$ whose analytic properties are tractable. We start in fact with a *family* of zeta-integrals $\mathscr{Z}(\phi, s)$, indexed by cusp forms ϕ in the space of π (and also other functions, some of which depend on s). In general, we can choose our data so that $\mathscr{Z}(\phi, s)$ is actually entire, or at least has finitely many known poles. For example, if π corresponds to a pair of classical cusp forms f_1, f_2 (i.e., an automorphic representation of $G = \mathrm{GL}(2) \times \mathrm{GL}(2)$), and r: $\mathrm{GL}_2(\mathbb{C}) \times \mathrm{GL}_2(\mathbb{C}) \to \mathrm{GL}_4(\mathbb{C})$ is the standard tensor product representation, then $\mathscr{Z}(\phi, s)$ looks like

$$\int_{Z_{\mathbb{A}}\mathrm{GL}_2(F)\backslash \mathrm{GL}_2(\mathbb{A})} \phi_{f_1}(g)\phi_{f_2}(g)E(g, F, s)\, dg, \tag{1.1}$$

with $E(s, F, g) = \sum_{B_F\backslash \mathrm{GL}_2(F)} F_s(\gamma g)$ an Eisenstein series on $\mathrm{GL}_2(\mathbb{A}_F)$. In this case, the zeta-integrals (1.1) generalize the integral representation of $\sum a_n b_n/n^s$ discovered almost 50 years ago by [Rankin] and [Selberg 1]; cf. Bump's survey talk in these proceedings ([Bump]).

In general, the strategy for using zeta-integrals to study automorphic L-functions can be broken down into four key steps:

Step I. Establish a "basic identity" for the "global" zeta-integrals, expressing them as a product of *local* zeta-integrals $\mathscr{Z}(\phi_v, s)$.

Step II. Analyze the meromorphic properties of the *global* zeta-integrals, including a functional equation of the form $\mathscr{Z}(\phi, s) = \tilde{\mathscr{Z}}(\tilde{\phi}, 1-s)$.

Step III. Verify that each *unramified* local zeta-integral—for $v \notin S$—coincides with the local Langlands factor $\det[I - r_v(t_{\pi_v})q_v^{-s}]^{-1}$.

Step IV. Complete the local theory of zeta-integrals. This includes the *local* functional equations and meromorphic properties of $\mathscr{Z}(\phi_v, s)$, and—ideally—a description of the g.c.d. of these local integrals, which—at least for finite v—should be of the form

$$L(s, \pi_v, r_v) = P_v(q_v^{-s})^{-1}. \tag{1.2}$$

In practice, Step III is the easiest and Step IV by far the hardest. In general, one can not always prove that the local zeta-integrals are of the form $L(s, \pi_v, r_v)\ R(q_v^{-s}, q_v^s)$, with $L(s, \pi_v, r_v)$ as above, and R arbitrary in $\mathbb{C}[q_v^{-s}, q_v^s]$. Fortunately, this last assertion is not always necessary; cf. [GeJa], where the local factor $L(s, \pi_v, r_v)$ is not introduced as the g.c.d. of local zeta-integrals.

For the sake of exposition, let us assume that Steps I through IV have been verified, and the (completed) L-function $L(s, \pi, r)$ has been defined as

$$L(s, \pi, r) = \left(\prod_{v \in S} L(s, \pi_v, r_v)\right) L_S(s, \pi, r).$$

Then the "local-global L-function machine," first introduced in [Tate], and then developed and perfected in [Jacquet–Langlands], works as follows to produce the desired analytic properties of $L_S(s, \pi, r)$ and $L(s, \pi, r)$. By Steps I-III we have

$$\begin{aligned} L_S(s, \pi, r) = \prod_{v \notin S} \mathscr{Z}(\phi_v, s) &= \left(\prod_{v \in S} \frac{1}{\mathscr{Z}(\phi_v, s)}\right) \mathscr{Z}(\phi, s) \\ &= \left(\prod_{v \in S} \frac{1}{\mathscr{Z}(\phi_v, s)}\right) \tilde{\mathscr{Z}}(\tilde{\phi}, 1 - s) \\ &= \left(\prod_{v \in S} \frac{\tilde{\mathscr{Z}}(\tilde{\phi}_v, 1 - s)}{\mathscr{Z}(\phi_v, s)}\right) L_S(1 - s, \pi, \tilde{r}). \end{aligned}$$

The first line of equations here, together with the fact that each local $\mathscr{Z}(\phi_v, s)$ is meromorphic in s, gives a meromorphic continuation for $L_S(s, \pi, r)$. Similarly, the local functional equation

$$\tilde{\mathscr{Z}}(\tilde{\phi}_v, 1 - s) = \gamma(s, \pi_v, r_v) \mathscr{Z}(\phi_v, s)$$

at least gives us the crude functional equation

$$L_S(s, \pi, r) = \left(\prod_{v \in S} \gamma(s, \pi_v, r_v)\right) L_S(1 - s, \pi, \tilde{r}). \tag{1.3}$$

To go further, we need more refined local results from Step IV. For example, if we can choose ϕ_v such that $\mathscr{Z}(\phi_v, s_0) \neq 0$ for each v in S, then we can conclude $L_S(s, \pi, r)$ has a pole at $s = s_0$ iff $\mathscr{Z}(\phi, s)$ does. More generally, if we can actually show that the family of zeta-integrals $\mathscr{Z}(\phi_v, s)$ admits a g.c.d. of the form $L(s, \pi_v, r_v)$ as explained above, then we may (find ϕ so that $\mathscr{Z}(\phi, s) = L(s, \pi, r)$ and) prove that $L(s, \pi, r)$ has the same analytic behavior as $\mathscr{Z}(\phi, s)$ and satisfies the functional equation

$$L(s, \pi, r) = \left(\prod_{v \in S} \varepsilon(s, \pi_v, r_v)\right) L(1 - s, \pi, \tilde{r}).$$

The key point here is that

$$\gamma(s, \pi_v, r_v) = \frac{\varepsilon(s, \pi_v, r_v) L(1 - s, \pi_v, \tilde{r}_v)}{L(s, \pi_v, r_v)} \tag{1.4}$$

with $\varepsilon(s, \pi_v, r_v)$ an exponential function of s. For details, see Section I.1.1 of [GeSh].

This L-function machine has been successfully applied—sometimes with modifications in the method just outlined—for the groups $\mathrm{GL}(n)$ and $\mathrm{GL}(n) \times \mathrm{GL}(m)$, with r the natural representation of ${}^LG$, or something close to it; cf. [GeSh] for a complete survey of results. For the classical groups $Sp(n)$ and $\mathcal{O}(n)$, or $G' \times \mathrm{GL}(n)$ with G' classical, there are the works of [PSR1] and [GePS]; here one encounters interesting obstacles to completing Step IV, which will be discussed below in Sections 3 and 4.

2. The Method of Eisenstein Series and Local Coefficients

This method hinges on the fact that $L_S(s, \pi, r)$ appears among the Fourier coefficients of Langlands' general Eisenstein series, and therefore (as explained in the lecture [Langlands3]) the analytic properties of $L_S(s, \pi, r)$ can be derived from the corresponding known properties of the Eisenstein series. The underlying assumption here is that G is the Levi component of a maximal parabolic subgroup P of some larger reductive group H, and $r\colon {}^LG \to \mathrm{GL}_d(\mathbb{C})$ appears in the decomposition of the adjoint action of ${}^LG \subset {}^LP$ on the Lie algebra of the unipotent radical of ${}^LP$.

In more detail: The Eisenstein series involved are "parabolically induced" from the cuspidal representation π on $G_{\mathbb{A}}$, twisted by a quasicharacter ω_s, and a "constant term" of these Eisenstein series is described by the intertwining operator

$$M(s, \pi)f = \left(\prod_{v \in S} A(s, \pi_v)f_v\right) \prod_{i=1}^{m} \frac{L_S(is, \pi, r^i)}{L_S(1 + is, \pi, r^i)}. \tag{2.1}$$

Here $f = \Pi f_v$ is an appropriately defined function in the space of $\mathrm{Ind}_{P_{\mathbb{A}}}^{H_{\mathbb{A}}} \pi \otimes \omega_s$, $A(s, \pi_v)$ is an intertwining operator for the *local* induced space $\mathrm{Ind}\, \pi_v \otimes \omega_{s,v}$, and $r^1, \ldots, r^m$ are the irreducible components of the aforementioned adjoint action of ${}^LG$; cf. [Langlands1]. The well-known theory of [Langlands2], giving the analytic continuation and functional equation of these Eisenstein series, then provides the analytic continuation and functional equation for each expression (2.1).

In particular, if $m = 1$, or if the remaining L-functions $L_S(s, \pi, r^i)$ are known, then (2.1) establishes the meromorphic continuation of $L(s, \pi, r^1)$. Langlands' theory of Eisenstein series also implies that $L_S(s, \pi, r^1)$ has only finitely many poles in the half-plane $\mathrm{Re}(s) \geq 0$, all simple and on the positive real axis. Thus a nice functional equation for $L_S(s, \pi, r^1)$, relating s to $1 - s$, is just what we need in order to conclude that the same nice properties hold in the left half-plane $\mathrm{Re}(s) \leq 0$.

In order to get a nice functional equation for $L_S(s, \pi, r)$, let us assume that π possesses a Whittaker model, i.e., π is "generic." In this case, Shahidi has defined certain "local coefficients" $c(s, \pi_v, r_v, \psi_v)$, which play the same role in the method of Eisenstein series as the gamma factors $\gamma(s, \pi_v, r_v, \psi_v)$ play in the theory of zeta-integrals. Indeed, using the identity

$$\left(\prod_{i=1}^{m} L_S(1 + is, \pi, r^i)\right) E_\psi(s, f, e) = \prod W_{f_v}(e),$$

relating $L_S(1 + is, \pi, r^i)$ to the "first" Fourier coefficients of $E(s, f, g)$ (and also the Whittaker functionals $W_{f_v}(e)$ on $\mathrm{Ind}\, \pi_v \otimes \omega_{v,s}$, $v \in S$), it is shown in [Sha2] that

$$\prod_{i=1}^{m} L_S(is, \pi, r^i) = \left(\prod_{v \in S} c(s, \pi_v, r_v, \psi_v)\right) \prod_{i=1}^{m} L_S(1 - is, \pi, \tilde{r}^i). \tag{2.2}$$

Note the strong analogy between this functional equation and the functional equation (1.3) obtained from the theory of zeta-integrals.

Now suppose $m \approx 1$, i.e., either $m = 1$, or $m = 2$, with r^2 a one-dimensional representation. Then Shahidi goes on (cf. [Sha3]) to express his local coefficients in the form

$$\prod_{v \in S} c(s, \pi_v, r_v) = \varepsilon_S(s, \pi, r) \prod_{v \in S} \frac{L(1 - s, \pi_v, \tilde{r}_v)}{L(s, \pi_v, r_v)} \tag{2.3}$$

where ε_S is an exponential function, and the L-functions $L(s, \pi_v, r_v)$ (and $L(s, \pi_v, \tilde{r}_v)$) are standard Euler factors of the form $P'_v(q_v^{-s})^{-1}$. Thus (2.1) is ultimately used to prove that the *completed* L-function $L(s, \pi, r) = \prod_v L(s, \pi_v, r_v)$ has only finitely many poles in $\mathrm{Re}(s) \geq 0$ and satisfies the functional equation

$$L(s, \pi, r) = \varepsilon(s, \pi, r) L(1 - s, \pi, \tilde{r}).$$

Note the analogy also between (2.3) and (1.4); in both cases, $1/L(s, \pi_v, r_v) = P_v(q_v^{-s})$ (or $P'_v(q_v^{-s})$) appears as the numerator of the "rational" gamma factor $\gamma(s, \pi_v, r_v)$ (or $c(s, \pi_v, r_v)$).

A complete list of "L-function data" (G, π, r) to which the above theory applies (in part or in whole) can be read off from the tables of [Sha3] (Section 4) and [Langlands1] (Section 6). Two particularly interesting examples are the following:

A. $H = G_2$ with Levi component G isomorphic to PGL_2, $r = r^1$ the symmetric *cube* representation of $\mathrm{SL}_2(\mathbb{C}) = {}^L G$, and π is any automorphic cuspidal representation of $\mathrm{PGL}_2(\mathbb{A}_F)$. In this case, the adjoint action of ${}^L G$ decomposes as the direct sum of (the four-dimensional representation) $r^1 = \mathrm{Sym}^3$ and the trivial representation. A slight modification of the above theory therefore allows us to still obtain the functional equation (and finiteness of poles result) for $L(s, \pi, \mathrm{Sym}^3)$; cf. [Sha1,4] and Section 4 below for more details.

B. From examples (iv) and (viii) on page 48 of [Langlands1], we obtain (via Shahidi's theory) the finiteness of poles and functional equation for $L(s, \pi, \mathrm{Sym}^2)$ and $L(s, \pi, \Lambda^2)$ on $\mathrm{GL}(n)$, provided π is *generic* cuspidal. (Here Λ^2 denotes the "exterior" or antisymmetric square representation of $\mathrm{GL}_n(\mathbb{C})$.)

3. Comparison of the Methods

Consider the following L-function data: $G = Sp(n)$, π is an automorphic cuspidal representation of $Sp_n(\mathbb{A}_F)$, and $r: {}^L G \to \mathrm{GL}_{2n+1}(\mathbb{C})$ corresponds to the standard embedding of ${}^LG = \mathrm{SO}_{2n+1}(\mathbb{C})$ in $\mathrm{GL}_{2n+1}(\mathbb{C})$. What can we prove then about $L(s, \pi, r)$?

In [PSR1], Piatetski–Shapiro and Rallis use zeta-integrals (generalizing those of Godement–Jacquet for GL(n)) in order to prove that $L_S(s, \pi, r)$ has a meromorphic continuation to $\mathbb{C}$ and satisfies a functional equation of the form $L_S(s, \pi, r) = \gamma_S(s, \pi, r)L_S(1 - s, \pi, \tilde{r})$; cf. (1.3). In this case, their basic identity (together with the local unramified computations) takes the form

$$\int_{G_f \times G_f \backslash G_{\mathbb{A}} \times G_{\mathbb{A}}} \phi_1(g_1)\phi_2(g_2)E(s, F_{\omega,s}, (g_1, g_2))\, dg_1\, dg_2 = \left(\prod_{v \in S} \mathscr{Z}(\phi_{1,v}, \phi_{2,v}, F_{v,s}, s)\right) L_S(s, \pi, r), \tag{3.1}$$

where ϕ_1 (resp. ϕ_2) belongs to the space of π (resp. $\tilde{\pi}$), and $E(s, F_{\omega,s}, h)$ is an Eisenstein series on the bigger symplectic group Sp_{2n} (which contains a copy of $Sp_n \times Sp_n$). Using the computations of [PSR2], Piatetski–Shapiro and Rallis go on to complete Step II of the L-function machine; in particular, they give precise information on the meromorphic properties of their (normalized) Eisenstein series.

The obstacle to using (3.1) to transfer analytic properties from $E(s)$ to $L_S(s, \pi, r)$ comes from the difficulty in completing Step IV for the local zeta-integrals $\mathscr{Z}(s, \phi_{1,v}, \phi_{2,v}, F_{v,s})$. Indeed, except for some special π (see Remark B below), knowledge of the (nonvanishing of the) local integrals in question is insufficient to conclude that $L_S(s, \pi, r)$ (let alone a suitably defined $L(s, \pi, r)$) has only finitely many poles (namely, those coming from $E(s, F, h)$).

Shahidi's theory, on the other hand, applies directly to $L(s, \pi, r)$, since $Sp(n)$ embeds nicely as the Levi component of a parabolic of Sp_{n+1}. (This corresponds to example (xx) on p.51 of [Langlands1]; in this case, the adjoint action of ${}^LG$ is irreducible, and so $r = r_1$.) The conclusion is that $L_S(s, \pi, r)$ may be "completed" to a meromorphic function $L(s, \pi, r)$, which has only finitely many poles in $\mathbb{C}$, and a functional equation of the desired type, *provided π is generic.*

Remark A. The assumption that π possesses a Whittaker model is definitely a restrictive one for such groups as $Sp(n)$ (where the automorphic representations corresponding to Siegel holomorphic cusp forms fail to have such models). Moreover, even when π is such that we can conclude $L(s, \pi, r)$ has only finitely many poles, the lack of an integral expression for $L(s, \pi, r)$ makes it unlikely we can locate these poles explicitly as in the theory of [PSR1] (except for some special groups and examples—see Section 4 below).

Remark B. Suppose π does possess a Whittaker model; suppose, moreover, that for each v in S, π is of the form $\mathrm{Ind}_{Q_v}^{G_v}(\rho_v \otimes 1)$, with ρ_v a supercuspidal or discrete series representation of the Levi component of a certain parabolic subgroup Q_v. In this case, the methods of both Piatetski–Shapiro–Rallis *and* Shahidi work independently to produce not only local L and ε factors for v in S, but also a completed L-function $L(s, \pi, r) = \prod_v L(s, \pi_v, r_v)$ that is meromorphic, satisfies a simple functional equation, and has at most finitely many (simple) poles. Thus it is natural to ask if the local factors—defined here in two entirely different ways—actually coincide. The answer must be "yes," and this has been checked by *global* arguments by S. Rallis and F. Shahidi (private communication); see Section III.2.5 of [GeSh] for more details.

Remark C. In very recent work, Piatetski–Shapiro and Rallis have introduced a completely new *global* zeta-integral (and basic identity) for $L_S(s, \pi, r)$ of the form

$$\int_{G_F \backslash G_{\mathbb{A}}} \phi_\pi(g)\Theta_T^\phi(g)E(g, s)\, dg = G^\phi(s)L_S(s', \pi, r). \tag{3.2}$$

Here Θ_T^ϕ is a theta function belonging to the Weil representation of G paired with the orthogonal group $\mathcal{O}_T$ of an n-dimensional quadratic form T, and $G^\phi(s)$ is a meromorphic function of s (depending on the Schwartz–Bruhat function ϕ). When $n = 1$, the zeta-integral in (3.2) reduces to the one used in [GeJa] to study $L(s, \pi, \mathrm{Sym}^2)$; in this case, the unipotent radical of the maximal parabolic subgroup $\left\{\begin{bmatrix} A & X \\ 0 & {}^tA^{-1} \end{bmatrix}\right\}$ coincides with the maximal unipotent subgroup of G, and therefore the local zeta-integrals involve "standard" Whittaker functions.

In general, let

$$U^P = \left\{u = \begin{bmatrix} I_n & X_u \\ 0 & I_n \end{bmatrix} : X_u^t = X_u\right\}$$

denote the unipotent radical of the maximal parabolic subgroup $P = \mathrm{GL}(n)U^P$ of $Sp(n)$, and identify the quadratic form T with the symmetric $n \times n$ matrix defining it; then T must be chosen in (3.2) so that the "degenerate" Fourier coefficient

$$\phi_T(g) = \int_{U_F^P \backslash U_{\mathbb{A}}^P} \phi(ug)\psi(\mathrm{tr}(TX_u))\, du$$

does not vanish, i.e., so that the degenerate Whittaker functional $\phi \to \phi_T(e)$, satisfying

$$\ell_T(\pi(u)v) = \psi(\mathrm{tr}\ TX_u)\ell_T(v)$$

is nontrivial. Such Whittaker functionals are not in general "unique;" in fact, the space of such functionals is usually infinite dimensional. Nevertheless, it is shown in [PSR3] that the Euler product (3.2) is valid. The key point here is that—for $v \notin S$—the same (unramified) local factors $L(s, \pi_v, r_v)$ arise regardless of which Whittaker functional is used. Moreover, it is shown that for any given $s = s_0$, ϕ may be chosen so that $G^\phi(s_0) \neq 0$. Thus $L_S(s', \pi, r)$ can at most inherit the (finitely many) poles of $E(g, s)$, and the problem of establishing nonvanishing properties for the local zeta-integrals in (3.1) is avoided.

Actually, a lot more can be concluded (globally) via this alternate basic identity. Because the oscillator representation appears in (3.2), this identity is ideally suited for relating the existence of poles for $L(s, \pi, r)$ to the fact that π might be the θ-series lift of a cusp form on $\mathcal{O}_T(\mathbb{A}_F)$, or a CAP representation in the sense of [PS1]; see [Soudry] for some interesting new results along these lines.

4. Combining the Methods

Consider first the example of

$$G = G' \times \mathrm{GL}(n)$$

where G' is a quasisplit reductive (classical) group of split rank n. This is the example treated in [GePS]. The idea is to use a generalization of the Rankin–Selberg method to reduce knowledge of the poles of $L(s, \Pi, r)$ to knowledge of a lower rank L-function whose properties are known (usually by Shahidi's theory). In this case, Π corresponds to a pair of cusp forms π and τ, one on $G'_{\mathbb{A}_F}$, the other on $\mathrm{GL}_n(\mathbb{A}_F)$. We denote this Π by $\pi \times \tau$ and consider the natural representation $r\colon {}^LG \to \mathrm{GL}_d(\mathbb{C})$ given by the tensor product of the "standard" representations of ${}^LG'$ and $\mathrm{GL}_n(\mathbb{C})$. The corresponding L-functions

$$L(s, \pi \times \tau, r)$$

indeed arise among the Fourier coefficients of certain Eisenstein series, but Shahidi's results on the finiteness of poles (and simple functional equation) do *not* apply here (since the relevant adjoint representation is too large).

For example, suppose $G' = SO_{2n+1}$, and $r: {}^LG \to GL_{2n^2}(\mathbb{C})$ is given by the tensor product of the standard representation of $Sp_n(\mathbb{C}) = {}^LG'$ and $GL_n(\mathbb{C})$. This case corresponds to example "B_{2n}" in the tables of Section 4 of [Sha3], according to which we conclude $L_S(s, \pi, r)$ appears (in some Eisenstein series Fourier coefficient) simultaneously with $L_S(s, \pi, r^2)$, where r^2 is a greater than one-dimensional representation of ${}^LG$. Hence Shahidi's theory neither provides local factors for v in S nor a finiteness of poles result for $L(s, \pi, r)$.

In [GePS], the L-function "machine" is applied to global zeta-integrals of the form

$$\mathscr{Z}(s, \phi, E) = \int_{H_F \backslash H_{\mathbb{A}}} \phi(h) E_\tau^f(h, s)\, dh$$

where ϕ belongs to the space of π, and $E_\tau^f(h, s) = \sum f(\gamma h, s)$ is an Eisenstein series, typically—but not always—defined on some reductive subgroup H of G' (say $H = SO_{2n}$ if $G' = SO_{2n+1}$) by inducing from the cuspidal representation $\tau \otimes |\det|^{s'}$ on the maximal parabolic subgroup $P = GL(n)U^P$ of H. The constant term of this Eisenstein series involves an automorphic L-function which *is* known (usually by Shahidi's theory) to be defined at all places and to have only finitely many poles. For the case $G' = SO_{2n+1}$, for example, we are dealing with the exterior square L-function $L(2s, \tau, \Lambda^2)$ for $GL(n)$. In any case, contrary to the situation encountered in Langlands' approach to automorphic L-functions, we are now confronted with some knowledge of the intertwining operators and wish to make conclusions about the Eisenstein series. In other words, we wish to *reverse* the usual strategy and conclude that—for appropriately chosen $f = \prod f_{v,s}$, $E_\tau^f(h, s)$ will inherit the analytic properties of some "known" L-function (like $L(2s, \tau, \Lambda^2)$. Granting the success of this procedure, we may conclude that $L(s, \pi \times \tau, r)$ itself has only finitely many poles in $\mathbb{C}$. Moreover, from the global functional equation for $E_\tau^f(s, h)$, together with the local functional equation and g.c.d. properties of the local Rankin–Selberg zeta-integrals (Step IV), we may also conclude that $L(s, \pi \times \tau, r)$ satisfies the expected functional equation.

Remark A. For the familiar Eisenstein series on $GL_2(\mathbb{A})$—say induced from the quasi-character $|a/b|^{s-1/2}\chi^{-1}(b)$ of the Borel subgroup—the procedure just outlined for proving the finiteness of poles of $E(h, s)$ reduces to the assertion that the *local* intertwining operators $M(s)f_v$ have the same meromorphic behavior in $\mathbb{C}$ as $L(2s-1, \chi_v)$ (at least for appropriately chosen f_v; for details, see Appendix A.2 to Section II.1.2 of [GeSh]. In general, it seems necessary to relate the relevant intertwining operators to higher dimensional analogues of $L(s', \chi_v)$ (such as $L(s', \tau_v, \Lambda^2)$ in the case of $\mathrm{SO}_{2n+1} \times \mathrm{GL}(n)$), which are defined thanks to Shahidi's theory of local coefficients. B. Tamir is in the process of investigating this problem for quasisplit groups, especially for the case of $U_{n+1,n} \times \mathrm{Re}\, s_F^E \,\mathrm{GL}(n)$. (In Shahidi's table of *quasisplit* groups, this corresponds to example ${}^2A_{n'} - 4$.)

Remark B. Ideally, one can study $L(s, \tau, \Lambda^2)$ itself via the method of zeta-integrals, so that the finitely many poles of $L(s, \tau, \Lambda^2)$ (and hence $L(s, \pi \times \tau, r)$) can be located explicitly. This is the subject matter of [Bump–Friedberg]. However, since only "completely unramified" r are considered there, Step IV is essentially bypassed.

On the other hand, an integral representation for $L(s, \tau, \Lambda^2)$ involving arbitrary cusp forms *on* GL(4) is analyzed in [JPSS1]. Here the focal point is the relation between cusp forms on $\mathrm{G}Sp(4)$ and GL(4), and the primary application requires understanding the pole of $L_S(s, \tau, \Lambda^2)$ at $s = 1$. Therefore the nonvanishing of certain local zeta-integrals is required at one point only, and Step IV is once again left virtually untouched.

Consider finally the example of

$$L(s, \pi, \mathrm{Sym}^3)$$

where π is any automorphic "nonmonomial" cuspidal representation of $\mathrm{PGL}_2(\mathbb{A}_F)$, and Sym^3 is the (four-dimensional) symmetric cube representation of $\mathrm{SL}_2(\mathbb{C})$. Recall from Section 2 that Shahidi's theory allows us to define $L(s, \pi_v, \mathrm{Sym}^3)$ at *every* local place and to prove that the completed L-function $\prod L(s, \pi_v, \mathrm{Sym}^3)$ satisfies the functional equation

$$L(s, \pi, \mathrm{Sym}^3) = \varepsilon(s, \pi, \mathrm{Sym}^3) L(1-s, \pi, \mathrm{Sym}^3) \tag{4.1}$$

and has only finitely many poles. By combining these results with the theory of zeta-integrals, one can actually prove that $L(s, \pi, \mathrm{Sym}^3)$ can be entire.

Whenever v is *unramified*, it is easy to check that

$$L(s, \pi_v, \mathrm{Sym}^3) = \frac{L(s, \pi_v \times \mathrm{Sym}^2(\pi_v), \rho_2 \otimes \rho_3)}{L(s, \pi_v, \rho_2)}, \tag{4.2}$$

with $\mathrm{Sym}^2(\pi_v)$ denoting the "symmetric square" of π_v (corresponding to the conjugacy class

$$\begin{pmatrix} \alpha_p^2 & 0 & 0 \\ 0 & 1 & 0 \\ 0 & 0 & \alpha_p^{-2} \end{pmatrix}$$

if

$$\pi_v \sim \begin{pmatrix} \alpha_p & 0 \\ 0 & \alpha_p^{-1} \end{pmatrix}.$$

The point is that this identity (4.2) can be shown to hold for *any* π_v, provided $\mathrm{Sym}^2(\pi_v)$ is understood to be the local Gelbart–Jacquet lifting to PGL_3, and $L(s, \pi_v \times \mathrm{Sym}^2(\pi_v), \rho_2 \otimes \rho_3)$ is the local factor on $\mathrm{GL}(2) \times \mathrm{GL}(3)$ studied in [JPSS2]; in other words, Shahidi's (completed) L-function $L(s, \pi, \mathrm{Sym}^3)$ may alternately be described (cf. [Sha1]) as the quotient

$$\frac{L(s, \pi \times \mathrm{Sym}^2(\pi), \rho_2 \otimes \rho_3)}{L(s, \pi, \rho_2)}. \tag{4.3}$$

N.B. We are assuming throughout that π is *nonmonomial*; equivalently, the Gelbart–Jacquet lifting $\mathrm{Sym}^2(\pi)$ is again a *cuspidal* representation.

Now by Shahidi's original theory, it already suffices (by the functional equation (4.1)) to understand the poles of $L(s, \pi, \mathrm{Sym}^3)$ in $\mathrm{Re}(s) \geq \frac{1}{2}$. On the other hand, by [JS1,2] we also know that the numerator and denominator in (4.3) are defined by absolutely convergent Euler products for $\mathrm{Re}(s) > 1$, and extend to continuous nonvanishing functions in $\mathrm{Re}(s) \geq 1$; cf. [Sha2]. Thus $L(s, \pi, \mathrm{Sym}^3)$ is entire (and nonzero) for $\mathrm{Re}(s) \geq 1$.

Finally, we appeal to an analog of the identity (2.1) that takes into account the trivial representation appearing along with Sym^3 in the adjoint action of ${}^LG$ inside ${}^LH = G_2(\mathbb{C})$. This identity reads as follows:

$$L(1 + s, \pi, \mathrm{Sym}^3) \frac{\zeta(1 + 2s)}{\zeta(2s)} M(s, \pi) f = \left(\prod_{v \in S} A^*(s, \pi_v) f_v \right) L(s, \pi, \mathrm{Sym}^3). \tag{4.4}$$

From this it follows that $L(s, \pi, \mathrm{Sym}^3)$ is entire at $s = \frac{1}{2}$ (since the simple pole of $\zeta(2s)$ at $s = \frac{1}{2}$ cancels any possible pole of $M(s, \pi)f$ at $s = \frac{1}{2}$, and $L(1 + s, \pi, \mathrm{Sym}^3)$ is entire at $s = \frac{1}{2}$ by the last paragraph). It also follows that the only possible poles of $L(s, \pi, \mathrm{Sym}^3)$ in $\mathrm{Re}(s) \geq \frac{1}{2}$ are then simple ones on the real interval $(\frac{1}{2}, 1)$. Indeed, the left-hand side of (4.4) has these properties, and—for any fixed s—we can choose f_v, $v \in S$, so that $A^*(s, \pi_v)f_v \neq 0$. From (4.3), it therefore allows that the only possible poles of $L(s, \pi, \mathrm{Sym}^3)$ in $\mathrm{Re}(s) \geq \frac{1}{2}$ come from the zeros of $L(s, \pi, \rho_2)$ in the interval $(\frac{1}{2}, 1)$. For "classical" π it is easy to check that $L(s, \pi, \rho_2)$ *never* vanishes in the interval, and hence $L(s, \pi, \mathrm{Sym}^3)$ is indeed entire (cf. [MorSha]).

5. Open Problems and Work in Progress

A

Automorphic L-functions attached to the exceptional groups G_2, F_4, or E_8 can *not* be analyzed via the method of Langlands–Shahidi explained in Section 2. Indeed, these groups can not be embedded as Levi components of any larger reductive group H. However, D. Ginsburg (in his Tel-Aviv University Ph.D. thesis) has recently made some progress finding local and global zeta-integrals that describe the "standard" automorphic L-function for G_2; similar *local* computations with a Hecke (as opposed to Rankin–Selberg) type zeta-integral go back to [Rodier].

B

A striking new development related to the L-functions discussed in this paper is a generalization to arbitrary groups of the "classical" theory of Poincaré series developed in [Selberg2]. In this latter paper, the Fourier and spectral expansions of the Poincaré series $P_n(z, s)$ are used to analyze its meromorphic continuation and to apply this information to the theory of cusp forms (real-analytic as well as holomorphic). In particular, estimates on the Kloosterman sums appearing in the Fourier expansion of $P_n(z, s)$ are used to provide a lower bound on the cuspidal spectrum of $\Delta = -y^2(\partial^2/\partial x^2 + \partial^2/\partial y^2)$ in $L^2(\Gamma\backslash H)$. In the context of $\mathrm{SO}(n, 1)$, this analysis has been generalized by the authors of [PSLS]; in particular, they obtain a lower bound for Δ on $\mathrm{SO}(n, 1)$ by analyzing certain generalized Kloosterman sums (see also [EGM]). For other developments in "Kloostermania," cf. [BFG], [Friedberg], [Sarnak], [Stevens], and reports at this Conference by Kuznetsov and others.

On the other hand, in the recent paper [PS2], a Poincaré series $P_r(g, s)$ is defined *for any reductive group G, and any representation r of its L-group*. The connection with our automorphic L-functions is provided by the identity

$$\prod_{i=1}^{t} \zeta_S(d_i s) \int_{G_F \backslash G_{\mathbb{A}}} P_r(g, s)\phi_\pi(g)\, dg = c(\phi_\pi) L_S(s, \pi, r), \tag{5.1}$$

where $c(\phi_\pi) \neq 0$. Thus the problem of analytically continuing $L_S(s, \pi, r)$ is reformulated in terms of a similar problem for $P_r(g, s)$. Unfortunately, this latter problem seems no more tractable than the former; indeed, $P_r(g, s)$ is defined in an almost artificial manner just so the identity (5.1) results. The strategy of directly establishing Langlands' conjecture on $L(s, \pi, r)$ by way of Poincaré series therefore seems a job for the distant future.

References

[Bump] Bump, D. "The Rankin–Selberg method: A survey," these proceedings.

[B,F] Bump, D. and Friedberg, S. "The exterior square automorphic L-functions on GL(n)," preprint (1987).

[BFG] Bump, D. Friedberg, S. and Goldfeld, D. "Poincaré series and Kloosterman sums for SL(3, $\mathbb{Z}$)," *Acta Arithmetica*, **50** (1987), pp. 31–89.

[EGM] Elstradt, J. Grunewald, F. and Mennicke, J. "Séries de Poincaré ...," *C.R. Acad. Sci. Paris*, t. 305, Série I (1987), 577–581.

[Friedberg] Friedberg, S. "Poincaré series for GL(n): Fourier expansion, Kloosterman sums, and algebreo-geometric estimates," *Math. Zeitschrift*, **196** (1987), pp. 165–188.

[GeJa] Gelbart, S. and Jacquet, H. "A relation between automorphic representations of GL(2) and GL(3)," *Ann. Sci. Ecole Normale Sup., 4e serie*, **11** (1978), 471–552.

[GePS] Gelbart, S. and Piatetski–Shapiro, I. "L-functions for $G \times$ GL(n)." In: *Explicit Constructions of Automorphic L-Functions, Lecture Notes in Mathematics* **1254**, Springer-Verlag, New York (1987).

[GeSh] Gelbart, S. and Shahidi, F. *Analytic properties of automorphic L-functions*, Perspectives in Mathematics, Vol. 6, Academic Press, Inc., Boston (1988).

[JL] Jacquet, H. and Langlands, R.P. "Automorphic forms on GL(2)," *Lecture Notes in Mathematics*, **114**, Springer-Verlag, New York (1970).

[JPSS1] Jacquet, H. Piatetski–Shapiro, I. and Shalika, J. in preparation.

[JPSS2] Jacquet, H. Piatetski–Shapiro, I. and Shalika, J. "Rankin–Selberg convolutions," *Amer. J. Math.*, **105** (1983), 367–464.

[JS1,2] Jacquet, H. and Shalika, J. "On Euler products and the classification of automorphic representations I, II," *Amer. J. Math.*, **103** (3) (1981), 499–558, 777–815.

[Langlands1] Langlands, R.P. *Euler Products*, James K. Whitmore Lectures Yale University Press, New Haven, (1967).

[Langlands2] Langlands, R.P. "On the functional equations satisfied by Eisenstein series," *Lecture Notes in Mathematics* **544**, Springer-Verlag, New York (1976).

[Langlands3] Langlands, R.P. "Eisenstein series, the trace formula, and the modern theory of automorphic forms," these proceedings.

[MorSha] Moreno, C.J. and Shahidi, F. "The L-function $L_3(s, \pi_\Delta)$ is entire," *Inventiones Math.* **79** (1985), 247–251.

[PS1] Piatetski–Shapiro, I. "Cuspidal representations associated to parabolic subgroups and Ramanujan's Conjecture." In: *Number Theory Related to Fermat's Last Theorem*, Progress in Mathematics Series, Birkhäuser Boston (1982).

[PS2] Piatetski–Shapiro, I. "Invariant theory and Kloosterman sums," in *Algebraic Groups, Utrecht 1986*, Lecture Notes in Mathematics, **1271**, Springer-Verlag, New York (1987), pp. 229–236.

[PSLS] Piatetski–Shapiro, I. Li, T.S. and Sarnak, P. "Poincaré Series for SO(n, 1)," Ramanujan Centennial Volume of the *Proceedings of the Indian Academy of Sciences*, to appear.

[PSR1] Piatetski–Shapiro, I. and Rallis, S. "ε-factors of representations of classical groups," *Proceedings of the National Academy of Sciences, U.S.A.* **83** (1986), 4589–4593.

[PSR2] Piatetski–Shapiro, I. and Rallis, S. "L-functions for the classical groups." In: *Explicit Constructions of Automorphic L-functions, Lecture Notes in Mathematics* **1254**, Springer-Verlag, New York (1987).

[PSR3] Piatetski–Shapiro, I. and Rallis, S. "A new way to get Euler products," preprint (1986).

[Rankin] Rankin, R. "Contributions to the theory of Ramanujan's function $\tau(n)$ and similar arithmetical functions, I and II," *Proc. Camb. Phil. Soc.* **35** (1939), 351–356 and 357–372.

[Rodier] Rodier, F. "Decomposition of principal series for reductive p-adic groups, and the Langlands' classification." In: *Operator Algebras and Group Representations*, Vol. II (Neptun, 1980), Monographs Stud. Math., **18**, Pitman, Boston, Mass., London (1984), 86–94.

[Sarnak] Sarnak, P. "Additive number theory and Maass forms." In: *Number Theory* (*New York* 1982), *Lecture Notes in Math.* **1052**, Springer-Verlag, New York (1984), 286–309.

[Selberg1] Selberg, A. *Arch. Math. Naturvid.* **43** (1940), 47–50.

[Selberg2] Selberg, A. "On the estimation of Fourier coefficients of modular forms." In: *Proc. Symp. Pure Math.* **VIII**, Amer. Math. Soc., Providence (1965), 1–15.

[Sha1] Shahidi, F. "Third Symmetric Power L-functions for GL(2)," to appear.

[Sha2] Shahidi, F. "On certain L-functions," *Amer. J. Math.* **103**, (2) (1981), 297–355.

[Sha3] Shahidi, F. "On the Ramanujan conjecture and finiteness of poles for certain L-functions," *Annals of Math.*, **127** (1988), pp. 547–584.

[Sha4] Shahidi, F. "Functional equation satisfied by certain L-functions," *Comp. Math.* **37** (1978), 171–208.

[Soudry] Soudry, D. "The CAP representations of $GSp(4, \mathbb{A})$," *J. reine angew. Math.* **383** (1988), pp. 87–108.

[Stevens] Stevens, G. "Poincaré Series on GL(r) and Kloosterman Sums," *Math. Ann.*, Band 277, Heft 1 (1987), 25–52.

[Tate] Tate, J. "Fourier analysis in number fields and Hecke'z zeta-function," Thesis, Princeton, 1950; also appears in *Algebraic Number Theory, Proceedings of the Brighton Conference*, J.W. Cassels and A. Frohlich, eds. Academic Press, New York (1968).

15 Explicit Formulae as Trace Formulae

DORIAN GOLDFELD[1]

1. Introduction

In his epoch-making paper [2], Selberg developed a general trace formula for discrete subgroups of $GL(2, \mathbb{R})$. The analogies with the explicit formulae of Weil [3] (relating very general sums over primes with corresponding sums over the critical zeroes of the zeta-function) are quite striking and have been the subject of much speculation over the years.[2]

It is the object of this note to show that Weil's explicit formula can in fact be interpreted as a trace formula on a suitable space. The simplest space we have been able to construct for this purpose, at present, is the semidirect product of the ideles of norm one with the adeles, factored by the discrete subgroup $\mathbb{Q}^* \ltimes \mathbb{Q}$, the semidirect product of the multiplicative group of rational numbers with the additive group of rational numbers. We will show that for a suitable kernel function on this space, the conjugacy class side of the Selberg trace formula, is precisely the sum over the primes occurring in Weil's explicit formula.

[1] This material is based upon work partially supported by the National Science Foundation under Grant No. NSF-DMS87-02169.

[2] In this connection see the article of Hejhal [1].

NUMBER THEORY,
TRACE FORMULAS and
DISCRETE GROUPS

ISBN 0-12-067570-6

This implies that the sum of the eigenvalues of the self-adjoint integral operator associated to the aforementioned kernel function is precisely the sum over the critical zeroes of the Riemann zeta-function occurring on the other side of Weil's formula. The relation between the eigenvalues of this integral operator and the zeroes of the zeta-function appears quite mysterious at present. What is lacking is a suitable generalization of the Selberg transform in this situation.

Finally, we should point out that our approach leads to various new equivalences to the Riemann Hypothesis, such as certain positivity hypotheses for the integral operators. Although we have worked over $\mathbb{Q}$, for simplicity of exposition, it is not hard to generalize our results to L-functions of arbitrary number fields.

2. Notation

Let A denote the adeles over $\mathbb{Q}$, I the ideles over $\mathbb{Q}$, and I_0 the subgroup of ideles of norm one. A general idele or adele will be written as

$$x = (x_\infty; x_2, x_3, \ldots, x_p, \ldots),$$

with $x_\infty \in \mathbb{R}$, $x_p \in \mathbb{Q}_p$ for all primes p. Since I_0 acts on A by left multiplication, say, we can consider the semidirect product $I_0 \ltimes A$ with group law

$$(x, x')(y, y') = (xy, x'y + y')$$

for all $(x, x'), (y, y') \in I_0 \ltimes A$.

Now $\mathbb{Q}^* \ltimes \mathbb{Q}$ acts discretely on $I_0 \propto A$ and the quotient space

$$\mathfrak{M} = \mathbb{Q}^* \times \mathbb{Q} \backslash I_0 \ltimes A$$

is compact with fundamental domain

$$\mathfrak{M} \cong \{\pm 1\} \times \left(\prod_p U_p\right) \times [0, 1] \times \left(\prod_p \mathbb{Z}_p\right),$$

where U_p denotes the units in $\mathbb{Q}_p$ and $\mathbb{Z}_p$ denotes the integers in $\mathbb{Q}_p$.

Let $|\ |_p$ denote the p-adic valuation on $\mathbb{Q}_p$ normalized so that $|p^n|_p = p^{-n}$. For $X = (x, x') \in I_0 \ltimes A$, we choose Haar measure

$$dX = \prod_p \frac{dx_p}{|x_p|_p} \cdot dx'_\infty \cdot \prod_p dx'_p$$

normalized so that

$$\int_{U_p} \frac{dx_p}{|x_p|_p} = 1$$

and

$$\int_{\mathbb{Z}_p} dx'_p = 1$$

$$\int_0^1 dx'_\infty = 1.$$

Given two continuous functions

$$F, G\colon \mathfrak{M} \to \mathbb{C}$$

which are $\mathbb{C}^\infty$ with respect to the variable x'_∞, we define the inner product

$$\langle F, G \rangle = \int_{\mathfrak{M}} F(X)\overline{G(X)}\, dX.$$

Let $\mathscr{L}^2(\mathfrak{M})$ denote the Hilbert space of square integrable functions of the above type. Then the operator

$$\nabla = \frac{\partial^2}{\partial x'^2_\infty}$$

is an unbounded operator on this space.

Let $K\colon I_0 \rtimes A \to \mathbb{C}$ be continuous and C^∞ with respect to the variable x'_∞. We also assume that the sum

$$\sum_{g \in \mathbb{Q}^* \rtimes \mathbb{Q}} \nabla^m K(X^{-1} g Y)$$

converges absolutely and uniformly for all $X, Y \in I_0 \rtimes A$ for $m = 0, 1, 2, 3, \ldots$. We can then define the kernel function

$$K(X, Y) = \sum_{g \in \mathbb{Q}^* \rtimes \mathbb{Q}} K(X^{-1} g Y).$$

it follows easily that the integral operator

$$F \to KF \qquad (F \in \mathscr{L}^2(\mathfrak{M}))$$

given by

$$KF(Y) = \int_{\mathfrak{M}} K(X, Y) F(X)\, dX$$

maps

$$\mathscr{L}^2(\mathfrak{M}) \to \mathscr{L}^2(\mathfrak{M})$$

and is of Hilbert Schmidt type. Furthermore, if

$$\frac{\partial^2}{\partial y_\infty'^2} K(X, Y) = \frac{\partial^2}{\partial x_\infty'^2} K(X, Y)$$

then the operators K and ∇ commute. Unfortunately, this is not usually the case.

3. The Explicit Formula

Let

$$\xi(s) = \pi^{-s/2}\Gamma\left(\frac{s}{2}\right)\zeta(s)$$

where $\zeta(s)$ is the Riemann zeta-function. Then we have the functional equation

$$\frac{\xi'}{\xi}(s) = -\frac{\xi'}{\xi}(1-s).$$

Now, let $f\colon \mathbb{R} \to \mathbb{C}$ be a $\mathbb{C}^\infty$-function satisfying

$$f(x) = f\left(\frac{1}{x}\right), \qquad f(x) = f(-x) \tag{3.1}$$

$$\left| x^m \frac{d}{dx^n} f(x) \right| = 0(1) \qquad (\forall m, n \geq 0, n \in \mathbb{Z}). \tag{3.2}$$

Then we have the Mellin transform pair

$$\tilde{f}(\lambda) = \int_0^\infty f(x) x^{-\lambda} \frac{dx}{x} \tag{3.3}$$

$$f(x) = \frac{1}{2\pi i} \int_{c - i\infty}^{c + i\infty} \tilde{f}(\lambda) x^\lambda \, d\lambda. \tag{3.4}$$

Clearly $f(\lambda) = f(-\lambda)$.

By the functional equation, we have for $c > \frac{1}{2}$

$$\frac{1}{2\pi i} \int_{c - i\infty}^{c + i\infty} -\frac{\xi'}{\xi}\left(\frac{1}{2} + \lambda\right)(x^\lambda + x^{-\lambda})\tilde{f}(\lambda)\, d\lambda$$

$$= (x^{1/2} + x^{-1/2})\tilde{f}\left(\frac{1}{2}\right) - \sum_{\substack{\rho \\ \xi(1/2 + \rho) = 0}} (x^\rho + x^{-\rho})\tilde{f}(\rho). \tag{3.5}$$

We now decompose the left side of (3.5) into three pieces I_1, I_2, I_3, where

$$I_1(x) = \sum_p \sum_{n=1}^{\infty} \frac{(\log p)}{p^{n/2}} \left(f(p^n x) + f\left(\frac{p^n}{x}\right)\right),$$

$$I_2(x) = (\log \sqrt{\pi})\left(f(x) + f\left(\frac{1}{x}\right)\right)$$

and

$$I_3(x) = -\frac{1}{4\pi i} \int_{c-i\infty}^{c+i\infty} \frac{\Gamma'}{\Gamma}\left(\frac{\frac{1}{2}+\lambda}{2}\right)(x^{\lambda} + x^{-\lambda})\tilde{f}(\lambda)\, d\lambda.$$

Using the formula

$$\frac{\Gamma'}{\Gamma}(s) = -\gamma - \frac{1}{s} + s \sum_{n=1}^{\infty} \frac{1}{n(s+n)},$$

we easily obtain

$$\begin{aligned} I_3(x) = {} & \frac{1}{2}\left(\gamma + \sum_{n=1}^{\infty} \frac{1}{n(4n+1)}\right)\left(f(x) + f\left(\frac{1}{x}\right)\right) + \int_0^1 \left(f\left(\frac{x}{t}\right) + f\left(\frac{1}{xt}\right)\right) \frac{dt}{\sqrt{t}} \\ & + \sum_{n=1}^{\infty} \frac{1}{2n+\frac{1}{2}}\left[f(x) + f\left(\frac{1}{x}\right) - (2n+\tfrac{1}{2}) \int_0^1 \left(f\left(\frac{x}{t}\right)\right.\right. \\ & \left.\left. + f\left(\frac{1}{xt}\right)\right) t^{2n+1/2} \frac{dt}{t}\right]. \end{aligned}$$

Finally, the explicit formula can be put in the form

$$\sum_{\substack{\rho \\ \xi(1/2+\rho)=0,\,\infty}} m_{\rho}[x^{\rho} + x^{-\rho}]\tilde{f}(\rho) = H(x) + H\left(\frac{1}{x}\right),$$

where m_{ρ} denotes the multiplicity of the zero or pole, taken positively if $\xi(\frac{1}{2}+\rho) = \infty$ and negatively if $\xi(\frac{1}{2}+\rho) = 0$. Here

$$\begin{aligned} H(x) = {} & c_0 f(x) + \sum_p \sum_{n=1}^{\infty} (\log p) p^{-n/2} f(p^n x) + \int_0^1 f\left(\frac{x}{t}\right) \frac{dt}{\sqrt{t}} \\ & + \sum_{n=1}^{\infty} \frac{1}{2n+\frac{1}{2}} \left[f(x) - \left(2n + \frac{1}{2}\right) \int_0^1 f\left(\frac{x}{t}\right) t^{2n+1/2} \frac{dt}{t}\right], \end{aligned}$$

where

$$c_0 = (\log \sqrt{\pi}) + \frac{1}{2}\left(\gamma + \sum_{n=1}^{\infty} \frac{1}{n(4n+1)}\right).$$

4. The Selberg Trace Formula

We now construct a special kernel function on $\mathfrak{M}$. First, we fix some notation, however.

Let $k_1, k_2: \mathbb{R} \to \mathbb{C}$ be $\mathbb{C}^\infty$-functions satisfying conditions (3.1) and (3.2). For $x_p \in Q_p$ and $B \subset Q_p$ let us define

$$\psi_B(x_p) = \begin{cases} 1 & x_p \in B \\ 0 & \text{otherwise} \end{cases}$$

to be the characteristic function of the set B. We also set for $x_\infty \in \mathbb{R}$

$$\delta(x_\infty) = \begin{cases} 1 & x_\infty \leq 0 \\ 0 & x_\infty > 0. \end{cases}$$

Let $X = (x, x') \in I_0 \bowtie A$. We now define a function

$$K^*: I_0 \bowtie A \to \mathbb{C}$$

by the absolutely convergent series

$$\begin{aligned} K^*(X) = {} & \delta(x_\infty) \sum_p (\log p)\, k(x_\infty, x'_\infty, x_p, x'_p) \prod_{q \neq \infty, p} \phi_q(x_q, x'_q) \\ & + \delta(x_\infty)\, H_1(x_\infty e^{x'_\infty(1 - x_\infty)}) k_2(e^{x'\infty/(1 - x_\infty)}) \prod_{q \neq \infty} \phi_q(x_q, x'_q), \end{aligned}$$

where we have put

$$\phi_q(x_q, x'_q) = \begin{cases} \psi_{U_q}(x_q)\, \psi_{\mathbb{Z}_q}\left(\dfrac{x'_q}{1 - x_q}\right) & q \mid (1 - x_\infty), \quad x_q \neq 1 \\ 0 & x_q = 1 \\ \psi_{U_q}(x_q)\, \psi_{\mathbb{Z}_q}(x'_q) & \text{otherwise} \end{cases}$$

$$\begin{aligned} k(x_\infty, x'_\infty, x_p, x'_p) = {} & |x_p|_p^{1/2}\, \psi_{p\mathbb{Z}_p}(x_p)\, \psi_{\mathbb{Z}_p}\left(\frac{x'_p}{1 - x_p}\right) \\ & \times k_1(x_\infty e^{x'\infty/(1 - x_\infty)})\, k_2(e^{x'_\infty(1 - x_\infty)}) \end{aligned}$$

and

$$H_1(x) = c_0 k_1(x) + \int_0^1 k_1\left(\frac{x}{t}\right) \frac{dt}{\sqrt{t}} + \sum_{n=1}^{\infty} \frac{(k_1(x) - (2n + \frac{1}{2}))}{2n + \frac{1}{2}} \int_0^1 k_1\left(\frac{x}{t}\right) t^{2n} \frac{dt}{\sqrt{t}}.$$

If we now let

$$K(X) = K^*(X) + \overline{K^*(X^{-1})},$$

a symmetric kernel function is then given in the form

$$K(X, Y) = \sum_{g \in \mathbb{Q}^* \rtimes \mathbb{Q}} K(X^{-1}gY),$$

and this gives rise to a self-adjoint integral operator of Hilbert-Schmidt type.

The Selberg trace formula arises by computing

$$\mathrm{Trace}(K) = \int_{\mathfrak{M}} K(X, X)\, dX,$$

the integral of the kernel function on the diagonal. We now briefly describe this calculation.

Firstly, for $X = (x, x')$, $Y = (y, y')$ in $I \rtimes A$ and $g = (\alpha, \beta) \in \mathbb{Q}^* \rtimes \mathbb{Q}$, we have

$$X^{-1}gY = \left(\alpha \frac{y}{x}, \frac{-\alpha x' y}{x} + \beta y + y'\right)$$

$$X^{-1}gX = (\alpha, (1 - \alpha)x' + \beta x).$$

It follows that

$$K(X, X) = 2\,\mathrm{Re}\left\{\sum_p (\log p) \sum_{n=1}^{\infty} p^{-n/2} \sum_{\beta \in \mathbb{Z}} k_1(p^n e^{x'_\infty + \beta})\, k_2(e^{x'_\infty + \beta}) + \sum_{\beta \in \mathbb{Z}} H_1(e^{x'_\infty + \beta})\, k_2(e^{x'_\infty + \beta})\right\}.$$

On integrating on the diagonal, we obtain

$$\mathrm{Trace}(K) = 2\,\mathrm{Re} \int_0^{\infty} \left[\sum_p (\log p) \sum_{n=1} p^{-n/2} k_1(p^n t) + H_1(t)\right] k_2(t) \frac{dt}{t},$$

which is precisely the sum over primes in Weil's explicit formula. It follows that we can also express

$$\mathrm{Trace}(K) = 2\,\mathrm{Re}\left\{\sum_{\substack{\rho \\ \xi(1/2+\rho) = 0, \infty}} m_\rho \tilde{k}_1(\rho)\, k_2(\rho)\right\}$$

as a sum over the zeroes and poles of the Riemann zeta-function. Further insights into the precise connection between the eigenvalues of the self-adjoint operator K and the critical values of $\xi(s)$ would be of considerable interest. It is hoped that further developments of the theory may shed some light on this problem.

In conclusion, it also might be worthwhile to remark that although our semidirect product group is not unimodular, it does exhibit many similarities with the classical GL(2) theory, not all of which have hitherto been elucidated. For example, every automorphic function in $\mathscr{L}^2(\mathfrak{M})$ has a Fourier expansion with respect to the additive component of the adele group.

References

[1] Hejhal, D.A. "The Selberg trace formula and the Riemann zeta function," *Duke Math. J.* **43** (1976), 441–482.

[2] Selberg, A. "Harmonic analysis and discontinuous groups in weakly symmetric Riemannian spaces," *J. Indian Math. Soc.* **20** (1956), 47–87.

[3] Weil, A. "Sur les 'formules explicites' de la théorie des nombres premiers," *Comm. Sém. Math. Univ. Lund* 1952, Tome supplémentaire (1952), 252–265.

16 Some Remarks on the Sieve Method

GEORGE GREAVES

1. Introduction

In this article we will survey the ideas underlying some recent, and some not so recent, published literature in the field described in the title.

As is well known, the first really effective sieve method was created by Viggo Brun [2] in the early 1920s. His method led to the following application, among others: If n is large enough then there is, between n and $n + \sqrt{n}$, a number having not more than 11 prime factors.

More generally, following the formulation provided by more recent authors [8], one may study sets $\mathscr{A}$ of integers, depending on a parameter X, which is to be considered as tending to $+\infty$, that satisfy

$$\sum_{a \in \mathscr{A};\, a \equiv 0, \bmod d} 1 = \frac{X\rho(d)}{d} + R(X, d) \qquad \text{if } |\mu(d)| = 1, \tag{1.1}$$

where ρ is a multiplicative function whose average value at primes is κ in some sense, such as

$$-L < \sum_{w < p < z} \frac{\rho(p) \log p}{p} - \kappa \log \frac{z}{w} < A_1, \tag{1.2}$$

NUMBER THEORY,
TRACE FORMULAS and
DISCRETE GROUPS

ISBN 0-12-067570-6

where

$$0 \le \frac{\rho(p)}{p} < A_2, \tag{1.3}$$

with $L > 0$, $A_1 > 0$. In practice one chooses ρ so that $R(X, d)$ is small, at least in some average sense. The formula (1.1) will then retain its significance for $d \le D$, say, in a sense to be made more precise below. Thus if $\mathscr{A}$ were the numbers in Brun's interval $[n, n + \sqrt{n}]$ we would take $X = \sqrt{n}$ and $\rho(d) = 1$, so that $R(X, d) = 0(1)$, for example. We could then take D close to X, e.g., $D = X/\log^2 X$, in the sequel.

In the linear sieve, which is the only case considered in this article, we restrict attention to those cases where the appropriate choice of the "dimension" κ is

$$\kappa = 1. \tag{1.4}$$

This covers the application made by Brun mentioned above, as well as many others.

In a sieve argument, one invents a suitable function $\lambda_D(d)$, defined over square-free d, that will satisfy

$$\lambda_D(d) \neq 0 \Rightarrow d \le D, \tag{1.5}$$

and then obtains from (1.1)

$$\sum_{a \in \mathscr{A}} \left\{ \sum_{d \mid a} \lambda_D(d) \right\} = X \sum_{d} \frac{\lambda_D(d)\rho(d)}{d} + \sum_{d} \lambda_D(d) R(X, d). \tag{1.6}$$

The object is to choose $\lambda_D(d)$ in such a way that the available information enables us to infer that the right side of (1.6) is positive and then draw an interesting inference from the conclusion that some a in $\mathscr{A}$ has

$$\sum_{d \mid a} \lambda_D(d) > 0. \tag{1.7}$$

Without loss of generality we may normalize the function λ so that it satisfies

$$|\lambda_D(d)| \le 1. \tag{1.8}$$

In the sieve arguments that we will consider, it will follow using (1.2) and (1.3) that

$$\sum_{d} \frac{\lambda_D(d)\rho(d)}{d} > \frac{c}{\log D} > 0, \tag{1.9}$$

for some absolute constant c. Accordingly, to make use of (1.6), one chooses a "level of distribution" $D = D(X)$ as large as possible consistently with some condition such as

$$\left|\sum_{d} \lambda_D(d) R(X, d)\right| \ll \frac{X}{\log^2 D}. \tag{1.10}$$

The conclusion (1.7) will then follow, if X is large enough. Of course D will in practice tend to ∞ with X.

The "trivial" way to establish (1.10) is via

$$\sum_{d \leq D} |\mu(d)|\,|R(X, d)| \ll \frac{X}{\log^2 D}, \tag{1.11}$$

but in recent years, following the work of Chen [3], Iwaniec [12], and others, it has become possible to perform such estimations in a better "nontrivial" way. At any rate, the procedure is to establish (1.9) with D chosen consistently with (1.10).

Let

$$S(a, z) = \sum_{d \mid (a, P(z))} \mu(d) \tag{1.12}$$

where

$$P(z) = \prod_{p < z} p$$

denotes the product of all primes p such that $p < z$, so that

$$S(a, z) = \begin{cases} 1 & \text{if } a \text{ has no prime factor } p < z \\ 0 & \text{otherwise.} \end{cases} \tag{1.13}$$

The "unweighted" sieve of Brun and his successors constructed functions $\lambda(d)$ with the property

$$\sum_{d \mid a} \lambda_D(d) \leq \sum_{d \mid (a, P(z))} \mu(d) = S(a, z). \tag{1.14}$$

Brun showed how to construct such λ so that z could be taken reasonably large compared with D. In the modern sieve method it is possible, in the "linear" context described by (1.4), to take

$$z = D^{1/2 - \varepsilon} \tag{1.15}$$

(where $\varepsilon > 0$) while retaining (1.9) so that (1.14) gives, for some a in $\mathscr{A}$,

$$S(a, z) > 0, \tag{1.16}$$

that is to say, some a in $\mathscr{A}$ has the property that all its prime factors exceed $D^{1/2-\varepsilon}$.

Let us now introduce a new parameter, the "degree" g, which is to satisfy

$$a \le D^g \qquad \text{for all } a \text{ in } \mathscr{A}. \tag{1.17}$$

With z given by (1.15), it now follows via (1.16) that some a in $\mathscr{A}$ has at most R prime factors,

$$\Omega(a) \le R, \tag{1.18}$$

wherever

$$R > 2g - 1. \tag{1.19}$$

The weighted sieve, of which the first example was introduced by Kuhn [14], has the different objective, not of establishing (1.16), but of aiming directly at the conclusion (1.18), with values of R somewhat smaller (i.e., better) than those given by (1.19). In the current state of the subject success is possible provided

$$g < R - \delta_R = \Lambda_R \tag{1.20}$$

for certain positive numbers δ_R satisfying

$$0 < \delta_R < \tfrac{1}{8}. \tag{1.21}$$

The second inequality in (1.21) depends upon a certain amount of numerical computation.

There is an irritating feature of current work on the weighted linear sieve problem, which we will describe in this article. In the unweighted sieve, an example provided by Selberg [17] shows that the choice $z = D^{1/2-\varepsilon}$ described in (1.15) is best possible in that any such result with z replaced by $D^{1/2+\varepsilon}$ would be false. Essentially the same example shows that a result of the type summarized by (1.20) could not hold for any $\delta_R < 0$ and leads to the conjecture that we ought to be able to take $\delta_R = 0$ in (1.21). We shall examine the reasons why current ideas fall short of leading to this objective.

2. The Linear Sieve

In this section we review some of the available knowledge of the linear sieve (with no reference to "weights" that will be relevant later on). We assume (1.1), (1.2), and (1.3) throughout, with $\kappa = 1$ as specified in (1.4).

The linear sieve lower bound may be summarized as follows. Under the hypotheses (1.1)–(1.4) we have, for $S(a, z)$ as defined in (1.12),

$$\sum_{a\in\mathscr{A}} S(a, z) > X\left\{\prod_{p<z}\left(1-\frac{\rho(p)}{p}\right)\right\}\left\{f\left(\frac{\log D}{\log z}\right)+O\left(\frac{1}{\log^{c} D}\right)\right\}$$
$$-\sum_{d<D}|\mu(d)|3^{\nu(d)}|R(X, d)|, \tag{2.1}$$

for a certain function f that satisfies

$$f(s) > 0 \qquad \text{if } s > 2. \tag{2.2}$$

For fuller versions of this theorem we refer the reader to the cited literature. Such a result, with $c = \frac{1}{14}$, appeared (under more restrictive assumptions about the multiplicative function ρ) in the paper [13] of Jurkat and Richert. At a conference in Stonybrook, Selberg [18] indicated how such results could be reached by a method originated by Rosser but never published by him. Meanwhile, the subject had been studied by Iwaniec, after seeing the paper of Jurkat and Richert. Iwaniec [10] paid particular attention to the attainable value of the constant c, returning to the case of arbitrary dimension κ in [11]. In Rosser's and Iwaniec's work, the factor $3^{\nu(d)}$ in the R-term did not appear; in the work of Jurkat and Richert it arose from a (multiple) use of the well-known λ^2 method of Selberg [16].

In the summary above, we have written what follows from use of the "trivial" method (1.11) of describing a suitable level of distribution D.

Next, let us review the salient features of the method of Rosser and Iwaniec that leads to a proof of the Linear Sieve Theorem. The method is, essentially, Brun's.

For square-free d write

$$d = p_1 p_2 \cdots p_\nu; \qquad p_1 > p_2 > \cdots > p_\nu,$$

p_i being prime. Let

$$B_{2i} = B_{2i}(p_{2i-1}, \ldots, p_1)$$

be, for the moment, arbitrary. Define χ_D by

$$\chi_D(d) = \begin{cases} 1 & \text{if } \; p_{2i} < B_{2i} \qquad \text{when } 1 \le 2i \le \nu \\ 0 & \text{otherwise.} \end{cases} \tag{2.3}$$

so that $\chi_D(1) = 1$. Then Brun observed, for the square-free number

$$A = (a, P(z)), \tag{2.4}$$

that

$$\sum_{d|A} \mu(d) \geq \sum_{d|A} \mu(d)\chi_D(d), \tag{2.5}$$

and proceeded by taking

$$\lambda_D(d) = \mu(d)\chi_D(d), \tag{2.6}$$

in the language of Section 1. The construction has to satisfy (1.5).

Brun's treatment of (2.5) was via an iteration of the well-known "Buchstab" identity, but it may perhaps be more quickly appreciated as follows. Write

$$\bar{\chi}_D(d) = \begin{cases} 1 & \text{if } \nu = 2j,\ B_{2j} \leq p_{2j},\ p_{2i} < B_{2i} (1 \leq 2i \leq \nu - 2) \\ 0 & \text{otherwise.} \end{cases} \tag{2.7}$$

Then we have, for an arbitrary arithmetic function ϕ, the "fundamental" identity (wherein $p(\delta)$ denotes the smallest prime factor of δ)

$$\sum_{d|A} \mu(d)\phi(d) = \sum_{d|A} \mu(d)\chi_D(d)\phi(d) + \sum_{\delta|A} \mu(\delta)\bar{\chi}_D(\delta) \sum_{t|(A/\delta,\, P(p(\delta)))} \mu(t)\phi(\delta t), \tag{2.8}$$

from which (2.5) immediately follows on taking $\phi(t) = 1$. On the other hand, (2.8) holds because the divisors d of A either satisfy $p_{2i} < B_{2i}$ for all i, or else there is a least j for which $B_{2j} \leq p_{2j}$, in which case we write $d = \delta t$ where $\bar{\chi}_D(\delta) = 1$ as in (2.7) and all prime factors of t are less than $p(\delta) = p_{2j}$.

Rosser's (and Iwaniec's) choice of the parameters B_{2i} was that given by

$$p_{2i} < B_{2i} \Leftrightarrow p_1 p_2 \cdots p_{2i-1} p_{2i}^3 < D. \tag{2.9}$$

By a rather nontrivial analysis, it has been shown (see [10], [11], for example) that this construction leads to a proof of the Linear Sieve Theorem summarized above. We make no attempt to discuss this analysis here, but we do have some comments to make on the construction (2.9) in the light of the instructive example provided by Selberg [17]. Before proceeding, note via (2.3), (2.6) that, even with the exponent 3 in (2.9) replaced by any number $\theta \geq 2$, the construction satisfies the conditions (1.5), (1.8) so that, given (1.11), the R-terms satisfy (1.10) as required.

Selberg's example is, nearly enough for present purposes, as follows. Take the set $\mathscr{A}$ to be the integers between $2D_1$ and $4D_1$ that have an even number of prime factors:

$$\mathscr{A} = \{a: 2D_1 < a < 4D_1, \Omega(a) \equiv 0, \bmod 2\}. \tag{2.10}$$

Take

$$4D_1 = D^{1+\varepsilon}, \tag{2.11}$$

where $\varepsilon > 0$ is a suitable decreasing function of D tending to 0 as $D \to \infty$. Then (because the Liouville function $\lambda(d) = (-1)^{\Omega(d)}$ satisfies

$$\sum_{d \le x} \lambda(d) = o(e^{-c\sqrt{\log x}}) \qquad \text{as } x \to \infty$$

for some $c > 0$), it is possible to show that $\mathscr{A}$ satisfies (1.1) with

$$X = D_1, \qquad \rho(d) = 1$$

(so that (1.2), (1.3) hold) and with $R(X, d)$ satisfying (1.11).

Possibly the simplest use of this example is to show that the "sieving limit" 2 appearing in (2.2) is best possible in that it cannot be replaced by a smaller number, $2 - \delta$, say. For if it were, then the resulting form of (2.1) would now show that some a in the set $\mathscr{A}$ defined in (2.10) has no prime factors smaller than $D^{1/(2-\delta)}$, which is false since each a in this $\mathscr{A}$ has at least two prime factors and does not exceed $4D_1 = 4D^{1+\varepsilon}$, where $\varepsilon > 0$ is arbitrarily small.

A more sophisticated use of Selberg's example is to show that the function f appearing at (2.1) in the Linear Sieve Theorem is best possible for all s. So far as the author is aware, the details of this have not been published.

The function can be defined by the system of equations

$$\frac{d}{ds}\{sf(s)\} = F(s-1) \qquad \text{if } s > 2; \qquad \frac{d}{ds}\{sF(s)\} = f(s-1) \qquad \text{if } s > 3;$$

$$f(s), F(s) = 1 + 0(e^{-s}) \qquad \text{if } s > 2; \qquad sF(s) = 2e^{\gamma} \qquad \text{if } 1 \le s \le 3.$$

Another instructive use of this example, which would incidentally lead to the same conclusion, is in examining the (negligibly small) values it gives to the terms in (2.8) containing a factor $\bar{\chi}_D$ that are cast out (as being non-negative) in an application of Brun's inequality (2.5). In this way one can gain an appreciation of why the details of the Rosser–Iwaniec construction (2.9) "ought" to be as they are.

We are working with the case $\phi(d) = 1$ of the identity (2.8) that, with the notations (2.4), (1.12), reads

$$S(a, z) = \sum_{d|(a, P(z))} \mu(d)\chi_D(d) + \sum_{\delta|(a, P(z))} \mu(\delta)\bar{\chi}_D(\delta)S\left(\frac{a}{\delta}, p(\delta)\right). \quad (2.12)$$

When (1.1) holds this leads to

$$\sum_{a\in\mathscr{A}} S(a, z) = X \sum_{d \leq D; d|P(z)} \frac{\mu(d)\chi_D(d)\rho(d)}{d} + \sum_{d \leq D} \mu(d)\chi_D(d)R(X, d)$$

$$+ \sum_{a\in\mathscr{A}} \sum_{\delta|(a, P(z))} \mu(\delta)\bar{\chi}_D(\delta)S\left(\frac{a}{\delta}, p(\delta)\right). \quad (2.13)$$

When the parameters B_{2i} are as specified by (2.9) the Rosser–Iwaniec proof of the Linear Sieve Theorem proceeds via showing

$$\sum_{d \leq D; d|P(z)} \frac{\mu(d)\chi_D(d)\rho(d)}{d} = \left\{\prod_{p<z}\left(1 - \frac{\rho(p)}{p}\right)\right\}\left\{f(s) + O\left(\frac{1}{\log^c D}\right)\right\}, \quad (2.14)$$

wherein $z = D^{1/s}$.

We have already observed that in Selberg's example, where $\mathscr{A}$ is given by (2.10), the R-term in (1.1) satisfies (1.11). Thus the assertion under discussion is that the "cast out" terms in (2.12) (as well as the R-terms) are negligibly small, in some sense such as

$$\sum_{a\in\mathscr{A}} \sum_{\delta|(a, P(z))} \mu(\delta)\bar{\chi}_D(\delta)S\left(\frac{a}{\delta}, p(\delta)\right) = O\left(\frac{X}{\log^{1+c} D}\right). \quad (2.15)$$

For example, let us examine the terms in (2.13) where the number $\nu(\delta)$ of prime factors of δ is 2 or 4. We may suppose $z < D^{1/2}$, the case $z \geq D^{1/2}$ having already been considered in a more direct way. We adopt standard notations that prime variables of summation satisfy

$$z > p_1 > p_2 > \cdots, \quad (2.16)$$

and that

$$S\left(\frac{a}{\delta}, \cdot\right) = 0 \qquad \text{if } \frac{a}{\delta} \text{ is not an integer}, \quad (2.17)$$

as might already be implied by (1.12). Thus we are considering

$$\Sigma_2 = \sum_{p_1, p_2; D \leq p_1 p_2^3} \sum_{a\in\mathscr{A}} S\left(\frac{a}{p_1 p_2}, p_2\right), \quad (2.18)$$

(the conditions $p_2 < p_1 < z$, $p_1p_2|a$ being implicit) and (more typically)

$$\Sigma_4 = \sum_{p_1p_2^3 < D \leq p_1p_2p_3p_4^3} \sum_{a \in \mathscr{A}} S\left(\frac{a}{p_1p_2p_3p_4}, p_4\right), \tag{2.19}$$

when

$$\frac{1}{2}D^{1+\varepsilon} < a < D^{1+\varepsilon}, \qquad \Omega(a) \equiv 0, \qquad \text{mod } 2 \tag{2.20}$$

as in (2.10), (2.11). For the purposes of a heuristic discussion, let us temporarily replace ε by zero. Then the only terms counted in Σ_2 would have $a = p_1p_2$, and those in Σ_4 have $a = p_1p_2p_3p_4$ (for the condition $D \leq p_1p_2p_3p_4^3$ would forbid $\Omega(a) \geq 6$, because any further prime factors of a have to exceed p_4). On the other hand, the condition $p_1p_2^3 < D$ in Σ_4 gives $p_1p_2p_3p_4 < D$, which is inconsistent with (2.20) when $\varepsilon > 0$. In Σ_2 we have $p_1p_2 < D$ whenever $z < D^{1/2}$, as a consequence of $p_2 < p_1 < z$.

It is, perhaps, a fairly routine exercise to elaborate this discussion into a proof of (2.15). Alternatively, one may simplify even further and require as a "test question" that the "cast out" terms in (2.13) should vanish when

$$\Omega(a) \equiv 0, \qquad \text{mod } 2; \qquad a = D. \tag{2.21}$$

We can also see how this argument would immediately break down if the exponent 3 in the Rosser–Iwaniec conditions (2.9) were replaced by any other value θ (although the resulting construction can be shown to lead to weaker but nontrivial results for any $\theta > e$).

3. The Weighted Linear Sieve

The object now is to show not that some a in $\mathscr{A}$ is free of small prime factors but, as remarked in the introduction, to show that some a in $\mathscr{A}$ has a suitably small number of prime factors,

$$\Omega(a) \leq R \qquad \text{for some } a \text{ in } \mathscr{A}, \tag{3.1}$$

when $\mathscr{A}$ satisfies (1.1)–(1.4) and (as in (1.9))

$$a < D^g \qquad \text{when } a \in \mathscr{A}. \tag{3.2}$$

Here, the "level of distribution" D is to satisfy (1.11).

As we remarked at (1.20), the available results all require the hypothesis

$$g < R - \delta_R = \Lambda_R \tag{3.3}$$

for a certain $\delta_R > 0$. On the other hand, Selberg observed, at a symposium on analytic number theory in Durham, England in 1979, that such results definitely do not hold for any $\delta_R < 0$. The example that shows this is, essentially, (2.10) again.

Suppose, for example, that we wish to deal with the case $R = 2$. Consider those $\mathscr{A}$ of the form

$$\mathscr{A} = \{a = pb: 2D_1 < b < 4D_1, \quad \Omega(b) \equiv 0 \bmod 2, \quad 4D_1 < p \leq 8D_1\}, \tag{3.4}$$

where p is prime. Write

$$4D_1 = D^{1+\varepsilon}$$

as in (2.11). Then, for the same reasons as before, we may take

$$X = D_1\{\pi(8D_1) - \pi(4D_1)\}$$

in (1.1), where π is the usual prime-counting function, and

$$\rho(d) = 1 \qquad \text{when } d \leq D.$$

Then (1.2), (1.3) hold and $R(X, d)$ satisfies (1.11). But $\mathscr{A}$ contains no elements a with $\Omega(a) \leq 2$, and $a \leq 32D_1^2$ for each a in $\mathscr{A}$ so that (for large enough D) (3.2) holds for each $g > 2$.

On the other hand, this example does, if anything, support the conjecture that the case $\delta_R = 0$; $\Lambda_R = R$ of (3.3) ought to provide a sufficient hypothesis from which to reach the desired conclusion (3.1). It appears, by analogy with the unweighted sieve discussed in Section 2, that "all" we have to do is construct an argument that is equally efficient in its dealing with the example given in (3.4). Moreover, in dealing with the cast-out terms of the type exemplified by (2.18), (2.19), the "large" prime p appearing in (3.4) and similar examples has no influence and we are, in effect, dealing with exactly the same example as was given in (2.10). For the purpose of heuristic guidance we may, then, adopt exactly the same test case (2.21) as before.

The basic machinery of a weighted sieve is, however, somewhat more involved. We again work with a parameter $z = D^{1/s}$, where now we will want to consider those values of s satisfying $s > 1$.

We will, in due course, introduce a constant $W(1) > 0$ and a function

$$w(p) = W\left(\frac{\log p}{\log D}\right) \qquad \text{for } p < z.$$

It will appear that the choice

$$W(t) = t(0 < t \leq 1) \tag{3.5}$$

would be a natural one and that it is only with reluctance that we are forced to work with modifications of this choice. Initially, then, let us restrict our discussion to this ideal case.

Denote

$$\log^+ x = \begin{cases} \log x & \text{if } x \geq 1 \\ 0 & \text{if } 0 < x < 1 \end{cases}$$

and retain the notational conventions (2.16), (2.17). We would like to be able to work successfully with the expression

$$E = S(a, z) \log D + \sum_{p_1} S\left(\frac{a}{p_1}, z\right) \log p_1 + \sum_{p_1, p_2} S\left(\frac{a}{p_1 p_2}, z\right) \log^+ \frac{p_1 p_2}{D}$$
$$+ \cdots \sum_{p_1, \ldots, p_r} S\left(\frac{a}{p_1 \ldots p_r}, z\right) \log^+ \frac{p_1 \ldots p_r}{D^{r-1}} + \cdots, \tag{3.6}$$

where $z = D^{1/s}$, for each $s > 1$. The expression S is as given in (1.12). The logarithms may, if desired, be taken to the base D, in accordance with the normalization (1.8) and consistently with (3.5). This expression E would count a number $a \in \mathscr{A}$ positively if it had at most one prime factor $p_1 < z$, or two prime factors $p_2 < p_1 < z$ whose product exceeded D, or (generally) r prime factors $p_r < \cdots < p_1 < z$ whose product exceeded D^{r-1} for some integer $r \geq 1$. If $a < D^g$ as in (3.2) with $g < R$, and if, for each $s > 1$, an expression (3.6) could be shown to be positive, then we should have $\Omega(a) \leq R$ as desired. This objective remains, however, an unachieved one at the time of writing.

Following Brun, we may decompose E by successive applications of the Buchstab identity. This can straightforwardly be shown to lead to

$$E = \log D - \sum_{p_1 | a;\, p_1 < z} \log\left(\frac{D}{p_1}\right) + \sum_{r \geq 2} \Sigma_r,$$

where now

$$\Sigma_2 = \sum_{p_1, p_2} S\left(\frac{a}{p_1 p_2}, p_2\right)\left\{\log D - \log p_1 - \log p_2 + \log^+ \frac{p_1 p_2}{D}\right\},$$

$$\Sigma_3 = \sum_{p_1, p_2, p_3} S\left(\frac{a}{p_1 p_2 p_3}, p_2\right)$$
$$\times \left\{\log p_3 - \log^+ \frac{p_1 p_3}{D} - \log^+ \frac{p_2 p_3}{D} + \log^+ \frac{p_1 p_2 p_3}{D^2}\right\},$$

and generally

$$\Sigma_r = \sum_{p_1,\ldots,p_r} \sum_{q|P(p_r)} S\left(\frac{a}{p_1p_2q}, p_2\right)\Delta_D(p_1p_2, q),$$

where

$$\Delta_D(p_1p_2, q) = \log^+ \frac{q}{D^{\nu(q)-1}} - \log^+ \frac{p_1q}{D^{\nu(q)}} - \log^+ \frac{p_2q}{D^{\nu(q)}} + \log^+ \frac{p_1p_2q}{D^{\nu(q)+1}},$$

$\nu(q)$ denoting the number of prime factors of q. It is also possible to derive this identity as a case of the "fundamental" identity (2.8).

It is straightforward to verify that $\Delta_D(p_1, p_2, q) \geq 0$ so that $\Sigma_r \geq 0$ when $r \geq 2$. Thus the expression defined in (3.6) satisfies

$$E \geq \log D - \sum_{p_1|a;\, p_1<z} \log \frac{D}{p_1} + \Sigma_2 + \Sigma_3', \tag{3.7}$$

where

$$\Sigma_3' = \sum_{p_1,p_2,p_3;\, p_1p_3<D} S\left(\frac{a}{p_1p_2p_3}, p_2\right) \log p_3.$$

Note that

$$\Sigma_2 = \sum_{p_1p_2<D} S\left(\frac{a}{p_1p_2}, p_2\right) \log \frac{D}{p_1p_2}.$$

The inequality (3.7) is implicit in more than one place in the literature (see [7], [4]), where, however, the discussions did not proceed along the lines suggested here. Observe that the inequality is sharp with respect to the test case (2.21) wherein $a = D$.

4. A Weighted Sieve Following Rosser

Of the possible approaches to our problem there are two that we wish to discuss. In the one more closely related to Rosser's construction, the first step is to invoke inequalities

$$\Sigma_2 \geq \sum_{p_1p_2^\beta<D} S\left(\frac{a}{p_1p_2}, p_2\right) \log \frac{D}{p_1p_2} = \Sigma_2^*, \tag{4.1}$$

$$\Sigma_3' \geq \sum_{p_1p_2^\beta<D} S\left(\frac{a}{p_1p_2p_3}, p_2\right) \log p_3 = \Sigma_3^*, \tag{4.2}$$

where $\theta = 3$. The role of the inequality $p_1 p_2^3 < D$ in (4.1) is essentially the same as in the unweighted sieve of Rosser discussed in Section 2. It appears also in (4.2) because a direct estimation of Σ_3' from below via the linear sieve would be inefficient; see [5] for a fuller discussion of this point. To attempt to avoid such inefficiencies, we treat Σ_2 and Σ_3 in a parallel fashion.

Note that the inequality (4.1) is sharp in the test case (2.21) because it is equivalent to

$$\sum_{p_1 p_2 < D \le p_1 p_2^3} S\left(\frac{a}{p_1 p_2}, p_2\right) \log \frac{D}{p_1 p_2} \ge 0.$$

Here the inequality $D \le p_1 p_2^3$ excludes the case $\Omega(a) \ge 4$, and when $p_1 p_2 = D$ we have $\log(D/p_1 p_2) = 0$. Observe that this remark depends on our current "ideal" choice of the weight functions in which

$$\frac{w(p)}{W(1)} = \frac{\log p}{\log D}.$$

The inequality (4.2) is, however, not sharp in the test case (2.21), since it says

$$\sum_{p_1 p_3 < D \le p_1 p_2^3} S\left(\frac{a}{p_1 p_2 p_3}, p_2\right) \log p_3 \ge 0.$$

Here the left side is, unfortunately, not negligible when

$$a = p_1 p_2 p_3 q = D,$$

where q is prime and $p_2 < q < p_2^2/p_3$, as may very well occur in our test case.

The unresolved difficulty in this subject is that of devising a method that does not, at some stage, suffer from this sort of defect.

Next, further applications of Buchstab's identity lead to

$$\Sigma_2^* + \Sigma_3^* = \sum_{p_1 p_2 | a;\, p_1 p_2^3 < D} \log \frac{D}{p_1 p_2} - \sum_{p_1 p_2 p_3 | a;\, p_1 p_2^3 < D} \log \frac{D}{p_1 p_2 p_3} + \Sigma_4 + \Sigma_5,$$

where

$$\Sigma_4 = \sum_{p_1 p_2^3 < D} S\left(\frac{a}{p_1 p_2 p_3 p_4}, p_4\right) \log \frac{D}{p_1 p_2 p_3 p_4},$$

$$\Sigma_5 = \sum_{p_1 p_2^3 < D} S\left(\frac{a}{p_1 p_2 p_3 p_4 p_5}, p_4\right) \log p_5.$$

In the sum Σ_4, we have

$$\Sigma_4 \geq \Sigma_4^* = \sum_{p_1 p_2^3 < D; p_1 p_2 p_3 p_4^3 < D} S\left(\frac{a}{p_1 p_2 p_3 p_4}\right) \log \frac{D}{p_1 p_2 p_3 p_4}, \tag{4.3}$$

because the condition $p_1 p_2^3 < D$ implies $p_1 p_2 p_3 p_4 < D$. This exemplifies a further important property of the choice of the exponent 3 in this construction. In addition, the inequality (4.3) is also sharp in the test case (2.21) for the same reasons as apply in the unweighted case at (2.19).

Continuing on the lines indicated leads to the following inequality for the expression defined in (3.6):

$$E \geq \sum_{d|a} \lambda_D(d),$$

where

$$\lambda_D(d) = \begin{cases} \mu(d)\chi_D(d) \log \dfrac{D}{d} & \text{if } d|P(z) \\ 0 & \text{if not.} \end{cases}$$

From the above remarks relating to the deficiencies of the current approach when measured against the test question (2.21) it appears we cannot expect to achieve a positive result of the desired shape (1.9) for arbitrary $z < D$. It is, however, possible to obtain a result of the desired type

$$\sum_{d|P(z)} \frac{\lambda_D(d)\rho(d)}{d} = \left\{\prod_{p<z}\left(1 - \frac{\rho(p)}{p}\right)\right\}\left\{g\left(\frac{\log D}{\log z}\right) + O\left(\frac{1}{\log^c D}\right)\right\},$$

where (setting $z = D^{1/s}$)$g(s) > 0$ when $s > 2/\sqrt{e}$. This is implicit in the results of [4], where the consequences of the idea we have just sketched were worked out.

The relationship between this approach and earlier ones was discussed in [5].

5. Weight-warping

Direct application of these methods in the form we have described in Section 4 does not, however, lead to the best available results. In fact we would, in the notation of the introduction, obtain only $\Omega(a) \leq R$ when $g < 2R/\sqrt{e}$. One can do better by adopting a modification of the natural logarithmic weights along lines introduced by Richert [15] in connection with a method introduced by Ankeny and Onishi [1].

Accordingly, we replace the expression in (3.6) by

$$\begin{aligned} E' = {} & W(1)S(a, z) + \sum w(p_1)S\left(\frac{a}{p_1}, z\right) \\ & + \sum \{w(p_1) + w(p_2) - W(1)\}^+ S\left(\frac{a}{p_1p_2}, z\right) \\ & + \sum \{w(p_1) + w(p_2) + \cdots + w(p_r) \\ & - (r-1)W(1)\}^+ S\left(\frac{a}{p_1 \dots p_r}, z\right) + \cdots \end{aligned} \tag{5.1}$$

where, as before, the superscript $^+$ indicates $f^+ = \max\{f, 0\}$, and we retain the notations (2.16), (2.17). The expression $w(p)$ is now to be nonzero only when $p > D^V$ (although V is allowed to be negative, in which case this restriction is vacuous). For the present, write $z = D^U$, where $1 \geq U > V$ and $U > \frac{1}{2}$.

Given that $a < D^g$ as in (3.2), we require that E' should be positive only when $\Omega(a) \leq R$, where R is to be specified. The term $w(p_1)S(a/p_1, z)$ counts the number $a \in \mathscr{A}$ positively if it has only one prime factor p_1 less than $z = D^U$, and if $p_1 > D^V$. This is acceptable provided

$$V + RU \geq g \tag{5.2}$$

so that $\Omega(a) \geq R + 1$ would be impossible. Furthermore, we would then have $g < (R+1)U$ so that the term $W(1)S(a, z)$ would also count positively only when $\Omega(a) \leq R$.

For our convenience let us write $W(1) = U - V$. It is perhaps not surprising that interpolation linear in $\log p/\log D$ is an appropriate procedure: Specify

$$w(p) \leq \frac{\log p}{\log D} - V \qquad \text{when } D^V < p < D^U$$

so that $w(p) \leq W(1)$. Then

$$\{w(p_1) + w(p_2) + \cdots + w(p_r) - (r-1)W(1)\}^+ S\left(\frac{a}{p_1 \dots p_r}, z\right)$$

cannot be nonzero unless each entry $w(p_i)$ is nonzero so that $p_i > D^V$. We would have

$$0 < \frac{\log(p_1 \dots p_r)}{\log D} - rV - (r-1)(U - V)$$

while all other prime factors of a exceed D^U. If the number of these were (strictly) greater than $R - r$ then we should have

$$\begin{aligned} g &\geq (R - r + 1)U + \frac{\log(p_1 \ldots p_r)}{\log D} \\ &> (R - r + 1)U + (r - 1)U + V \\ &= RU + V, \end{aligned}$$

contrary to the hypothesis (5.2). Hence $\Omega(a) \leq R$ as required.

One now proceeds to estimate the expression E' from below in the style described in Sections 3 and 4. There is (for some choices of the parameters U, V) a further restriction on the function w needed to make this analogous process a legitimate one. We shall not follow this process in detail here, save to note that the quantity analogous to (2.18) is now estimated via

$$\sum_{D \leq p_1 p_2^3} \{W(1) - w(p_1) - w(p_2)\}^+ \geq 0.$$

This inequality is, in general, not sharp in the test case (2.21), since the quantity

$$W(1) - w(p_1) - w(p_2) = U + V - \frac{\log(p_1 p_2)}{\log D}$$

is not zero when $p_1 p_2 = D$, unless $U + V = 1$.

It appears, therefore, that "weight-warping" of the type discussed in this section is not going to feature substantially in any possible future treatment of the case $g = R$. For we should need $U + V = 1$ (at least approximately) for the reasons discussed above, and also $R = g \leq RU + V$ from (5.2). Since $U \leq 1$ this implies $U = 1$, $V = 0$, which returns us to the ideal logarithmic weights

$$w(p)/W(1) = \frac{\log p}{\log D}$$

of the previous section.

Full details of the construction introduced in the last two sections were given in [5]. The matter has been taken further in [9], where the function λ_d that is involved was given a bilinear form suitable for use in applications with nontrivial treatment of the R-terms appearing in (1.10).

6. Selberg's λ^2 Method

The second approach to the weighted linear sieve that we wish to discuss uses the well-known λ^2 method [16] in the following way. Details have appeared in [6].

We start with the analogue, for the expression E' defined in (5.1), of the inequalities (3.7), (4.1), (4.2), where we now choose $\theta = 2$. Since we will not be able to use the ideal weights $\log p/\log D$, these inequalities will already be not sharp in the test case (2.21).

We make further use of Buchstab's identity, but proceeding in the opposite direction to that used previously, on the terms in which $p_1^3 > D$. This leads to

$$E' \geq W(1) - \sum_{p_1|a;p_1^3<D} \{W(1) - w(p_1)\} + \Sigma_1 + \Sigma_2'' + \Sigma_3'', \tag{6.1}$$

where

$$\Sigma_1 = \sum_{D \leq p_1^3} \{W(1) - w(p_1)\} S\left(\frac{a}{p_1}, \sqrt{\frac{D}{p_1}}\right)$$

$$+ \sum_{D \leq p_1^3; p_2 < \sqrt{(D/p_1)}} w(p_2) S\left(\frac{a}{p_1 p_2}, \sqrt{\frac{D}{p_1}}\right),$$

$$\Sigma_2'' = \sum_{p_1^3 < D} S\left(\frac{a}{p_1 p_2}, p_2\right)\{W(1) - w(p_1) - w(p_2)\},$$

$$\Sigma_3'' = \sum_{p_1^3 < D} S\left(\frac{a}{p_1 p_2 p_3}, p_2\right) w(p_3).$$

The parameter $\sqrt{(D/p_1)}$ arises because it is the upper bound for p_2 implied by the inequalities $p_2 < p_1$, $p_2^2 p_1 < D$ appearing in (4.1), (4.2) when $\theta = 2$. This choice of θ is appropriate because the sum Σ_1 is to be bounded from above using the upper bound method of Selberg. Thus we require

$$\Sigma_1 \leq \sum_{D \leq p_1^3; p_1|a; p_1 < z} \{W(1) - w(p_1)\} \left\{ \sum_{d|a/p_1; d < \sqrt{(D/p_1)}} \lambda(d) \right\}^2,$$

where $\lambda(1) = 1$ and $\lambda(d)$ is chosen as is usual in Selberg's method. For this to be a valid inequality we will require

$$w(p_2) \leq \{W(1) - w(p_1)\}\{\lambda(1) + \lambda(p_2)\}^2 \qquad \text{when } p_2^2 p_1 < D \leq p_1^3. \tag{6.2}$$

The inequality (6.2) constitutes a new requirement on the "warping" of the weight function w required by the method under discussion. On the other hand, it is instructive to compare the requirement with that which would be needed if we were to employ, instead of the λ^2 method, the other well-known method, Rosser's, for obtaining an upper bound for the sum

$$\Sigma_1' = \sum_{D \le p_1^3} \{W(1) - w(p_1)\} S\left(\frac{a}{p_1}, \sqrt{\frac{D}{p_1}}\right)$$

that contributes to Σ_1. The two methods lead to bounds for

$$\sum_{a \in \mathscr{A}} \Sigma_1'$$

that have the same leading term. However, if Rosser's method were to be used in an analogous way to the current proposal to bound the sum Σ_1 from above, then we should require $w(p_2) = 0$ if $p_2^3 p_1 < D$. Thus the use of the λ^2 method offers, at least when $p_2^3 p_1 < D \le p_1^3$, a flexibility in the choice of w that is not offered by the method of Rosser, and does so at no cost.

The remaining contributions Σ_2'', Σ_3'' to (6.1) are dealt with, initially, by the combinatorial methods described earlier. In the first place we have

$$\Sigma_2'' \ge \sum_{p_1^3 < D; p_1 p_2^3 < D} S\left(\frac{a}{p_1 p_2}, p_2\right)\{W(1) - w(p_1) - w(p_2)\} = \Sigma_2^{**},$$

$$\Sigma_3'' \ge \sum_{p_1^3 < D; p_1 p_2^3 < D} S\left(\frac{a}{p_1 p_2 p_3}, p_2\right) w(p_3) = \Sigma_3^{**},$$

provided the condition $p_1^3 < D$ guarantees $w(p_1) + w(p_2) \le W(1)$. The inequality for Σ_3'' is, like the corresponding inequality (4.2) in the method of Section 4, not sharp in the test case (2.21). Nor is the inequality for Σ_2'', because the inequality (6.2) necessarily involves a "warping" of the logarithmic weights $\log p/\log D$, introduced for reasons additional to those of the type discussed in Section 5. However, these deficiencies are now damped by the presence of the inequality $p_1^3 < D$.

Next, further applications of Buchstab's identity give

$$\Sigma_2^{**} + \Sigma_3^{**} = \sum_{p_1^3 < D; p_1 p_2^3 < D; p_1 p_2 | a} \{W(1) - w(p_1) - w(p_2)\}$$

$$- \sum_{p_1^3 < D; p_1 p_2^3 < D} \left[S\left(\frac{a}{p_1 p_2 p_3}, p_3\right)\left\{W(1) - \sum_{i=1}^{3} w(p_i)\right\} \right.$$

$$\left. + S\left(\frac{a}{p_1 p_2 p_3 p_4}, p_3\right) w(p_4) \right].$$

The contribution to the last sum in which $p_1p_2p_3^3 \geq D$ is bounded by the λ^2 method and the remaining contribution is expressed via Buchstab's identity in terms of sums

$$\Sigma_4'' + \Sigma_5' = \sum_{p_1^3 < D; p_1p_2^3 < D; p_1p_2p_3^3 < D} \left[S\left(\frac{a}{p_1p_2p_3p_4}, p_4\right)\left\{W(1) - \sum_{i=1}^{4} w(p_i)\right\} + S\left(\frac{a}{p_1p_2p_3p_4p_5}, p_4\right)w(p_5)\right],$$

whose treatment is then analogous to that of $\Sigma_2'' + \Sigma_3''$.

The role of the inequalities $p_1p_2^3 < D \leq p_1p_2p_3^3$ is to keep the application of the λ^2 method within a range in which it is sharp, in its treatment of the $S(a/p_1p_2p_3)$ term, in the test case (2.21). This feature is taken from the well-known paper [13] of Jurkat and Richert on the linear sieve (without "weights"); see also Chapter 8 of the text [8] of Halberstam and Richert.

Probably the most that can be said for our remarks concerning the role of the λ^2 method in this construction is that the reader should not be surprised to learn that it leads, in certain cases, to results superior to those given by the more combinatorial methods of Sections 4 and 5. Full details are given in [6].

For applications, numerical approximations are needed to the numbers δ_R that may legitimately appear in inequalities of the type (1.20). For the combinatorial method of Section 5 it appears (see [4]) that

$$\delta_2 = 0.06373.., \qquad \delta_3 = 0.09999..,$$
$$\delta_4 = 0.11647.., \ldots, \qquad D_\infty = 0.12482...$$

For the method of Section 6 we have (see [6])

$$\delta_2 = 0.04456.., \qquad \delta_3 = 0.07426.., \qquad \delta_4 = 0.10397...$$

For values of R greater than 4 the value found for δ_R is worse than that following via the method of Section 5. It seems that this is because the new inequality (6.2) becomes, for the choice of the function w appropriate for large R, an excessively restrictive requirement. It seems probable that the ideas of Section 6 could be adapted to obtain small improvements in the known values of δ_R for $R > 4$, but the details would appear likely to be excessively tiresome in view of the small improvement likely to be obtained.

It is, it is hoped, clear that in any case the methods discussed in this article are, at least in their present form, unsuitable for a satisfactory solution of the problem under consideration, for which the answer is, presumably,

$$\delta_R = 0.$$

References

[1] Ankeny, N.C. and Onishi, H. "The general sieve," *Acta Arith.* **10** (1964–65), 31–62.

[2] Brun, V. "Le crible d'Eratosthene et le theoreme de Goldbach," *Skr. Norske Vid. Akad. Kristiania.* **I** No. 3 (1920), 36 pp.

[3] Chen, J.-R. "On the distribution of almost-primes in an interval," *Sci. Sinica.* **28** (5) (1975), 611–627.

[4] Greaves, G. "A weighted sieve of Brun's type," *Acta Arith.* **40** (1982), 297–332.

[5] Greaves, G. "A comparison of some weighted sieves," *Banach Centre Publ.* **17** (1985), 143–153.

[6] Greaves, G. "The weighted linear sieve and Selberg's λ^2 method," *Acta Arith.* **47** (1986), 71–96.

[7] Halberstam, H., Heath-Brown, D.R., and Richert, H.-E. "Almost-primes in short intervals." In: *Recent Progress in Analytic Number Theory* **I**, Halberstam, H. and Hooley, C., eds., Academic Press, London (1981), 69–101.

[8] Halberstam, H. and Richert, H.-E. *Sieve Methods.* Academic Press, London (1974).

[9] Halberstam, H. and Richert, H.-E. "A weighted sieve of Greaves' type I, II," *Banach Centre Publ.* **17** (1985), 155–215.

[10] Iwaniec, H. "On the error term in the linear sieve," *Acta Arith.* **29** (1971), 1–30.

[11] Iwaniec, H. "Rosser's sieve," *Acta Arith.* **36** (1980), 171–202.

[12] Iwaniec, H. "A new form of the error term in the linear sieve," *Acta Arith.* **37** (1980), 307–320.

[13] Jurkat, W.B. and Richert, H.-E. "An improvement of Selberg's sieve method I," *Acta Arith.* **11** (1965), 217–240.

[14] Kuhn, P. "Zur Viggo Brun'schen Siebmethode. I," *Norske Vid. Selske Forh. Trondhjem.* **14** (39) (1941), 145–148.

[15] Richert, H.-E. "Selberg's sieve with weights," *Mathematika.* **16** (1969), 1–22.

[16] Selberg, A. "On an elementary method in the theory of primes," *Norske Vid. Selske. Forh. Trondhjem.* **19**(18) (1947), 64–67.

[17] Selberg, A. "The general sieve method and its place in prime number theory," *Proc. Inter. Congress. Math., Cambridge* (Mass.) **I** (1950), 286–292.

[18] Selberg, A. "Sieve methods," *Amer. Math. Soc. Proc. Sympos. Pure Math.* **20** (1971), 311–351.

17 Critical Zeros of GL(2) L-Functions

JAMES LEE HAFNER

1. Introduction

What I would like to describe is the basic ingredients that go into a proof that a positive proportion of the zeros of certain Dirichlet series lie on their critical line. The basic ideas have their germs in the methods of Hardy and Littlewood, and others, but it was Selberg who brought the ideas to fruition, adding the major ingredients that, in the case of the Riemann zeta-function, pushed the result from a good lower bound to a positive proportion result. It is also the case that a number of the ideas that are crucial in the other cases for which the theorem is known have their origin in Selberg's work in harmonic analysis and modular forms. Hence it is only fitting that this be a topic of discussion at this conference in Selberg's honor.

There are actually two different methods that can be used. The first is due to Selberg [22], and it is the one we describe here. Titchmarsh [24] (or [25], Chapter X) also gave a proof of Selberg's result by simplifying some of the technical details, but the essential ingredients were Selberg's. We will try to point out where these two variations on a theme differ. The second fundamentally different method, due to Levinson [18], provides a bit more information but does not seem to be

NUMBER THEORY,
TRACE FORMULAS and
DISCRETE GROUPS

ISBN 0-12-067570-6

applicable to any case other than (essentially) the Riemann zeta-function, so we do not discuss it here.

The general plan of a proof of this type is actually quite simple, but to carry out the details in particular cases requires (at times) significant amounts of machinery. As a result, the theorem has only been shown for a small class of all Dirichlet series for which it is expected a Riemann Hypothesis should hold. This class includes only the Riemann zeta-function itself (Selberg [22]), Dirichlet L-functions (Parmankulov [19] and Zhuravlev [26]), and L-functions attached to GL(2) cusp forms (Hafner [11,12,13]). This last group is actually subdivided into those L-series arising from holomorphic cusp forms ([11,12]) and those arising from real-analytic cusp forms, the so-called Maass wave forms ([13]). In both of these last cases we assume the cusp forms are also Hecke eigenforms. Note that all these cases arise from modular forms.

There are two things I would like to emphasize in this paper. The first is the general outline of a proof, i.e., the basic ingredients common to all three cases. Secondly, I would like to point out the major differences between the actual implementation of the method in the various known cases.

Let me now set up a bit of notation. Our Dirichlet series will be denoted by

$$L(s) = \sum_{n=1}^{\infty} \frac{a(n)}{n^s}.$$

This will be the L-series attached to some modular form $F(z)$ of weight k on a congruence subgroup of $\Gamma = \mathrm{PSL}(2, \mathbb{Z})$. We assume the coefficients $a(n)$ are multiplicative so that $L(s)$ has an Euler product

$$L(s) = \prod_p \left(1 + \frac{a(p)}{p^s} + \frac{a(p^2)}{p^{2s}} + \cdots\right).$$

We do not really need to assume that each Euler factor has a nice form as a rational function in p^{-s}, though that is the case for all the examples we consider. We also assume that the coefficients $a(n)$ (these are also the Hecke eigenvalues in the case of cusp forms) are real valued. Without a functional equation we would not even expect a Riemann Hypothesis to hold. Hence, we also assume that the L-series satisfies a functional equation of the type

$$\xi(s) \overset{\text{def}}{=} G(s)L(s) = \varepsilon\xi(k - s)$$

where

$$\begin{aligned} \varepsilon &= \pm 1, \\ G(s) &= B^s \Gamma^*(s), \\ \Gamma^*(s) &= \text{product of gamma functions}, \\ k &= \text{weight of associated modular form}. \end{aligned}$$

We take as normalization, that $G(s)/\Gamma(s)$ has a nonzero limit as $\operatorname{Im} s \to \infty$.

Let me give three examples to help crystallize the situation:

(A) The Riemann Zeta-Function.

$$F(z) = \theta(z) = \sum_{m=-\infty}^{\infty} e^{\pi i m^2 z},$$

$$a(n) = \begin{cases} 1 & \text{if } n = m^2, \\ 0 & \text{otherwise}, \end{cases} \qquad k = \frac{1}{2}, \qquad \varepsilon = 1, \qquad G(s) = \pi^{-s}\Gamma(s);$$

(B) Ramanujan's Tau Function.

$$F(z) = \Delta(z) = q \prod_{n=1}^{\infty} (1 - q^n)^{24} = \sum_{n=1}^{\infty} \tau(n) q^n, \qquad q = e^{2\pi i z},$$

$$a(n) = \tau(n), \qquad k = 12, \qquad \varepsilon = (-1)^{k/2} = 1, \qquad G(s) = (2\pi)^{-s}\Gamma(s);$$

(C) Maass wave form f for Γ with Laplacian eigenvalue $\frac{1}{4} + r^2$.

$$F(z) = f(z) = \sum_{n=1}^{\infty} a(n) W_{0,ir}(4\pi n y) \times \begin{cases} \cos(2\pi n x) & \text{for } f \text{ even}, \\ \sin(2\pi n x) & \text{for } f \text{ odd}, \end{cases}$$

$$a(n) = \text{Hecke eigenvalue}, \qquad k = 0, \qquad \varepsilon = \begin{cases} +1 & \text{for } f \text{ even}, \\ -1 & \text{for } f \text{ odd}, \end{cases}$$

and

$$G(s) = \begin{cases} \pi^{-s}\Gamma\left(\dfrac{s + ir}{2} + \dfrac{1}{4}\right)\Gamma\left(\dfrac{s - ir}{2} + \dfrac{1}{4}\right) & \text{for } f \text{ even}, \\[2ex] \pi^{-s}\Gamma\left(\dfrac{s + ir}{2} + \dfrac{3}{4}\right)\Gamma\left(\dfrac{s - ir}{2} + \dfrac{3}{4}\right) & \text{for } f \text{ odd}. \end{cases}$$

(Here, $W_{0,ir}(y)$ is the standard Whittaker function.) Note that in the first and last cases we have used a nonstandard normalization. We include Dirichlet L-functions in case (A) as well. Of course case (B) includes the L-series attached to any holomorphic cusp form for $\mathrm{PSL}(2, \mathbb{Z})$ and its congruence subgroups.

With this setup it is easy to formalize the questions about zeros. As we have it, the critical strip is centered on the line $\sigma = k/2$ and has width $\frac{1}{2}$ in the case of the Riemann zeta-function and width one in the other cases. So we let

$$N(T) = \#\left\{\rho = \beta + i\gamma \colon \left|\beta - \frac{k}{2}\right| < \frac{1}{2}, \quad 0 < \gamma \leq T, \quad L(\rho) = 0\right\}$$

and

$$N_0(T) = \#\left\{\rho = \frac{k}{2} + i\gamma \colon \ 0 < \gamma \leq T, \quad L(\rho) = 0\right\},$$

It is easy to show that $N(T) \sim 1/\pi \times T \log T$ as T tends to infinity. The Riemann Hypothesis is of course $N_0(T) = N(T)$. Note that we have formulated this hypothesis to allow for zeros off the line but on the real axis (which is also a subject of debate). The best that can be shown, however, is only

$$N_0(T) > AT \log T$$

for some positive constant A (depending heavily on L), and this only for the examples mentioned above. This is, of course, the positive proportion result.

As an aside, we note that Levinson's method (applied to the Riemann zeta-function) gives a numerical value for the constant A. The best result to date is $0.3658 \times 1/\pi$ (see Conrey [2]). Selberg's method can be used to get numerical values for this constant, but the results are very weak and so not very interesting. As far as I know no (successful) attempt has been made to get a value in the case of cusp forms either holomorphic or real analytic.

2. The General Plan of the Proof

We have $\xi(s) = \varepsilon\xi(k - s)$ so that on the critical line

$$\varepsilon^{1/2}\xi\left(\frac{k}{2} + it\right)$$

is real for real values of t. Thus to count zeros (of odd order) we need only count sign changes of this function. This is done by comparing, for $0 < t < T$,

$$\left|\int_t^{t+h} \xi\left(\frac{k}{2} + iu\right) du\right| \quad \text{and} \quad \int_t^{t+h} \left|\xi\left(\frac{k}{2} + iu\right)\right| du. \tag{1}$$

When these two integrals differ we have found a zero in the interval $[t, t+h]$.

To simplify the discussion we make note of two factors:

(a) We can "factor out" the function $G(s)$ from $\xi(s)$ in these integrals because its asymptotics are well understood and it contributes no extraneous zeros.

(b) To measure the frequency of having the two integrals in (1) differ we estimate the measure of the set $E \subseteq [0, T]$ of t where this occurs.

As a consequence of these remarks, we need only compare the two integrals

$$\int_0^T \left|\int_t^{t+h} L\left(\frac{k}{2} + iu\right) du\right| dt \quad \text{and} \quad \int_0^T \int_t^{t+h} \left|L\left(\frac{k}{2} + iu\right)\right| du\, dt. \tag{2}$$

We show that the measure of E satisfies $m(E) \gg T$, which indicates frequent sign changes, and deduce

$$N_0(T) \gg \frac{m(E)}{h} \gg \frac{T}{h}.$$

The smaller we can take h the better the lower bound result. Qualitatively, the smaller we take h the more sensitive is our comparison in (1) (or (2)) to detection of zeros.

Comparing the integrals in (1) as they stand, the best that has been achieved is h a constant. This was Hardy and Littlewood's result [14] in the case of the Riemann zeta-function, Lekkerkerker's [17] for holomorphic cusp forms, and Epstein–Hafner–Sarnak's [5] for Maass forms. The essential difficulty in getting a smaller h is the following. Consider the case of the Riemann zeta-function. It satisfies

$$\int_0^T \left|\zeta\left(\frac{1}{2} + it\right)\right| dt \asymp T(\log T)^{1/4} \quad \text{and} \quad \int_0^T \left|\zeta\left(\frac{1}{2} + it\right)\right|^2 dt \asymp T(\log T).$$

This means that there is significant fluctuation in the size of $\zeta(\frac{1}{2} + it)$ even in short intervals. Thus our technique for trying to detect zeros is hindered if the zero is near (within h, say) of a point of large size (e.g., if it has a steep slope through the zero). The functions $L(k/2 + it)$ exhibit similar behavior. If we could replace L by $L/|L|$ then this would not be a problem. However this is unreasonable as this effectively requires complete knowledge of the zeros. Thus we seek a way to "mollify" the large fluctuations in L. This is done using a Dirichlet polynomial, which should approximate $|L|^{-1}$. Let

$$\phi(s) = \sum_{\nu \leq X} \frac{\beta_\nu}{\nu^s},$$

where X will be chosen later (as a function of T) and the coefficients β_ν are chosen by the following scheme. Write for Re $s > k$

$$\begin{aligned} L(s)^{-1/2} &= \prod_p \left(1 + \frac{a(p)}{p^s} + \frac{a(p^2)}{p^{2s}} + \cdots\right)^{-1/2} \\ &= \prod_p \left(1 + \frac{\alpha(p)}{p^s} + \frac{\alpha(p^2)}{p^{2s}} + \cdots\right) \\ &= \sum_{\nu=1}^{\infty} \frac{\alpha(\nu)}{\nu^s}, \end{aligned}$$

and then put

$$\beta_\nu = \begin{cases} \alpha(\nu)\left(1 - \dfrac{\log \nu}{\log X}\right) & \text{if } \nu \leq X, \\ 0 & \text{otherwise.} \end{cases}$$

We can see immediately that

(a) $\phi(s)$ is entire;

(b) $\phi(s)\phi(k - s)$ is invariant under $s \to k - s$;

(c) $0 \leq \phi(s)\phi(k - s)|_{s = k/2 + it} = |\phi(k/2 + it)|^2 \approx |L|^{-1}$ as required; and

(d) $|\phi(k/2 + it)|^2$ has only even order zeros.

Now we will replace $L(s)$ by $L(s)\phi(s)\phi(k - s)$ in the above discussion. This satisfies the same functional equation (with the same gamma factor $G(s)$) and on the critical line has no new zeros of odd order. Thus those we detect must come from $L(s)$.

For simplicity we let $M(u) = L|\phi|^2(k/2 + iu)$. With this notation and the plan outlined above our job is reduced to the following two tasks:

(a) produce a good lower bound for

$$\int_0^T \int_t^{t+h} |M(u)|\, du\, dt$$

and

(b) produce a good upper bound for

$$\int_0^T \left| \int_t^{t+h} M(u)\, du \right| dt.$$

These estimates will provide a tool for getting a lower bound on $m(E)$. (We will not go into the details of this aspect of the proof, however. Rather, we concentrate on the deeper problems involved.)

It is very easy to show that a good lower bound in the first case is hT, so we leave out the details and add just a few remarks. First, if our scheme of choosing ϕ to mollify the fluctuations in L is successful, then this should be the correct order of magnitude (the integrand should be effectively constant). Second, the proof of this lower bound can be outlined as follows. Interpret $|L|\phi|^2|$ as $|L\phi^2|$, invert orders of integration, and pull the absolute value signs outside the integral. Then use Cauchy's theorem to replace the integral with one in the right half-plane of absolute convergence of the series for $L\phi^2(s)$. At this point, the evaluation is trivial. This procedure works for any Dirichlet series and so (a) is not where the difficult parts of the proof lie.

We can get a good upper bound for item (b) by first applying the Cauchy–Schwarz inequality (again we should not lose much if our integrand is essentially constant, i.e., if we chose ϕ appropriately) and then bounding

$$\int_0^h \int_0^h \left\{ \int_0^T M(t+u)\overline{M(t+v)}\, dt \right\} du\, dv.$$

The inner integral is rather delicate to evaluate because of the dependence on u and v, but the basic idea can be illustrated with the special case $u = v = 0$. If we can show that

$$\int_0^T |M(t)|^2\, dt \ll T \tag{3}$$

for some choice of X, then we will in fact have chosen ϕ correctly and we will have the machinery in place to fill in the details of the proof in the general case. Without the mollifier ϕ, this integral should satisfy

$$\int_0^T \left| L\left(\frac{k}{2} + it\right) \right|^2 dt = c_1 T \log T + c_2 T + o(T).$$

Introducing the mollifier to pull down the main term may make it difficult to control the effect in the secondary and error terms. This is the fundamental problem that needs to be dealt with.

The difference between Selberg's version of this method and Titchmarsh's can be described briefly here. What we need to do is estimate the mean-square of a Dirichlet series that satisfies a functional equation. This can be done by the direct approach that we outline below or through a Tauberian argument as follows. Instead of estimating the integral directly, one first analyzes the integral

$$\int_{-\infty}^{\infty} |M(u)|^2 e^{-|u|/T}\, du.$$

This can be done by realizing the integral as the L^2 norm of some function and then applying Parseval's identity. This, in effect, reintroduces the modular forms from which the L-series came from and makes some of the calculations simpler. This idea of using Fourier analysis is due to Titchmarsh. The plan we describe here is more in the spirit of Selberg's original proof and so seems more appropriate in this forum.

In the rest of this section we will outline how the evaluation of the integral in (3) breaks up into main terms and error terms. In the next section we will indicate how the main terms can be dealt with. We will point out what the key ingredients are in this calculation. It will turn out that these ingredients are present in essentially any example where we expect the Riemann Hypothesis to hold, not just in the examples for which our theorem is known to hold. Thus the proof can be carried out up to this point for many more cases. In the following sections we will indicate how the error terms are dealt with. It is here where the proof breaks up into cases corresponding to the three examples we gave earlier. Each case requires a different approach, though there are some parallels in how the two types of cusp forms are handled.

The integral in (3) is the mean square of the L function on its critical line multiplied by two Dirichlet polynomials of length X. To evaluate this integral, we replace L by a Dirichlet polynomial of length (necessarily) T and then expand the products and square modulus. This can be done in practice by using an approximate functional equation. Of course the actual calculations are much more complicated than outlined here, but our purpose is merely expositional and so we leave out a greal deal of detail. This leaves us with six sums and an integral of the type

$$\sum_{\kappa, \lambda, \mu, \nu \leq X} \frac{\beta_\kappa \beta_\lambda \beta_\mu \beta_\nu}{(\kappa\lambda\mu\nu)^{k/2}} \sum_{n, m \leq T} \frac{a(n)a(m)}{(nm)^{k/2}} \int_0^T \left(\frac{n\kappa\lambda}{m\mu\nu}\right)^{it} dt. \tag{4}$$

The main terms should clearly be those where the integrand is equal to 1 (whence the integral is exactly T). All other terms have integral much smaller than T and should be considered as error terms.

We can separate the main terms in the following way. Let

$$q = gcd(\kappa\lambda, \mu\nu), \qquad b = \frac{\kappa\lambda}{q}, \qquad d = \frac{\mu\nu}{q}, \qquad gcd(b, d) = 1,$$

and then put

$$dm = bn + N.$$

We can then rewrite the expression in (4) (using symmetry) as

$$\sum_{\kappa, \lambda, \mu, \nu \leq X} \frac{\beta_\kappa \beta_\lambda \beta_\mu \beta_\nu}{(\kappa\lambda\mu\nu)^{k/2}} \sum_{N \geq 0} \sum_{n \leq T} \frac{a(n)a\left(\frac{bn+N}{d}\right)}{\left[n\left(\frac{bn+N}{d}\right)\right]^{k/2}} \int_0^T \left(\frac{bn}{bn+N}\right)^{it} dt. \tag{5}$$

(We use the convention that $a(x) = 0$ if x is not an integer.) In this form the main terms are when $N = 0$ and the error terms are when $N \geq 1$.

3. The Main Terms

The terms in (5) where $N = 0$ give

$$T \sum_{\kappa, \lambda, \mu, \nu \leq X} \frac{\beta_\kappa \beta_\lambda \beta_\mu \beta_\nu}{(\kappa\lambda\mu\nu)^{k/2}} \left(\frac{d}{b}\right)^{k/2} \sum_{n \leq T} \frac{a(n)a(bn/d)}{n^k}$$

$$= T \sum_{\kappa, \lambda, \mu, \nu \leq X} \frac{\beta_\kappa \beta_\lambda \beta_\mu \beta_\nu}{(\kappa\lambda\mu\nu bd)^{k/2}} \sum_{n \leq T} \frac{a(bn)a(dn)}{n^k}. \tag{6}$$

(This is not strictly correct; we have ignored some subtle points concerning the range of summation in n.) If the coefficients $a(n)$ were completely multiplicative, this could be expressed as

$$T \sum_{\kappa, \lambda, \mu, \nu \leq X} \frac{\beta_\kappa \beta_\lambda \beta_\mu \beta_\nu a(b)a(d)}{(\kappa\lambda\mu\nu bd)^{k/2}} \left(\sum_{n \leq T} \frac{a^2(n)}{n^k}\right). \tag{7}$$

This last sum in parentheses can be estimated classically as $\log T$. That is, one studies the Dirichlet series

$$D(s) = \sum_{n=1}^{\infty} \frac{a^2(n)}{n^s} \tag{8}$$

and discovers that this has a simple pole at $s = k$ and other effectively described analytic properties. For the case of the Riemann zeta-function, $a(n) = 1$ if n is a square and zero otherwise, so this is trivial. In the case of cusp forms, either holomorphic or Maass, this is done via the Rankin–Selberg convolution. (See Rankin [20] and Selberg [21]. Also some details of this kind of calculation, in a more general setting, can be found in Section 4.) From this the estimate for the innermost sum is easily derived.

The fact that the coefficients are not completely multiplicative is more a nuisance than a serious impediment to the proof. The effect of the b and d in the inner sum on the right side of (6) can be dealt with by considering the Dirichlet series.

$$\sum_{n=1}^{\infty} \frac{a(bn)a(dn)}{n^s}$$

and realizing that this is a finite Euler product over the prime divisors of bd times the series in (8). Hence its analytic behavior is easily determined. The result is effectively the same as that given in (7).

Finally we are left with the sum

$$\sum_{\kappa, \lambda, \mu, \nu \leq X} \frac{\beta_\kappa \beta_\lambda \beta_\mu \beta_\nu a(b)a(d)}{(\kappa\lambda\mu\nu bd)^{k/2}}. \tag{9}$$

This can again be evaluated by classical methods. We first write the sum as (again assuming complete multiplicativity)

$$\sum_{n \leq X^2} \sum_{m|n} \left(\frac{n}{m}\right)^k \mu(m)a^2(m)\left(\sum_{\kappa\lambda \equiv 0(n)} \frac{\beta_\kappa \beta_\lambda}{(\kappa\lambda)^k} a\left(\frac{\kappa\lambda}{n}\right)\right)^2.$$

Replacing λ with $l\lambda$, μ with $l_1\mu$ where $n|ll_1$ and $gcd(\lambda\mu, n) = 1$ (so that ll_1 is composed only of primes dividing n, indicated by an asterisk on the summation), the inner sum can be put in the form

$$\sum_{n|ll_1}^{*} \frac{a(ll_1/n)}{(ll_1)^k} \sum_{gcd(\lambda,n)=1} \frac{\beta_{l\lambda}a(\lambda)}{\lambda^k} \sum_{gcd(\mu,n)=1} \frac{\beta_{l_1\mu}a(\mu)}{\mu^k}.$$

The last two sums are essentially the same, namely, recalling the definition of β_v,

$$S_l = \frac{\alpha_l}{\log X} \sum_{\substack{gcd(\lambda, n) = 1 \\ \lambda \leq X/l}} \frac{\alpha_\lambda a(\lambda)}{\lambda^k} \log \frac{X}{\lambda l}.$$

This is a weighted average of the coefficients of the Dirichlet series

$$\begin{aligned} \sum_{gcd(\lambda, n) = 1} \frac{\alpha_\lambda a(\lambda)}{\lambda^s} &= \prod_{p \nmid n} \left(1 + \frac{\alpha_p a(p)}{p^s} + \cdots\right) \qquad (10) \\ &\approx \prod_{p \nmid n} \left(1 + \frac{a^2(p)}{p^s} + \cdots\right)^{-1/2} \\ &= D(s)^{-1/2} \times \text{finite Euler product,} \end{aligned}$$

where $D(s)$ is the Dirichlet series in (8). Consequently, we deduce that the series in (10) has a fractional order zero at $s = k$. From this the estimates for S_l, and so all the other sums can be derived. We find that the sum in (9) is bounded by $(\log X)^{-1}$.

Putting all the estimates together, we find that the main term is estimated by

$$T \frac{\log T}{\log X} \ll T$$

as required (in (3)) if we can take X to be any fixed power of T. There has been no limitation on X so far so that the feasibility of choosing X in this way is completely determined by our ability to show that the error terms in (5) contribute no more than T in order of magnitude.

We summarize what has gone into the proof so far and what the essential ingredients have been. First, we have used the functional equation in two ways. One way is to get a function that is real on the critical line and whose (odd order) zeros are those of L. The other way is in the form of the approximate functional equation used to replace L in the initial problem with a Dirichlet polynomial. The length of this polynomial, T in our case, was determined by the size of the gamma factors in the functional equation. Up to this point in the proof, a Dirichlet polynomial of any length that is a power of T would have sufficed. In Titchmarsh's variation of this proof, the functional equation is implicit in the transformation (via Parseval's identity) back to modular forms.

The next ingredient was the Euler product for the Dirichlet series. This was essential to enable us to choose the mollifier (the coefficients α_ν) so that they were multiplicative and so that the last sum we studied could be effectively estimated using our knowledge of $a(n)$. In fact, only average data about these coefficients was used.

Finally, we used deeply the analytic properties of the series $D(s)$. The main tool here was the fact that this series has an analytic continuation beyond the point of absolute convergence, except for a simple pole at that point, and moderate growth in vertical strips (used implicitly in the estimation of the above sums). Of these points, the fact that the pole is simple is not so important; any higher order pole could be dealt with by modifying the definition of the mollifier slightly.

Each of these ingredients is present in essentially any example where we expect a Riemann Hypothesis to hold. Hence the stumbling blocks to extending this theorem to a more general class of functions lies in the problem of estimating the error terms in (5). As we will see this problem gets exceedingly more difficult as we move through the examples. The case of the Riemann zeta-function, though difficult, requires little machinery beyond real and complex analysis. For the L-series coming from holomorphic forms, the Selberg trace formula seems to be an essential tool. To prove the theorem in the case of Maass wave forms, the current proof requires the Kuznetsov–Bruggeman trace formula. In the next section we will see how these tools play a role in the problem of estimating the error term.

4. The Error Terms

Recall that the error terms are

$$\sum_{\kappa,\lambda,\mu,\nu\le X}\frac{\beta_\kappa\beta_\lambda\beta_\mu\beta_\nu}{(\kappa\lambda\mu\nu)^{k/2}}\sum_{N\ge 1}\sum_{n\le T}\frac{a(n)a\left(\dfrac{bn+N}{d}\right)}{\left[n\left(\dfrac{bn+N}{d}\right)\right]^{k/2}}\int_0^T\left(\frac{bn}{bn+N}\right)^{it}dt. \tag{11}$$

Now N can be assumed small (less than an arbitrarily small power of T) because the integral decays rapidly as N goes to infinity. (In practice, the actual integral is more complicated and shows this decay more readily.) Also, since κ, λ, μ, $\nu\le X$ and we want to choose X to be a power of T, our problem is reduced to an estimate of the type

$$\sum_{n\le T}a(n)a\left(\frac{bn+N}{d}\right)w(n,T;b,d,N)=O_{b,d,N}(T^A) \tag{12}$$

with explicit (and no worse than polynomial) bounds in b, d, N, and some fixed $A < 1$. Here $w(n, T; b, d, N)$ is some weight function. As an aside, we note that in Titchmarsh's variation on the proof we must deal with a similar sum, but with a different w. Thus, the rest of our discussion applies to both versions of Selberg's method. With this estimate our error terms contribute

$$X^B T^{A+\varepsilon} = o(T),$$

provided X is a small enough (fixed) power of T, and this is indeed smaller than our main terms.

Note that a trivial estimate for the sum in (11) (using the Cauchy–Schwarz inequality) is T (showing only this dependence). Hence for an estimate of the type (12) to hold, some significant cancellation must occur.

We break up into the various examples now to point out how each case is handled separately.

Case (A). The Riemann Zeta-function. In this case the terms in the sum in (11) vanish unless both n and $(bn + N)/d$ are integral squares. The infrequency of this occurring makes the estimates in this case fairly straightforward. The tools required are some classical estimates for exponential integrals. It was essential that something like this occur. The coefficients are non-negative, so we can not expect any cancellation from the arithmetical structure of the sum; it all must occur in the weight function (the exponential integral). We refer the reader to Selberg [22], Titchmarsh [24] or Chapter X of [25], or Edwards [4], Chapter 11 for more details.

Cases (B) and (C). GL(2) Cusp Forms. One should ask now why cancellation should be expected in the sum in (11) in these cases, especially since (on average) none of the coefficients are zero. The answer lies in the following estimates:

$$\sum_{n \le x} a(n) \ll x^{k/2+\varepsilon}, \qquad \text{and} \qquad \sum_{n \le x} |a(n)| \gg x^{(k+1)/2-\varepsilon}.$$

This shows that the distribution of the signs of the coefficients is somewhat random so that we expect no significant correlation between $a(n)$ and $a(n + N)$, for example. Thus built in to the sum in (11) should be some fundamental cancellation among the terms themselves. This is what we need to exploit.

The problem of getting a good bound in T for the sum (12) is not the only aspect we have to consider. Equally important is the quality of the estimates in b, d, and N. Thus whatever technique we use must show this dependence explicitly, and it must be quantitatively good.

We will see that there are two basic ingredients that go into our study of this problem but that they show themselves in different ways in each of these two cases. The ingredients are:

(a) Spectral data
(b) Sums of Kloosterman sums

Let us now break up our discussion into the separate cases to see how these two ideas contribute to the solution of our problem.

Case (B). Holomorphic Cusp Forms. We will take the simpler case $b = d = 1$ to illustrate the ideas and then add a few remarks on what is needed to extend to the general case.

We first write the sum in question as

$$\sum_{n \le T} a(n)a(n+N)w(n, T; N) = \int_{k+1-i\infty}^{k+1+i\infty} D(s, N)\Phi(s)T^{s-k+1}\, ds, \quad (13)$$

where $\Phi(s)$ depends on w and

$$D(s, N) = \sum_{n=1}^{\infty} \frac{a(n)a(n+N)}{(n+N)^s}.$$

This is a generalization of the series $D(s)$ given in (8). From the information about $D(s)$ obtained from the Rankin–Selberg convolution, we easily see that this series is absolutely convergent if Re $s > k$. This means we can move the path of integration to Re $s = k + \varepsilon$ and get the trivial estimate for our sum $O(T^{1+\varepsilon})$. This is, of course, not good enough, but it does suggest that if we can move the path of integration to the left of the line Re $s = k$ we would be able to get the kind of estimate (at least in T) that we want. To do this path shift requires two things. First, we need an explicit analytic continuation of this series beyond the line of absolute convergence (encountering no poles on this line), and second we need good bounds on the series in a vertical strip to the left of this line. We must also maintain explicit dependence on the parameter N (and b, d in the general case).

If $F(z)$ is the cusp form in question, then we can realize the series $D(s, N)$ as a Rankin–Selberg convolution (as we did for $D(s)$):

$$D(s, N) = \text{elementary factors} \times \int_{\Gamma\backslash\mathscr{H}} y^k F(z)\overline{F(z)} P(z, s-k+1, N) \frac{dx\,dy}{y^2}, \tag{14}$$

where $\Gamma\backslash\mathscr{H}$ is a fundamental domain for the action of Γ on the upper half-plane $\mathscr{H}$ and $P(z, s, N)$ is the (nonholomorphic) Poincaré series defined in Re $s > 1$ by

$$P(z, s, N) = \sum_{M\in\Gamma_\infty\backslash\Gamma} \frac{e^{2\pi i N M(z)}}{|\gamma z + \delta|^{2s}}.$$

Here $M = \left(\begin{smallmatrix}\alpha & \beta\\ \gamma & \delta\end{smallmatrix}\right)$ runs over a complete set of coset representatives for $\Gamma_\infty = \{\left(\begin{smallmatrix}1 & n\\ 0 & 1\end{smallmatrix}\right) : n \in \mathbb{Z}\}$ in Γ. (Also, $M(z) = (\alpha z + \beta)/(\gamma z + \delta)$.) Of course, the case $N = 0$ is just the Eisenstein series.

As we remarked before, since we expect no correlation between the signs of the coefficients of $D(s, N)$, we should not expect this to have any poles on the line of absolute convergence. It was Selberg [23] who first showed that this was the case. Using the fact that the Poincaré series (when $N > 0$) is square-integrable over $\Gamma\backslash\mathscr{H}$, we can expand it using the spectral decomposition of the space $L^2(\Gamma\backslash\mathscr{H})$. We get

$$D(s, N) = \text{elementary factors} \times \sum_{j>0} \gamma_j(N)\Gamma(s-k-s_j)\Gamma(s-k+s_j)\langle F, Fe_j\rangle_k + \text{managable terms}. \tag{15}$$

Here the sum is over the discrete spectrum of the Laplacian for the group Γ, the $e_j(z)$ are the Maass wave forms (the eigenfunctions of this Laplacian), $\gamma_j(N)$ are their Fourier coefficients, and $s_j = \frac{1}{2} + ir_j$ are determined by their eigenvalues $\lambda_j = \frac{1}{4} + r_j^2$. Also, the factor in angle brackets is the Petersson inner product

$$\langle F, Fe_j\rangle_k = \int_{\Gamma\backslash\mathscr{H}} y^k F(z)\overline{F(z)e_j(z)} \frac{dx\,dy}{y^2}. \tag{16}$$

One then easily sees that $D(s, N)$ has no poles to the right of the line Re $s = k - \frac{1}{2}$ and poles on this line at $s = k - \frac{1}{2} \pm ir_j$.

In more general cases (either when b and d are not both equal to one or when $F(z)$ is a holomorphic cusp form on a congruence subgroup), there is the possibility of so-called exceptional eigenvalues, those less than $\frac{1}{4}$. These introduce poles on the real axis to the right of the point $k - \frac{1}{2}$. However, Selberg has shown [23] that for congruence groups these poles lie to the left of the point $k - \frac{1}{4}$, i.e., the eigenvalues are all greater than $\frac{3}{16}$, so they cause no problem.

We see already that spectral data has provided us with good qualitative information. But in order to proceed with our application, we must have quantitative information. Any trivial attempt to estimate $D(s, N)$ to the left of the line of absolute convergence will not succeed (the resulting bounds will be exponential in $|s|$.) Ideally, to keep the qualitative information already obtained explicit, we would like to be able to estimate the series in (15) term by term. This is possible only if we can bound the inner product in (16), as a function of the eigenvalue, by

$$\langle F, Fe_j \rangle_k \ll e^{-\pi r_j/2} r_j^{k+\varepsilon}.$$

The constant $\pi/2$ in the exponential is crucial to get term-wise convergence of the series in (15).

Good [6,7,8] provides us with the necessary ideas on how to proceed. The key is to find another representation for the integral in (16). Since $F(z)$ is a holomorphic cusp form of weight k, we can write $F(z)$ as a (finite) linear combination of holomorphic Poincaré series

$$P_k(z, m) = \sum_{M \in \Gamma_\infty \backslash \Gamma} \frac{e^{2\pi i m M(z)}}{(\gamma z + \delta)^k}.$$

We replace $F(z)$ in (16) by one of these series, and then the resulting integral can be evaluated as a complicated series involving Legendre functions. This can be estimated (carefully) and the resulting bound is of the required type. Using this bound (16) in our expression (15) gives us usable bounds on $D(s, N)$ in $\operatorname{Im} s$. This in turn fits neatly into (13) and allows us to conclude an estimate like (12).

There are a number of points that should be made at this time. The first is that we have implicitly used the Selberg trace formula in a number of places. The most obvious are in the spectral decomposition of $P(z, s, N)$ and in the spectral sum in (15). The second point is that this procedure has provided us with the necessary estimates in s (and actually in N as well) to deal with (13) in the case $b = d = 1$. To extend to the general case, to maintain explicit dependence in these parameters, more work is required. (See Hafner [10] for details.) The analogue of (14) for the generalization of $D(s, N)$ can be given in a couple of different ways, but only one seems most suited for the purposes at hand. That is, we replace the factor $\overline{F(z)}$ and $\overline{F(dz/b)}$, and then we are forced to use a Poincaré series and a fundamental domain for the congruence subgroup

$$\Gamma_0(d, b) = \left\{\begin{pmatrix} \alpha & \beta \\ \gamma & \delta \end{pmatrix} \in \Gamma : d|\gamma, b|\beta\right\}.$$

Now the trace formula for this group becomes the essential tool, and all spectral estimates (particularly the general version of (16) that entails the Maass wave forms for this subgroup) must show the dependence on b and d. A second problem, though a very subtle one, is in the step where we replace the form $F(z)$ in the integral (16) by a linear combination of Poincaré series. In this general case, this step must either be independent of the extra parameters or show this dependence explicitly. Luckily, we have set it up to achieve the first hope, i.e., we can do this substitution on the group on which the form F is defined independent of b and d, not on the subgroup $\Gamma_0(d, b)$.

The final point to be made here concerns the two major ingredients that we said earlier lie at the heart of the method. The first, spectral data, has shown itself in a fundamental and obvious way. The second, sums of Kloosterman sums, has not been made explicit. It is, however, very implicit in this key step of replacing the form F by a sum of Poincaré series. The Fourier coefficients of these Poincaré series are (infinite) sums of Kloosterman sums. Hence we have the equivalence:

$$F = \sum_{\text{finite}} P\text{-series} \Leftrightarrow a(n) = \sum_{\substack{m \\ \text{finite}}} \sum_{c} S(m, n, c) \times \text{J-Bessel function}.$$

The point I wish to make here is that we have (implicitly) replaced the coefficients $a(n)$ in (13), which involve deep arithmetic concepts, by sums of Kloosterman sums, which are much more mysterious in this regard.

This concludes our discussion of the holomorphic case. Note that our procedure has been essentially classical, obtaining bounds on arithmetical sums via the associated Dirichlet series ($D(s, N)$). In some sense, this is the most natural and satisfying method. It also has the advantage that the information we obtained about these Dirichlet series becomes a basic tool that can be used in other problems. (See, e.g., Good [9].) In the next (and last) case these aesthetically pleasing aspects disappear.

Case (C). Maass wave forms. We now let $f(z)$ be a Maass wave form for Γ with eigenvalue $\lambda = \frac{1}{4} + r^2$. We would like to estimate

$$\sum_{n \leq T} a(n) a\left(\frac{bn + N}{d}\right) w(n, T; b, d, N), \tag{17}$$

where $a(n)$ are the Fourier coefficients of f and w is some weight function. We would also like to be able to proceed in this case exactly as was done above, but we are thwarted in the key step where we replace the form f in the integral in (16) by the holomorphic Poincaré series. No such analogue exists for the Maass forms. Hence a new approach is required. This is provided by Kuznetsov [16]. We still rely on the two ingredients, spectra and Kloosterman sums, but in a different way. We return to our sum (17) and try to estimate it directly. Again we will take the special case where $b = d = 1$ and add remarks later on how the general case is dealt with.

Our difficulty is that we do not have an exact formula for the coefficients $a(n)$ in terms of Kloosterman sums. But the nonholomorphic Poincaré series $P(z, s, n)$ have the property that

$$a(n) = \text{elementary factors} \times \langle P(\cdot, s, n), f \rangle_0,$$

where the elementary factors depend on s and the eigenvalue $\lambda = \frac{1}{4} + r^2$. Furthermore, the Fourier coefficients of $P(z, s, n)$ involve sums of Kloosterman sums $S(m, n, c)$ (like the holomorphic Poincaré series) with weight functions that are Bessel-like, though more complicated. With these two ideas, it is possible to get an approximate formula for $a(n)$ as sums of Kloosterman sums. But how is this to be used?

The main steps can now be summarized as follows. We replace $a(n + N)$ in (17) (when $b = d = 1$) with this approximate formula as sums of Kloosterman sums of the type $S(m, n + N, c)$. Inverting orders of summation, we see that the inner sum (over n) involves $a(n)$ twisted by an additive character. This sum can be transformed by the Poisson summation formula for $a(n)$. After this is done, we re-invert orders of summation and reconstruct a new series of Kloosterman sums of the type $S(N, -n + m, c)$. We get an expression like

$$\sum_m \sum_n a(n) \sum_N \sum_c S(N, -n + m, c) W(c; N, n, m) \tag{18}$$

with some new weight function W. It might be useful to point out that the sum on m is finite and essentially independent of the other parameters, so it should cause no problems. The sums on n and c are infinite and the sum on N is short (at most T^{ε} long). This step of decomposing and then reconstructing the Kloosterman sums may seem unnecessary, but it actually has a very subtle value. First, a large part of the n-dependence in the initial sum has been "smoothed" into a continuous variable by the transformations of the Poisson summation formula. Secondly, we have separated the N- and n-dependence; the advantage of this is not obvious at this point but will be indicated later.

Now we can appeal to spectral data in the form of the Kuznetsov–Bruggeman Trace Formula. (See Kuznetsov [15], Bruggeman [1] or Deshoullier and Iwaniec [3].) This very deep formula is difficult to state in any detail, so for purposes of this exposition we describe it in very vague terms. The formula can be outlined by the graphic:

$$\begin{aligned}\sum_c S(m, n, c)\phi\left(\frac{4\pi\sqrt{mn}}{c}\right) &= \sum_j \gamma_j(m)\overline{\gamma_j(n)}\tilde{\phi}(r_j)\\ &\quad + \sum_k \text{holomorphic forms of weight } k\\ &\quad + \int_t \text{Eisenstein series at } \tfrac{1}{2} + it. \qquad (19)\end{aligned}$$

On the left-hand side of this formula are Kloosterman sums times some fairly general weight function. The first term on the right is a sum over all Maass forms (the discrete spectrum), the $\gamma_j(m)$ and $\gamma_j(m)$ are the Fourier coefficients of the forms, and $\tilde{\phi}$ is a certain transform of ϕ. The second and third terms on the right are of a similar type but with the m-th and n-th Fourier coefficients of the holomorphic forms and the Eisenstein series, respectively, replacing those of the Maass forms. The transforms of ϕ in these two cases are also somewhat different.

This formula has a number of applications, not the least of which is our problem at hand. Its value is that it expresses the arithmetic information of Kloosterman sums in terms of analytic objects in a very effective manner. It is more general than the Selberg Trace Formula in this regard (and, it seems implies it as well). In any case, we use this formula to re-express (18) in the form (considering only the first term on the right)

$$\sum_m \sum_n a(n) \sum_j \gamma_j(-n + m)h_j(n; m),$$

where

$$h_j(n; m) = \sum_N \gamma_j(N) \times \text{transforms of weight functions.}$$

We can see now the second reason why we went through the earlier step involving the Kloosterman sums and Poisson summation. By separating the dependence on n and N, we can exploit in h_j the cancellation over the sum on N. Furthermore, the final transformed sums, though very complicated, have a much more exploitable form. That is, they can be estimated effectively to give us our desired bound for the sum in (17).

There are still a few points that we should add here. First, the other terms in (19) are dealt with in essentially the same manner as above. Secondly, the procedure to estimate these sums is, in some sense, more complicated than what we started with, since now we have to track dependencies on the complete spectrum. This is not so fundamentally different, however, from what was required in Case (B) of the holomorphic forms, where (15) and (16) had to be dealt with.

Thirdly, the extension to the more general case (arbitrary b and d) is achieved in much the same way as before. That is, we need to appeal to the spectral and Kloosterman sum structures of the congruence subgroup $\Gamma_0(b, d)$. Again, dependencies on b and d in the associated Kuznetsov–Bruggeman formula must be tracked. These complications turn out to be tedious, but no more fundamentally difficult than in the earlier case.

We would like to add one final comment. In the case of the holomorphic forms, by studying the appropriate generating function, $D(s, N)$, we developed a tool that can be applied directly, with classical techniques, to a number of related problems. In this case, the "tool," is the Kuznetsov–Bruggeman Trace Formula. Though very valuable and deep, it does not lend itself to ease of application; each example of a sum of type (17) must be handled separately, with all the integral and sum estimates dealt with as special cases. This is a disappointing state of affairs. One solution to this dilemma would be to provide an estimate of the type

$$\langle f, fe_j \rangle_0 \ll e^{-\pi r_j/2} r_j^A,$$

for some constant A. As we saw before, this would provide the necessary quantitative information about the series $D(s, N)$ and so provide a more pleasant tool for other applications. As far as I know, this integral with three Maaß forms has not been effectively estimated, despite significant efforts by a number of people (including the author).

References

[1] Bruggeman, R.W. "Fourier coefficients of cusp forms," *Invent. Math.* **45** (1978), 1–18.

[2] Conrey, B. "On the distribution of the zeros of the Riemann zeta-function." In: *Topics in Analytic Number Theory*, S. Graham and J. Vaaler, eds., University of Texas Press, Austin (1985).

[3] Deshouillers, J.-M., and Iwaniec, H. "Kloosterman sums and Fourier coefficients of cusp forms," *Invent. Math.* **70** (1982/83), 219–288.

[4] Edwards, H.M. *Riemann's Zeta Function.* Associated Press, New York (1974).

[5] Epstein, C., Hafner, J.L., and Sarnak, P. "Zeros of L-functions attached to Maass forms," *Math. Zeit.* **190** (1985), 113–128.

[6] Good, A. "Beiträge zur Theorie den Dirichletreihen, die Spitzenformen zugeordnet sind," *J. Number Theory* **13**, (1981), 18–65.

[7] Good, A. "Cusp forms and eigenfunctions of the Laplacian," *Math. Ann.* **255** (1981), 523–548.

[8] Good, A. "The square mean of Dirichlet series associated with cusp forms," *Mathematika.* **29** (1982), 278–295.

[9] Good, A. "On various means involving the Fourier coefficients of cusp forms," *Math. Zeit.* **183** (1983), 95–129.

[10] Hafner, J.L. "Explicit estimates in the arithmetic theory of cusp forms," *Math. Ann.* **264** (1983), 9–20.

[11] Hafner, J.L. "Zeros on the critical line for Dirichlet series attached to certain cusp forms," *Math. Ann.* **264** (1983), 21–37.

[12] Hafner, J.L. "On the zeros (à la Selberg) of Dirichlet series attached to certain cusp forms." In: *Topics in Analytic Number Theory*, S. Graham and J. Vaaler, eds., University of Texas Press, Austin (1985).

[13] Hafner, J.L. "Zeros on the critical line for Maass wave form L-functions," *J. Reine Angew. Math.*, to appear.

[14] Hardy, G.H., and Littlewood, J.E. "The zeros of the Riemann's zeta function on the critical line," *Math. Zeit.* **10** (1921), 283–317.

[15] Kuznetsov, N.V. "Petersson hypothesis for parabolic forms of weight zero and Linnik hypothesis. Sums of Kloosterman sums," *Math. USSR Sbornik.* **39** (1981), 299–342.

[16] Kuznetsov, N.V. "Mean value of the Hecke series associated with cusp forms of weight zero," *J. Soviet Math.* **24** (1984), 215–238.

[17] Lekkerkerker, C.G. "On the zeros of a class of Dirichlet series," Dissertation, Utrecht (1955).

[18] Levinson, N. "More than one-third of zeros of Riemann's zeta-function are on $\sigma = \frac{1}{2}$," *Adv. Math.* **13** (1974), 383–436.

[19] Parmankulov, S. "On the number of zeros of Dirichlet L-functions on the line $\sigma = \frac{1}{2}$," *Izv. Akad. Nauk UzSSR Ser. Fiz.-Mat. Nauk.* **16** (1972), 32–37.

[20] Rankin, R.A. "Contributions to the theory of Ramanujan's function $\tau(n)$ and similar arithmetic functions, I and II," *Proc. Cambridge Phil. Soc.* **35** (1939), 351–372.

[21] Selberg, A. "Bemerkungen über eine Dirichletsche Reihe, die mit der Theorie der Modulformen nahe verbunden ist," *Arch. Math. Naturvid.* **43** (1940), 47–50.

[22] Selberg, A. "On the zeros of Riemann's zeta-function," *Skr. Norske. Vid.-Akad. Oslo I.* **10** (1942), 1–59.

[23] Selberg, A. "On the estimation of Fourier coefficients of modular forms," *Proc. Symp. Pure Math., AMS,* **VIII**, *Theory of Numbers* (1965), 1–15.

[24] Titchmarsh, E.C. "On the zeros of the Riemann zeta function," *Quart. J. Math.* (Oxford) **18** (1947), 4–16.

[25] Titchmarsh, E.C. *The Theory of the Riemann Zeta-function*, 2nd ed., Oxford University Press (Clarendon), London, New York (1987).

[26] Zhuravlev, V.G. "The zeros of a Dirichlet L-function on the critical line," *Math Notes.* **19** (1976), 341–346.

18 A New Upper Bound in the Linear Sieve*

H. HALBERSTAM, LOU SHITUO
AND *YAO QI*

1. Introduction

Let $\mathscr{A}$ be a finite set of integers (that are not necessarily distinct or positive) and $\mathscr{P}$ be a set of primes. As usual, we write

$$P(z) = \prod_{\substack{p \in \mathscr{P} \\ p < z}} p(z \geq 2), \qquad P(z_1, z) = \prod_{\substack{p \in \mathscr{P} \\ z_1 \leq p < z}} p = \frac{P(z)}{P(z_1)} \qquad (2 \leq z_1 < z),$$

$$\mathscr{A}_d = \{a \in \mathscr{A}, a \equiv 0 \bmod d\}, \qquad A = \max_{a \in \mathscr{A}} |a|,$$

and

$$S(\mathscr{A}, z) = S(\mathscr{A}, \mathscr{P}, z) = |\{a \in \mathscr{A}, (a, P(z)) = 1\}|,$$

where $|\{\ldots\}|$ denotes the cardinality of the set described in the parentheses. We suppose that there exist a non-negative arithmetic function $w(\cdot)$, defined on the sequence of square-free integers, such that $w(p) = 0 (p \notin \mathscr{P})$ and $0 < w(p) < p (p \in \mathscr{P})$, and a convenient approximation X to $|\mathscr{A}|$, such that the quantities

$$R_d := |\mathscr{A}_d| - \frac{w(d)}{d} X, \qquad \mu(d) \neq 0, \tag{1.1}$$

* This work was supported in part by a grant from the National Science Foundation.

ISBN 0-12-067570-6

may be viewed as remainders, at least on average, in a sense that will be made apparent below. We write

$$V(z) = \prod_{p<z}\left(1 - \frac{w(p)}{p}\right), \qquad z \geq 2,$$

and impose the Iwaniec condition

$$\frac{V(z_1)}{V(z_2)} \leq \frac{\log z_2}{\log z_1}\left(1 + O\left(\frac{1}{\log z_1}\right)\right), \qquad 2 \leq z_1 < z; \tag{I}$$

thus the values taken by $w(p)$ may be said to be at most 1 on average, by reference to Mertens' prime number theory. Accordingly it has become customary to refer to the problem of estimating $S(\mathscr{A}, \mathscr{P}, z)$ in these circumstances as the linear sieve. We place on record one form of the fundamental upper inequality in the linear sieve:

Theorem A. *Suppose that* $y > z > 3$. *Then*

$$S(\mathscr{A}, z) \leq XV(z)\left\{F\left(\frac{\log y}{\log z}\right) + O\left(\frac{1}{\log\log z}\right)^{3/10}\right\}$$
$$+ \sum_{D\in\mathscr{D}} \mu(d)\chi_y^+(D) \sum_{\substack{d\in D\\ d<y^{1-\varepsilon}}} \sum_{\substack{m|P(z)\\ m<A^{\varepsilon}}} \alpha_{md} R_{md},$$

where $|\alpha_{md}| \leq 1$, *and* $F(\cdot)$ *and* $f(\cdot)$ *are the standard upper bound and lower bound functions (see, e.g., "Sieve Methods," Chapter* 8.2 *and* [5]).

Writing

$$\phi^- = f, \qquad \phi^+ = F,$$

we have the well-known relations

$$u\phi^{\pm}(u) - v\phi^{\pm}(v) = \int_v^u \phi^{\mp}(t-1)\,dt, \qquad 2 \leq v \leq u.$$

Put $z_1 < z$, where z_1 is large enough but

$$\frac{\log z_1}{\log z} \to 0.$$

Introduce the partition of the interval $[z_1, z)$:

$$z_1 < z_1 z_2 < z_1 z_2^2 < \cdots < z_1 z_2^{j-1} < z_1 z_2^j < \cdots < z_1 z_2^{J-1} < z \leq z_1 z_2^J, \tag{J}$$

where $z_2 < z_1$ and J is a large integer about $\log(z/z_1)/\log z_2$ in size.

Denote the generic subinterval by I. If $I_1 = [z_1 z_2^{j_1-1}, z_1 z_2^{j_1})$ and $I_2 = [z_1 z_2^{j_2-1}, z_1 z_2^{j_2})$ write $I_1 > I_2$ if $j_1 > j_2$. In a typical direct product

$$D = I_1 \otimes \cdots \otimes I_r = I_1 \ldots I_r$$

for short, with

$$I_1 > \cdots > I_r,$$

write $\nu(D) = r$, $\mu(D) = (-1)^{\nu(D)}$, $q(D) = I_1$ and $p(D) = I_r$, and say that a divisor d of $P(z_1, z)$, $d = p_1 p_2 \cdots p_r (p_i \text{ prime}, p_1 > \cdots > p_r)$ belongs to D, $d \in D$, if and only if $p_j \in I_j$, for $1 \le j \le r$. Let D_0 denote the empty product and adopt the natural convention that $1 \in D_0$. Let $\mathscr{D}$ denote the set of all direct products D. We may say further, unambiguously, that $d \in \mathscr{D}$ if $d \in D$ for some D in $\mathscr{D}$. Define

$$\chi_y^+(D) = \eta_y^+(I_1)\eta_y^+(I_1 I_2 I_3)\ldots, \qquad \text{if } D = I_1 I_2 \cdots I_r (I_1 > I_2 > \cdots > I_r),$$

where

$$\eta_y^+(D) = \begin{cases} 1 & r \text{ even} \\ 1 & r \text{ odd and } i_r^3 i_{r-1} \cdots i_1 < y \\ 0 & \text{otherwise,} \end{cases}$$

and i_j is the right-hand end point of I_j, and let

$$\bar{\chi}_y^+(D) = \chi_y^+(I_1 \cdots I_{r-1}) - \chi_y^+(I_1 \cdots I_r).$$

Later on we shall take $\mathscr{A} = \{n, x - x^\theta < n \le x\}$, $\frac{1}{2} < \theta < 1$, and $\mathscr{P}$ the set of all primes. Here $X = x^\theta$, $w(p) = 1$ for all primes p (so that (I) holds even with equality) and

$$R_d = \left[\frac{x}{d}\right] - \left[\frac{x - x^\theta}{d}\right] - \frac{x^\theta}{d},$$

so that $|R_d| \le 1$.

2. A New Upper Bound Method

Consider the upper bound in Theorem A. Iwaniec gave the remainder term in a bilinear form in order to estimate it as accurately as existing analytic methods permit (cf. [4]) and hence to choose y as large as possible (consistent with having the remainder be $o(XV(z))$. Since $F(u)$ is decreasing, a larger admissible choice of y would decrease, for a fixed z, $F(\log y/\log z)$ and hence improve the upper bound for $S(\mathscr{A}, \mathscr{P}, z)$. For example, the trivial estimate $|R_d| \leq 1$ would allow us to estimate the remainder sum by y. In the specific instance when

$$\mathscr{A} = \{n, x - x^\theta < n \leq x\},$$

$\theta = \frac{11}{20} + \varepsilon$ and $\mathscr{P}$ is the set of primes, this would make the choice $y = x^{0.92 - \varepsilon}$ admissible. Using his bilinear form of the remainder sum, Iwaniec found that then

$$\pi(x) - \pi(x - x^{11/20 + \varepsilon}) \leq \frac{2(1 + c)x^{11/20 + \varepsilon}}{\log x} \qquad c = \frac{1}{0.92 - \varepsilon} - 1. \quad (2.1)$$

Given the size of θ, the "perfect" result would be without the multiplier 2 on the right (use of a sieve method makes the factor 2 inevitable so far); therefore c can be viewed as measuring the "defect" of this sieve estimate, in other words, the defect in this method of handling the remainders. With the method we are about to describe, we can choose even $y = x^{1 - \varepsilon}$ and then obtain (2.1) with $c = 0.001$. This would be the simplest application of our method, but we shall not publish the details because in the meantime Heath-Brown has been able to do much better by an asymptotic method (not using the sieve).

The gist of our approach is as follows. A careful study of [3] reveals that the bilinear remainder sum could be better estimated if, among the bilinear forms

$$\sum_{M < m \leq 2M} \sum_{N < n \leq 2N} a_m b_n R_{mn}$$

that one studied there, a relatively small numbers of length-pairs (M, N) did not occur. A first step to avoid these "bad" pairs is given in [2]. Ours is a more systematic approach in which we go back to the original form of the remainder sum, focus on a typical constituent sum

$$\sum_{p_i \in I_i (i = 1, \ldots, r)} R_{p_1, \ldots, p_r} \qquad (I_1 > \cdots > I_r) \tag{2.2}$$

and identify "bad" products D of I-intervals. It will be shown below that we can avoid the occurrence of "bad" products D in the remainder sum by removing from $\mathscr{A}$, at the very outset, the set

$$\mathscr{A}^* = \bigcup_{\substack{d \in D \\ D \in \mathscr{D}^*}} \mathscr{A}_d, \tag{2.3}$$

where $\mathscr{D}^* \subset \mathscr{D}$ and will be described later. Form the collection

$$\mathscr{D}_2 = \{DT; D \in \mathscr{D}^*, T \in \mathscr{D}, T < D\}. \tag{2.4}$$

We have then that

$$\mathscr{A}^* = \bigcup_{\substack{d \in D \\ D \in \mathscr{D}_2}} \mathscr{A}_d,$$

and we now remove the products from the remainder sum that are not only in $\mathscr{D}^*$, but in $\mathscr{D}_2$ also. For applications, we always assume that $\{D \in \mathscr{D}_2; \chi_y^+(D) \neq 0\}$ covers all the "bad" products. Writing

$$\tilde{\mathscr{A}} = \mathscr{A} \backslash \mathscr{A}^*,$$

$S(\mathscr{A}, \mathscr{P}, z) = S(\tilde{\mathscr{A}}, \mathscr{P}, z)$ because $\mathscr{D}^*$ is such that if $a \in \mathscr{A}^*$ then automatically $(a, P(z)) > 1$. However, when the sieve mechanism is applied, the removal of $\mathscr{A}^*$ from $\mathscr{A}$ exacts a price. The price, fortunately, is small; in the example (2.1) it is measured by the "defect" $c = 0.001$ and derives precisely from two terms in (2.7) below.

We shall describe $\mathscr{D}^*$ later.

We write $T|D$ if T is the direct product of a subset of I-intervals making up D, and D/T for the complementary direct product. $T < D$ means $q(T) < p(D)$.

We now state our main result.

Theorem 1. *Suppose that $y > z > 3$ and $\mathscr{A}$ is a finite set of integers and satisfies condition (I). $\mathscr{D}^*$ is a subset of $\mathscr{D}$ such that $\mu(D^*) = 1$ if $D^* \in \mathscr{D}^*$, $\mathscr{D}_2$ is a subset of $\mathscr{D}$, $D \in \mathscr{D}_2$ if and only if $D = D^*T$, where $D^* \in \mathscr{D}^*$ and $T < D^*$, and $\mathscr{D}_1 = \mathscr{D} \backslash \mathscr{D}_2$. Then*

$$\begin{aligned} S(\mathscr{A}, z) \leq{} & XV(z)\left\{F\left(\frac{\log y}{\log z}\right) + O\left(\frac{1}{\log\log z)^{3/10}}\right)\right\} \\ & + \sum_{D \in \mathscr{D}_1} \mu(D)\chi_y^+(D) \sum_{\substack{d \in D \\ d < y^{1-\varepsilon}}} \sum_{\substack{m | p(z) \\ m < A^{\varepsilon}}} \alpha_{dm} R_{dm} \\ & + \sum_{D \in \mathscr{D}^*} \chi_y^+(D) \sum_{\substack{d \in D \\ d < y^{1-\varepsilon}}} S(\mathscr{A}_d, z(D)) \\ & - \sum_{D \in \mathscr{D}^*} \chi_y^+(D) \sum_{\substack{d \in D \\ d < y^{1-\varepsilon}}} \frac{w(d)}{d} XV(z(D)) F\left(\frac{\log(y/d)}{\log z(D)}\right), \end{aligned} \tag{2.5}$$

where $|\alpha_{md}| \leq 1$, and $z(D)$ is the left-hand end point of $p(D)$.

In all applications, we define $\mathscr{D}_1$ and $\mathscr{D}_2 = \mathscr{D}\backslash\mathscr{D}_1$, where $\{D \in \mathscr{D}_2; \chi_y^+(D) \neq 0\}$ covers $\mathscr{D}'_2$, and $\mathscr{D}'_2$ is a subset of $\mathscr{D}$ that contains all those $D \in \mathscr{D}$ for which

$$\chi_y^+(D) \sum_{\substack{d \in D \\ d < y^{1-\varepsilon}}} \sum_{\substack{m \mid P(z) \\ m < A^{\varepsilon}}} \alpha_{dm} R_{dm} \tag{2.6}$$

cannot be estimated to the requisite order. Of course, for every $D \in \mathscr{D}_1$, (2.6) can be estimated. Then $\mathscr{D}^*$ is (cf. (2.4)) such that an element D of $\mathscr{D}^*$ is a direct product of the largest $2r$ intervals of an element DT of $\mathscr{D}_2$ (with r suitably chosen—see later discussion).

We usually apply Theorem A to estimate $S(\mathscr{A}_d, z(D))(d \in D,\ D \in \mathscr{D}^*$ and $\chi_y^+(D) \neq 0)$ appearing in the third expression on the right of (2.5). First, we identify $\mathscr{D}'_2$, and then we define $\mathscr{D}^*$ and $\mathscr{D}_2$ (if there exist T_1 and $T_2 \in \mathscr{D}$, such that $T_1 < D^* < T_2$ and $T_1 D^* T_2 \in \mathscr{D}_2$, then there exist $T_3 \in \mathscr{D}$ and $D_1^* \in \mathscr{D}^*$ with $T_3 < D_1^*$ and $D_1^* T_3 = T_1 D^* T_2$). In practice, $\mathscr{D}^*$ has a composite structure

$$\mathscr{D}^* = \bigcup_j \mathscr{D}_j^*.$$

and if Theorem 1 is to improve on Theorem A then it is necessary to use for at least one $\mathscr{D}_j^*$ a parameter y_j different from y. Then the third expression on the right of (2.5)

$$\sum_{D \in \mathscr{D}^*} \chi_y^+(D) \sum_{\substack{d \in D \\ d < y^{1-\varepsilon}}} S(\mathscr{A}_d, z(D))$$

$$= \sum_j \left\{ \sum_{D \in \mathscr{D}^*_j} \chi_y^+(D) \sum_{\substack{d \in D \\ d < y^{1-\varepsilon}}} \frac{w(d)}{d} XV(z(D)) F\left(\frac{\log(y_j/d)}{\log z(D)}\right) \right.$$

$$\times \left\{1 + O\left(\frac{1}{(\log\log z(D))^{-3/10}}\right)\right\}$$

$$\left. + \sum_{D \in \mathscr{D}^*_j} \chi_y^+(D) \sum_{\substack{d \in D \\ d < y^{1-\varepsilon}}} \sum_{\substack{T \in \mathscr{D} \\ T < D}} \mu(T) \chi^+_{y_j/i(D)}(T) \sum_{\substack{t \in T \\ m < A^{\varepsilon}}} \alpha_{dmt} R_{dmt} \right\},$$

where $D = I_1 \cdots I_r$, $i(D)$ is the product of the right-hand end points of the I's, and y_j satisfies

$$\{D = D^* T;\ D^* \in \mathscr{D}_j^*,\ T \in \mathscr{D},\ T < D^*,\ \chi_y^+(D^*) \chi^+_{y_j/i(D^*)}(T) \neq 0\} \cap \mathscr{D}'_2 = \phi.$$

(If $T = \phi$ and therefore $\chi^+_{y_j/i(D)}(\phi) = 1$, we have

$$\{D \in \mathscr{D}_j^*; \chi_y^+(D) \neq 0\}$$
$$\subset \{D = D^*T;\ D^* \in \mathscr{D}_j^*,\ T \in \mathscr{D},\ \chi_y^+(D^*)\chi^+_{y_j/i(D^*)}(T) \neq 0\},$$

whence

$$\mathscr{D}^* \cap \mathscr{D}_2' = \phi;$$

in other words, $\mathscr{D}^*$ does not always consist of "bad" direct products.)

In [6], one of the authors discusses the error term that appears in the example (2.1). Applying the results of [6], we find that

$$D_2' = \mathscr{D}_{2,1}' \cup \mathscr{D}_{2,2}',$$

where

$$\mathscr{D}_{2,1}' = \{D \in \mathscr{D};\ \nu(D) = 4;\ D = I_1I_2I_3I_4;\ I_i \subset [x^{a_{i1}}, x^{b_{i1}}),\ i = 1, 2, 3, 4\}$$

and $a_{11} = 1 - 8t_0/5$, $b_{11} = \frac{1}{3}$, $a_{21} = (1 - 2t_0/5 - \log i_1/\log x)/3$, $b_{21} = 2t_0/5$, $a_{31} = (1 - 2t_0/5 - \log\ i_1/\log\ x - \log\ i_2/\log\ x)/2$, $b_{31} = 2t_0/5$, $a_{41} = (1 - 2t_0/5 - \log i_1/\log x - \log i_2/\log x - \log i_3/\log x)$; $t_0 = 1 - \theta + \eta < \frac{9}{20}$, $\theta = \frac{11}{20} + \varepsilon$, and ε and η are suitable small numbers; and

$$\mathscr{D}_{2,2}' = \{D \in \mathscr{D},\ \nu(D) = 6 \text{ and } D = I_1 \cdots I_6,\ I_i \subset [x^{a_{i2}}, x^{b_{i2}}),\ 1 \leq i \leq 6\},$$

where $a_{12} = \max(t_0/2,\ 1 - 6t_0/7 - \log\ i_2/\log\ x - \log\ i_3/\log\ x - \log i_4/\log x)$, $b_{12} = 0.4 - \log i_2/\log x$, $a_{22} = \log i_3/\log x$, $b_{22} = 2t_0/7$, $a_{32} = \log\ i_4/\log\ x$, $b_{32} = \log\ i_2/\log\ x$, $a_{42} = (1 - 10t_0/7)/3$, $b_{42} = \log\ i_3/\log\ x$, $a_{52} = 0.5 - t_0/7 - \sum_{j=1}^{4} \log i_j/\log x$, $b_{52} = \log i_4/\log x$, $a_{62} = 1 - 2t_0$, and $b_{62} = \log i_5/\log x$. We take

$$\mathscr{D}^* = \mathscr{D}_1^* \cup \mathscr{D}_2^*, \tag{2.7}$$

$\mathscr{D}_1^* = \{D \in \mathscr{D},\ \nu(D) = 2,\ \text{and}\ D = I_1I_2,\ I_i \subset [x^{a_{i1}}, x^{b_{i1}}),\ i = 1, 2\}$, $\mathscr{D}_2^* = \{D \in \mathscr{D}, \nu(D) = 4 \text{ and } D = I_1 \cdots I_4, I_i \subset [x^{a_{i2}}, x^{b_{i2}}), 1 \leq i \leq 4\}$ where a_{ij} and b_{ij} are as given above. Then we have $\mathscr{D}_2 \supset \mathscr{D}_2'$.

We apply Theorem A with $y = y_1 = x^{\beta_1}$, $\beta_1 = 1.23 - \log d/\log x - \varepsilon$ for $d \in D$, $D \in \mathscr{D}_1^*$ and $y = y_2 = x^{\beta_2}$, $\beta_2 = 1.5 - 3t_0/7 - \log\ d/2\ \log\ x - \varepsilon$ for $d \in D$, $D \in \mathscr{D}_2{}^*$. We can see that

$$\chi^+_{y_j/i(D)}(T) = 0$$

if $DT \in \mathscr{D}_{2,j}'$ and $D \in \mathscr{D}_j^*$ (here $a_{31} \not< (\beta_1 - \log d/\log x)/3$, $d \in D^*$, $D^* \in \mathscr{D}_1^*$, and $a_{52} \not< (\beta_2 - \log d/\log x)/3$, $d \in D^*$, $D^* \in \mathscr{D}_2^*$, and therefore $i_1i_2i_3^3 \not< y$ for $D \in \mathscr{D}_{2,1}'$, $D = I_1I_2I_3I_4$, and $i_1i_2i_3i_4i_5^3 \not< y_2$ for $D \in \mathscr{D}_{2,2}'$, $D = I_1I_2I_3I_4I_5I_6$.)

Therefore

$$\{D^*T;\ D^* \in \mathscr{D}_j^*,\ T \in \mathscr{D},\ T < D^*,\ \chi_y^+(D^*)\chi_{y_j/i(D^*)}^+(T) \neq 0\} \cap \mathscr{D}_2' = \phi.$$

On computing the function F, we obtain $c = 0.001$.

3. Proof of Theorem 1

$\mathscr{A}^*$ is a subset of $\mathscr{A}$ given by (2.3) so that

$$S(\mathscr{A}^*, z) = 0,$$

and if $\tilde{\mathscr{A}} = \mathscr{A} \backslash \mathscr{A}^*$ then $\tilde{\mathscr{A}}_d = \varnothing$ if $d \in D$ and $D \in \mathscr{D}_2$. These imply $S(\mathscr{A}, z) = S(\tilde{\mathscr{A}}, z)$. We have the fundamental identity (cf. [4])

$$\begin{aligned} S(\mathscr{A}, z) = & \sum_{D \in \mathscr{D}} \mu(D)\chi_y^+(D) \sum_{d \in D} S(\tilde{\mathscr{A}}_d, z_1) \\ & + \sum_I \sum_{\substack{D \in \mathscr{D} \\ I < D}} \mu(D)\chi_y^+(ID) \sum_{\substack{p' < p \\ p, p' \in I}} \sum_{d \in D} S(\tilde{\mathscr{A}}_{p'pd}, p') \\ & + \sum_{D \in \mathscr{D}} \mu(D)\bar{\chi}_y^+(D) \sum_{d \in D} S(\tilde{\mathscr{A}}_d, p(d)) := \sum_1 + \sum_2 + \sum_3, \end{aligned} \tag{3.1}$$

and there is, of course, a corresponding identity for $S(\mathscr{A}^*, z)$. Now $\mathscr{D} = \mathscr{D}_1 \cup \mathscr{D}_2, \mathscr{D}_1 \cap \mathscr{D}_2 = \varnothing$. Hence, starting with the latter identity (for $S(\mathscr{A}^*, z)$)

$$\begin{aligned} - & \sum_{D \in \mathscr{D}_1} \mu(D)\chi_y^+(D) \sum_{d \in D} S(\mathscr{A}_d^*, z_1) \\ & = \sum_{D \in \mathscr{D}_2} \mu(D)\chi_y^+(D) \sum_{d \in D} S(\mathscr{A}_d^*, z_1) \\ & \quad + \sum_I \sum_{\substack{D \in \mathscr{D}_2 \\ I < D}} \mu(D)\chi_y^+(ID) \sum_{\substack{p' < p \\ p, p' \in I}} \sum_{d \in D} S(A_{pp'd}^*, p') \\ & \quad + \sum_I \sum_{\substack{D \in \mathscr{D}_1 \\ I < D}} \mu(D)\chi_y^+(ID) \sum_{\substack{p' < p \\ p, p' \in I}} \sum_{d \in D} S(A_{pp'd}^*, p') \\ & \quad + \sum_{D \in \mathscr{D}_2} \mu(D)\bar{\chi}_y^+(D) \sum_{d \in D} S(\mathscr{A}_d^*, p(D)) \\ & \quad + \sum_{D \in \mathscr{D}_1} \mu(D)\bar{\chi}_y^+(D) \sum_{d \in D} S(\mathscr{A}_d^*, p(D)) \\ := & \sum_4 + \sum_5 + \sum_6 + \sum_7 + \sum_8, \end{aligned} \tag{3.2}$$

say. Moreover, by (3.2) we have (using the fact that $\tilde{\mathscr{A}}_d = \varnothing$ if $d \in D$, $D \in \mathscr{D}_2$)

$$\sum_1 = \sum_{D \in \mathscr{D}_1} \mu(D)\chi_y^+(D) \sum_{d \in D} S(\tilde{\mathscr{A}}_d, z_1)$$

$$= \sum_{D \in \mathscr{D}_1} \mu(D)\chi_y^+(D) \sum_{d \in D} \{S(\mathscr{A}_d, z_1) - S(\mathscr{A}_d^*, z_1)\}$$

$$= \sum_{D \in \mathscr{D}_1} \mu(D)\chi_y^+(D) \sum_{d \in D} S(\mathscr{A}_d, z_1) + \sum_4 + \sum_5 + \sum_6 + \sum_7 + \sum_8.$$

Consider the sum of the three $\sum$'s over $\mathscr{D}_2$,

$$\sum_4 + \sum_5 + \sum_7,$$

remembering that $D \in \mathscr{D}_2$ if and only if $D = D^*T$, $D^* \in \mathscr{D}^*$ and $T < D^*$. Since $\mu(D^*) = 1$, we have

$$\sum_4 + \sum_5 + \sum_7 = \sum_{D \in \mathscr{D}^*} \sum_{\substack{T \in \mathscr{D} \\ T < D}} \mu(TD)\chi_y^+(TD) \sum_{d \in D} \sum_{t \in T} S(\mathscr{A}_{td}^*, z_1)$$

$$+ \sum_{D \in \mathscr{D}^*} \sum_I \sum_{\substack{T \in \mathscr{D} \\ I < T < D}} \mu(TD)\chi_y^+(ITD) \sum_{\substack{p' < p \\ p, p' \in I}} \sum_{d \in D} \sum_{t \in T} S(\mathscr{A}_{pp'td}^*, p')$$

$$+ \sum_{D \in \mathscr{D}^*} \sum_{\substack{T \in \mathscr{D} \\ T < D}} \mu(TD)\bar{\chi}_y^+(TD) \sum_{d \in D^*} \sum_{t \in T} S(\mathscr{A}_{td}^*, p(t))$$

$$= \sum_{D \in \mathscr{D}^*} \sum_{d \in D} \left\{ \sum_{\substack{T \in \mathscr{D} \\ T < D}} \mu(T)\chi_y^+(TD) \sum_{t \in T} S(\mathscr{A}_{dt}^*, z_1) \right.$$

$$+ \sum_I \sum_{\substack{T \in \mathscr{D} \\ I < T < D}} \mu(T)\chi_y^+(ITD) \sum_{\substack{p' < p \\ p, p' \in I}} \sum_{t \in T} S(\mathscr{A}_{dp'pt}^*, p')$$

$$\left. + \sum_{\substack{T \in \mathscr{D} \\ T < D}} \mu(T)\bar{\chi}_y^+(TD) \sum_{t \in T} S(\mathscr{A}_{dt}^*, p(t)) \right\}$$

$$= \sum_{D \in \mathscr{D}^*} \chi_y^+(D) \sum_{d \in D} \left\{ \sum_{\substack{T \in \mathscr{D} \\ T < D}} \mu(T)\chi_{y/i(D)}^+(T) \sum_{t \in T} S(\mathscr{A}_{dt}^*, z_1) \right.$$

$$+ \sum_I \sum_{\substack{T \in \mathscr{D} \\ I < T < D}} \mu(T)\chi_{y/i(D)}^+(IT) \sum_{\substack{p' < p \\ p', p \in I}} \sum_{t \in T} S(\mathscr{A}_{dp'pt}^*, p')$$

$$\left. + \sum_{\substack{T \in \mathscr{D} \\ T < D}} \mu(T)\bar{\chi}_{y/i(D)}^+(T) \sum_{t \in T} S(\mathscr{A}_{dt}^*, p(t)) \right\},$$

and the expression in parentheses equals, by the fundamental identity,

$$\sum_{D\in\mathscr{D}^*} \chi_y^+(D) \sum_{d\in D} S(\mathscr{A}_d^*, z(D)).$$

In this argument we employed the notations $i(D) = \prod_{j=1}^{r} i_j$ if $D = I_1 \ldots I_r$, and $z(D)$ for the left-hand end point of $p(D)$. Clearly

$$z(D) \geq \frac{p(d)}{z_2}$$

so that

$$S(\mathscr{A}_d^*, z(D)) \leq S\left(\mathscr{A}_d^*, \frac{p(d)}{z_2}\right).$$

Also,

$$\sum_3 + \sum_8 \leq 0$$

by the definition of χ_y^+. Hence

$$\begin{aligned} S(\tilde{\mathscr{A}}, z) \leq & \sum_{D\in\mathscr{D}_1} \mu(D)\chi_y^+(D) \sum_{d\in D} S(\mathscr{A}_d, z_1) \\ & + \sum_{D\in\mathscr{D}^*} \chi_y^+(D) \sum_{d\in D} S(\mathscr{A}_d^*, z(D)) + \sum_2 + \sum_6, \end{aligned} \tag{3.3}$$

and $\sum_2$, $\sum_6$ are $o(XV(z))$ in magnitude because of the two primes p, p' that are close to one another. Let $\sum_0$ denote the first sum on the right. We apply the fundamental lemma (cf. [4])

$$S(\mathscr{A}_d, z_1) = XV(z_1)\frac{w(d)}{d}(1 + O(e^{-L})) - \sum_{\nu | P(z_1)} \alpha_{\nu d} R_{\nu d}$$

where $L = u \log \frac{1}{2}u$, $u = \log y_1/\log z_1$, $z_1 = z^{\tau^{-2}}$, $\tau = (\log \log z)^{1/10}$, $y_1 = z^{\tau^{-1}}$, $z_2 = z^{\tau^{-9}}$. Therefore $\sum_0$ is

$$\begin{aligned} XV(z_1) & \sum_{D\in\mathscr{D}_1} \mu(D)\chi_y^+(D) \sum_{d\in D} \frac{w(d)}{d} \\ & \leq XV(z_1) \sum_{D\in\mathscr{D}_1} \mu(D)\chi_y^+(D) \sum_{d\in D} \left(\frac{w(d)}{d}\right)\phi^{(-)^{\nu(D)}}\left(\frac{\log(y/d)}{\log z_1}\right) \\ & = XV(z_1)\left\{\sum_{D\in\mathscr{D}} - \sum_{D\in\mathscr{D}_2}\right\}\mu(D)\chi_y^+(D) \sum_{d\in D} \frac{w(d)}{d}\, \phi^{(-)^{\nu(D)}}\left(\frac{\log(y/d)}{\log z_1}\right). \end{aligned} \tag{3.4}$$

The first sum we handle as in [4]. The second sum is equal to (the factor $XV(z_1)$ apart, and using Motohashi's result [5, Lemma 15])

$$\sum_{D\in\mathscr{D}^*} \chi_y^+(D) \sum_{d\in D} w(d)/d \sum_{\substack{T\in\mathscr{D} \\ T<D}} \mu(T)\chi_{y/i(D)}^+(T) \sum_{t\in T} (w(t)/t)\phi^{(-)^{\nu(T)}}\left(\frac{\log(y/dt)}{\log z_1}\right)$$

$$= \sum_{D\in\mathscr{D}^*} \chi_y^+(D) \sum_{d\in D} \frac{w(d)}{d} \frac{V(z(D))}{V(z_1)} F\left(\frac{\log(y/d)}{\log z(D)}\right)$$

$$\times (1 + O(1/(\log\log z(D))^{3/10})). \tag{3.5}$$

Combining (3.3), (3.4), and (3.5) we have proved Theorem 1.

Using a more complicated version of this method, the last named two authors have proved that for any $\theta > \frac{6}{11}$ and $x \geq x_0(\theta)$,

$$\pi(x) - \pi(x - x^\theta) > \frac{0.037x^\theta}{\log x}.$$

In other words,

$$p_{n+1} - p_n \ll p_n^\theta.$$

This result represents only a small improvement on [1] and [2]. However, our method introduces a novel refinement of the linear sieve that is of independent interest and has other applications.

References

[1] Iwaniec, H. and Pintz, J. "Primes in short intervals," *Mh. Math.* **98** (1984), 115–143.

[2] Heath-Brown, D.R. and Iwaniec, H. "On the difference between consecutive primes," *Inventiones Math.* **55** (1979), 49–69.

[3] Iwaniec, H. "A new form of the error term in the linear sieve," *Acta Arith.* **37** (1980), 307–320.

[4] Halberstam, H. "Lectures on the linear sieve," *Topics in Analytic Number Theory* Univ. of Texas Press (1985), 165–220.

[5] Motohashi, Y. *Lectures on Sieve Method and Prime Number Theory.* Tata Institute of Fundamental Research (1983).

[6] Lou, S. "On a sum that occurs in the error term of the linear sieve applied to an interval," to appear.

19 On the Distribution of $\log|\zeta'(\frac{1}{2}+it)|$

*DENNIS A. HEJHAL**

1

In probability theory, convergence of moments typically leads to convergence of the (underlying) probability distributions themselves (cf. [3, pp. 406, 408, 416(6)]).

It is not too surprising then that A. Selberg was able to use his moment formalism (cf. [20, 21, 24]) to obtain a number of *probabilistic* results concerning the (asymptotic) value distribution of functions like $\log \zeta(s)$ [1949, unpublished].

In the case at hand, one finds, for instance, that

$$\lim_{T\to\infty} \frac{1}{T} m\left\{T \le t \le 2T: \frac{\log \zeta(\frac{1}{2}+it)}{\sqrt{\pi \log\log T}} \in E\right\} = \iint_E e^{-\pi(x^2+y^2)}\, dx\, dy. \tag{1.1}$$

Selberg's formalism also leads to several other results (concerning zeta-functions) that have a distinctly probabilistic "feel" to them.

For example, as shown in [4], the methods of [20, 21] can be combined with analytic function theory (in the form of Hadamard's product formula) to yield a kind of "travelling polynomial approximation" for

* Supported in part by NSF Grant DMS 86-07958.

NUMBER THEORY,
TRACE FORMULAS and
DISCRETE GROUPS

ISBN 0-12-067570-6

a wide variety of (multiplicative) zeta-functions along $\mathrm{Re}(s) = \frac{1}{2}$. These "travelling" approximations are valid except for (certain) t-sets of small *relative* measure (cf. the LHS of (1.1)).

Under a modest[2] hypothesis concerning the spacing of successive zeros of $L_j(s)$, one is *then* able to prove that

$$\{GRH \text{ for } L_j\}$$
$$\Rightarrow \left\{ \sum_{j=1}^{N} c_j L_j(s) \text{ has almost all of its zeros along the critical line} \right\} \quad (1.2)$$

for $(c_j) \in \mathbb{R}^N$ and a rather wide class of "basic" Euler products L_j. The essential restriction is that the L_j should all satisfy the *same* functional equation. (It is tacitly assumed that $c_1 L_1(s) + \cdots + c_N L_N(s) \not\equiv 0$ in (1.2).)

The techniques of [4] are quite general. The aim of this paper is to show *how* (with a few minor changes) they can be employed to investigate the asymptotic value distribution of functions like $\log |L_j'(\frac{1}{2} + it)|$. The results we obtain will serve to complement (1.1). (Note that a Dirichlet series expansion for $\log L_j'(s)$ is essentially out of the question!!)

To illustrate the main ideas, it's enough to restrict ourselves to the case of $\zeta(s)$. We shall also assume that RH holds.[3]

See [5, 24] for the basic properties of $\zeta(s)$.

2

We begin with some notation. Write

$$A(t) \equiv \frac{t}{2\pi} \log\left(\frac{t}{2\pi e}\right)$$

and define $\phi(s)$ on $\{0 < \mathrm{Re}(s) < 1\}$ by the equations

$$e^{i\phi(s)} = \pi^{1/4 - s/2} \sqrt{\frac{\Gamma\left(\frac{s}{2}\right)}{\Gamma\left(\frac{1-s}{2}\right)}}, \quad \phi\left(\frac{1}{2}\right) = 0.$$

[2] statistical

[3] Cf. Section 10 however.

Set:

$$\xi(s) = \pi^{-s/2}\Gamma\left(\frac{s}{2}\right)\zeta(s)$$

$$V(s) = e^{i\phi(s)}\zeta(s).$$

One knows that

(a) $\phi(s)$ and $V(s)$ are both real along $\left\{\mathrm{Re}(s) = \frac{1}{2}\right\}$;

(b) $\xi(s) \equiv \xi(1-s)$;

(c) $V(s) \equiv V(1-s)$;

(d) $V(s) \equiv \dfrac{\pi^{1/4}}{\sqrt{\Gamma\left(\frac{s}{2}\right)\Gamma\left(\frac{1-s}{2}\right)}}\,\xi(s)$.

In addition

$$\phi\left(\frac{1}{2}+it\right) = \pi A(t) - \frac{\pi}{8} + O\left(\frac{1}{t}\right)$$

$$\frac{d\phi(\frac{1}{2}+it)}{dt} = \pi A'(t) + O\left(\frac{1}{t}\right)$$

for large t. Cf. [6, p. 47(1)(7)] and [24, pp. 79(2), 329, 221, 179].

The function $\xi(s)$ is holomorphic except for simple poles at $s = 0, 1$. We denote its zeros by $\rho = \frac{1}{2} + i\gamma$ and recall that Hadamard factorization yields

$$\xi(s) = \frac{e^{as}}{s(s-1)}\prod_{\rho}\left(1-\frac{s}{\rho}\right)e^{s/\rho},$$

(cf. [24, p. 30]).

3

The two functions whose value distribution we propose to consider are

$$\log\left|\frac{\zeta'(\frac{1}{2}+it)}{A'(t)}\right| \qquad \text{and} \qquad \log\left|\frac{V'(\frac{1}{2}+it)}{A'(t)}\right|.$$

Heuristically speaking, the idea is very simple. Let M_1 be any large constant. Take $T \leq t \leq 2T$ and keep T sufficiently large (as determined by M_1). For $|u - t| \leq 9M_1/\log T$, write (*exactly* as in [4, equation (8)])

$$V\left(\frac{1}{2} + iu\right) \equiv \exp[\Omega_t(u)] \cdot \prod_{|\gamma - t| \leq 10M_1/\log T} [A(u) - A(\gamma)] \ .^{4}$$

As mentioned in [4], the total variation of $\Omega_t(u)$ on $[t - M_1/\log T,\ t + M_1/\log T]$ is $O_{M_1}(1)$ for "most" t. Bear in mind here that $-\Omega_t(u)$ is essentially convex (cf. lemmas H and J in Section 6 below).

To save writing, we now define

$$\mathscr{F}_t = \left[t - \frac{10M_1}{\log T}, t + \frac{10M_1}{\log T}\right] \qquad (\mathscr{F} \text{ for "frame"})$$

$$\mathscr{W}_t = \left[t - \frac{M_1}{\log T}, t + \frac{M_1}{\log T}\right] \qquad (\mathscr{W} \text{ for "window"})$$

$$x = A(u), \qquad \theta(u) = \phi(\tfrac{1}{2} + iu)$$

$$\mathscr{P}_t(x) = \prod_{\gamma \in \mathscr{F}_t} [x - A(\gamma)]$$

and then observe that

$$V\left(\frac{1}{2} + iu\right) = \exp[\Omega_t(u)]\mathscr{P}_t(x)$$

$$\zeta\left(\frac{1}{2} + iu\right) = \exp[\Omega_t(u) - i\theta(u)]\mathscr{P}_t(x)$$

$$i\,\frac{\zeta'(\frac{1}{2} + iu)}{A'(u)} = \zeta\left(\frac{1}{2} + iu\right)\left[\frac{\Omega_t'(u)}{A'(u)} + \frac{\mathscr{P}_t'(x)}{\mathscr{P}_t(x)} - i\,\frac{\theta'(u)}{A'(u)}\right]$$

$$\frac{\log\left|\dfrac{\zeta'(\frac{1}{2} + iu)}{A'(u)}\right|}{\sqrt{\pi \log\log T}} = \frac{\log|\zeta(\frac{1}{2} + iu)|}{\sqrt{\pi \log\log T}} + \frac{\log\left|\dfrac{\Omega_t'(u)}{A'(u)} + \dfrac{\mathscr{P}_t'(x)}{\mathscr{P}_t(x)} - i\,\dfrac{\theta'(u)}{A'(u)}\right|}{\sqrt{\pi \log\log T}}.$$

Recall, however, that

$$N[0 \leq \gamma \leq T] = A(T) + O(\log T) \qquad [24, \text{p. } 181]$$

$$\theta'(u) = \pi A'(u) + O(u^{-1})$$

$$\frac{dA}{du} = \frac{1}{2\pi}\log\left(\frac{u}{2\pi}\right) = \frac{1}{2\pi}\log T + O(1).$$

[4] Note that $\Omega_t(u)$ is real-valued mod πi.

For most t

$$\int_{\mathscr{W}_t} |\Omega_t'(u)|\, du = O_{M_1}(1).$$

This suggests that, on *most* windows $\mathscr{W}_t$, we have

$$|\Omega_t'(u)| = O_{M_1}(\log T).$$

At the same time

$$\frac{\mathscr{P}_t'(x)}{\mathscr{P}_t(x)} \equiv \sum_{\gamma \in \mathscr{F}_t} \frac{1}{x - A(\gamma)}, \tag{3.1}$$

while the average spacing between successive $A(\gamma)$ is 1. Since $A(\mathscr{F}_t)$ has length $\sim 10M_1/\pi$, the number of terms on the right side of (3.1) is typically $O_{M_1}(1)$. We conclude (*at least heuristically*) that

$$\log\left|\frac{\Omega_t'(u)}{A'(u)} + \frac{\mathscr{P}_t'(x)}{\mathscr{P}_t(x)} - i\frac{\theta'(u)}{A'(u)}\right| \tag{3.2}$$

tends to be bounded by $O_{M_1}(1)$ except for a $[T, 2T]$ subset having small relative measure.

We now let $T \to \infty$ and remember that [on $T \le u \le 2T$]

$$\frac{\log|\zeta(\frac{1}{2}+iu)|}{\sqrt{\pi \log\log T}}$$

behaves (distributionally) like a Gaussian variable with density $\exp(-\pi\xi^2)$. That is,

$$\lim_{T\to\infty} \frac{1}{T} m\left\{T \le t \le 2T; \frac{\log|\zeta(\frac{1}{2}+iu)|}{\sqrt{\pi \log\log T}} \in E\right\} = \int_E \exp(-\pi x^2)\, dx, \tag{3.3}$$

(cf. (1.1) and [4, equation (6)], [15], [25].)

Since

$$\frac{\log\left|\dfrac{\zeta'(\frac{1}{2}+iu)}{A'(u)}\right|}{\sqrt{\pi \log\log T}} = \frac{\log|\zeta(\frac{1}{2}+iu)|}{\sqrt{\pi \log\log T}} + \frac{O_{M_1}(1)}{\sqrt{\pi \log\log T}} \quad \text{“generally”},$$

we conclude that

$$\frac{\log|\zeta(\frac{1}{2}+iu)|}{\sqrt{\pi \log\log T}} \quad \text{and} \quad \frac{\log\left|\dfrac{\zeta'(\frac{1}{2}+iu)}{A'(u)}\right|}{\sqrt{\pi \log\log T}}$$

are (in effect) the *same* random variable. Equation (3.3) should therefore hold with $\log|\zeta'(\frac{1}{2}+iu)/A'(u)|$ in place of $\log|\zeta(\frac{1}{2}+iu)|$.

One naturally expects the exact same result for $\log|V'(\frac{1}{2}+iu)/A'(u)|$. The only difficulty here is that

$$\log\left|\frac{\Omega_t'(u)}{A'(u)}+\frac{\mathscr{P}_t'(x)}{\mathscr{P}_t(x)}\right|$$

now appears in place of (3.2), and the absolute value could conceivably be very small. A *non*removable difficulty on this score would (of course) be most unusual indeed!!

4

In stating the following theorems, we define

$$F_1(t) \equiv \log\left|\frac{\zeta'(\frac{1}{2}+it)}{A'(t)}\right|, \qquad F_2(t) \equiv \log\left|\frac{V'(\frac{1}{2}+it)}{A'(t)}\right|$$

$$D_1(t) \equiv F_1(t) - \log\left|\zeta\left(\frac{1}{2}+it\right)\right|, \qquad D_2(t) \equiv F_2(t) - \log\left|\zeta\left(\frac{1}{2}+it\right)\right|$$

$$\beta_k \equiv \begin{cases} 0 & \text{for } k \text{ odd} \\ \dfrac{(2l)!}{(4\pi)^l l!} & \text{for } k = 2l \end{cases}.$$

We continue to assume the truth of RH.

Theorem 1. *Let k be any positive integer and $T \geq 4$. Then*

$$\int_T^{2T} |D_j(t)|^{2k}\,dt = O(T) \qquad \text{for } j = 1, 2.$$

The implied constant depends solely on k.

Theorem 2. *We also have*

$$\int_T^{2T} [F_j(t)]^k\,dt = \beta_k T(\pi \log\log T)^{k/2} + O[T(\log\log T)^{(k-1)/2}].$$

Theorem 3. *For any Borel measurable E having Jordan content*[5]

$$\lim_{T\to\infty} \frac{1}{T} m\left\{T \leq t \leq 2T\colon \frac{F_j(t)}{\sqrt{\pi\log\log T}} \in E\right\} = \int_E \exp(-\pi x^2)\,dx.$$

[5] Possession of Jordan content is equivalent to assuming that ∂E has exterior (Lebesgue) measure zero (cf. [2, pp. 226, 230] and [3, p. 231(5)(6)]).

In connection with Theorem 3, observe that Theorem 1 immediately yields

$$\lim_{T\to\infty}\frac{1}{T}m\left\{T\le t\le 2T:\left|\frac{\log|\zeta(\frac{1}{2}+it)|}{\sqrt{\pi\log\log T}}-\frac{F_j(t)}{\sqrt{\pi\log\log T}}\right|>\delta\right\}=0 \quad (4.1)$$

for any positive δ. This equation completely justifies the expectations expressed at the end of Section 3.

5

We now turn to the proofs. First of all, note that Theorem 2 is an immediate consequence of Theorem 1. In fact

$$\int_T^{2T}\left[\log\left|\zeta\left(\frac{1}{2}+it\right)\right|\right]^k dt=\beta_k T(\pi\log\log T)^{k/2}+O[T(\log\log T)^{(k-1)/2}] \quad (5.1)$$

is already known from the proof of (1.1). Compare [15], [16, pp. 148, 150], [25], [26, Lemma 2]. The numbers β_k are just the moments of $\exp(-\pi x^2)\,dx$. We can now write

$$\left[\log\left|\zeta\left(\frac{1}{2}+it\right)\right|+D_j(t)\right]^k\equiv\left[\log\left|\zeta\left(\frac{1}{2}+it\right)\right|\right]^k$$
$$+\sum_{\iota=0}^{k-1}\binom{k}{\iota}[\log|\zeta|]^{\iota}[L_\tau(\varepsilon)]^{\phi-\iota}$$

and apply Cauchy–Schwarz.

Theorem 3 follows from Theorem 2 by standard probabilistic techniques (cf. [3]).[6]

It remains *then* to prove Theorem 1.

[6] To avoid any misunderstanding, the following facts should be kept in mind.

(I) Let $\{F_n\}_{n=1}^{\infty}$ be any sequence of probability distributions on $\mathbb{R}$ converging pointwise to a *continuous* probability distribution $G(x)$. Then $F_n(x)\to G(x)$ uniformly on $\mathbb{R}$.

(II) Let E be any Borel measurable subset of $\mathbb{R}$ having Jordan content. Assume [*in (I)*] that $G(x)$ has continuous density. Then

$$F_n\{E\}\to G\{E\}.$$

Cf. [3, pp. 390, 176, 177].

(III) Facts I and II have immediate analogs when $\{F_n\}_{n=1}^{\infty}$ is replaced by $\{F_T\colon T\ge 1\}$.

(IV) Let $r(x, T)$ be given on $\mathbb{R}\times[1,\infty)$. Suppose that, for any sequence $T_n\uparrow\infty$, we can select a *sub*sequence $n_j\uparrow\infty$ such that

$$\lim_{j\to\infty} r(x, T_{n_j})=0 \qquad \text{uniformly in } x.$$

Then: $\lim_{T\to\infty} r(x,T)=0$ uniformly on $\mathbb{R}$. The obvious choice for $r(x,T)$ is $F_T(x)-G(x)$. Cf. the Helly selection theorem [3, p. 345].

6

For this purpose, a variety of lemmas will be necessary. We'll collect them all in this section and say a few words about each of their proofs.

To begin with, recall that

$$N[0 \le \gamma \le T] = A(T) + \frac{7}{8} + S(T) + 0\left(\frac{1}{T}\right), \tag{6.1}$$

where $S(t) \equiv \pi^{-1} \arg \zeta(\frac{1}{2} + it)$ (cf. [24, p. 179]). For large t, write

$$\eta_t = \inf_{\rho} |t - \gamma|.$$

In accordance with [24, pp. 190–191], one knows that

$$\lim_{t \to \infty} \eta_t = 0. \tag{6.2}$$

Let k be any positive integer and T be any number bigger than 10^{12} (say). Set

$$T_1 = \frac{3}{4} T, \qquad T_2 = \frac{9}{4} T. \tag{6.3}$$

Lemma A. *Given any $B \ge 1$. We then have*

$$\int_T^{2T} \left[\sum_{|\gamma - t| \le B/\log T} 1 \right]^{2k} dt = O_{kB}(1).$$

This lemma emerges quite early in the Selberg moment formalism [20, 21]. There are *two* steps. The first is to remember that

$$\sum_{\text{all } \rho} \frac{\sigma_{xt} - \frac{1}{2}}{(\sigma_{xt} - \frac{1}{2})^2 + (\gamma - t)^2} = O_{k\varepsilon}[F(t)] \tag{6.4}$$

for $T_1 \le t \le T_2$, where

$$x \equiv T^{\varepsilon/k}, \qquad 0 < \varepsilon < \frac{1}{100}, \qquad \sigma_{xt} \equiv \frac{1}{2} + \frac{4}{\log x}$$

$$F(t) \equiv \log t + \left| \sum_{n \le x^3} \Lambda_x(n) n^{-\sigma_{xt} - it} \right|.$$

Cf. [20, p. 7], [21, p. 26], [24, p. 310]. The second step is to use Montgomery-Vaughan [18] as in [15], [25], [26, Lemmas 1 and 2] to show that

$$\int_{T_1}^{T_2} \left[F(t) \left(\sigma_{xt} - \frac{1}{2} \right) \right]^{4k} dt = O_{k\varepsilon}(T).$$

Alternatively, see [20, Lemma 3], [11, p. 30], [24, p. 312(bot)]. Note that

$$1 \le (1 + B^2) \frac{(\sigma_{xt} - \frac{1}{2})^2}{(\sigma_{xt} - \frac{1}{2})^2 + (t - \gamma)^2} \qquad \text{for } |\gamma - t| \le \frac{B}{\log T}.$$

For real-valued f, we now write

$$f^+(x) = \max[0, f(x)], \qquad f^-(x) = \max[0, -f(x)]$$

so that

$$|f(x)| = f^+(x) + f^-(x) \qquad \text{and} \qquad f(x) = f^+(x) - f^-(x).$$

Lemma B. *One has*

$$\int_T^{2T} \left[\log^+\left(\frac{1}{\eta_t \log T}\right)\right]^{2k} dt = O_k(T).$$

The trick here is to observe that the LHS does not exceed

$$\sum_{T-1 \le \gamma \le 2T+1} \int_{\gamma - 1/\log T}^{\gamma + 1/\log T} \left[\log\left(\frac{1}{|u - \gamma| \log T}\right)\right]^{2k} du$$

$$= \sum_{T-1 \le \gamma \le 2T+1} \frac{2(2k)!}{\log T} = O_k(T).$$

Lemma C. *For $0 < h < 1$, we have*

$$\int_T^{2T} |S(t + h) - S(t)|^{2k} dt = O[T \log^k(2 + h \log T)],$$

with an implied constant depending solely on k.[7]

This result corresponds to [7], [25], [26, Theorem 4]. In essence, one merely combines [20, Theorem 2], [21, Theorem 4], [24, p. 311] with Montgomery-Vaughan [18].

Lemma D. *One also has*

$$\int_T^{2T} [\log^+(\eta_t \log T)]^{2k} dt = O_k(T).$$

The LHS is not more than

$$2 \sum_{T_1 \le \gamma_n \le T_2} \frac{1}{\log T} \int_0^{(\gamma_{n+1} - \gamma_n)\log T} (\log^+ |v|)^{2k} dv.$$

[7] In particular, it is independent of h.

It is now *sufficient* to prove that

$$\frac{1}{\log T} \sum_{T_1 \le \gamma_n \le T_2} [(\gamma_{n+1} - \gamma_n) \log T]^2 = O(T). \tag{6.5}$$

But, (6.1) and Lemma C (with $k = 1$) immediately yield

$$\sum_{\substack{T_1 \le \gamma_n \le T_2 \\ \gamma_{n+1} - \gamma_n \ge 2H}} (\gamma_{n+1} - \gamma_n) = O\left[T \frac{\log(H \log T)}{(H \log T)^2}\right]$$

for $2/\log T \le H \le 1$. This implies that

$$N[T_1 \le \gamma_n \le T_2, \gamma_{n+1} - \gamma_n \ge 2H] = O\left[T \log T \cdot \frac{\log(H \log T)}{(H \log T)^3}\right]. \tag{6.6}$$

A straightforward integration by parts then yields (6.5). Equation (6.6) is similar *in spirit* to [8, p. 35 (top)] and [19, p. 329].

Lemma E. *Take* $x = T^{\varepsilon/k}$ *and* $0 < \varepsilon < \frac{1}{100}$. *Then*

$$\int_{T_1}^{T_2} \left[\log\left|\zeta\left(\frac{1}{2} + it\right)\right| - \operatorname{Re} \sum_{p \le x} p^{-1/2 - it}\right]^{2k} dt = O_{k\varepsilon}(T).$$

This is just the real counterpart of [20, Theorem 2], [21, Theorem 4], [24, p. 311]. There are (of course) a few modifications here and there, but no serious difficulties occur. Cf. [15, 25] for further details. The fundamental identity is

$$\log\left|\zeta\left(\frac{1}{2} + it\right)\right| = \operatorname{Re} \sum_{n \le x^3} \frac{\Lambda_x(n)}{\log n} n^{-\sigma_{xt} - it} + O[F(t)(\sigma_{xt} - \tfrac{1}{2})]$$
$$- \sum_{\rho} \operatorname{Re} \int_{1/2}^{\sigma_{xt}} \left[\frac{1}{u + it - \rho} - \frac{1}{\sigma_{xt} + it - \rho}\right] du$$

in the notation of (6.4).

Lemma E plays (as one might expect) a key role in the proof of (3.3) and (5.1).

Lemma F. *Given any* $R > 0$. *With* $x = T^{\varepsilon/k}$ *and* $0 < \varepsilon < \frac{1}{100}$,

$$\int_{T_1}^{T_2} \left[\log\left|\zeta\left(\frac{1}{2} + it\right)\right| - \operatorname{Re} \sum_{p \le x} p^{-1/2 - it} - \sum_{|\gamma - t| \le R/\log T} \log|A(t) - A(\gamma)|\right]^{2k} dt = O_{kR\varepsilon}(T).$$

This corresponds to [4, equation (11)]. In view of Lemma E, it suffices to check that

$$\int_{T_1}^{T_2}\left[\sum_{|\gamma-t|\le R/\log T}\log|A(t)-A(\gamma)|\right]^{2k}dt = O_{kR}(T).$$

By Lemma A, we can take $R=1$ WLOG. But, then,

$$\frac{\log T}{4\pi}|t-\gamma| \le |A(t)-A(\gamma)| \le \frac{\log T}{2\pi}|t-\gamma| \le \frac{1}{2}$$

$$|\log|A(t)-A(\gamma)|| \le \log\left[\frac{4\pi}{(\log T)|t-\gamma|}\right]$$

$$\left|\sum_{|\gamma-t|\le R/\log T}\log|A(t)-A(\gamma)|\right| \le \log^{+}\left[\frac{4\pi}{\eta_t\log T}\right]\cdot\sum_{|\gamma-t|\le R/\log T}1.$$

We can now exploit Lemmas A and B together with Cauchy–Schwarz.

In the next lemma, let M_1 be any number bigger than 1.[8] Let n_1 and n_2 be distinct integers of absolute value ≤ 5. Define $\Omega_t(u)$ as in Section 3 and keep T sufficiently large as determined by M_1.[9] Put $h_j = n_jM_1/\log T$.

Lemma G. *For $T \ge T_0(M_1)$*

$$\int_T^{2T}|\Omega_t(t+h_1)-\Omega_t(t+h_2)|^{2k}\,dt = 0_{kM_1}(T).$$

This corresponds to [4, eq. (10)]. Take $\varepsilon = 1/500$, $R = 20M_1$, and temporarily write

$$\log\left|\zeta\left(\frac{1}{2}+iu\right)\right| = \operatorname{Re}\sum_{p\le x}p^{-1/2-iu}$$
$$+\sum_{|\gamma-u|\le R/\log T}\log|A(u)-A(\gamma)| + \mathscr{U}_R(u)$$

for $T_1 \le u \le T_2$. By virtue of Lemma F

$$\int_{T_1}^{T_2}|\mathscr{U}_R(u)|^{2k}\,du = O_{kM_1}(T). \tag{6.7}$$

[8] Though M_1 can actually be fixed at the outset for purposes of proving Theorem 1, the case of adjustable M_1 is essential for both [4] and Section 8.

[9] We indicate this restriction (more or less symbolically) by writing $T \ge T_0(M_1)$.

On the other hand,

$$\Omega_t(t+h_1)-\Omega_t(t+h_2)=\operatorname{Re}\sum_{p\le x}\frac{1}{\sqrt{p}}(p^{-ih_1}-p^{-ih_2})p^{-it}$$
$$+\mathscr{U}_R(t+h_1)-\mathscr{U}_R(t+h_2)$$
$$+O_{M_1}(1)\sum_{|\gamma-t|\le 25M_1/\log T}1$$

for $T\le t\le 2T$ as seen by elementary manipulation. The necessary moments are now (entirely) straightforward. Cf. (6.7), Lemma A, and [7], [25], [26, Equation (4.1)].

Before turning to the last four lemmas, it is convenient to record a few *standard* (or else, readily verified) formulas. Recall that

$$\mathscr{F}_t\equiv\left[t-\frac{10M_1}{\log T},\quad t+\frac{10M_1}{\log T}\right],\qquad \mathscr{W}_t\equiv\left[t-\frac{M_1}{\log T},\quad t+\frac{M_1}{\log T}\right].$$

We then have

$$\Psi(z)\equiv\frac{\Gamma'(z)}{\Gamma(z)}=\operatorname{Log}z-\frac{1}{2z}-\frac{1}{12z^2}+O(z^{-4})$$
$$\Psi'(z)=\frac{1}{z}+\frac{1}{2z^2}+\frac{1}{6z^3}+O(z^{-4})$$
$$\text{for } |z|\ge 1,\ |\operatorname{Arg}(z)|\le\pi-\delta\qquad[6,\text{ p. }47(7)];\tag{6.8}$$

$$\frac{\xi'(s)}{\xi(s)}=\frac{1}{2}\sum_\rho\left(\frac{1}{s-\rho}+\frac{1}{s+\rho-1}\right)-\left[\frac{1}{s}+\frac{1}{s-1}\right]$$
$$\frac{V'(s)}{V(s)}=-\frac{1}{4}\Psi\left(\frac{s}{2}\right)+\frac{1}{4}\Psi\left(\frac{1-s}{2}\right)+\frac{\xi'(s)}{\xi(s)}\qquad\text{(cf. Section 2)};\tag{6.9}$$

$A(u)$ is holomorphic + schlicht on $|u-t|\le 1$ for $t\ge 50000$

$$\frac{A(u_2)-A(u_1)}{u_2-u_1}=\frac{1}{2\pi}\log\left(\frac{t}{2\pi}\right)+O\left(\frac{1}{t}\right)$$
$$\log\left[\frac{A(u_2)-A(u_1)}{u_2-u_1}\right]=\log\left[\frac{1}{2\pi}\log\left(\frac{t}{2\pi}\right)\right]+O\left[\frac{1}{t\log t}\right]$$
$$\text{for } u_j\in\mathbb{C},\ t\ge 50000,\ |u_j-t|\le 1;\tag{6.10}$$

$$\Omega_t'(u)=\frac{i}{4}\left[\Psi\left(\frac{1}{4}-\frac{1}{2}iu\right)-\Psi\left(\frac{1}{4}+\frac{1}{2}iu\right)\right]-\frac{2u}{u^2+\frac{1}{4}}$$
$$+\frac{1}{2}\sum_\rho\left[\frac{1}{u-\gamma}+\frac{1}{u+\gamma}\right]-\sum_{\gamma\in\mathscr{F}_t}\frac{1}{u-\gamma}$$
$$-\sum_{\gamma\in\mathscr{F}_t}\frac{\partial}{\partial u}\log\left[\frac{A(u)-A(\gamma)}{u-\gamma}\right]$$
$$\text{for } |u-t|\le\frac{9M_1}{\log T},\ T_1\le t\le T_2,\ T\ge T_0(M_1).\tag{6.11}$$

Lemma H. *For $T_1 \le t \le T_2$, $T \ge T_0(M_1)$, $|u-t| \le 9M_1/\log T$, we have*

$$\Omega_t''(u) = O\left(\frac{1}{T}\right) - \sum_{\gamma \notin \mathscr{F}_t} \frac{1}{(u-\gamma)^2}.$$

The implied constant is absolute *so long as $T \ge T_0(M_1)$.*

This follows easily from (6.8)–(6.11). Bear in mind here that

$$\left(\frac{\partial}{\partial u}\right)^2 \log\left[\frac{A(u)-A(\gamma)}{u-\gamma}\right] = O\left[\frac{1}{T\log T}\right]$$

(for $\gamma \in \mathscr{F}_t$) by a standard Cauchy estimate and that

$$N[t-1 \le \gamma \le t+1] = O(\log t).$$

Cf. [1] regarding the Cauchy estimate.

Lemma I. *Given any $h > 0$, $A \ge 3$. Let $G(v)$ be C^2 and convex on $[-Ah, Ah]$. Then*

$$V_G[-h,h] \le |G(Ah) - G(Ah-2h)| + |G(-Ah) - G(-Ah+2h)|.$$

The symbol V_G means total variation of G.

This is a trivial calculus exercise.

Lemma J. *For $T \ge T_0(M_1)$, we have*

$$\int_T^{2T} |V_{\Omega_t}[W_t]|^{2k}\, dt = O_{kM_1}(T).$$

Since $\lim_{t\to\infty} \eta_t = 0$,[10] Lemma H shows that $-\Omega_t(u)$ is *eventually* convex. We can now combine Lemmas G and I in order to obtain the desired result.

Lemma K. *For $T_1 \le t \le T_2$, $T \ge T_0(M_1)$, $|u-t| \le 9M_1/\log T$, the function*

$$f_t(u) \equiv \Omega_t'(u) + \sum_{\gamma \in \mathscr{F}_t} \frac{A'(u)}{A(u)-A(\gamma)}$$

is strictly decreasing except for the obvious *discontinuities.*

[10] i.e., $\lim_{n\to\infty}(\gamma_{n+1}-\gamma_n) = 0$

In fact, equations (6.8) and (6.11) immediately yield

$$f_t'(u) = O\left(\frac{1}{T}\right) - \sum_{\rho} \frac{1}{(u-\gamma)^2}. \tag{6.12}$$

Recall (once again) that $\lim_{t\to\infty} \eta_t = 0$.

7

We can now (at long last) prove Theorem 1.

It *suffices* to consider $j = 2$, i.e., the more difficult case. Since D_2 has nothing to do (per se) with either $\mathscr{F}_t$ or $\log T$ [cf. Section 4], we need only verify that

$$\int_{(1.1)T}^{(1.9)T} |D_2(t)|^{2k}\, dt = O_k(T).$$

However,

$$\int_{(1.1)T}^{(1.9)T} |D_2(t)|^{2k}\, dt \le \int_T^{2T} \left[\frac{\log T}{2M_1} \int_{t-M_1/\log T}^{t+M_1/\log T} |D_2(u)|^{2k}\, du\right] dt$$

for $T \ge T_0(M_1)$. That is,

$$\int_{(1.1)T}^{(1.9)T} |D_2(t)|^{2k}\, dt \le \frac{1}{2M_1} \int_T^{2T} \left[\log T \int_{\mathscr{W}_t} |D_2(u)|^{2k}\, du\right] dt.$$

As in Section 3,

$$\left|\frac{V'(\frac{1}{2}+iu)}{A'(u)}\right| = \left|V\left(\frac{1}{2}+iu\right)\right| \left|\frac{\Omega_t'(u)}{A'(u)} + \frac{\mathscr{P}_t'[A(u)]}{\mathscr{P}_t[A(u)]}\right| \qquad (u \in \mathscr{W}_t)$$

$$\log\left|\frac{V'(\frac{1}{2}+iu)}{A'(u)}\right| = \log\left|\zeta\left(\frac{1}{2}+iu\right)\right| + \log\left|\frac{\Omega_t'(u)}{A'(u)} + \frac{\mathscr{P}_t'[A(u)]}{\mathscr{P}_t[A(u)]}\right|$$

$$D_2(u) = \log\left|\frac{\Omega_t'(u)}{A'(u)} + \sum_{\gamma\in\mathscr{F}_t} \frac{1}{A(u)-A(\gamma)}\right|.$$

Let

$$D_2^+(u) = \log^+\left|\frac{\Omega_t'(u)}{A'(u)} + \sum_{\gamma\in\mathscr{F}_t} \frac{1}{A(u)-A(\gamma)}\right|$$

$$D_2^-(u) = \log^-\left|\frac{\Omega_t'(u)}{A'(u)} + \sum_{\gamma\in\mathscr{F}_t} \frac{1}{A(u)-A(\gamma)}\right|.$$

Since

$$|D_2(u)|^{2k} \equiv |D_2^+(u)|^{2k} + |D_2^-(u)|^{2k},$$

it suffices to prove that

$$\text{(i)} \quad \int_T^{2T} \left[\log T \int_{\mathscr{W}_t} |D_2^+(u)|^{2k}\, du \right] dt = O_{kM_1}(T)$$

$$\text{(ii)} \quad \int_T^{2T} \left[\log T \int_{\mathscr{W}_t} |D_2^-(u)|^{2k}\, du \right] dt = O_{kM_1}(T)$$

for $T \geq T_0(M_1)$.

We begin with case (i). Let $b_1, b_2, \ldots$ denote positive constants depending solely on k and M_1. Keep $u \in \mathscr{W}_t$ and $T \leq t \leq 2T$. Then

$$|D_2^+(u)| \leq \log^+ \left[1 + \left| \frac{\Omega_t'(u)}{A'(u)} \right| + \sum_{\gamma \in \mathscr{F}_t} \frac{1}{|A(u) - A(\gamma)|} \right]$$

$$|D_2^+(u)| \leq \log \left[1 + \left| \frac{\Omega_t'(u)}{A'(u)} \right| \right] + \sum_{\gamma \in \mathscr{F}_t} \log \left[1 + \frac{1}{|A(u) - A(\gamma)|} \right]$$

$$|D_2^+(u)|^{2k} \leq b_1 \left[\log\left(1 + \left| \frac{\Omega_t'(u)}{A'(u)} \right| \right) \right]^{2k} + b_1 \left[\sum_{\gamma \in \mathscr{F}_t} \log\left(1 + \frac{1}{|A(u) - A(\gamma)|} \right) \right]^{2k}$$

$$|D_2^+(u)|^{2k} \leq b_1 \left[\log\left(1 + \left| \frac{\Omega_t'(u)}{A'(u)} \right| \right) \right]^{2k} + b_1 N[\gamma \in \mathscr{F}_t]^{2k-1} \sum_{\gamma \in \mathscr{F}_t} \log^{2k}\left(1 + \frac{1}{|A(u) - A(\gamma)|} \right)$$

{by Hölder's inequality}

$$|D_2^+(u)|^{2k} \leq b_2 + b_3 \left| \frac{\Omega_t'(u)}{\log T} \right| + b_1 N[\gamma \in \mathscr{F}_t]^{2k-1} \sum_{\gamma \in \mathscr{F}_t} \log^{2k}\left(1 + \frac{1}{|A(u) - A(\gamma)|} \right)$$

$$\log T \int_{\mathscr{W}_t} |D_2^+(u)|^{2k}\, du \leq \log T \left\{ b_2 \frac{2M_1}{\log T} + b_3 \frac{V_{\Omega_t}[\mathscr{W}_t]}{\log T} \right\} + b_1 \log T \cdot N[\gamma \in \mathscr{F}_t]^{2k-1} \times \sum_{\gamma \in \mathscr{F}_t} \int_{\mathscr{W}_t} \log^{2k}\left(1 + \frac{1}{|A(u) - A(\gamma)|} \right) du$$

$$\log T \int_{\mathscr{W}_t} |D_2^+(u)|^{2k}\, du \leq b_4 + b_5 V_{\Omega_t}[\mathscr{W}_t] + b_6 \log T \cdot N[\gamma \in \mathscr{F}_t]^{2k-1} \times \sum_{\gamma \in \mathscr{F}_t} \int_0^{10M_1} \log^{2k}\left(1 + \frac{1}{w}\right) \frac{dw}{\log T}$$

$$\log T \int_{\mathscr{W}_t} |D_2^+(u)|^{2k}\, du \leq b_7 + b_7 V_{\Omega_t}(\mathscr{W}_t) + b_7 N[\gamma \in \mathscr{F}_t]^{2k}$$

$$\log T \int_{\mathscr{W}_t} |D_2^+(u)|^{2k}\, du \leq 2b_7 + b_7 V_{\Omega_t}(\mathscr{W}_t)^2 + b_7 N[\gamma \in \mathscr{F}_t]^{2k}.$$

Lemmas A and J immediately yield

$$\int_T^{2T} \left[\log T \int_{\mathscr{W}_t} |D_2^+(u)|^{2k}\, du \right] dt \leq O_{kM_1}(T).$$

To treat case (ii), it is convenient to divide things into two subcases. Specifically,

$$\text{(A)} \quad \eta_t \leq \frac{M_1}{\log T}$$

$$\text{(B)} \quad \eta_t > \frac{M_1}{\log T}$$

Case (B) goes as follows. Keep $u \in \mathscr{W}_t$ and $T \geq T_0(M_1)$. Note that

$$D_2^-(u) = \log^- \left| \frac{f_t(u)}{A'(u)} \right|.$$

Since $A'(u) \sim (1/2\pi) \log T$, we immediately get

$$|D_2^-(u)| \leq \log^- \left| \frac{f_t(u)}{\log T} \right|.$$

The function $f_t(u)$ is continuous and strictly decreasing on $\mathscr{W}_t$ by Lemma K. Since $\eta_t > M_1/\log T$, equation (6.12) readily shows that

$$|f_t'(u)| \geq \frac{1}{4\eta_t^2} \qquad \text{for } u \in \mathscr{W}_t.$$

A little thought will now yield

$$\int_{\mathscr{W}_t} |D_2^-(u)|^{2k}\, du \leq 2 \int_0^{2M_1/\log T} \left[\log^- \left(\frac{w}{4\eta_t^2 \log T} \right) \right]^{2k} dw,$$

whereupon

$$\log T \int_{\mathscr{W}_t} |D_2^-(u)|^{2k}\, du \le 8\eta_t^2(\log T)^2 \int_0^{M_1/2\eta_t^2\log^2 T} [\log^-(v)]^{2k}\, dv.$$

Observe, however, that

$$\frac{M_1}{2\eta_t^2 \log^2 T} < \frac{1}{2M_1} < 1.$$

This gives

$$\begin{aligned} \log T \int_{\mathscr{W}_t} |D_2^-(u)|^{2k}\, du &\le b_8\eta_t^2(\log T)^2 \left[\log\left(\frac{M_1}{2\eta_t^2 \log^2 T}\right)\right]^{2k} \frac{M_1}{2\eta_t^2 \log^2 T} \\ &\le b_9 \left[\log\left(\frac{M_1}{2\eta_t^2 \log^2 T}\right)\right]^{2k} \\ &\le b_{10}[\log(\eta_t \log T)]^{2k} + b_{10}, \end{aligned}$$

where the constants b_i *continue* to depend solely on k and M_1.

Case (A) exploits a similar estimate. Just as before

$$|D_2^-(u)| \le \log^- \left|\frac{f_t(u)}{\log T}\right|.$$

Since $\eta_t \le M_1/\log T$, the function $f_t(u)$ now has *at least one* discontinuity on $\mathscr{W}_t$. Equation (6.12) readily shows, however, that

$$|f_t'(u)| \ge \left[\frac{\log T}{2M_1}\right]^2 \qquad \text{for } u \in \mathscr{W}_t,\ u \ne \gamma.$$

By dividing $\mathscr{W}_t$ into the obvious subintervals, we get

$$\int_{\mathscr{W}_t} |D_2^-(u)|^{2k}\, du \le 2\{1 + N[\gamma \in \mathscr{W}_t]\} \int_0^{2M_1/\log T} \left[\log^-\left(\frac{\log T}{4M_1^2} w\right)\right]^{2k} dw$$

$$\int_{\mathscr{W}_t} |D_2^-(u)|^{2k}\, du \le \frac{16M_1^2}{\log T} N[\gamma \in \mathscr{W}_t] \int_0^{1/2M_1} [\log^-(v)]^{2k}\, dv,$$

whereupon

$$\log T \int_{\mathscr{W}_t} |D_2^-(u)|^{2k}\, du \le b_{11} N[\gamma \in \mathscr{W}_t].$$

Taken together, cases (A) and (B) now yield

$$\log T \int_{\mathscr{W}_t} |D_2^-(u)|^{2k}\, du \le b_{10}[\log^+(\eta_t \log T)]^{2k} + b_{11} N[\gamma \in \mathscr{W}_t] + b_{10}.$$

By applying Lemmas A and D, we finally establish that

$$\int_T^{2T} \left[\log T \int_{\mathscr{W}_t} |D_2^-(u)|^{2k}\, du \right] dt \le O_{kM_1}(T).$$

This completes the proof of Theorem 1.

8

In this section, we combine the ideas of [4] with those of Sections 3 and 6 to say something about the "statistics" of

$$\log \left| \frac{\zeta'(\frac{1}{2} + i\gamma)}{A'(\gamma)} \right| \qquad \text{[i.e., } F_1(\gamma)\text{]}.$$

We continue to assume that RH holds. Note, incidentally, that

$$\left| \zeta'\left(\frac{1}{2} + i\gamma\right) \right| \equiv \left| V'\left(\frac{1}{2} + i\gamma\right) \right|. \tag{8.1}$$

In the context of $\zeta(s)$, hypothesis $(\mathscr{H}_\alpha)$ asserts that

$$\limsup_{T \to \infty} \frac{N\left[T \le \gamma \le 2T\colon 0 \le \gamma' - \gamma \le \dfrac{c}{\log T} \right]}{A(2T) - A(T)} \le \mathscr{M} c^\alpha$$

for $0 < c < 1$, with suitably chosen $\mathscr{M}$. The zero $\frac{1}{2} + i\gamma'$ is (just) the immediate successor of $\frac{1}{2} + i\gamma$ (with an obvious convention at multiple zeros). Cf. [4,14,15]. We also recall that the pair correlation conjecture automatically *implies* $(\mathscr{H}_1)$.

Theorem 4 [assuming both RH and $(\mathscr{H}_\alpha)$]. *Let E be any Borel measurable subset of $\mathbb{R}$ having Jordan content. Then*

$$\lim_{T \to \infty} \frac{N\left[T \le \gamma \le 2T\colon \dfrac{F_1(\gamma)}{\sqrt{\pi \log\log T}} \in E \right]}{A(2T) - A(T)} = \int_E e^{-\pi x^2}\, dx.$$

Proof. By standard probability theory, it *suffices* to prove this for $E = (a, b)$ with $-\infty < a < b < \infty$. Cf. [3] and the footnote for Section 5. The argument that follows is closely related to [4, 15]. To get started, let k be any positive integer and Q be any number bigger than 50000.

Think of M_1 and ω as being *explicit* functions of Q, k, α subject to the condition that

$$0 < \omega < 10^{-6}, \qquad Q \leq M_1 \leq \exp(\sqrt{Q}).$$

The quantities a, b, $\mathscr{M}$, α, k are fixed once and for all. Let η be *any* positive number less than $\min((6 - a)/10, 1)$. Let $[\![X]\!]$ signify a real number of absolute value $\leq X$.

In everything that follows, it is understood that

$$T \geq T_0(a, b, \mathscr{M}, \alpha, k, \eta, Q). \tag{8.2}$$

It is also understood that B_l is a positive constant depending solely on $\mathscr{M}$, α, k.

Exactly as in [4, p. 215], we can construct a large set $\mathscr{G}_Q$ of "good" $t \in [T, 2T]$ such that

(i) the relative complement of $\mathscr{G}_Q$ has measure $\leq B_1 T Q^{-2k}$.
(ii) the number of γ in $[t, t \pm M_j/\log T]$ is $M_j/2\pi + [\![Q\sqrt{\log M_j}]\!]$.
(iii) the total variation of $\Omega_t(u)$ on $\mathscr{W}_t$ is $\leq \Omega$.
(iv) $\zeta(s)$ has mingap at t exceeding $e^{-Q}/\log T$.
(v) the distance from t to the closest γ exceeds $e^{-Q}/\log T$.

Here: $M_2 \equiv 10M_1$ and Ω is a positive constant depending *solely* on Q, k, α. The word "mingap" refers to the minimum γ–spacing on window-frame $\mathscr{F}_t$.

Let $t_n = nM_1/\log T$. We say that n is "good" iff

(a) $[t_{n-100}, t_{n+100}] \subseteqq [T, 2T]$.
(b) $(t_n, t_n + \omega/\log T)$ contains at least one point of $\mathscr{G}_Q$ (call it τ_n).
(c) no two γ_μ and γ_ν can be found within $5\omega/\log T$ units of the *same* t_{n+q} with $0 \leq q \leq 10$.

An elementary consideration shows that the number of "good" n is

$$\frac{T \log T}{M_1} + [\![1000]\!] + B_2[\![\omega^\alpha T \log T]\!] + B_3\left[\!\!\left[\frac{T \log T}{\omega Q^{2k}}\right]\!\!\right].$$

With a slight abuse of language, this means that the number of "bad" intervals $[t_n, t_{n+1})$ is

$$B_4\left[\!\!\left[\omega^\alpha T \log T + \frac{T \log T}{\omega Q^{2k}}\right]\!\!\right]. \tag{8.3}$$

Recall, however, that

$$\lim_{T\to\infty}\frac{1}{T}m\left\{T\le t\le 2T\colon \frac{\log|\mathrm{V}(\frac{1}{2}+it)|}{\sqrt{\pi\log\log T}}\in(a-2\eta,\, a+2\eta)\right\}$$

$$=\int_{a-2\eta}^{a+2\eta}e^{-\pi x^2}\,dx.$$

For large T, we therefore have

$$m\left\{T\le t\le 2T\colon \frac{\log|V(\frac{1}{2}+it)|}{\sqrt{\pi\log\log T}}\in(a-2\eta,\, a+2\eta)\right\}=[\![5\eta]\!]T.$$

Let $N_{a\eta}$ be the number of "good" n for which

$$a-\eta\le\frac{\operatorname{Re}\Omega_{\tau_n}(\tau_n)}{\sqrt{\pi\log\log T}}\le a+\eta.$$

For *each* such n, form $[t_n, t_{n+1})$ and consider the equation

$$\frac{\log|V(\frac{1}{2}+it)|}{\sqrt{\pi\log\log T}}=\frac{\operatorname{Re}\Omega_{\tau_n}(t)+\sum_{\gamma\in\mathscr{F}_{\tau_n}}\log|A(t)-A(\gamma)|}{\sqrt{\pi\log\log T}}$$

$$=\frac{\operatorname{Re}\Omega_{\tau_n}(\tau_n)+[\![\Omega]\!]+\sum_{\gamma\in\mathscr{F}_{\tau_n}}\log|A(t)-A(\gamma)|}{\sqrt{\pi\log\log T}}.$$

Note that $N[\gamma\in\mathscr{F}_{\tau_n}]=M_2/\pi+2[\![Q\sqrt{\log M_2}]\!]$.

By deleting the appropriate $e^{-Q}/\log T$ neighborhoods, we immediately discover that

$$\frac{\log|V(\frac{1}{2}+it)|}{\sqrt{\pi\log\log T}}=\frac{\operatorname{Re}\Omega_{\tau_n}(\tau_n)+[\![\Omega]\!]+O_{M_2Q}(1)}{\sqrt{\pi\log\log T}}$$

$$\frac{\log|V(\frac{1}{2}+it)|}{\sqrt{\pi\log\log T}}=\frac{\operatorname{Re}\Omega_{\tau_n}(\tau_n)+O_{Qk\alpha}(1)}{\sqrt{\pi\log\log T}}$$

$$\frac{\log|V(\frac{1}{2}+it)|}{\sqrt{\pi\log\log T}}\in(a-2\eta,\, a+2\eta)$$

for

$$t\in[t_n, t_{n+1})-\left\{\begin{array}{l}\text{a set of measure}\\ \left[\!\!\left[\dfrac{20M_1}{\pi}+8Q\sqrt{\log M_1}\right]\!\!\right]\dfrac{e^{-Q}}{\log T}\end{array}\right\}.$$

The *essential* point here is that T is kept sufficiently large, cf. (8.2). It follows that

$$N_{a\eta}\left[\frac{M_1}{\log T} - 8[M_1 + Q\sqrt{\log M_1}]\frac{e^{-Q}}{\log T}\right] \leq 5\eta T$$

$$N_{a\eta}\frac{M_1}{\log T}\left[1 - 8\left(1 + \frac{Q\sqrt{\log M_1}}{M_1}\right)e^{-Q}\right] \leq 5\eta T$$

$$N_{a\eta} \leq 10\eta\frac{T\log T}{M_1}. \tag{8.5}$$

Similarly for $N_{b\eta}$.

Let $N_{ab\eta}$ now be the number of "good" n for which

$$\frac{\operatorname{Re}\Omega_{\tau_n}(\tau_n)}{\sqrt{\pi\log\log T}} \in (a + \eta, b - \eta).$$

Since T is large, we already know that

$$m\left\{T \leq t \leq 2T: \frac{\log|V(\frac{1}{2} + it)|}{\sqrt{\pi\log\log T}} \in [a, b]\right\} = T\left[\int_a^b e^{-\tau x^2}\,dx + [\![\eta]\!]\right].$$

Divide the left-hand t-set into six pieces corresponding to the (side)conditions

(A) $t \in \text{bad}\,[t_n, t_{n+1})$;

(B) $t \in \text{good}\,[t_n, t_{n+1})$, $\quad \mathscr{U}_n \in (a + \eta, b - \eta)$, $\quad \min_\gamma|\gamma - t| \geq \dfrac{e^{-Q}}{\log T}$;

(C) $t \in \text{good}\,[t_n, t_{n+1})$, $\quad \mathscr{U}_n < a - \eta$, $\quad \min_\gamma|\gamma - t| \geq \dfrac{e^{-Q}}{\log T}$;

(D) $t \in \text{good}\,[t_n, t_{n+1})$, $\quad \mathscr{U}_n > b + \eta$, $\quad \min_\gamma|\gamma - t| \geq \dfrac{e^{-Q}}{\log T}$;

(E) $t \in \text{good}\,[t_n, t_{n+1})$, $\quad \mathscr{U}_n \in [a - \eta, a + \eta] \cup [b - \eta, b + \eta]$,

$\min_\gamma|\gamma - t| \geq \dfrac{e^{-Q}}{\log T}$;

(F) $t \in \text{good}\,[t_n, t_{n+1})$, $\quad \min_\gamma|\gamma - t| < \dfrac{e^{-Q}}{\log T}$,

where

$$\mathscr{U}_n \equiv \frac{\operatorname{Re}\Omega_{\tau_n}(\tau_n)}{\sqrt{\pi\log\log T}}. \tag{8.6}$$

Let $\ell_A, \ldots, \ell_F$ be the associated Lebesgue measures.

By virtue of (8.2), (8.3), (8.4), and (8.5), we immediately see that

$$\ell_A = \frac{M_1}{\log T} B_4 \left[\!\left[\omega^\alpha T \log T + \frac{T \log T}{\omega Q^{2k}} \right]\!\right]$$

$$\ell_C = 0$$

$$\ell_D = 0$$

$$\ell_E \leq 2\left(10\eta \frac{T \log T}{M_1} \right) \frac{M_1}{\log T} = 20\eta T$$

$$\ell_F \leq T \log T \cdot \frac{2e^{-Q}}{\log T} = 2Te^{-Q}$$

This shows that

$$\ell_B = T\left(\int_a^b e^{-\pi x^2} dx + [\![\eta]\!] \right) + B_4 M_1 T \left[\!\left[\omega^\alpha + \frac{1}{\omega Q^{2k}} \right]\!\right] + [\![20\eta T]\!] + [\![2Te^{-Q}]\!]. \tag{8.7}$$

On the other hand, note that

$$\ell_B = N_{ab\eta}\left[\frac{M_1}{\log T} - 20[\![M_1 + Q\sqrt{\log M_1}]\!] \frac{e^{-Q}}{\log T} \right]$$

yields

$$\ell_B = N_{ab\eta} \frac{M_1}{\log T} + [\![40T(1+Q)e^{-Q}]\!] \tag{8.8}$$

By *combining* (8.7) and (8.8), we find that

$$N_{ab\eta} = \frac{T \log T}{M_1} \int_a^b e^{-\pi x^2} dx + 43 \frac{T \log T}{M_1} [\![Qe^{-Q} + \eta]\!] + B_4 T \log T \left[\omega^\alpha + \frac{1}{\omega Q^{2k}} \right]. \tag{8.9}$$

It *remains* for us to compute

$$N\left[T \leq \gamma \leq 2T \colon \frac{\log \left| \dfrac{V'(\frac{1}{2} + i\gamma)}{A'(\gamma)} \right|}{\sqrt{\pi \log \log T}} \in (a, b) \right].$$

Cf. (8.1). To this end, we first divide the associated γ–set into five pieces corresponding to the (side)conditions

(A) $\gamma \in \text{bad } [t_n, t_{n+1})$;

(B) $\gamma \in \text{good } [t_n, t_{n+1})$, $\quad \mathscr{U}_n \in (a+\eta, b-\eta)$;

(C) $\gamma \in \text{good } [t_n, t_{n+1})$, $\quad \mathscr{U}_n < a-\eta$;

(D) $\gamma \in \text{good } [t_n, t_{n+1})$, $\quad \mathscr{U}_n > b+\eta$;

(E) $\gamma \in \text{good } [t_n, t_{n+1})$, $\quad \mathscr{U}_n \in [a-\eta, a+\eta] \cup [b-\eta, b+\eta]$.

Cf. (8.6). The associated cardinalities will be denoted by $N_A, \ldots, N_E$.

A trivial manipulation shows that

$$\left|\frac{V'(\frac{1}{2}+i\gamma)}{A'(\gamma)}\right| = \exp[\operatorname{Re}\Omega_{\tau_n}(\gamma)] \cdot \prod_{\gamma^* \in \mathscr{F}_{\tau_n}}{}' |A(\gamma) - A(\gamma^*)| \tag{8.10}$$

in cases (B)-(E) [where $\prod'$ means omit γ].

In such cases, it emerges that

$$\frac{\log\left|\dfrac{V'(\frac{1}{2}+i\gamma)}{A'(\gamma)}\right|}{\sqrt{\pi \log\log T}} = \frac{\operatorname{Re}\Omega_{\tau_n}(\tau_n) + [\![\Omega]\!] + \sum'_{\gamma^* \in \mathscr{F}_{\tau_n}} \log|A(\gamma) - A(\gamma^*)|}{\sqrt{\pi \log\log T}}$$

$$= \mathscr{U}_n + \frac{[\![\Omega]\!] + O_{M_2 Q}(1)}{\sqrt{\pi \log\log T}} \text{ (cf. the definition of } \mathscr{G}_Q)$$

$$= \mathscr{U}_n + \frac{O_{Qk\alpha}(1)}{\sqrt{\pi \log\log T}}.$$

Compare (8.4).

The preceding equation [combined with (8.2) and (8.5)] immediately shows that

$$N_C = N_D = 0$$

$$N_E \leq 20\eta \frac{T \log T}{M_1}\left\{\frac{M_1}{2\pi} + [\![2]\!] + [\![Q\sqrt{\log M_1}]\!]\right\}$$

$$\leq 20\eta T \log T\left[1 + \frac{Q\sqrt{\log M_1}}{M_1}\right].$$

In a similar manner

$$N_B = N_{ab\eta}\left\{\frac{M_1}{2\pi} + 2[\![Q\sqrt{\log M_1}]\!]\right\}.$$

To control N_A, we simply observe that

$$\begin{aligned}
N_A &\le N\{T \le \gamma \le 2T: \gamma \in \text{bad } [t_n, t_{n+1})\} \\
&= N[T \le \gamma \le 2T] - N\{T \le \gamma \le 2T: \gamma \in \text{good } [t_n, t_{n+1})\} \\
&= N[T \le \gamma \le 2T] - \left(\frac{T \log T}{M_1} + B_5 T \log T \left[\!\left[\omega^\alpha + \frac{1}{\omega Q^{2k}} \right]\!\right]\right) \\
&\quad \times \left(\frac{M_1}{2\pi} + [\![2]\!] + [\![Q\sqrt{\log M_1}]\!]\right)
\end{aligned}$$

cf. (8.3) and the definition of $\mathscr{G}_Q$

$$\begin{aligned}
&= N[T \le \gamma \le 2T] - \frac{T \log T}{M_1}\left(1 + B_5 M_1 \left[\!\left[\omega^\alpha + \frac{1}{\omega Q^{2k}} \right]\!\right]\right) \\
&\quad \times \frac{M_1}{2\pi}\left(1 + 4\pi \left[\!\left[\frac{Q\sqrt{\log M_1}}{M_1} \right]\!\right]\right) \\
&= N[T \le \gamma \le 2T] - \frac{T \log T}{2\pi}\left(1 + B_5 M_1 \left[\!\left[\omega^\alpha + \frac{1}{\omega Q^{2k}} \right]\!\right]\right) \\
&\quad \times \left(1 + B_6 \left[\!\left[\frac{Q\sqrt{\log M_1}}{M_1} \right]\!\right]\right).
\end{aligned}$$

But

$$\begin{aligned}
N[T \le \gamma \le 2T] &= A(2T) - A(T) + O(\log T) \\
&= \int_T^{2T} \left[\frac{1}{2\pi} \log\left(\frac{u}{2\pi}\right)\right] du + O(\log T) \\
&= \int_T^{2T} \left[\frac{\log T}{2\pi} + O(1)\right] du + O(\log T).
\end{aligned}$$

Hence

$$N[T \le \gamma \le 2T] = \frac{T \log T}{2\pi}\left[1 + O\left(\frac{1}{\log T}\right)\right] \tag{8.11}$$

with an *absolute* implied constant.

It is now apparent that we should try to minimize

$$\left|\left(1 + B_5 M_1 \left[\!\left[\omega^\alpha + \frac{1}{\omega Q^{2k}} \right]\!\right]\right)\left(1 + B_6 \left[\!\left[\frac{Q\sqrt{\log M_1}}{M_1} \right]\!\right]\right) - 1\right|.$$

This leads to the following basic choices

$$\omega^\alpha = \frac{1}{\omega Q^{2k}} \Rightarrow \omega = Q^{-2k/(1+\alpha)}$$

$$M_1 Q^{-2k\alpha/(1+\alpha)} = \frac{Q}{M_1} \Rightarrow M_1 = Q^{1/2 + k\alpha/(1+\alpha)}.$$

To satisfy the various constraints, we need to take

$$k > \frac{1+\alpha}{2\alpha} \qquad \text{and} \qquad Q \geq Q_0(k, \alpha). \tag{8.12}$$

Note that these *choices* are *exactly* the same as in [4, p. 216].

With $\psi \equiv k\alpha/(1+\alpha) - \frac{1}{2}$, we immediately obtain

$$
\begin{aligned}
N&\left[T \leq \gamma \leq 2T\colon \frac{F_1(\gamma)}{\sqrt{\pi \log\log T}} \in (a, b)\right] \\
&= N_A + N_B + N_C + N_D + N_E \\
&= \frac{T\log T}{2\pi}\left\{O\left(\frac{1}{\log T}\right) + B_7[\![Q^{-\psi}\sqrt{\log Q}]\!]\right\} \\
&\quad + N_{ab\eta}\frac{M_1}{2\pi}\{1 + B_8[\![Q^{-\psi}\sqrt{\log Q}]\!]\} \\
&\quad + B_9[\![\eta T\log T]\!] \\
&= B_7 T\log T[\![Q^{-\psi}\sqrt{\log Q}]\!] + N_{ab\eta}\frac{M_1}{2\pi}\{1 + B_8[\![Q^{-\psi}\sqrt{\log Q}]\!]\} \\
&\qquad\qquad \text{cf. (8.2)} \\
&\quad + B_9[\![\eta T\log T]\!] \\
&= B_{10} T\log T[\![Q^{-\psi}\sqrt{\log Q}]\!] + N_{ab\eta}\frac{M_1}{2\pi} \\
&\qquad\qquad \left(\text{since } N_{ab\eta} \leq 2\frac{T\log T}{M_1}\right) \\
&\quad + B_9[\![\eta T\log T]\!] \\
&= B_{10} T\log T[\![Q^{-\psi}\sqrt{\log Q}]\!] \\
&\quad + \frac{T\log T}{2\pi}\int_a^b e^{-\pi x^2}\,dx + B_{11} T\log T[\![Qe^{-Q} + \eta]\!] \\
&\quad + B_{12} T\log T[\![Q^{-\psi}]\!] \\
&\quad + B_9[\![\eta T\log T]\!] \qquad \text{via (8.9)} \\
&= \frac{T\log T}{2\pi}\int_a^b e^{-\pi x^2}\,dx + B_{13} T\log T[\![Qe^{-Q} + Q^{-\psi}\sqrt{\log Q} + \eta]\!] \\
&= \frac{T\log T}{2\pi}\int_a^b e^{-\pi x^2}\,dx + B_{14} T\log T[\![Q^{-\psi}\sqrt{\log Q} + \eta]\!].
\end{aligned}
$$

By fixing k and then letting $\eta \to 0^+$, $Q \to \infty$, we finally deduce that

$$\lim_{T \to \infty} \frac{N\left[T \le \gamma \le 2T \colon \dfrac{F_1(\gamma)}{\sqrt{\pi \log \log T}} \in (a, b)\right]}{A(2T) - A(T)} = \int_a^b e^{-\pi x^2}\, dx$$

This completes the proof of Theorem 4.

9

There is little doubt that Theorems 1-4 can be extended to include higher derivatives, at least if we impose *both* RH and $(\mathscr{H}_\alpha)$.

A heuristic computation (similar to Section 3) shows that

$$\frac{\log|\zeta(\tfrac{1}{2} + iu)|}{\sqrt{\pi \log \log T}} \quad \text{and} \quad \frac{\log\left|\dfrac{\zeta^{(N)}(\tfrac{1}{2} + iu)}{A'(u)^N}\right|}{\sqrt{\pi \log \log T}} \tag{9.1}$$

are effectively the "same" random variable[11] as $T \to \infty$; similarly, for $V(\tfrac{1}{2} + iu)$. The higher derivatives of $\Omega_t(u)$ are controlled on $\mathscr{W}_t$ by

(A) observing that $\Omega_t'(u)$, $\Omega_t^{(3)}(u)$, $\Omega_t^{(5)}(u)$, ... are monotonic

(B) passing to finite differences ala [17, pp. 6 (top), 56]

(C) applying Lemma G (in a suitably extended form)

Cf. (6.11)(6.8)(6.10) and Lemma H regarding (A).

10

In a somewhat different direction, there is *no* question that *all* these results can be formulated for much more general L-functions (cf. [7, 10, 15, 16, 22], for instance).

It is (also) reasonable to expect that GRH can be replaced by a somewhat weaker hypothesis (at the expense of calling for lengthier proofs). Compare [4, footnote 3], [15], [21], [24, p. 204(C)].

11

Theorem 4 shows that, on average, 50% of all ρ satisfy

$$|\zeta'(\rho)| \le A'(\gamma).$$

[11] on $[T, 2T]$

This *implies* that $\sum|\rho\zeta'(\rho)|^{-1}$ diverges. To the extent that large $|\omega|$ can be neglected, we also have

$$\sum_{T\le\gamma\le 2T}\frac{1}{|\rho\zeta'(\rho)|^{\lambda}}\approx\frac{1}{[T\log T]^{\lambda-1}}\int_{\mathbb{R}}e^{-\pi\omega^2-\lambda\omega\sqrt{\pi\log\log T}}\,d\omega.$$

The RHS behaves like $T^{1-\lambda}(\log T)^{\lambda^2/4+1-\lambda}$. This *suggests* that $\sum|\rho\zeta'(\rho)|^{-\lambda}$ converges iff $\lambda>1$. Compare [24, pp. 319, 323].

Cf. [12, 13] for some related ideas.

12

Finally, note that Theorem 4 is a kind of "real counterpart" to [23, p. 199 (lines 7-17)]. See [9] for additional results along these lines.

References

[1] Ahlfors, L.V. *Complex Analysis*, 2nd edition. McGraw-Hill (1966).

[2] Apostol, T. *Mathematical Analysis*. Addison-Wesley (1957).

[3] Billingsley, P. *Probability and Measure*, 2nd edition. John Wiley (1986), especially pp. 390, 406, 408, 416(6), 600, 335, 350(13).

[4] Bombieri, E. and Hejhal, D.A. "Sur les zéros des fonctions zêta d'Epstein," *C.R. Acad. Sci. Paris* **304** (1987), 213–217.

[5] Davenport, H. *Multiplicative Number Theory*, 2nd edition. Springer-Verlag (1980).

[6] Erdélyi, A. et al. *Higher Transcendental Functions*, vol. 1. McGraw-Hill (1953).

[7] Fujii, A. "On the zeros of Dirichlet *L*-functions I," *Trans. Amer. Math. Soc.* **196** (1974), 225–235. See also the *errata* in **267** (1981), 33 (para 3), 38, 39.

[8] Fujii, A. "On the zeros of Dirichlet *L*-functions II," *Trans. Amer. Math. Soc.* **267** (1981), 33–40.

[9] Fujii, A. "On the zeros of Dirichlet *L*-functions V, VI," *Acta Arith.* **28** (1975), 395–403; **29** (1976), 49–58.

[10] Fujii, A. "A remark on the zeros of some *L*-functions," *Tôhoku Math. J.* **29** (1977), 417–426.

[11] Ghosh, A. "On Riemann's zeta function—sign changes of *S(T)*." In: *Recent Progress in Analytic Number Theory*, Vol. 1, H. Halberstam and C. Hooley eds., Academic Press (1981), 25–46.

[12] Ghosh, A. and Conrey, B. "A mean value theorem for the Riemann zeta function at its relative extrema on the critical line," *J. Lond. Math. Soc.* **32** (1985), 193–202.

[13] Gonek, S.M. "Mean values of the Riemann zeta function and its derivatives," *Inventiones Math.* **75** (1984), 123–141.

[14] Hejhal, D.A. "Zeros of Epstein zeta functions and supercomputers," *Proceedings of the International Congress of Mathematicians* 1986 (*Berkeley*), 1362–1384.

[15] Hejhal, D.A. manuscript on the zeros of Epstein zeta functions, in preparation.

[16] Joyner, D. *Distribution Theorems of L-Functions.* Longman Scientific & Technical (1986). Note that p. 117 requires the *analog* of A. Selberg's $N(\sigma, T, T + H)$ estimate.

[17] Milne-Thomson, L.M. *The Calculus of Finite Differences.* MacMillan (1951).

[18] Montgomery, H.L. and Vaughan, R.C. "Hilbert's inequality," *J. Lond. Math. Soc.* **8** (1974), 73–82, especially equations (1.9)(1.11). See also: Ivić, A. *The Riemann Zeta Function.* Wiley (1985), 130–135.

[19] Mueller, J. "On the difference between consecutive zeros of the Riemann zeta function," *J. Number Th.* **14** (1982), 327–331.

[20] Selberg, A. "On the remainder in the formula for $N(T)$, the number of zeros of $\zeta(s)$ in the strip $0 < t < T$," *Avhand. Norske Vid. Akad. Oslo* No. 1, (1944) especially Sections 2-4.

[21] Selberg, A. "Contributions to the theory of the Riemann zeta function," *Archiv. for Math. og Naturvid.* **48**(5) (1946), especially Sections 4, 5, 3 (thm 1).

[22] Selberg, A. "Contributions to the theory of Dirichlet's L-functions," *Skr. Norske Vid. Akad. Oslo* No. 3, (1946), especially Sections 5, 6.

[23] Selberg, A. "The zeta-function and the Riemann hypothesis," *Skandinaviske Matematiker Kongres.* (1946), 187–200.

[24] Titchmarsh, E.C. *The Theory of the Riemann Zeta-Function.* Oxford Univ. Press (1951). The moment formalism is discussed on pp. 308–314.

[25] Tsang, K.M. "The distribution of the values of the Riemann zeta function," Ph.D. thesis, Princeton University (1984), 179 pp.

[26] Tsang, K.M. "Some Ω-theorems for the Riemann zeta-function," *Acta Arith.* **46** (1986), 369–395.

20 Selberg's Lower Bound of the First Eigenvalue for Congruence Groups

H. IWANIEC

1. Introduction

Let Γ be a congruence subgroup of the modular group $\mathrm{SL}(2,\mathbb{Z})$ and let

$$\lambda_0 = 0 < \lambda_1 \leq \lambda_2 \leq \cdots \tag{1}$$

be the point spectrum of the Laplace operator

$$\Delta = y^2\left(\frac{\partial^2}{\partial x^2} + \frac{\partial^2}{\partial y^2}\right)$$

acting on Γ-automorphic functions. A. Selberg [3] conjectured that

$$\lambda_1 \geq \frac{1}{4},$$

and he was able to show that

$$\lambda_1 \geq \frac{3}{16} \tag{2}$$

by using A. Weil's estimate for the Kloosterman sums

$$|S(m, n; c)| \leq (m, n, c)^{1/2} c^{1/2} \tau(c), \tag{3}$$

NUMBER THEORY,
TRACE FORMULAS and
DISCRETE GROUPS

ISBN 0-12-067570-6

thus using indirectly the Riemann Hypothesis for curves over finite fields. S. S. Gelbart and H. Jacquet [2] have proved the strict inequality

$$\lambda_1 > \frac{3}{16}$$

by means of a certain lift from GL(2) to GL(3). Their method does not depend on bounds for Kloosterman sums.

The spectrum of Δ is related to the Kloosterman sums $S(n, n; c)$ through the spectral representation of the zeta-function

$$Z(s) = \sum_{c \equiv 0 (\operatorname{mod} q)} c^{-2s} S(n, n; c). \tag{4}$$

A. Selberg [3] has shown that $Z(s)$ has meromorphic continuation to the whole complex s-plane. In the half-plane $s = \sigma + it$, $\sigma > \frac{1}{2}$ the zeta-function $Z(s)$ has only a finite number of simple poles at the points s_j of the segment $\frac{1}{2} < s_j < 1$ corresponding to the exceptional eigenvalues

$$0 < \lambda_j = s_j(1 - s_j) < \frac{1}{4}.$$

However, (3) implies that the series (4) converges absolutely in $\sigma > \frac{3}{4}$, so $Z(s)$ is holomorphic in $\sigma > \frac{3}{4}$, whence $\lambda_1 \geq \frac{3}{16}$.

In this article we show a simple proof of (2) without using Weil's celebrated estimate. For clarity we restrict our arguments to the Hecke congruence groups $\Gamma = \Gamma_0(N)$.

2. An Estimate for Sums of Kloosterman Sums

In place of (3) we establish an upper bound for Kloosterman sums on average of the following type

Lemma 1. *Let $n \geq 1$, $Q \geq 1$, and $R \geq 1$. We have*

$$\sum_{q \leq Q} \sum_{r \leq R} |S(n, n; qr)| \leq \tau_3(n)(Q + R)QR(\log 4QR)^2.$$

Proof. First we estimate the partial sum

$$S_1(Q, R) = \sum_{\substack{q \leq Q \\ (q,r)=1}} \sum_{r \leq R} |S(n, n; qr)|.$$

The key property of the Kloosterman sum $S(n, n; qr)$ we are going to use is the "twisted" multiplicativity

$$S(n, n; qr) = S(n\bar{q}, n\bar{q}; r)S(n\bar{r}, n\bar{r}; q),$$

where $q\bar{q} \equiv 1(\text{mod } r)$ and $r\bar{r} \equiv 1(\text{mod } q)$. By Cauchy's inequality we get

$$S_1^2(Q, R) \leq S_2(Q, R)S_2(R, Q),$$

say, where

$$S_2(Q, R) = \sum_{\substack{q \leq Q \\ (q,r)=1}} \sum_{r \leq R} |S(n\bar{q}, n\bar{q}; r)|^2.$$

We have

$$\begin{aligned} S_2(Q, R) &\leq \sum_{r \leq R} \left(1 + \frac{Q}{r}\right) \sum_{a(\text{mod } r)} |S(an, an; r)|^2 \\ &= \sum_{r \leq R} (r + Q)\nu_n(r), \end{aligned}$$

where $\nu_n(r)$ stands for the number of solutions to

$$n(x + \bar{x}) \equiv n(y + \bar{y})(\text{mod } r)$$

in $x(\text{mod } r)$, $y(\text{mod } r)$, $(xy, r) = 1$. Elementarily we infer

$$\nu_n(r) \leq (n, r)r\tau(r).$$

Hence

$$S_2(Q, R) \leq (Q + R)R \sum_{r \leq R} (n, r)\tau(r) \leq \tau_3(n)(Q + R)R^2 \log 4R.$$

This and the analogous result for $S_2(R, Q)$ yield

$$S_1(Q, R) \leq \tau_3(n)(Q + R)QR \log 4QR.$$

Finally the constraint $(q, r) = 1$ in $S_1(Q, R)$ can be relaxed as follows

$$\begin{aligned} S(Q, R) &= \sum_{q \leq Q} \sum_{r \leq R} |S(n, n; qr)| \\ &\leq \sum_k \sum_\ell S_1\left(\frac{Q}{k}, \frac{R}{\ell}\right). \end{aligned}$$

Applying the estimate for S_1 we complete the proof of Lemma 1.

3. Kloosterman Sums Formula

It is essential for our arguments to work with a finite sum of Kloosterman sums rather than with the zeta-function $Z(s)$. Set $\sigma = \frac{1}{2} + (\log 4X)^{-1}$,

$$F_q(X) = \sum_{c \equiv 0(\text{mod } q)} c^{-1} f\left(\frac{2\pi nX}{c}\right) S(n, n; c)$$

and

$$G_q(X) = \sum_{\sigma < s_j < 1} \Gamma(2s_j - 1)\hat{f}(1 - 2s_j)X^{2s_j - 1}|\rho_j(n)|^2,$$

where f is a smooth function compactly supported in $(0, \infty)$, $\hat{f}$ is the Mellin transform, and $\rho_j(n)$ stands for the n-th Fourier coefficient of the orthonormalized Maass cusp forms for $\Gamma = \Gamma_0(q)$:

$$u_j(z) = y^{1/2} \sum_{n \neq 0} \rho_j(n)K_{s_j - 1/2}(2\pi|n|y)e(nx),$$

$$(\Delta + \lambda_j)u_j = 0, \langle u_j, u_k \rangle = \delta_{jk}.$$

Lemma 2. *For $n \geq 1$, $q \geq 1$, and $X \geq 2$ we have*

$$F_q(X) = G_q(X) + 0(n \log qX),$$

where the constant implied in 0 depends on the test function f alone.

Proof. It follows from Theorems 1 and 2 of [1].

4 Proof of Selberg's Bound

Notice that $\Gamma(2s_j - 1) > 0$ and $\hat{f}(1 - 2s_j) > 0$ if $f \geq 0$ and $f \neq 0$, which we henceforth assume. Moreover, if u_j is a normalized Maass cusp form for $\Gamma_0(N)$ then $\ell(q)^{-1/2}u_j$, where

$$\ell(q) = [\Gamma_0(qN) \colon \Gamma_0(N)] \leq q \prod_{p|q} \left(1 + \frac{1}{p}\right),$$

is a normalized cusp form for $\Gamma_0(qN)$. From this observation it follows (by positivity argument)

$$\ell(q)^{-1}G_N(X) \leq G_{qN}(X).$$

Sum up this inequality over q with $Q < q < 2Q$ getting

$$G_N(X) \ll \sum_{Q < q \leq 2Q} G_{qN}(X).$$

Hence by Lemmas 1 and 2 we obtain

$$\begin{aligned} G_N(X) &\ll \sum_{Q < q \leq 2Q} F_{qN}(X) + nQ \log QX \\ &\ll (Q + XQ^{-1})(n \log QX)^2 \\ &\ll X^{1/2}(n \log X)^2 \end{aligned}$$

on taking $Q = X^{1/2}$. Finally, letting $X \to \infty$, this implies $2s_j - 1 \le \frac{1}{2}$, $s_j \le \frac{3}{4}$, and $\lambda_j \ge \frac{3}{16}$.

Remark. Our method yields the holomorphic continuation of the Kloosterman-Selberg zeta-function $Z(s)$ to Re $s > \frac{3}{4}$, but it does not show that the series (4) is absolutely convergent in Re $s > \frac{3}{4}$.

References

[1] Deshouillers, J-M. and Iwaniec, H. "Kloosterman sums and Fourier coefficients of cusp forms," *Invent. Math.* **70** (1982), 219–288.
[2] Gelbart, S. S. and Jacquet, H. "A relation between automorphic representations of GL(2) and GL(3)," *Ann. Ecole Norm. Sup.* **11** (1978), 471–542.
[3] Selberg, A. "On the estimation of Fourier coefficients of modular forms," *Proc. Symposia in Pure Math.* **VIII**, A.M.S., Providence (1965), 1–15.

21 Discrete Subgroups and Ergodic Theory*

G. A. MARGULIS

In spite of the quite general title, we only give results on the behavior of "individual" orbits and the description of invariant measures for actions of groups on homogeneous spaces. Some number theoretic corollaries of these results will also be provided. In particular, we prove Oppenheim's conjecture on values of indefinite quadratic forms.

1. Formulation of Results and Conjectures

1.1

Let B be a real nondegenerate indefinite quadratic form in n variables. It is well known that if $n \geq 5$ and the coefficients of B are rational, then B represents zero nontrivially, i.e., there exist integers $x_1, \ldots, x_n$ not

* Editorial note. After consultation with G. A. Margulis it was agreed to make some changes directly on the original manuscript, without returning it to him, in order to save time. A. Borel kindly provided the necessary revision, including an additional argument supplied by G. Prasad in the proof of Lemma A. The appendix in the original manuscript has been transformed and reorganized into Section 4; the historical background material and the bibliography have been corrected and updated.

The editors wish to thank A. Borel and G. Prasad for their help, still taking ultimate responsibility for making the changes.

NUMBER THEORY,
TRACE FORMULAS and
DISCRETE GROUPS

ISBN 0-12-067570-6

all equal to 0 such that $B(x_1, \ldots, x_n) = 0$. Theorem 1 stated below can be considered as an analogue of this assertion in the case where B is not proportional to a form with rational coefficients. Note that in Theorem 1 the condition "$n \geq 5$" is replaced by the weaker one "$n \geq 3$".

Theorem 1. *Suppose that $n \geq 3$ and that B is not proportional to a form with rational coefficients, or equivalently the ratio of some two coefficients of B is irrational. Then for any $\varepsilon > 0$ there exist integers $x_1, \ldots, x_n$ not all equal to 0 such that $|B(x_1, \ldots, x_n)| < \varepsilon$.*

In 1.3 Theorem 1 will be deduced from a result on the action of the group preserving B on the space of lattices. It should be noted that this result (Theorem 2) is essentially equivalent to Theorem 1.

Theorems 1 and 2 were announced in [10], with sketches of proofs. Complete proofs are given in [11]. In the present paper we shall repeat them with small modifications.

In [10] and [11], I had stated that Theorem 1 gave a proof of a conjecture of H. Davenport. After the first version of the present paper was submitted, A. Borel informed me that the conjecture was due to A. Oppenheim [13, 14] and that Oppenheim, in fact, had a stronger one in mind, which he formulated explicitly in [15]: If B is not proportional to a form with rational coefficients, then, for any $\varepsilon > 0$ there should exist integers $x_1, \ldots, x_n$ such that

$$0 < |B(x_1, \ldots, x_n)| < \varepsilon. \tag{1}$$

(The two conditions are of course equivalent for forms that do not represent zero rationally.) I then completed the earlier argument and proved a result, to be called Theorem 1′, the statement of which differs from that of Theorem 1 only in that the condition $|B(x_1, \ldots, x_n)| < \varepsilon$ is replaced by (1). The proof is also a reduction to Theorem 2. It is contained in Section 4, while Section 2 establishes Theorem 2, modulo some lemmas that are proved in Section 3.

1.2. Remarks on Theorem 1

(a) One can easily understand that if Theorem 1 is proved for some n_0, then it is proved for all $n \geq n_0$. So it is enough to prove this theorem for $n = 3$.

(b) If $n = 2$ then the assertion analogous to Theorem 1 is not true. To see this it is enough to consider the form

$$x_1^2 - \lambda x_2^2 = (x_1 - \sqrt{\lambda}x_2)(x_1 + \sqrt{\lambda}x_2),$$

where λ is an irrational positive number such that $\sqrt{\lambda}$ has a continued fraction development with bounded partial quotients.

(c) Theorem 1′ had been proved earlier in the following cases: For diagonal forms in five variables [8], for forms representing zero rationally in $n \geq 5$ variables [16], and in four variables [17], and for forms in $n \geq 21$ variables [9]. These proofs are in the context of analytic number theory and reduction theory. In [15], it is shown for $n \geq 3$ that if the set $(0, \varepsilon] \cap B(\mathbb{Z}^n)$ is not empty for any $\varepsilon > 0$, then the same is true for the form $-B$. On the other hand, it is clear that $B(\mathbb{Z}^n)$ is invariant under multiplication by square of integers. Thus Theorem 1′ implies, under the conditions of that theorem, that $B(\mathbb{Z}^n)$ is dense in $\mathbb{R}$, a consequence of [15], which had already been pointed out in [9].

1.3

As usual let $\mathbb{C}$, $\mathbb{R}$, $\mathbb{Q}$, $\mathbb{Z}$, and $\mathbb{N}^+$ denote sets of complex, real, rational, integer and positive integer numbers, respectively, and let $\mathrm{SL}(n, \mathbb{R})$ (resp. $\mathrm{SL}(n, \mathbb{Z})$) denote the group of unimodular matrices of order n with real (resp. integer) coefficients.

Theorem 2. *Let $G = \mathrm{SL}(3, \mathbb{R})$ and $\Gamma = \mathrm{SL}(3, \mathbb{Z})$. Let us denote by H the group of elements of G preserving the form $2x_1x_3 - x_2^2$ and by Ω the space of lattices in $\mathbb{R}^3$ having determinant* 1. *(The quotient space G/Γ can be naturally identified with Ω. Under this identification the* coset $g\Gamma$ *goes to the lattice $g\mathbb{Z}^3$.) Let G_y denote the stabilizer $\{g \in G | gy = y\}$ of $y \in \Omega$. If $z \in \Omega = G/\Gamma$ and the orbit Hz is relatively compact in Ω, then the quotient space $H/H \cap G_z$ is compact.*

Theorem 2 will be proved in Section 2. Now we give the reduction of Theorem 1 to Theorem 2. According to the remark (a) in 1.2, it is enough to prove Theorem 1 for $n = 3$. Let H_B denote the group of elements of G preserving B. Since $n = 3$ and the form B is indefinite, in some basis of $\mathbb{R}^3$ the form B has the type $\lambda(2x_1x_3 - x_2^2)$ where $\lambda = \pm 1$. So $H = g_B H_B g_B^{-1}$ for some $g_B \in G$.

Suppose now that the assertion of Theorem 1 is not true, i.e., $|B(x)| > \varepsilon$ for some $\varepsilon > 0$ and all $x \in \mathbb{Z}^3$, $x \neq 0$. Then, since H_B preserves B, we have $|B(x)| > \varepsilon$ for any $h \in H_B$ and $x \in h\mathbb{Z}^3$, $x \neq 0$. In view of the Mahler compactness criterion, this implies that the set $H_B\mathbb{Z}^3$ is relatively compact in Ω. By Theorem 2 for $z = g_B\mathbb{Z}^3$, it follows that the quotient space $H/H \cap G_z$ is compact. But $H = g_B H_B g_B^{-1}$ and $G_z = g_B \Gamma g_B^{-1}$. So the quotient space $H_B/H_B \cap \Gamma$ is compact. Then, in view of Borel's density theorem (see [1]), $H_B \cap \Gamma$ is Zariski dense in H_B. But $\Gamma = \mathrm{SL}(3, \mathbb{Z})$. Furthermore, if the $\mathbb{Q}$-rational points in an affine manifold are Zariski dense, then that manifold is defined over $\mathbb{Q}$ (see [2] chapter AG, 14.4). So H_B is a $\mathbb{Q}$-subgroup of G. Hence, $H_B = H_{\sigma B}$ for any automorphism σ of $\mathbb{C}$ over $\mathbb{Q}$, where σB is obtained from B by applying σ to the coefficients of B. It follows that the forms σB and B are proportional for any σ. But this contradicts the assumption that "the ratio of some two coefficients of B is irrational."

1.4

Let G be a Lie group. An element u of G will be called *unipotent* if the transformation Ad u of the Lie algebra of G is unipotent. A subgroup $U \subset G$ will be called *unipotent* if it consists of unipotent elements.

Let us denote a right-invariant Haar measure on G by μ_G. For any discrete subgroup $\Gamma \subset G$, the measure μ_G induces a measure on G/Γ that will also be denoted by μ_G. A discrete subgroup $\Gamma \subset G$ is said to be a *lattice* if $\mu_G(G/\Gamma) < \infty$. A lattice Γ is called *uniform* if G/Γ is compact, and *nonuniform* otherwise. It is well known that $\mathrm{SL}(n, \mathbb{Z})$ is a nonuniform lattice in $\mathrm{SL}(n, \mathbb{R})$ (see [18] Corollary 10.5).

The statement of Theorem 2 is a very special case of the following

Conjecture 1. *Let G be a connected Lie group, Γ be a lattice in G, and H be a subgroup of G. Suppose that H is generated by unipotent elements. Then for any point $x \in G/\Gamma$, there exists a closed subgroup $P \subset G$ containing H such that the closure of Hx coincides with Px.*

The following conjecture, due to Raghunathan, is a special case of Conjecture 1.

Conjecture 2. *Let G be a connected Lie group, Γ be a lattice in G, and U be a unipotent subgroup of G. Then for any point $x \in G/\Gamma$, there exists a closed subgroup $P \subset G$ containing U such that the closure of Ux coincides with Px.*

Raghunathan also noted the connection of his conjecture with Theorem 1 of the present paper.

1.5. Remarks on Conjectures 1 and 2

a) It is not possible to assume in Conjecture 1 that H is an arbitrary subgroup of G. For example, if

$$G = \mathrm{SL}(2, \mathbb{R}),\ \Gamma = \mathrm{SL}(2, \mathbb{Z}),\ D = \left\{ \begin{pmatrix} \lambda & 0 \\ 0 & \lambda^{-1} \end{pmatrix} \middle| \lambda \in \mathbb{R},\ \lambda \neq 0 \right\},$$

then there exists $x \in G/\Gamma$ such that the closure of Dx is not a manifold. On the other hand, in Conjecture 1 the condition "H is generated by unipotent elements" is not the most general of all possible conditions, as shown by the following fact: Let $G = \mathrm{SL}(2, \mathbb{R})$, Γ be a lattice in G and

$$P = \left\{ \begin{pmatrix} a & b \\ 0 & a^{-1} \end{pmatrix} \middle| a, b \in \mathbb{R},\ a \neq 0 \right\}.$$

Then for any point $x \in G/\Gamma$, the orbit Px is dense in G/Γ. (See 2.4 for a more general statement.)

b) Let $P \subset G$ be a closed subgroup, $x \in G/\Gamma$ and $G_x = \{g \in G | gx = x\}$. Then the orbit Px is closed in G/Γ iff the natural map $P/P \cap G_x \to G/\Gamma$ is proper.

c) It is not difficult to prove that if $G = \mathrm{SL}(n, \mathbb{R})$ and $\Gamma = \mathrm{SL}(n, \mathbb{Z})$, then in Conjectures 1 and 2 the subgroup P is algebraic and has a conjugate defined over $\mathbb{Q}$.

d) It seems that Conjecture 1 should follow from Conjecture 2. At least this is so if Γ is an arithmetic subgroup of G. (In this case, as Raghunathan communicated to the author, the above-mentioned reduction follows from remark (c) and the countability of the set of $\mathbb{Q}$-subgroups.)

1.6

Using the method of the proof of Theorem 1 presented in this paper along with some additional arguments, one can prove Conjecture 2 when $G = \mathrm{SL}(3, \mathbb{R})$ and the orbit Ux is relatively compact in G/Γ. This makes it possible to prove the following theorem, which can be considered as a generalization of Theorem 1.

Theorem 3. *Let B_1 and B_2 be two real quadratic forms in three-variables. Suppose that* (1) *there exists a basis of the space $\mathbb{R}^3$ in which B_1 and B_2 have the form $2x_1x_3 - x_2^2$ and x_1^2, respectively*; (2) *every nonzero linear combination of B_1 and B_2 is not proportional to a form with rational coefficients. Then for any $\varepsilon > 0$, there exist integers x_1, x_2, x_3, not all equal to 0 such that*

$$|B_1(x_1, x_2, x_3)| < \varepsilon \quad \text{and} \quad |B_2(x_1, x_2, x_3)| < \varepsilon.$$

1.7

Conjecture 2 is connected with

Conjecture 3. *Let, as in Conjecture 2, G be a connected Lie group, Γ be a lattice in G, and U be a unipotent subgroup of G. Then for any U-invariant, U-ergodic locally finite Borel measure σ on G/Γ, there exist $x \in G/\Gamma$ and a closed subgroup $P \subset G$ containing U such that the set Px is closed in G/Γ and σ is a finite P-invariant measure the support of which is Px.*

1.8

We mention the cases in which Conjectures 2 and 3 are proved.

Conjecture 3 is proved in [3] for reductive G and maximal unipotent subgroups U, and Conjecture 2 is proved in [6] for reductive G and horospherical U (the definition of a horospherical subgroup will be given below in 2.4; recall that any maximal unipotent subgroup of a reductive Lie group is horospherical). For the case $G = \mathrm{SL}(2, \mathbb{R})$, Conjectures 2 and 3 were proved in [4] for $\Gamma = \mathrm{SL}(2, \mathbb{Z})$ and in [7] for arbitrary Γ; the main difference between the case $G = \mathrm{SL}(2, \mathbb{R})$ and the general one is that every connected nontrivial unipotent subgroup of $\mathrm{SL}(2, \mathbb{R})$ is horospherical.

1.9

The following theorem was also proved in [4] and [7].

Theorem 4. *Let $G = \mathrm{SL}(2, \mathbb{R})$, Γ be a lattice in G, and $x \in G/\Gamma$. We set*

$$u_t = \begin{pmatrix} 1 & t \\ 0 & 1 \end{pmatrix} \quad \text{and} \quad u = u_1 = \begin{pmatrix} 1 & 1 \\ 0 & 1 \end{pmatrix}.$$

Suppose that the orbit $\{u_t x \mid t \in \mathbb{R}\}$ *is not periodic, i.e., that* $u_t x \neq x$ *for every* $t \neq 0$. *Then*

(i) *the orbit* $\{u_t x \mid t \in \mathbb{R}\}$ *is uniformly distributed in* G/Γ, *i.e., for any bounded continuous function* f *on* G/Γ

$$\lim_{T\to\infty} \frac{1}{T}\int_0^T f(u_t x)\,dt = \int_{G/\Gamma} f\,d\mu_G;$$

(ii) *the sequence* $\{u^n x \mid n \in \mathbb{Z}\}$ *is uniformly distributed in* G/Γ, *i.e., for any bounded continuous function* f *on* G/Γ

$$\lim_{N\to\infty} \frac{1}{N}\sum_{n=0}^{N-1} f(u^n x) = \int_{G/\Gamma} f\,d\mu_G.$$

1.10

For $t \in \mathbb{R}$, let $[t]$ denote the largest integer x with $x \leq t$ and let

$$\{t\} = t - [t].$$

For $m, n \in \mathbb{N}^+$, let (m, n) denote g.c.d. of m and n. The following theorem is proved in [4] with the help of Theorem 4(ii).

Theorem 5. *For any irrational* θ,

$$\lim_{T\to\infty} \frac{1}{T} \sum_{\substack{0 < m \leq T\{m\theta\} \\ (m, [m\theta]) = 1}} \{m\theta\}^{-1} = \frac{1}{\zeta(2)} = \frac{6}{\pi^2},$$

where ζ *denotes Riemann zeta-function.*

1.11

Suppose that G is a locally compact group acting continuously on a locally compact space X. We say that a subgroup H of G *has property (D) with respect to* X if for every H-invariant locally finite Borel measure μ on X, there exist Borel subsets $X_i \subset X$, $i \in \mathbb{N}^+$, such that

1) $\mu(X_i) < \infty$

2) $HX_i = X_i$

3) $X = \bigcup X_i$.

If H has property (D) with respect to X, then, as one can easily see, every H-ergodic, H-invariant locally finite Borel measure on X is finite. (When G and X are separable and metrizable, the converse is also true.)

The following theorem should be mentioned in connection with Conjecture 3.

Theorem 6 (see [5]). *Let G be a connected Lie group and let Γ be a lattice in G. Then any unipotent subgroup U of G has property (D) with respect to G/Γ.*

Using results of Moore on the Mautner phenomenon (see [12]), one can deduce from Theorem 6 the following

Theorem 7. *Let G be a connected Lie group, Γ be a lattice in G, and H be a connected subgroup of G. Suppose that the quotient of H by its unipotent radical is semisimple. Then H has property (D) with respect to G/Γ.*

(By the unipotent radical of H we mean, as usual, the greatest connected normal unipotent subgroup of H.)

The following is a special case of Theorem 7.

Theorem 8. *Let G, Γ, and H be the same as in Theorem 7 and let $x \in G/\Gamma$. Suppose that the orbit Hx is closed in G/Γ. Then $H \cap G_x$ is a lattice in H, where $G_x = \{g \in G | gx = x\}$ is the stabilizer of x.*

Applying Theorem 7 to the case where $G = \mathrm{SL}(n, \mathbb{R})$, $\Gamma = \mathrm{SL}(n, \mathbb{Z})$, $x = e\Gamma$, and $H \subset G$ is the set of $\mathbb{R}$-rational points of a $\mathbb{Q}$-subgroup of G, we obtain the following theorem of Borel and Harish-Chandra.

Theorem 9. *Let H be a connected $\mathbb{Q}$-group. Suppose that the quotient of H by its unipotent radical is semisimple. Then $H(\mathbb{Z})$ is a lattice in $H(\mathbb{R})$.*

2. Proof of Theorem 2

In this and following sections we denote by $\bar{A}$ the closure of a subset A of a topological space, by $\mathbb{N}_G(F)$ the normalizer of a subgroup F in a group G, and by e the identity element of G.

The proof of Theorem 2 itself will be given in 2.5. Before this, in 2.1, some simple lemmas on actions of groups on homogeneous spaces will be formulated and proved. In 2.2 we shall formulate lemmas about closures of subsets of the form $P'MP$, where P' and P are subgroups of a real algebraic group G and M is a subset of G with $e \in M$. These lemmas will be proved in Section 3. In 2.3 and 2.4 some more lemmas used in the proof of Theorem 2 will be proved.

2.1

In Lemmas 1-4, G is an arbitrary second countable locally compact group and Ω is a homogeneous space of G. As in Section 1, $G_y = \{g \in G | gy = y\}$ denotes the stabilizer of $y \in \Omega$.

Lemma 1. *Let F, P, and $P'(F \subset P, F \subset P')$ be closed subgroups of G, and let Y and Y' be closed subsets of Ω, and let $M \subset G$. Suppose that*

(i) $PY = Y$ *and* $P'Y' = Y'$

(ii) $mY \cap Y' \neq \emptyset$ *for any* $m \in M$

(iii) *Y is a compact minimal F-invariant subset (minimality means that Fy is dense in Y for any $y \in Y$).*

Then $hY \subset Y'$ for any $h \in \mathbb{N}_G(F) \cap P'MP$.

Proof. Set $S = \{g \in G | gY \cap Y' \neq \emptyset\}$. Conditions (i) and (ii) imply that $S \supset P'MP$. On the other hand, since Y is compact and Y' is closed, the set S is closed in G. So $S \supset \overline{P'MP}$. Hence, $hY \cap Y' \neq \emptyset$, i.e., $hy = y'$ for some $y \in Y$ and $y' \in Y'$. But $h \in \mathcal{N}_G(F)$, $FY = Y$, and $FY' = Y'$. Therefore,

$$hgy = (hgh^{-1})hy \in Fy' \subset FY' = Y'$$

for any $g \in F$. In other words, $hFy \subset Y'$. But the orbit Fy is dense in Y, and Y' is closed. So $hY \subset Y'$.

Lemma 2. *Let F be a closed subgroup of G, let $Y \subset \Omega$ be a closed F-invariant subset, and let $g \in \mathcal{N}_G(F)$. If $gY \cap Y \neq \emptyset$, then $gY = Y$.*

Proof. Since $g \in \mathcal{N}_G(F)$ and $FY = Y$, we have $FgY = gY$, and consequently $F(gY \cap Y) = gY \cap Y$. But $gY \cap Y \neq \emptyset$ and Y is a minimal closed F-invariant subset. So $gY \cap Y = Y$. Hence, $gY = Y$.

Lemma 3. *Let F and P $(F \subset P)$ be closed subgroups of G and let $Y \subset \Omega$ and $M \subset G$. Suppose that* (i) $PY = Y$; (ii) $mY \cap Y \neq \varnothing$ *for any* $m \in M$; (iii) *Y is a compact minimal F-invariant subset. Then $hY = Y$ for any $h \in \mathcal{N}_G(F) \cap \overline{PMP}$.*

Lemma 4. *Let F be closed subgroup of G and let $y \in \Omega$. Suppose that the quotient space $F/F \cap G_y$ is not compact and that Fy is a compact minimal F-invariant subset. Then the closure of the subset $\{g \in G - F | gy \in Fy\}$ contains e.*

Proof. Suppose the contrary. Then there exists a relatively compact neighborhood $Y \subset G$ of the identity such that $Uy \cap Fy = (U \cap F)y$. Let us represent F as the union of an increasing sequence of compact subsets K_n, $n \in \mathbb{N}^+$. Since $F/F \cap G_y$ is not compact, for any $n \in \mathbb{N}^+$ there exists $z_n \in Fy$ such that $K_n z_n \cap (U \cap F)y = \varnothing$. But $Uy \cap Fy = (U \cap F)y$ and $K_n z_n \subset Fy$. So

$$K_n z_n \cap Uy = \varnothing. \tag{1}$$

Since $\overline{Fy}$ is compact, we can assume (replacing $\{z_n\}$ by a subsequence) that the sequence $\{z_n\}$ tends to $z \in \overline{Fy}$ as $n \to \infty$. Now Uy is open in Ω, $K_n \subset K_{n+1}$, and $F = \bigcup_{n \in \mathbb{N}^+} K_n$. Thus (1) implies that $\overline{Fz} \cap Uy = \varnothing$. In view of the inclusion $z \in \overline{Fy}$, this contradicts the fact that $\overline{Fy}$ is a minimal closed F-invariant subset.

2.2

We define *a real algebraic group* to be a group of $\mathbb{R}$-rational points of an $\mathbb{R}$-group or, in other words, an algebraic subgroup of $\mathrm{GL}(n, \mathbb{R})$. A subgroup U of a real algebraic group G will be called *unipotent* if U consists of unipotent matrices. (This definition of a unipotent subgroup differs slightly from the definition given in 1.4. These two definitions are equivalent iff the kernel of the adjoint representation of G consists of unipotent matrices.) Recall that a unipotent subgroup of a real algebraic group is connected iff it is algebraic.

Lemma 5. *Let G be a real algebraic group, U be a connected unipotent subgroup of G, and $M \subset G$. Suppose that $e \in \overline{M} - M$ and $M \subset G - \mathcal{N}_G(U)$. Then the connected component of e in the set $\mathcal{N}_G(U) \cap \overline{UMU}$ differs from U.*

For formulation of following two lemmas, we have to introduce notation for some elements and subsets of SL(3, $\mathbb{R}$). Let

$$d(t) = \begin{pmatrix} t & 0 & 0 \\ 0 & 1 & 0 \\ 0 & 0 & t^{-1} \end{pmatrix}, \qquad v_1(t) = \begin{pmatrix} 1 & t & t^2/2 \\ 0 & 1 & t \\ 0 & 0 & 1 \end{pmatrix} = \exp t \begin{pmatrix} 0 & 1 & 0 \\ 0 & 0 & 1 \\ 0 & 0 & 0 \end{pmatrix},$$

$$v_2(t) = \begin{pmatrix} 1 & 0 & t \\ 0 & 1 & 0 \\ 0 & 0 & 1 \end{pmatrix} = \exp t \begin{pmatrix} 0 & 0 & 1 \\ 0 & 0 & 0 \\ 0 & 0 & 0 \end{pmatrix}.$$

Let $D = \{d(t) | t > 0\}$, $V_1 = \{v_1(t) | t \in \mathbb{R}\}$, $V_2 = \{v_2(t) | t \in \mathbb{R}\}$, $V_2^+ = \{v_2(t) | t > 0\}$, and $V_2^- = \{v_2(t) | t < 0\}$. Further we set

$$V = V_1 \cdot V_2 = \left\{ \begin{pmatrix} 1 & v_1 & v_2 \\ 0 & 1 & v_1 \\ 0 & 0 & 1 \end{pmatrix} \middle| \, v_1, v_2 \in \mathbb{R} \right\}$$

and denote by $W \subset \mathrm{SL}(3, \mathbb{R})$ the group of unipotent upper triangular matrices. As in the formulation of Theorem 2, let H denote the group of elements of SL(3, $\mathbb{R}$) preserving the form $2x_1x_3 - x_2^2$. It is clear that D normalizes each of subgroups V_1, V_2, V, and W, and that V is commutative. Let us also note that $D \subset H$ and $V_1 = H \cap W$.

Lemma 6. *Let $G = \mathrm{SL}(3, \mathbb{R})$. If $M \subset G - \mathcal{N}_G(V)$ and $e \in \overline{M}$, then $W \cap \overline{VMV}$ generates W.*

Lemma 7. *Let $G = \mathrm{SL}(3, \mathbb{R})$. If $M \subset G - H$ and $e \in \overline{M}$, then $\overline{HMDV_1}$ contains either V_2^+ or V_2^-.*

2.3

Lemma 8. *Let G be a real algebraic group and $M \subset G$. Suppose that $e \in \overline{M} - M$.*

(i) *If U is a connected unipotent subgroup of G and $M \subset G - U$, then the closure of the subgroup generated by the set $\mathcal{N}_G(U) \cap \overline{UMU}$ contains a one-parameter subgroup that is not contained in U.*

(ii) *Let $G = \mathrm{SL}(3, \mathbb{R})$ and let D, V_1, and V be as in 2.2. If $M \subset G - V_1$, then the closure of the subgroup generated by the set $\mathcal{N}_G(V_1) \cap \overline{V_1MV_1}$ contains either V or a subgroup of the form vDV_1v^{-1}, where $v \in V$.*

Proof. (i) Let us denote by Ψ the closure of the subgroup generated by $\mathcal{N}_G(U) \cap \overline{UMU}$ and by Ψ° the connected component of the identity of the Lie group Ψ. It is enough to prove $\Psi^\circ \neq U$. In view of Lemma 5, one can assume $M \subset \mathcal{N}_G(U)$. Then $\Psi \supset M$. But $e \in \overline{M} - M$. So $\Psi^\circ \cap M \neq \varnothing$ and, since $M \subset G - U$, we have $\Psi^\circ \neq U$.

(ii) One can directly check that the connected component of the identity of the group $\mathcal{N}_G(V_1)$ coincides with DV. This easily implies that, for any one-parameter subgroup $S \subset \mathcal{N}_G(V_1)$, $S \not\subset V_1$, the subgroup SV_1 contains either V or a subgroup of the form vDV_1v^{-1}, where $v \in V$. Now it remains to use the assertion (i).

2.4

From now until the end of Section 2, G, Γ, H, and Ω will be as in the formulation of Theorem 2, and $d(t)$, D, V_1, V_2, V_2^+, V_2^-, V, and W will be as in 2.2.

As in [6], by the *horospherical subgroup* corresponding to an element $g \in G$ we mean the subgroup

$$\{u \in G | g^j u g^{-j} \to e \qquad \text{as } j \to \infty\}.$$

This subgroup will be denoted by U_g. It is well known that U_g is a connected closed unipotent subgroup and g normalizes U_g. If $U_g \neq \{e\}$, F denotes the subgroup generated by U_g and g, and Λ is a lattice in G, then the orbit Fz is dense in G/Λ for any $z \in G/\Lambda$ (see Propositions 1.1 and 1.2 in [6]). However, as it was noted in 1.4, $\Gamma = \mathrm{SL}(3, \mathbb{Z})$ is a nonuniform lattice in $G = \mathrm{SL}(3, \mathbb{R})$ and, as can be easily checked, $W = U_{d(t)}$ for any $0 < t < 1$. So we have

Lemma 9. *For any $y \in \Omega = G/\Gamma$, the orbit DWy is dense in Ω and consequently is not relatively compact in Ω.*

As it was noted, $W = U_{d(t)}$ for any $0 < t < 1$. On the other hand (see [18] Theorem 1.12), if $y \in \Omega = G/\Gamma$, $\gamma \in G_y$, $\gamma \neq e$, $\{g_n | n \in \mathbb{N}^+\} \subset G$, and $g_n \gamma g_n^{-1} \to e$ as $n \to \infty$, then the set $\{g_n y | n \in \mathbb{N}^+\}$ is not relatively compact in Ω. So we have

Lemma 10. *Let $y \in \Omega$. Suppose that the orbit Dy is relatively compact in Ω. Then $W \cap G_y = \{e\}$, and consequently the quotient space $U/U \cap G_y$ is not compact for any nontrivial closed subgroup U of W.*

Lemma 11. *For any $y \in \Omega$, the orbit DVy is not relatively compact in Ω.*

Proof. Suppose the contrary, i.e., that $\overline{DVy}$ is compact. Then the DV-invariant subset $\overline{DVy}$ contains a minimal closed V-invariant subset Y. One can assume that $y \in Y$. Let us denote the subgroup $\{g \in \mathcal{N}_G(V) | gY = Y\}$ by F. As Y is closed, F is closed. Let us denote the connected component of the identity of the group F by F^0. One can easily check that the connected component of the identity of the group $\mathcal{N}_G(V)$ coincides with DW. This easily implies that any unimodular connected closed subgroup of $N_G(V)$ containing V is either W or V. Thus either $F^0 \subset W$ or F^0 is nonunimodular. But a nonunimodular group doesn't contain a lattice (see [18] Remark 1.9). In view of Lemma 10, this implies that the quotient space $F^0/F^0 \cap G_y$ (and consequently the quotient space $F/F \cap G_y$) is noncompact. Now we use Lemma 4 and see that the closure of $\phi \overset{\text{def}}{=} \{g \in G - F | gY \cap Y \neq \emptyset\}$ contains e. On the other hand, it follows from Lemma 2 that $\phi \cap \mathcal{N}_G(V) = \emptyset$. So the closure of the set $M \overset{\text{def}}{=} \{g \in G - \mathcal{N}_G(V) | gY \cap Y \neq \emptyset\}$ contains e. Then in view of Lemma 6, $\overline{VMV} \cap W$ generates W. This fact, Lemma 3, and the inclusion $W \subset \mathcal{N}_G(V)$ imply $WY = Y$. Now in view of Lemma 9, the set $\overline{DVy} \supset \overline{DY} = \overline{DWY}$ is not compact.

Lemma 12. *For any $y \in \Omega$, the sets $DV_1V_2^+ y$ and $DV_1V_2^- y$ are not relatively compact in Ω.*

Proof. If $n \in \mathbb{N}^+$, $t \in \mathbb{R}$, and $|t| \leq n$, then $v_2(t)v_2(n)y \subset V_2^+ y$. This easily implies that if z is a limit point of $\{v_2(n)y | n \in \mathbb{N}^+\}$, then $V_2 z \subset V_2^+ y$, and consequently $DVz = DV_1V_2z \subset \overline{DV_1V_2^+ y}$. But if $V_2^+ y$ is relatively compact in Ω, then the set of limit points of $\{v_2(n)y | n \in \mathbb{N}^+\}$ is not empty. Therefore, in view of Lemma 11, $DV_1V_2^+ y$ is not relatively compact in Ω. The fact that $DV_1V_2^- y$ is not relatively compact can be proved analogously.

2.5. Proof of Theorem 2

Since the set $\overline{Hz}$ is compact and H-invariant, it contains a minimal closed H-invariant set X. Then, since $H \supset V_1$ and X is compact, the set X contains a minimal closed V_1-invariant set Y. Let us choose $y \in Y$. As

$Hy \subset X$ is relatively compact and $D \subset H$, Lemma 10 implies that $V_1/V_1 \cap G_y$ is not compact. Therefore, in view of Lemma 4, the closure of $M_1 \overset{\text{def}}{=} \{g \in G - V_1 | gY \cap Y \neq \varnothing\}$ contains e. Let us denote by Ψ the closure of the subgroup generated by $\mathcal{N}_G(V_1) \cap \overline{V_1 M_1 V_1}$. Lemma 3 implies

$$\Psi Y = Y. \tag{1}$$

As $HX = X$, $Y \subset X$, and $DV_1 \subset H$, we have $DV_1 Y \subset X$. So, in view of (1), $DV_1 \Psi Y \subset X$. Using this fact, the compactness of X and Lemma 12, we obtain

$$DV_1 \Psi \not\supset V_1 V_2^+ \text{ and } DV_1 \Psi \not\supset V_1 V_2^- . \tag{2}$$

According to Lemma 8(ii), Ψ contains either V or $vDV_1 v^{-1}$, where $v \in V$. But if $v \in V - V_1$, one can easily check that $DV_1(vDV_1 v^{-1})$ contains either $DV_1 V_2^+$ or $DV_1 V_2^-$. Therefore, in view of (2), $\Psi \supset DV_1$. This fact and (1) imply

$$DV_1 Y = Y. \tag{3}$$

Set

$$M = \{g \in G - H | gy \in \overline{Hz}\}. \tag{4}$$

Suppose that $e \in \overline{M}$. Then, in view of Lemma 7, $\overline{HMDV_1}$ contains either V_2^+ or V_2^-. On the other hand, in view of (3) and Lemma 1, $gY \subset \overline{Hz}$ for any $g \in \mathcal{N}_G(V_1) \cap \overline{HMDV_1}$. So either $V_2^+ Y \subset \overline{Hz}$ or $V_2^- Y \subset \overline{Hz}$. This fact, the equalities $DV_1 V_2^+ = V_2^+ DV_1$ and $DV_1 V_2^- = V_2^- DV_1$, and the equality (3) imply that $\overline{Hz}$ contains either $DV_1 V_2^+ Y$ or $DV_1 V_2^- Y$. In view of Lemma 12, this contradicts the compactness of Hz. Thus we have shown

$$e \notin \overline{M}. \tag{5}$$

We have $y \in Y \subset X \subset \overline{Hz}$ and X is a minimal closed H-invariant subset. So, in view of Lemma 4, (4) and (5) imply that $H/H \cap G_y$ is compact. On the other hand, as $y \in \overline{Hz}$, then, in view of (4) and (5), $y \in Hz$. So $H/H \cap G_z$ is compact.

Remark. In Theorem 2, one can replace Γ by any nonuniform lattice in $SL(3, \mathbb{R})$. The proof goes in exactly the same way.

3. Proofs of Lemmas 5–7

3.1

In the proofs of Lemmas 5–7, we shall need some assertions about unipotent groups of linear transformations. These are contained in the following lemma. We recall beforehand that all orbits are closed for linear actions of connected unipotent groups (see [19]).

Lemma 13. *Let U be a connected unipotent group of linear transformations of $\mathbb{R}^n$ and let $Y \subset \mathbb{R}^n$. Let $L = \{x \in \mathbb{R}^n \mid Ux = x\}$.*

(i) *If $p \in L \cap Y$ and $L \cap Y = \varnothing$, then the connected component of p in the set $\overline{UY} \cap L$ is not compact.*

(ii) *Let $X \subset \mathbb{R}^n$ be a closed subset. Suppose that for any $x \in (\overline{UY} \cap X) - L$, the connected component of x in the set $Ux \cap X$ is not compact. If $p \in L \cap \overline{Y \cap X}$ and $L \cap Y = \varnothing$, then the connected component of p in the set $\overline{UY} \cap X \cap L$ is not compact.*

Proof. The connected unipotent group U is isomorphic to its Lie algebra $\mathfrak{u}$ as an algebraic manifold (this isomorphism can be realized by the logarithmical mapping $\ln: U \to \mathfrak{u}$). On the other hand, (a) for any $x \in \mathbb{R}^n$, the coordinates of ux are regular functions of $u \in U$; (b) the set of values of any nonconstant regular function on a linear space is noncompact. Therefore, for any $x \in \mathbb{R}^n - L$, the connected set Ux is not compact. Hence, (ii) implies (i).

Let us prove (ii) using induction on n. Since U is unipotent, the Lie–Kolchin theorem implies that there exists a U-invariant linear subspace S in $\mathbb{R}^n$ of codimension 1 such that $S \supset L$ and U acts on R^n/S trivially. Further, let us choose a sequence $\{y_i \in Y \cap X \mid i \in \mathbb{N}^+\}$ tending to p. Let us denote by Ψ the upper topological limit of sets $Uy_i \cap X$, i.e., Ψ is the set of limit points of all sequences of the form $\{z_i \in Uy_i \cap X \mid i \in \mathbb{N}^+\}$. Since $p \in L \subset S$, $y_i \to p$ as $i \to \infty$ and U acts on R^n/S trivially, we have $\Psi \subset S$. On the other hand, since $y_i \to p$ as $i \to \infty$ and the connected component of y_i in the set $Uy_i \cap X$ is not compact for any $i \in \mathbb{N}^+$, it follows that the connected component of p in the set Ψ is not compact. So, in view of the inclusion $\Psi \subset \overline{UY} \cap X$, the connected component of p in the set $\overline{UY} \cap X \cap S$ is not compact. Then either the connected component of p in the set $\overline{UY} \cap X \cap L$ is not compact or contains a point q that belongs to the closure of the set $(\overline{UY} \cap X \cap S) - L$. But $\overline{UA} \cap X \cap S \subset \overline{UY} \cap X \cap S$ for any $A \subset \overline{UY}$. So we can replace X by $X \cap S$ and Y by $\overline{UY} \cap X \cap S - L$ and assume that $X \subset S$ and $Y \subset S$. The desired induction step follows easily from this.

3.2

Proof of Lemma 5. (i) Let $\pi: G \to G/U$ be the natural projection and $T = \{x \in G/U \mid Ux = x\}$. Since $\mathcal{N}_G(U) = \{g \in G \mid UgU = gU\}$, we have $\pi(N_G(U)) = T$. So the desired assertion follows from

(*) the connected component of $\pi(e)$ in the set $T \cap \overline{U\pi(M)}$ is not compact and consequently differs from $\pi(e)$.

The connected unipotent subgroup U is algebraic and has no rational characters. So, in view of Chevalley's theorem (see [2] Theorem 5.1), there exists $m \in \mathbb{N}^+$, a faithful rational representation $\alpha: G \to \mathrm{GL}(m, \mathbb{R})$ and $x_0 \in \mathbb{R}^m$ such that $U = \{g \in G \mid \alpha(g)x_0 = x_0\}$. According to the lemma on orbit closures (see [2] Proposition 1.8), $\alpha(G)x_0$ is a smooth manifold that is open in its closure. This implies that the map $gU \to \alpha(g)x_0$ is a homomeomorphism of G/U onto $\alpha(G)x_0$. Now to prove (*), it remains to apply Lemma 13(i) for $Y = \alpha(M)x_0$ and $p = x_0$. (We should only note that (1) since $e \in \overline{M}$, $M \subset G - \mathcal{N}_G(U)$, and $T = \pi(\mathcal{N}_G(U))$, we have $\pi(e) \in T \cap \overline{\pi(M)}$ and $T \cap \pi(M) = \varnothing$; (2) since U is connected and unipotent and the representation α is rational, $\alpha(U)$ is connected and unipotent.)

3.3

Proof of Lemma 6. Let us denote by $a_{ij}(g)$ the coefficient of a matrix g in the intersection of i-th row and j-th column. Set $W^+ = \{w \in W \mid a_{12}(w) \geq a_{23}(w)$ and $W^- = \{w \in W \mid a_{12}(w) \leq a_{23}(w)\}$. One can easily see that each of the sets W^+ and W^- generates W. So it is enough to prove

$$\text{(A1)} \quad \overline{VMV} \text{ contains either } W^+ \text{ or } W^-.$$

The Lie algebra $\mathfrak{g}$ of G is naturally identified with the space of real matrices of order 3 having trace zero. Let E_{ij} denote the matrix such that $a_{ij}(E_{ij}) = 1$ and $a_{kl}(E_{ij}) = 0$ if $k \neq i$ or $l \neq j$. Set $x_0 = E_{12} + E_{23} \in \mathfrak{g}$. Let us denote by Ad the adjoint representation of G and note that $(\mathrm{Ad}\, g)x = gxg^{-1}$ for any $g \in G$ and $x \in \mathfrak{g}$. According to the lemma on orbit closures, $(\mathrm{Ad}\, G)x_0$ is a smooth manifold that is open in its closure. This implies that the map sending gV onto $(\mathrm{Ad}\, g)x_0$ is a homeomorphism of G/V onto $(\mathrm{Ad}\, G)x_0$. So (A1) is equivalent to

$$\text{(A2)} \quad \overline{(\mathrm{Ad}\ V)(\mathrm{Ad}\ M)x_0} \text{ contains either } (\mathrm{Ad}\ W^+)x_0 \text{ or } (\mathrm{Ad}\ W^-)x_0.$$

Let $\mathfrak{v}$, $\mathfrak{v}_1$, and $\mathfrak{v}_2$ denote the Lie algebras of V, V_1, and V_2, respectively. Note that $\mathfrak{v}_1 = \{tx_0 | t \in \mathbb{R}\}$, $\mathfrak{v}_2 = \{tE_{13} | t \in \mathbb{R}\}$, and $\mathfrak{v} = \mathfrak{v}_1 + \mathfrak{v}_2$. Set $\mathfrak{v}_2^+ = \{tE_{13} | t > 0\}$ and $\mathfrak{v}_2^- = \{tE_{13} | t < 0\}$. Direct calculations show that $(\mathrm{Ad}\, W^+)x_0 = x_0 + \mathfrak{v}_2^+$ and $(\mathrm{Ad}\, W^-)x_0 = x_0 + \mathfrak{v}_2^-$: Further, one can easily check that $\mathcal{N}_G(V) = \{g \in G | \mathrm{Ad}\, g)x_0 \in \mathfrak{v}\}$. But $M \subset G - \mathcal{N}_G(V)$. So $\mathfrak{v} \cap (\mathrm{Ad}\, M)x_0 = \varnothing$. Since $e \in \overline{M}$, we have $x_0 \in \overline{(\mathrm{Ad}\, M)x_0}$. In view of aforesaid, (A2) is a particular case of

(B1) If $Y \subset (\mathrm{Ad}\, G)x_0$, $x_0 \in \overline{Y}$, and $\mathfrak{v} \cap Y = \varnothing$, then $\overline{(\mathrm{Ad}\, V)Y}$ contains either $x_0 + \mathfrak{v}_2^+$ or $x_0 + \mathfrak{v}_2^-$.

Set $X = \{x \in \mathfrak{g} | a_{12}(x) = 1\}$ and denote by N the set of nilpotent elements of $\mathfrak{g}$. We have $x_0 \in X \cap N$, $(\mathrm{Ad}\, G)x_0 \subset N$, and $y/a_{12}(y) \in X \cap N$ if $y \in N$ and $a_{12}(y) \neq 0$. So (B1) can be easily deduced from

(B2) If $Y \subset X \cap N$, $x_0 \in Y$, and $\mathfrak{v} \cap Y = \varnothing$, then $\overline{(\mathrm{Ad}\, V)Y}$ contains either $x_0 + \mathfrak{v}_2^+$ or $x_0 + \mathfrak{v}_2^-$.

Since N is closed and invariant under Ad G, we have $\overline{(\mathrm{Ad}\, V)Y} \subset N$ for any $Y \subset N$. On the other hand, one can easily check that

(a) $X \cap \mathfrak{v} = x_0 + \mathfrak{v}_2$, and if $S \subset x_0 + \mathfrak{v}_2$ is a connected closed noncompact subset and $x_0 \in S$, then either $S \supset x_0 + \mathfrak{v}_2^+$ or $S \supset x_0 + \mathfrak{v}_2^-$
(b) $\mathfrak{v} = \{x \in \mathfrak{g} | (\mathrm{Ad}\, V)x = x\}$
(c) the group V, and consequently the group Ad V, are connected and unipotent.

Therefore, in view of Lemma 13(ii), (B2) follows from

(C) For any $x \in (X \cap N) - \mathfrak{v}$, the connected component of x in the set $(\mathrm{Ad}\, V)x \cap X$ is not compact.

Let us denote by $\mathfrak{t} \subset \mathfrak{g}$ the space of upper triangular matrices with trace zero. If $x \in (\mathfrak{t} \cap X \cap N) - \mathfrak{v}$, then $x = x_0 + sE_{23} + tE_{13}$, where $s \neq 0$. In this case, direct calculations show that $(\mathrm{Ad}\, V)x = x + \mathfrak{v}_2 \subset X$. So one can assume that $x \notin \mathfrak{t}$. One can easily check that the isotropy subgroup $V_y = \{v \in V | \mathrm{Ad}\, v)y = y\}$ is trivial for any $y \in g - \mathfrak{t}$. In particular, $V_x = \{e\}$. On the other hand, since (see [19]) all orbits are closed for linear actions of connected unipotent groups, the set $(\mathrm{Ad}\, V)x$ is closed. So the orbit mapping $v \to (\mathrm{Ad}\, v)x$, $v \in V$, is proper. Now to prove (C), it remains to show that the connected component of x in the set

$$\phi \overset{\mathrm{def}}{=} \{v \in V | a_{12}(vxv^{-1}) = 1\}$$

is not compact. A direct calculation shows that the polynomial

$$P(s,t) \overset{\text{def}}{=} a_{12}\left(\begin{pmatrix}1 & s & t\\ 0 & 1 & s\\ 0 & 0 & 1\end{pmatrix} \times \begin{pmatrix}1 & s & t\\ 0 & 1 & s\\ 0 & 0 & 1\end{pmatrix}^{-1}\right)$$

is linear with respect to t for any s. But $P(0,0) = 1$. So either $P(0,t) = 1$ for any $t \in \mathbb{R}$ or, for all but a finite number of $s \in \mathbb{R}$, there exists $t(s) \in \mathbb{R}$ such that $P(s, t(s)) = 1$. Hence, ϕ is not compact.

3.4

Proof of Lemma 7. Let $a_{ij}(g)$, $\mathfrak{g}$, Ad, $\mathfrak{v}$, $\mathfrak{v}_1$, $\mathfrak{v}_2$, $\mathfrak{v}_2^+$, and $\mathfrak{v}_2^-$ be as in the proof of Lemma 6. Further, let us denote by $\mathfrak{h} \subset \mathfrak{g}$ the Lie algebra of H and by $P \subset \mathfrak{g}$ the orthogonal complement to $\mathfrak{h}$ with respect to Killing form. One can easily check

$$P = \{x \in \mathfrak{g} \mid a_{11}(x) = a_{33}(x),\ a_{12}(x) = -a_{23}(x), \quad \text{and} \quad a_{21}(x) = -a_{32}(x)\}. \tag{1}$$

We have $\mathfrak{h} + P = \mathfrak{g}$ and $\mathfrak{h} \cap P = 0$. There exists, therefore, a neighborhood A of e in H (resp. B of 0 in P) such that $(a,b) \to a \exp b$ is a homeomorphism of $A \times B$ onto a neighborhood of e in G. On the other hand, the conclusion is not altered if we replace M by a set of elements $h(m) \cdot m (m \in M,\ h(m) \in H)$. It follows, therefore, that it suffices to consider the case where $M \subset \exp B$. Let $M_0 = \log M$. Let us consider two cases:

$$\text{(a)}\quad M \cap V_2 = \varnothing;\ \text{(b)}\ M \cap V_2 \neq \varnothing.$$

(i) Set $M_0 = \{m - e \mid m \in M\}$. Since $e \in \overline{M}$ and $e \notin M$, we have

$$0 \in \overline{M_0} \qquad \text{and} \qquad 0 \notin M_0. \tag{2}$$

It is immediately checked that $\mathfrak{v} = \{x \in \mathfrak{g} \mid (\mathrm{Ad}\ V_1)x = x\}$. But $\mathfrak{v}_2 = \mathfrak{v} \cap P$. Thus

$$\mathfrak{v}_2 = \{x \in P \mid (\mathrm{Ad}\ V_1)x = x\}. \tag{3}$$

The subalgebra $\mathfrak{h}$ and the Killing form on $\mathfrak{g}$ are invariant under Ad H. Therefore, P is invariant under Ad H. In particular, $(\mathrm{Ad}\ V_1)P = P$. Now applying Lemma 13(i) and using (2), (3), and the equality $M \cap V_2 = \varnothing$, we see that the connected component of zero in the set $\overline{(\mathrm{Ad}\ V_1)M_0} \cap \mathfrak{v}_2$ is not compact. Consequently, $\overline{(\mathrm{Ad}\ V_1)M_0}$ contains either $\mathfrak{v}_2^+$ or $\mathfrak{v}_2^-$. However, $e + \mathfrak{v}_2^+ = V_2^+$, $e + \mathfrak{v}_2^- = V_2^-$ and

$$HMDV_1 \supset V_1MV_1 \supset \bigcup_{v \in V_1} (vMv^{-1}) = e + (\mathrm{Ad}\ V_1)M_0.$$

So $\overline{HMDV_1}$ contains either V_2^+ or V_2^-.

(ii) Let $v_2(t_0) \in M \cap V_2$. Then

$$HMDV_1 \supset DMD \supset Dv_2(t_0)D \supset \{d(t)v_2(t_0)\, d(t)^{-1} | t > 0\}$$
$$= \{v_2(t^2 t_0) | t > 0\}.$$

But $t_0 \neq 0$ (because $e \notin M$). Thus $HMDV_1$ contains either V_2^+ or V_2^-.

4. Proof of Theorem 1′

As stated in the introduction, we prove here the strengthened Oppenheim conjecture [15]:

Theorem 1′. *Let B be a real nondegenerate indefinite quadratic form in n variables. Suppose that $N \geq 3$ and that B is not proportional to a form with rational coefficients. Then, for any $\varepsilon > 0$, there exist integers $x_1, \ldots, x_n$ such that*

$$0 < |B(x_1, \ldots, x_n)| < \varepsilon.$$

As in the case of Theorem 1, the proof will be a reduction to Theorem 2. To carry it out, we need some lemmas, contained in the next section.

4.1

We use notation introduced in Sections 1.3 and 2.2. For $y \in \Omega = G/\Gamma$, let us denote by $\phi(y) \subset \Omega$ the set of limits of all sequences of the form $\{v_1(t_k) \cdot y | k \in \mathbb{N}^+, t_k > 0\}$, where $t_k \to +\infty$ as $k \to +\infty$.

Lemma A. *Let $z \in \Omega$ and $y \in \Omega$. Suppose that the following conditions are satisfied*: (i) *the orbit Hz is compact (equivalently, the quotient space $H/H \cap G_z$ is compact)*; (ii) *$y \notin Hz$ and the orbit $V_1 y$ is relatively compact in Ω. Then $\phi(y) \not\subset Hz$.*

Proof. Suppose the contrary. Then since $V_1 y$ is relatively compact in Ω, we have that $v_1(t)y$ tends to the compact orbit Hz as $t \to +\infty$ (it means that the distance from $v_1(t)y$ to Hz tends to 0 as $t \to \infty$). This implies the existence of g, $g_t \in G$ and $h_t \in H$ such that

$$y = g \cdot z, \qquad v_1(t) \cdot g \cdot z = g_t \cdot h_t \cdot z, \qquad \text{with } g_t \to e \text{ as } t \to \infty.$$

There exists therefore $u_t \in G_z$ such that

$$v_1(t) \cdot g = g_t \cdot h_t \cdot u_t. \tag{1}$$

Since $H \cap G_z$ is cocompact in H, there exists a neighborhood U_0 of e in G such that $HU_0H \cap G_z \subset H \cap G_z$ (cf. Lemma 1.16 in [18]). From this follows the existence of a neighborhood U of e in G such that we have $UHu \cap UHv \neq \varnothing$ $(u, v \in G_z)$ if and only if $v \cdot u^{-1} \in H$. Since $V_1 \cdot g$ is connected, (1) then shows that if $g_t \in U$, there exists a constant $t(U)$ such that the element u_t may be assumed to be constant for $t \geq t(U)$. Replacing g by $g \cdot u_t^{-1}$, we have then $v_1(t) \cdot gH = g_t \cdot H$ for $t \geq t(U)$. Letting U run through a fundamental set of neighborhoods of e, we see that there exists a sequence $t \to +\infty$, and elements $g_t \to e$ such that $v_1(t) \cdot g \cdot H = g_t \cdot H$. Then $v_1(t)gH$ tends to eH in G/H as $t \to +\infty$. Thus we have found in G/H points $p = gH$ and $q = eH$ such that $p \neq q$ and $v_1(t)p \to q$ as $t \to +\infty$. However, it is impossible because the manifold G/H is affine and the group $V_1 = \{v_1(t) | t \in \mathbb{R}\}$ is unipotent.

Lemma B. *Let $z = g\mathbb{Z}^3 \in \Omega$, $g \in G$. Then the following conditions are equivalent*: (a) *the orbit Hz is compact*; (b) *there exists a rational form B such that B doesn't represent* 0 *over $\mathbb{Q}$ and $H = gH_Bg^{-1}$, where H_B denotes the group of elements of G preserving B.*

Proof. The implication (a) $\Rightarrow$ (b) was in fact proved in the process of the reduction of Theorem 1 to Theorem 2. The implication (b) $\Rightarrow$ (a) is a special case of the theorem on the compactness of the quotients by arithmetic subgroups of $\mathbb{Q}$-anisotropic groups.

Lemma C. *Let $y \in \Omega$. Suppose that the orbit V_1y is relatively compact in Ω and that the orbit Hz is compact for any $z \in \phi(y)$. Then the orbit Hy is compact.*

Proof. $\phi(y)$ is obviously compact. We note next that it is connected. In fact, if this is not the case, $\phi(y)$ is the union of two non-empty disjoint closed and open subsets ϕ_1, ϕ_2, and there exists a continuous function with compact support equal to 1 on ϕ_1 and to 0 on ϕ_2. This is then easily seen to contradict the fact that $V_1 \cdot y$ is connected.

Lemma B implies that the set of compact orbits of H in Ω is countable. However, Hz is compact for any $z \in \phi(y)$. Since a compact connected set is not the union of at least two, and at most countably many, non-empty disjoint closed subsets, it follows that $\phi(y) \subset H \cdot z$ for some $z \in \phi(y)$. Now there remains to use Lemma A.

4.2

Proof of Theorem 1′. As in the proof of Theorem 1, one can assume that $n = 3$. Let H_B and g_B denote the same as in the reduction of Theorem 1 to Theorem 2.

Let us set $S = \{x \in \mathbb{R}^3 \mid B(x) = 0\}$, $S_{\mathbb{Q}} = S \cap \mathbb{Q}^3$, and $S_{\mathbb{Z}} = S \cap \mathbb{Z}^3$. Since B is not proportional to a rational form, the set $S_{\mathbb{Q}}$ is not Zariski dense in the two-dimensional cone S. But $S_{\mathbb{Q}}$ is invariant under multiplications by rationals. Thus $S_{\mathbb{Q}}$ belongs to the union of a finite set of lines. Then after an integral change of variables, we can assume that $g_B S_{\mathbb{Q}} - \{0\}$ does not intersect the coordinate axes. In this case

$$\lim_{t \to +\infty} (\inf\{\|v_1(t)x\| \mid x \in g_B S_{\mathbb{Z}}, \quad x \neq 0\}) = \infty, \tag{1}$$

where $\| \ \|$ denotes the Euclidean norm.

Suppose now that the assertion of Theorem 1′ is not true, i.e., that $|B(x)| > \varepsilon$ for some $\varepsilon > 0$ and all $x \in \mathbb{Z}^3 - S_{\mathbb{Z}}$. Then since H_B preserves B and $H = g_B H_B g_B^{-1}$, we have

$$\inf\{\|hx\| \mid x \in g_B(\mathbb{Z}^3 - S_{\mathbb{Z}}), h \in H\} > 0. \tag{2}$$

It follows from the Mahler compactness criterion, (1), and (2) that the set $V_1 g_B \mathbb{Z}^3$ is relatively compact in Ω and that

$$\inf\{\|hx\| \mid h \in H, x \in z, x \neq 0, z \in \phi(g_B \mathbb{Z}^3)\} > 0. \tag{3}$$

In view of the Mahler compactness criterion, (3) implies that, for any $z \in \phi(g_B \mathbb{Z}^3)$, the set Hz is relatively compact in Ω and consequently (see Theorem 2) is compact. Now applying Lemma C, we get that the orbit $Hg_B\mathbb{Z}^3$ is compact. This and Lemma B imply that B is proportional to a rational form. Contradiction.

References

[1] Borel, A. "Density properties for certain subgroups of semisimple groups without compact components," *Ann. Math.* **72** (1960), 179–188.

[2] Borel, A. *Linear Algebraic Groups*. Benjamin, New York (1969).

[3] Dani, S.G. "Invariant measures and minimal sets of horospherical flows," *Invent. Math.* **64** (1981), 357–385.

[4] Dani, S.G. "On uniformly distributed orbits of certain horocycle flows," *Ergod. Th. and Dynam. Syst.* **2** (1981), 139–158.

[5] Dani, S.G. "On orbits of unipotent flows on homogeneous spaces," *Ergod. Th. and Dynam. Syst.* **4** (1984), 25–34.

[6] Dani, S.G. "Orbits of horospherical flows," *Duke Math. J.* **53** (1986), 177–188.

[7] Dani, S.G. and Smillie, J. "Uniform distribution of horocycle flows for Fuchsian groups," *Duke Math. J.* **51** (1984), 185–194.

[8] Davenport, H. and Heilbronn H. "On indefinite quadratic forms in five variables," *J. London Math. Soc.*, Ser. II **21** (1946), 185–193.

[9] Davenport, H., Ridout H. "Indefinite quadratic forms," *Proc. London Math. Soc.*, Ser. III **9** (1959), 544–555.

[10] Margulis, G.A. "Formes quadratiques indéfinies et flots unipotents sur les espaces homogènes," *C.R. Acad. Sci., Paris*, Ser. I **304** (1987), 249–253.

[11] Margulis, G.A. "Indefinite quadratic forms and unipotent flows on homogeneous spaces." In: *Semester on Dynamical Systems and Ergodic Theory*; *Warsaw* 1986, Banach Center Publications, to appear.

[12] Moore, C.C. "The Mautner phenomenon for general unitary representations," *Pac. J. Math.* **86** (1980), 155–169.

[13] Oppenheim, A. "The minimum of indefinite quaternary quadratic forms of signature zero," *Proc. Nat. Acad. Sci. USA* **15** (1929), 724–727.

[14] Oppenheim, A. "The minima of indefinite quaternary quadratic forms," *Ann. Math.* **32** (1931), 271–298.

[15] Oppenheim, A. "Values of quadratic forms, I," *Quarterly Jour. Math.* (2) **4** (1953), 54–59.

[16] Oppenheim, A. "Values of quadratic forms, II," *Quarterly Jour. Math.* **2** (4) (1953), 60–66.

[17] Oppenheim, A. "Values of quadratic forms, III," *Monatshefte f. Math.* (N.F.) **57** (1954), 97–101.

[18] Raghunathan, M.S. "Discrete subgroups of Lie groups," Springer, Berlin, Heidelberg, New York (1972).

[19] Rosenlicht, M. "On quotient varieties and the affine embeddings of certain homogeneous spaces," *Trans. Amer. Math. Soc.* **101** (1961), 211–223.

22 Good Rational Approximation Derived from Thue's Inequality

*JULIA MUELLER**

1. Introduction

It was proved independently by Dyson and by Gelfond (see [1]) that when α is algebraic of degree $r \geq 3$ and $\mu > \sqrt{2r}$ then there are only finitely many rationals x/y with $(x, y) = 1$ and

$$\left|\alpha - \frac{x}{y}\right| < \frac{1}{|y|^{\mu}}. \tag{1.1}$$

This improves on earlier estimates due to Liouville, to Thue, and to Siegel and is in turn superseded by work of Roth, who showed this to be true for any $\mu > 2$. However, the exponent $\sqrt{2r}$ deserves attention, since the proof of the above assertions for $\mu > \sqrt{2r}$ is accomplished by using auxiliary polynomials in two variables, whereas stronger assertions require auxiliary polynomials in more variables. Hence a rational number x/y that is a solution of (1.1) will be called a *good rational approximation to* α. Moreover, although we are at present unable to bound the size of the denominator y of possible solutions x/y

* Research partially supported by NSF grant DMS-8604568.

ISBN 0-12-067570-6

to (1.1), it may be shown (see, e.g., Bombieri [1]) that when $\mu > \sqrt{2r}$, the denominators y lie in at most two intervals of the following types:

$$1 \leq y \leq (2H)^{c_1(\mu)r}, \tag{1.2}$$

where H is the height of α, i.e., the maximum modulus of the coefficients of the minimal polynomial for α, and

$$B \leq y \leq B^{c_2(\mu)r^2}. \tag{1.3}$$

Unfortunately, we have no information on B, and in fact it may well be that all the solutions satisfy (1.2). We call such a solution x/y "small" or "large" when y satisfies (1.2) or (1.3). Now if $S = \{x_1/y_1, \ldots, x_n/y_n\}$ is a set of solutions in reduced form of (1.1) with $0 < y_1 \leq \cdots \leq y_n$, then we have the easily derived *Gap Principle* that

$$y_i > \frac{1}{2} y_{i-1}^{\mu - 1} \qquad (i = 1, \ldots, n).$$

It is then easy to give a bound for the number of both small and large solutions.

Suppose now we are given a Thue equation

$$|F(x, y)| = h. \tag{1.4}$$

Here F is a form of degree $r \geq 3$, say

$$F(x, y) = a_r x^r + a_{r-1} x^{r-1} y + \cdots + a_0 y^r, \tag{1.5}$$

where the coefficients lie in Z and where F is irreducible over Q. Thue discovered in 1909 that the number of solutions of (1.4) in integers x, y is finite. The first important results on the number of solutions of (1.4) were obtained by C.L. Siegel in the case of a binomial form $F(x, y) = ax^r - by^r$. In 1983, J.-H. Evertse [3] obtained a bound on the number of primitive solutions of (1.4) that depends only on r and h, but is independent of F. Let t be the number of distinct prime factors of h, then Evertse's bound depends exponentially on r and also on t. This result was superseded by Bombieri and Schmidt [2] in 1985. They showed that the number of primitive solutions of (1.4) does not exceed cr^{1+t}. In fact, the absolute constant c can be taken to be 215 if r is sufficiently large.

Siegel [8] had raised a general question that in the special context of Thue equations asks whether the number of solutions of (1.4) may be bounded in terms of h and s only where $s + 1$ is the number of nonzero coefficients of F. In this note, we will refer to this as *Siegel's conjecture*. When F is a binomial form, i.e., $s = 1$, Mueller [5] showed that the number of primitive solutions of $|ax^r - by^r| \leq h$ is at most five provided r is sufficiently large. When F is a trinomial form, i.e., $s = 2$, Mueller and Schmidt [6] showed that the number of solutions of $|ax^r + bx^s y^{r-s} + cy^r| \leq h$, $r \geq 9$, does not exceed $c_3 h^{2/r}$ where c_3 is an absolute constant. The general case of this conjecture was proved by Mueller and Schmidt [7] in 1986. Let $N(F, h)$ be the number of solutions of the Thue inequality

$$|F(x, y)| \leq h, \tag{1.6}$$

where F is given by (1.5). Then the main result in [7] is

$$N(F, h) \leq c_4 s^2 h^{2/r}(1 + \log h^{1/r}). \tag{1.7}$$

The factor $1 + \log h^{1/r}$ can be eliminated when $r \geq \max(4s, s\log^3 s)$. The constant c_4 is absolute.

Let $\alpha_1, \ldots, \alpha_r$ be the roots of $f(z) = F(z, 1)$ and let (x, y), $y \neq 0$ be a solution of (1.6). Let α be a root α_j with

$$|\alpha y - x| = \min_{1 \leq i \leq r} |\alpha_i y - x|. \tag{1.8}$$

In 1969 Lewis and Mahler [4] obtained the following estimate, which has recently been refined by Bombieri and Schmidt (see Lemma 1 of [2]):

$$\left|\alpha - \frac{x}{y}\right| \leq \frac{(c_5(r)H)^r h}{H(x, y)^r}, \tag{1.9}$$

where $H(x, y) = \max(|x|, |y|)$ is the height of x/y. It is clear from (1.9) that if $H(x, y)$ is larger than a power of H and also of h, then x/y is a solution of (1.1) with $\mu > \sqrt{2r}$. In fact, it is a solution of

$$\left|\alpha - \frac{x}{y}\right| < \frac{1}{Ky^\mu}, \tag{1.10}$$

where $\mu > \sqrt{2r}$ and K is a power of H. We call such solutions of (1.6) *large*. As we see, such a large solution of (1.6) gives rise to a large solution of (1.1) that is a good rational approximation to α derived from Thue's inequality. By applying (1.3) and the Gap Principle to this set of large solutions, we may then bound their number. This was the method used by Bombieri and Schmidt [2] and also by Mueller and Schmidt [7] for bounding the number of large solutions. Amazingly enough, the number of solutions of (1.10) has been shown to be bounded by an absolute constant. Taking into account the number of roots α, this gives a bound $\ll r$ for the number of large solutions.

We have seen that the large solutions can be counted by the method of Thue and Siegel, which has a long history, going back to 1909. At the present state, the counting of small solutions is perhaps the more difficult task. The method of Bombieri and Schmidt [2] for bounding the number of small solutions gives a bound that intrinsically depends on r.

In trying to find a bound that depends on s rather than on r, the more difficult case again turned out the case of small solutions. In [7], Mueller and Schmidt established a new estimate that shows that small solutions of the Thue inequality (1.6) are solutions of (1.1), and in fact of (1.10), where again K is large.

We begin with the following easily proved

Proposition 1. *Let* (x, y) *with* $y \neq 0$ *be a solution of* (1.6) *and let* α *be defined by* (1.8). *Suppose* $f^{(u)}(\alpha) \neq 0$ *for some* u *in* $1 \leq u \leq r$, then

$$\left|\alpha - \frac{x}{y}\right| \leq \frac{r}{2}\left(\frac{2^r h}{f^{(u)}(\alpha) y^r}\right)^{1/u}.$$

We refer to Lemma 10 of [7] for a proof of this proposition. Clearly the value of this proposition rests on the assertion that at least one of the derivative $f^{(u)}(\alpha)$ is "large." We have

Proposition 2. *Let* $\hat{f}(z) = F(1, z)$ *be the reciprocal polynomial to* $f(z) = F(z, 1)$. *For each root* α *of* $f(z)$ *there is either a* u *in* $1 \leq u \leq s$ *such that*

$$|f^{(u)}(\alpha)| \geq c_6(r, s) H^{1 - u/r} \tag{1.11}$$

or a v *in* $1 \leq v \leq s$ *such that*

$$|\hat{f}^{(v)}(\alpha^{-1})| \geq c_6(r, s) H^{1 - v/r}. \tag{1.12}$$

We remark first that α^{-1} is a root of $f(z) = z^{-r} f(z)$ if and only if α is a root of $f(z)$. The proposition is symmetric in f and $\hat{f}$. Our second remark is that (1.11) will happen when $|\alpha|$ is relatively large, and (1.12) will happen when $|\alpha^{-1}|$ is relatively large.

The two propositions do not quite give the next theorem. A difficulty is that if α is defined by (1.8), and $\hat{\alpha}$ is the root of $\hat{f}$ with $|\alpha - y/x| = \min_{1 \leq i \leq r} |1/\alpha_i - y/x|$, then not necessarily $\hat{\alpha} = 1/\alpha$. However, certain modifications lead to the following:

Theorem. *Let (x, y) be a solution of (1.6) with $x \neq 0$ and $y \gg h^{1/r}$ and let α be defined by (1.8). Then we have either*

$$\left|\alpha - \frac{x}{y}\right| \leq c_7(r, s)\left(\frac{h}{H^{1-s/r}|y|^r}\right)^{1/s}$$

or

$$\left|\alpha^{-1} - \frac{y}{x}\right| \leq c_7(r, s)\left(\frac{h}{H^{1-s/r}|x|^r}\right)^{1/s}.$$

We remark that the constants $c_6(r, s)$ in Proposition 2 and $c_7(r, s)$ in the above theorem are harmless as far as our application to counting the number of small solutions of (1.6) is concerned. We refer to the original paper [7] for details.

In the next section we will describe briefly the main ideas in our proof of Proposition 2.

We hope very much that the expository approach of this paper will serve as a useful introduction to [7], which is rather technical in nature.

2. On the Proof of Proposition 2

To prove (1.11) we will show first that for a given root α of f, there is an $i(K)$ in $1 \leq i(K) \leq s$ and a u in $1 \leq u \leq i(K)$ such that

$$|f^{(u)}(\alpha)| \geq c_8(r, s)|a_{i(K)}\alpha^{r_{i(K)}-u}|. \tag{2.1}$$

Afterwards we will show that

$$|a_{i(K)}\alpha^{r_{i(K)}-u}| \geq c_9(r, s)H^{1-u/r}. \tag{2.2}$$

In what follows we will come upon the same idea again and again, and that is to use the convexity property of the Newton polygon associated with the polynomial f. We begin with some definitions. Write

$$f(z) = a_o + a_1 z^{r_1} + \cdots + a_s z^{r_s},$$

with $0 = r_o < r_1 < \cdots < r_s = r$ and with nonzero coefficients. Construct the points

$$P_i = (r_i, -\log|a_i|) \qquad (i = 0, \ldots, s)$$

and their convex hull C. The *Newton polygon* consists of the lower boundary of C, i.e., the elements (x, y) of C such that $(x, y') \notin C$ when $y' < y$. The Newton polygon then consists of certain vertices

$$P_0 = P_{i(o)}, P_{i(1)}, \ldots, P_{i(\ell)} = P_s$$

with $0 = i(o) < i(1) < \cdots i(\ell) = s$, and the segments between adjacent vertices. We define $\sigma[k]$ as the slope of the k-th segment, i.e., the segment

$$P_{i(k-1)}P_{i(k)} \qquad (k = 1, \ldots, \ell). \tag{2.3}$$

Then by the convexity of the Newton polygon, we have

$$\sigma[1] < \sigma[2] < \cdots < \sigma[\ell].$$

In this exposition we will make the simplifying assumption that the Newton polygon has rather sharp angles at its vertics; more precisely, we will suppose that

$$\sigma[k+1] - \sigma[k] > \log 9 \qquad (k = 1, \ldots, \ell - 1). \tag{2.4}$$

With the segment (2.3) we associate the annulus A_k given by

$$\frac{1}{3}e^{\sigma[k]} < |z| < 3e^{\sigma[k]}.$$

These annuli are disjoint by (2.4), and it may be shown that each root of f lies in an annulus A_k. In other words, for each root α there is a k with

$$|\log|\alpha| - \sigma[k]| < \log 3. \tag{2.5}$$

More precisely, for given k in $1 \le k \le \ell$, there are precisely $r_{i(k)} - r_{i(k-1)}$ such roots.

Now for given α, the system of equations

$$0 = f(\alpha) = f'(\alpha) = \cdots = f^{(s)}(\alpha) \tag{2.6}$$

is a system of $s + 1$ linear homogeneous equations in the coefficients a_0, $a_1, \ldots, a_s$ of f. This system is easily seen to have nonzero determinant, so that we would have $a_0 = \cdots = a_s = 0$ as the only solution. Since the coefficients a_i of f are nonzero by hypothesis, (2.6), cannot happen. It may be seen that we can derive a lower bound for $\min(|f(\alpha)|, |f'(\alpha)|, \ldots, |f^{(s)}(\alpha)|)$ in terms of the coefficients of f. In other words, when α is a root, one of the $|f'(\alpha)|, \ldots, |f^{(s)}(\alpha)|$ will be large.

It turns out to be better to apply this argument not directly to f, but to a truncated polynomial

$$g(z) = \sum_{i \le i(K)} a_i z^{r_i}, \tag{2.7}$$

where K will be determined depending on α. K will be chosen such that the last term $|a_{i(K)}\alpha^{r_{i(K)}}|$ in (2.7) that corresponds to a vertex $P_{i(K)}$ of the

Newton polygon is relatively large in comparison to the terms $|a_i\alpha^{r_i}|$ with $i > i(K)$, which do not occur in (2.7). More about the choice of K later.

Now $z^j g^{(j)}(z)$ is again a linear combination of the monomials z^{r_i} with $i \leq i(K)$. More precisely, we have

$$z^j g^{(j)}(z) = \sum_{i \leq i(K)} c_{ij} a_i z^{r_i} \qquad (j = 0, \ldots, i(K)), \tag{2.8}$$

with certain coefficients c_{ij} depending only on i, j. There is a nontrivial combination of the $i(K) + 1$ expressions (2.8) such that the terms $a_i z^{r_i}$ with $i < i(K)$ cancel and only the last term $a_{i(K)} z^{r_{i(K)}}$ remains

$$\sum_{j=0}^{i(K)} c_j z^j g^{(j)}(z) = c a_{i(K)} z^{r_{i(K)}}.$$

Thus we have "isolated" the term $a_{i(K)} z^{r_{i(K)}}$ associated with the vertex $P_{i(K)}$. It turns out that $c \neq 0$. Thus if $|a_{i(K)}\alpha^{r_{i(K)}}|$ is large, then some $|\alpha^j g^{(j)}(\alpha)|$ is large.

Now going back from g to f and substituting α with $f(\alpha) = 0$ we obtain

$$\sum_{j=1}^{i(K)} c_j \alpha^j f^{(j)}(\alpha) = c a_{i(K)} \alpha^{r_{i(K)}} + \sum_{i > i(K)} d_i a_i \alpha^{r_i}. \tag{2.9}$$

How are we to choose K? We note first that in the summation on the left-hand side of (2.9), we need to have $K > 0$, so that $i(K) > 0$. The assertion $K > 0$ will be provided by Lemma 2. Aside from this, our choice of K will be such that the first term on the right-hand side of (2.9) cannot be cancelled by the other terms.

Lemma 1. *Let K be such that $\sigma[K+1] > \log|\alpha|$, then $|a_i\alpha^{r_i}| < |a_{i(K)}\alpha^{r_{i(K)}}|$ when $i > i(K)$.*

Proof. Let

$$\begin{aligned} T(i) &= \log\left|\frac{a_i\alpha^{r_i}}{a_{i(K)}\alpha^{r_{i(K)}}}\right| \\ &= \log|a_i| - \log|a_{i(K)}| + (r_i - r_{i(K)}) \log|\alpha| \\ &= (r_i - r_{i(K)})(\log|\alpha| - \tau(i(K), i)) \end{aligned}$$

where

$$\tau(i, j) = -\frac{(\log|a_i| - \log|a_j|)}{r_i - r_j}$$

is the slope of the segment P_iP_j. Note that $\tau(i,j) = \tau(j,i)$. By the convexity of the Newton polygon,

$$\tau(i(K), i) \geq \tau(i(K), i(K+1)) = \sigma[K+1] \qquad \text{for } i > i(K).$$

Thus by choosing K with $\sigma[K+1] > \log|\alpha|$, we obtain $T(i) < 0$ for $i > i(K)$. Lemma 1 is proved.

Now in view of the presence of the coefficients c, d_i in (2.9), which can be estimated by routine arguments, in order that the first term on the right-hand side of (2.9) dominate the other terms, we need a little more than $\sigma[K+1] > \log|\alpha|$. A good choice of K is the following: Set $K = K(\alpha) < \ell$ to be the *least* index with

$$\sigma[K+1] \geq \log|\alpha| + \log s + 3 \tag{2.10}$$

or set $K = \ell$ if there is no such K with (2.10). We remark that the definition of K as *least* with (2.10) will be used in the proof of (2.2).

With this choice of $K = K(\alpha)$, and after a somewhat technical estimation of the coefficients c_j, c, d_i in (2.9), which can be expressed in terms of certain determinants, it then follows that there is an u in $1 \leq u \leq i(K)$ with

$$|\alpha|^u |f^{(u)}(\alpha)| \geq c_8(r, s)|a_{i(K)}\alpha^{r_{i(K)}}|$$

so that

$$|f^{(u)}(\alpha)| \geq c_8(r, s)|a_{i(K)}\alpha^{r_{i(K)} - u}|,$$

which is (2.1).

Our next object is to establish the claim that $K \neq 0$.

Lemma 2. *Define* $K = K(\alpha)$ *by* (2.10), *then* $K > 0$.

Proof. Suppose $K = 0$, then (2.10) gives $\sigma[1] - \log|\alpha| \geq \log s + 3$. By the convexity of the Newton polygon, we have $\sigma[k] - \log|\alpha| \geq \log s + 3$ for any k in $1 \leq k \leq \ell$. Since $\log s + 3 > \log 3$, this contradicts (2.5) and Lemma 2 follows.

To establish (2.2) we will again use the convexity argument. Let q be the smallest index with $|a_q| = H$. Then $P_q = (r_q, -\log|a_q|)$ is the lowest vertex (or the left one of two equally low vertices) of the Newton polygon.

Lemma 3. *Suppose*

$$q < i(K) \tag{2.11}$$

where K is defined as before. Then (2.2) holds.

Proof. Denote the left-hand side of (2.2) by $\Delta(\alpha, u)$. By construction of $K(\alpha)$ we have

$$\sigma[K] < \log|\alpha| + s + 3 \tag{2.12}$$

It is at this point that we use the definition of $K < \ell$ as least with (2.10). If there is no K with (2.10), then (2.12) again holds by definition of $K = K(\alpha)$, which gives $K = \ell$ in this case. We have

$$\begin{aligned}\log \Delta(\alpha, u) &= (r_{i(K)} - u) \log|\alpha| + \log|a_{i(K)}| \\ &\geq (r_{i(K)} - u)(\sigma[K] - \log s - 3) + \log|a_{i(K)}|.\end{aligned}$$

In view of $q < i(K)$ and the convexity of the Newton polygon, $\sigma[K] \geq \tau(q, i(K))$, so that

$$\log \Delta(\alpha, u) \geq (r_{i(K)} - u)\tau(q, i(K)) + \log|a_{i(K)}| - c_{10}(r, s).$$

Since the slope

$$\tau(q, i(K)) = -\frac{\log|a_q| - \log|a_{i(K)}|}{r_q - r_{i(K)}},$$

this is the same as

$$\log \Delta(\alpha, u) \geq (r_q - u)\tau(q, i(K)) + \log|a_q| - c_{10}(r, s).$$

Since P_q is the lowest (or one of the lowest) vertics of the Newton polygon, (2.11) give $\tau(q, i(K)) \geq 0$. Therefore, in the case when $r_q \geq u$, we get

$$\log \Delta(\alpha, u) \geq \log|a_q| - c_{10}(r, s) = \log H - c_{10}(r, s),$$

which is rather stronger than (2.10). But when $r_q < u$, another glance at the Newton polygon gives $\tau(q, i(K)) \leq \tau(q, s)$, so that

$$\log \Delta(\alpha, u) \geq (r_q - u)\tau(q, s) + \log|a_q| - c_{10}(r, s).$$

Substituting the quotient defining $\tau(q, s)$ and using $r_q < u \leq s$, one deduces in a few lines that

$$\log \Delta(\alpha, u) \geq \left(1 - \frac{u}{r}\right) \log H - c_{10}(r, s),$$

which gives (2.2).

All this was derived assuming (2.11). When (2.11) is violated we turn to the reciprocal polynomial $\hat{f}$ and its root α^{-1}. The exponents, instead of $r_0, \ldots, r_s$, now are $\hat{r}_0, \ldots, \hat{r}_s$ with $\hat{r}_i = r - r_{s-i}$, and the Newton polygon has slopes $-\sigma[\ell] < \cdots < -\sigma[1]$. We define $\hat{K}(\alpha^{-1})$ and $\hat{q}$, and the analogue of (2.11) becomes

$$\hat{q} < \hat{\imath}(\hat{K}), \tag{2.13}$$

with $\hat{\imath}(t) = s - i(\ell - t)$. It may then be seen that (2.13) is equivalent with

$$i(\hat{k}) < q, \tag{2.14}$$

where $\hat{k} = \hat{k}(\alpha)$ is largest with $\sigma[\hat{k}] \leq \log|\alpha| - \log s - 3$. Since $K(\alpha)$ was smallest with $\sigma[K+1] \geq \log|\alpha| + \log s + 3$, we have $\hat{k} \leq K$. In fact, since there is a k with (2.5), we have $\hat{k} < K$, and therefore $i(\hat{k}) < i(K)$. Hence either (2.11) and therefore (1.11) holds, or (2.13) and therefore (1.12) holds.

References

[1] Bombieri, E. "On the Thue-Siegel Dyson theorem," *Acta Math.* **148** (1982), 255–296.

[2] Bombieri, E. and Schmidt, W.M. "On Thue's equation," *Inv. Math.* **88** (1987), 69–81.

[3] Evertse, J.-H. "Upper bounds for the number of solutions of diophantine equations," *Math. Centrum.* Amsterdam (1983), 1–127.

[4] Lewis, D. and Mahler, K. "Representation of integers by binary forms," *Acta. Arith.* **6** (1961), 333–363.

[5] Mueller, J. "Counting solutions of $|ax^r - by^r| \leq h$," *Quarterly J. Oxford* **38** (152) (1987).

[6] Mueller, J. and Schmidt, W.M. "Trinomial Thue equations and inequalities," *J.f. Math.* **379** (1987), 76–99.

[7] Mueller, J. and Schmidt, W.M. "Thue's equation and a conjecture of Siegel," *Acta Math.*, **160** (1988), 207–247.

[8] Siegel, C.L. "Über einige anuendungen diophantischer approximationen," *Abh. Preuss. Akad. Wiss. Phys.-Math. Kl.* No. 1 (1929).

23 The Selberg Zeta-Function of a Kleinian Group

S. J. PATTERSON

1. Introduction

In [20] Selberg introduced the Selberg trace formula for cocompact discrete groups as an infinite analogue of the Kronecker class-number formula. It describes a very precise relationship between the geometric and spectral properties of such a group. Selberg also showed how it was possible in certain cases to extend the formula to cases where the discrete group acts on a symmetric space of rank 1 and has a quotient of finite volume, but not necessarily cocompact. In this case the spectral theory is complicated by the existence of a continuous component of the spectrum.

Selberg also showed how the trace formula could be reformulated in the properties of a certain analytic function, the Selberg zeta-function. In [11], III, Section 6 we showed how such a Selberg zeta-function could be defined and investigated for a convex cocompact Fuchsian group. This discussion made use of very special properties of Fuchsian groups and does not extend to higher dimensional cases. Another approach, more directly based on scattering theory, is given in [7].

In this paper we shall study this question for a certain class of convex cocompact discrete groups acting on hyperbolic spaces of

NUMBER THEORY,
TRACE FORMULAS and
DISCRETE GROUPS

ISBN 0-12-067570-6

arbitrary dimension. To explain this in more detail we must introduce some notation. Let $N \geq 1$ and take as a model of $(N+1)$-dimensional hyperbolic space the unit ball

$$B^{N+1} = \{x \in \mathbb{R}^{N+1} \mid \|x\| < 1\},$$

where $\| \ \|$ denotes the Euclidean norm. Let $L\colon B^{N+1} \times B^{N+1} \to \mathbb{R}$ be defined by

$$L(x, x') = 1 + \frac{\|x - x'\|^2}{(1 - \|x\|^2)(1 - \|x'\|^2)},$$

and then the hyperbolic structure of B^{N+1} can be described through the group $\mathrm{Con}(N)$ of diffeomorphisms $g\colon B^{N+1} \to B^{N+1}$ satisfying for all $x, x' \in B^{N+1}$

$$L(g(x), g(x')) = L(x, x').$$

This preserves the metric $ds^2 = (1 - \|x\|^2)^{-2}\|dx\|^2$ and therefore acts conformally; the conformal dilation is

$$j(g, x) = \frac{(1 - \|g(x)\|^2)}{(1 - \|x\|^2)}.$$

This function extends to the closed ball and hence to the sphere

$$S^N = \{x \in \mathbb{R}^{N+1} \mid \|x\| = 1\}.$$

Recall that an element γ of $\mathrm{Con}(N)$ is called loxodromic (or hyperbolic) if it fixes precisely two points of S^N. If γ is loxodromic then there exists $\delta \in \mathrm{Con}(N)$ so that $\delta^{-1}\gamma\delta$ fixes $(1, 0, \ldots, 0)$ and $(-1, 0, \ldots, 0)$. The subgroup of $\mathrm{Con}(N)$ fixing these two points is isomorphic to $\mathbb{R}_+^\times \times O(N)$ where $\lambda \in \mathbb{R}_+^\times$ maps $(w_1, \ldots, w_{N+1})$ into

$$((1 - \|w\|^2) + a(1 + \|w\|^2) + 2bw_1)^{-1} \times (b(1 + \|w\|^2) + 2aw_1, 2w_2, \ldots, 2w_{N+1}),$$

with

$$a = \frac{1}{2}(\lambda + \lambda^{-1})$$

and

$$b = \frac{1}{2}(\lambda - \lambda^{-1}).$$

This corresponds to a translation of $\frac{1}{2}\log\lambda$ along the geodesic joining the fixed points of γ.

The action of $A\in O(N)$ is the linear one trivial on the axis joining $(1,0,\ldots,0)$ to $(-1,0,\ldots,0)$ and the natural action on the orthogonal complement to the complete line of which this is a segment. We note that the conjugacy class of A depends only on the conjugacy class of γ.

Let us now take $\gamma\in\mathrm{Con}(N)$, γ loxodromic, and define the zeta-function associated with the group $<\gamma>$ generated by γ to be (for $\mathrm{Re}(s)>0$)

$$Z_{<\gamma>}(s)=\prod_{1_1\geq 0,\ldots,1_N\geq 0}(1-\Lambda^{-(s+1_1+\cdots+1_N)}\alpha_1^{1_1}\cdots\alpha_N^{1_N})^2,$$

where $\Lambda=\max(\lambda,\lambda^{-1})$, λ is as above, and $\alpha_1,\ldots,\alpha_N$ are the eigenvalues of A. Note that this depends only on the conjugacy class of $<\gamma>$ in $\mathrm{Con}(N)$.

For Γ a discrete subgroup of $\mathrm{Con}(N)$ we define

$$Z_\Gamma(s)=\prod_H Z_H(s),$$

where H runs through a set of representatives of conjugacy classes of maximal hyperbolic subgroups of Γ. It will turn out that this converges for $\mathrm{Re}(s)>\delta(\Gamma)$ where $\delta(\Gamma)$ is the exponent of convergence. $Z_\Gamma(s)$ is the Selberg zeta-function of Γ.

Next we need the concept of the S-matrix. Let $L(\Gamma)$ denote the limit set of Γ and let $\Omega(\Gamma)=S^N-L(\Gamma)$ be the set of ordinary points on which Γ acts discontinuously. Let for $\mathrm{Re}(s)>\delta(\Gamma)$, $\zeta_1,\zeta_2\in\Omega(\Gamma)$

$$S(\zeta_1,\zeta_2;s)=\sum_{\gamma\in\Gamma}\frac{j(\gamma,\zeta_1)^s}{\|\gamma(\zeta_1)-\zeta_2\|^{2s}}.$$

Let us denote by $L(s)$ the set of smooth functions $f\colon\Omega(\Gamma)\to\mathbb{C}$ satisfying

$$f(g(\zeta))=j(g,\zeta)^{s-N}f(\zeta).$$

This is in a natural way the fiber of a holomorphic vector bundle over $\mathbb{C}$. It is possible to regard the $S(\zeta_1,\zeta_2;s)$ as the kernel of a pseudodifferential operator $S(s)\colon L(s)\to L(N-s)$; this will be explained in Section 3. Let $c_N(s)$ be the Harish-Chandra c-function for $\mathrm{Con}(N)$, i.e.,

$$c_N(s)=\pi^{N/2}\cdot 2^{2s-N}\frac{\Gamma\left(s-\dfrac{N}{2}\right)}{\Gamma(s)}.$$

If Γ is convex cocompact $S(s)$ can be analytically continued to a meromorphic family of pseudodifferential operators on $\mathbb{C}$, and one has the functional equation

$$S(N-s)\cdot S(s) = c_N(s)c_N(N-s)Id_{L(s)};$$

for more details see Section 3. The S-matrix is the analogue of the "constant term" of Eisenstein series in the classical setting.

One of our main objectives is to describe a relationship between the analytic properties of $S(s)$ and of $Z_\Gamma(s)$. This we shall do under in Section 5 under the restriction

$$\Gamma \text{ is convex cocompact and } \delta(\Gamma) < \frac{N}{2}.$$

This restriction can probably be relaxed to demanding merely that Γ be geometrically finite, but there are many technical difficulties to be resolved in so doing.

In the analysis that we carry out here an important role is played by a geometric construction of a "truncated" fundamental domain for Γ. Let $A\colon \Omega(\Gamma) \to \mathbb{R}_+^\times$ be smooth and satisfy

$$A(g(\zeta)) = j(g, \zeta)A(\zeta) \qquad (g \in \Gamma, \zeta \in \Omega(\Gamma)).$$

One can regard $A(\zeta)^{-2}\|d\zeta\|^2$ as giving a Γ-invariant metric on $\Omega(\Gamma)$ conformal to the standard one. For each $\zeta \in \Omega(\Gamma)$ we consider the horosphere

$$H_\zeta(A) = \{w \in B^{N+1} | P(w, \zeta)A(\zeta) > 1\},$$

where

$$P(w, \zeta) = \frac{(1 - \|w\|^2)}{\|w - \zeta\|^2}$$

is the Poisson kernel. One has, for $\gamma \in \Gamma$,

$$H_{\gamma(\zeta)}(A) = \gamma(H_\zeta(A)).$$

Thus $\bigcup_{\zeta \in \Omega(\Gamma)} H_\zeta(A)$ is a Γ-invariant neighborhood of $\Omega(\Gamma)$ in $\overline{B^{N+1}}$; let

$$F(A) = B^{N+1} - \bigcup_{\zeta \in \Omega(\Gamma)} H_\zeta(A).$$

When Γ is convex cocompact, $\Gamma\backslash F(A)$ is relatively compact; that is, there exists a relatively compact subset K of B^{N+1} so that

$$F(A) \subset \bigcup_{\gamma \in \Gamma} \gamma(K).$$

When A is "small" one can set up a natural bijection from $\Omega(\Gamma)$ to $\partial F(A)$, and this is for us the most significant fact. For details see Section 2. As a consequence of investigation here, we begin to study the asymptotic geometry of $F(A)$ and $\Gamma\backslash F(A)$, which is of interest in itself for a study of such Γ. For the formulation of these results see Section 5. C. Epstein has informed me that he has used a similar technique in a related context.

Our work here is based on that of Mandouvalos [8], [9]. We have taken over many of his concepts and results. It is closely related to other recent studies of Eisenstein series associated with discrete subgroups of Con(N) of infinite covolume, and in particular to [1], [10], [15]. The approach to the Selberg Trace Formula is related to that of [2].

We remark here that the method described here could also be used to prove a Selberg Trace Formula directly. Since there is little to be gained from this we shall not discuss it further here.

Finally I would like to thank C.L. Epstein, L.D. Faddeev, R. Melrose, and C. Series for very helpful conversations about the subject of this paper.

2. Geometrical Considerations

In this section we shall summarize some of the geometrical facts that we shall need later. First of all, let $J(g, w) \in \mathrm{GL}_{N+1}(\mathbb{R})$ be the Jacobian of $g: B^{N+1} \to B^{N+1}$. As g is conformal, $j(g, w)^{-1}J(g, w)$ is an orthogonal matrix, which we denote by $J_0(g, w)$.

Let ∇ be the Euclidean gradient operator on $\mathbb{R}^{N+1}$. Then for $g \in \mathrm{Con}(N)$ one has $\nabla(f \circ g) = J(g, \cdot)(\nabla f) \circ g$. In particular, if f is continuously differentiable one has

$$(1 - \|w\|^2)\|\nabla(f \circ g)(w)\|^2 = (1 - \|g(w)\|^2)^2 \cdot \|(\nabla f)(g(w))\|^2.$$

It follows from this that the Laplace operator on B^{N+1} with respect to the hyperbolic metric is given by

$$\Delta f(w) = \frac{1}{4} \cdot (1 - \|w\|^2)^{N+1}\nabla \cdot ((1 - \|w\|^2)^{-(N-1)}\nabla f)$$

where $\cdot$ indicates the Euclidean inner product.

We shall also need the gradient operator on S^N, which we define as follows. Let f be a differentiable function on S^N. We extend it to a neighborhood of S^N by demanding that it be homogenous of degree 0.

Then form ∇f and restrict this to S^N; this defines a gradient $\nabla_S f$; note that $w \cdot (\nabla_S f)(w) = 0$ by Euler's theorem.

Now let Γ be a discrete subgroup of Con(N) and let A be as in Section 1. One example of such an A, which is continuous but not necessarily differentiable, is

$$A_{FN}(\zeta) = \operatorname*{Inf}_{\zeta_1, \zeta_2 \in L(\Gamma)} \frac{\|\zeta - \zeta_1\| \cdot \|\zeta - \zeta_2\|}{\|\zeta_1 - \zeta_2\|}.$$

We shall assume that $L(\Gamma)$ consists of at least two points; from this we see that A_{FN} is bounded and that

$$A_{FN}(\zeta) \geq c \cdot d(L(\Gamma), \zeta)$$

where $d(L(\Gamma), \zeta)$ denotes the Euclidean distance and c is an absolute constant. This inequality is valid for any bounded metric space and will be left as an exercise for the reader. Also, there exists $c' > 0$, depending on $L(\Gamma)$ we have

$$A_{FN}(\zeta) \leq c' d(L(\Gamma), \zeta).$$

Suppose now that Γ is convex cocompact, so that $\Gamma \backslash \Omega(\Gamma)$ is compact. As A/A_{FN} is Γ-invariant and continuous, we see that A is bounded and satisfies also

$$c_A \, d(L(\Gamma), \zeta) \leq A(\zeta) \leq c'_A \, d(L(\Gamma), \zeta)$$

for some constants c_A, c'_A depending on A.

Now let $F(A)$ be as in Section 1. We shall now consider, for suitable A, the envelope of the horospheres $H_\zeta(A)$. This will be the boundary of $F(A)$ in B^{N+1}, and each $H_\zeta(A)$ will touch the envelope in one point, which we denote by $r_A(\zeta)$. Thus we demand that

$$P(r_A(\zeta), \zeta)A(\zeta) = 1$$

and

$$P(r_A(\zeta), \eta)A(\eta) \geq 1, \qquad \text{for } \eta \in \Omega(\Gamma).$$

We suppose the boundary of $F(A)$ to be smooth; then the second condition implies that

$$\nabla_S P(r_A(\zeta), \eta)A(\eta)|_{\eta = \zeta} = 0,$$

where the gradient is taken with respect to η. These equations can easily be solved, and we obtain for $r_A(\zeta)$ the following expression

$$r_A(\zeta) = \zeta - 2\frac{((1 + A(\zeta)) \cdot \zeta + (\nabla_S A)(\zeta))A(\zeta)}{(1 + A(\zeta))^2 + \|\nabla_S A(\zeta)\|^2}$$

so that

$$1 - \|r_A(\zeta)\|^2 = \frac{4A(\zeta)}{((1 + A(\zeta))^2 + \|\nabla_S A(\zeta)\|^2)}.$$

Since our equations are invariant under conjugation by an element γ of Γ, and since we have obtained a unique solution, it follows that

$$r_A(\gamma(\zeta)) = \gamma(r_A(\zeta)).$$

We shall now show that for suitable A the map r_A is a bijection from $\Omega(\Gamma)$ to $\partial F(A)$. First we show that it is surjective. Suppose $w_0 \in \partial F(A)$; then there exist $w_j \to w_0$, $\zeta_j \in \Omega(\Gamma)$ so that $P(w_j, \zeta_j)A(\zeta_j) \geq 1$. Since $H_{\zeta_j}(A)$ has diameter $2A(\zeta_j)/(1 + A(\zeta_j))$, the sequence ζ_j cannot accumulate at a point of $L(\Gamma)$. Let ζ_0 be an accumulation point of this sequence, $\zeta_0 \in \Omega(\Gamma)$. Then $P(w_0, \zeta_0)A(\zeta_0) \geq 1$. If $w_0 \in \partial F(A)$ then there has to exist ζ^* with $P(w_0, \zeta^*)A(\zeta^*) = 1$. Thus w_0 will satisfy precisely the conditions that we imposed on $r_A(\zeta^*)$ and so $w_0 = r_A(\zeta^*)$, i.e., r_A is surjective.

It is not true without restriction that $r_A(\Omega(\Gamma)) \subset \partial F(A)$ or that r_A is injective. Let $Y \in \mathbb{R}_+^\times$; we will show that these conditions are satisfied for A/Y and Y sufficiently large. First we show that $r_{A/Y}$ is injective. Suppose that $K \subset \Omega(\Gamma)$ is a compact region so that $\Gamma K = \Omega(\Gamma)$. Suppose Y_j is a sequence, $Y_j \to \infty$, so that r_{A/Y_j} is not injective and let η_j, $\eta_j' \in \Omega(\Gamma)$, $\eta_j \neq \eta_j'$ be so that $r_{A/Y_j}(\eta_j) = r_{A/Y_j}(\eta_j')$. By the covariance of r_{A/Y_j} we can assume that $\eta_j \in K$. As A is bounded, and as the diameter of $H_\zeta(A)$ is $2A(\zeta)/(1 + A(\zeta))$, we have $\|\eta_j - \eta_j'\| = O(Y_j^{-1})$ as $j \to \infty$. However, by the Mean Value Theorem we have

$$r_{A/Y_j}(\eta_j) - r_{A/Y_j}(\eta_j') = \eta_j - \eta_j' + O\left(\frac{\|\eta_j - \eta_j'\|}{Y_j}\right).$$

Thus we would have a contradiction if $\eta_j \neq \eta_j'$ for sufficiently large j. This shows that $r_{A/Y}$ is injective for sufficiently large Y.

We shall now show that for Y sufficiently large the image of $r_{A/Y}$ is the boundary of $F(A)$. To do this we note that we have to prove that for $\eta \neq \zeta$

$$P(r_{A/Y}(\eta), \zeta)A(\zeta) > Y.$$

This would mean that $r_{A/Y}(\eta)$ was in the horosphere at ζ of diameter $2A(\zeta)/(Y + A(\zeta))$, and so we would need that

$$\|r(\eta) - \zeta\| \leq 2A(\zeta)/(Y + A(\zeta)).$$

But

$$\|r_{A/Y}(\eta) - \eta\| = 2A(\eta)/((Y + A(\eta))^2 + \|\nabla_S A(\eta)\|^2)^{1/2}$$

so that we would have, for a certain $c > 0$,

$$\|\zeta - \eta\| \leq \frac{c}{Y}.$$

Let now $W = (Y + A(\eta))^2 + \|\mathrm{B}_S A(\eta)\|^2$. Then

$$\|r_{A/Y}(\eta) - \zeta\|^2 = \|\zeta - \eta\|^2\left(1 - 2A(\eta)\frac{(Y + A(\eta))}{W}\right) + \frac{4A(\eta)^2}{W} - 4A\left((\eta - \zeta)\cdot\frac{\nabla_S A(\eta)}{W}\right).$$

From this it follows that

$$\frac{P(r_{A/Y}(\eta), \eta)A(\eta)}{P(r_{A/Y}(\eta), \zeta)A(\zeta)} = 1 + \frac{W - 2A(\eta)(Y + A(\eta))}{4A^2}\|\eta - \zeta\|^2 + O(\|\zeta - \eta\|^3 \cdot Y^2 + \|\zeta - \eta\|^2)$$

so that under the assumption that

$$\|\zeta - \eta\| < \frac{c}{Y}$$

we would then have, for sufficiently large Y and ζ, η in a relatively compact subset of $\Omega(\Gamma)$,

$$\frac{P(r_{A/Y}(\eta), \eta)A(\eta)}{P(r_{A/Y}(\eta), \zeta)A(\zeta)} \geq 1 + \frac{Y^2}{8A(\eta)^2}\|\eta - \zeta\|^2,$$

and the assertion follows.

We summarize the conclusions of these considerations in the following proposition:

Proposition 2.1. *Let A be as above and let*

$$F(A) = \{w \in B^{N+1} \mid P(w, \zeta)A(\zeta) < 1 \quad \text{for all } \zeta \in \Omega(\Gamma)\}.$$

Let

$$r_A(\zeta) = \zeta - 2\frac{(((1 + A(\zeta))\zeta + (\nabla_S A)(\zeta))A(\zeta)}{(1 + A(\zeta))^2 + \|\nabla_S A(\zeta)\|^2}.$$

Then $r_A(\Omega(\Gamma)) \supset \partial F(A)$ and if $Y \in \mathbb{R}_+$ is sufficiently large then $r_{A/Y}$ is a bijection from $\Omega(\Gamma)$ to $\partial F(A/Y)$. The horosphere

$$H_\zeta(A/Y) = \{w \in B^{N+1} \mid P(w, \zeta)A(\zeta)/Y \geq 1\}$$

is tangent to $\partial F(A/Y)$ *at* $r_{A/Y}(\zeta)$. *Also*

$$r_{A/Y}(g(\zeta)) = g(r_{A/Y}(\zeta)).$$

We need one further result concerning the r_A; we shall assume that A has been replaced by a suitable A/Y so that r_A is a bijection, and that r_A is nonsingular. We need the "surface element" of $\partial F(A)$ which, as we have just seen, is parametrized by r_A. The surface element along the outward normal will be denoted by $\mathbf{dS}$, and a standard exercise in differential geometry shows that

$$\mathbf{a} \cdot \mathbf{dS} = \det\left[\frac{\partial r_A}{\partial \zeta_1}, \ldots, \frac{\partial r_A}{\partial \zeta_N}, \ldots, \frac{\mathbf{a}}{\zeta_{N+1}}\right] dm(\zeta),$$

the element being taken at $r_A(\zeta)$, and m being the "area" of S^N. One has an expansion

$$\mathbf{a} \cdot \mathbf{dS}(r_{A/Y}(\zeta)) = \left(\zeta \cdot \mathbf{a} + \sum_{k>0} P_k(\zeta) \cdot \mathbf{a} Y^{-k}\right) dm(\zeta),$$

where the P_k can be given explicitly in terms of the A and $\nabla_S A$.

There is one final fact that we shall need later and that is best recorded here, although it is not related directly to what has gone before. It is:

Proposition 2.2. *Let* Γ *be a convex cocompact subgroup of* Con(N) *without elliptic elements. Then there exists a smooth function* F *defined on* $\Omega(\Gamma) \times \Omega(\Gamma)$ *so that*

(i) $F(\zeta_1, \zeta_2) \geq 0$

(ii) $\sum_{g_1, g_2 \in \Gamma} F(g_1(\zeta_1), g_2(\zeta_2)) = 1$

(iii) There exists a neighborhood U of the diagonal of $\Omega(\Gamma) \times \Omega(\Gamma)$ so that if $(\zeta_1, \zeta_2) \in U$ and $F(g_1(\zeta_1), g_2(\zeta_2)) \neq 0$ then $g_1 = g_2$.

Proof. Let $\Delta\colon \Omega(\Gamma) \to \Omega(\Gamma) \times \Omega(\Gamma)$ be the diagonal embedding. This induces a map $\Delta\colon \Gamma\backslash\Omega(\Gamma) \to \Gamma\backslash\Omega(\Gamma) \times \Gamma\backslash\Omega(\Gamma)$. In view of our assumptions, $\Gamma\backslash\Omega(\Gamma)$ is compact and the sets $\gamma\Delta(\Omega(\Gamma))$ for $\gamma \in \Gamma$ are disjoint and do not accumulate in $\Omega(\Gamma) \times \Omega(\Gamma)$. In view of these facts the assertion follows by a standard "partition of unity" argument, which we leave to the reader.

3. Eisenstein Series

In this section we shall summarize the basic properties of Eisenstein series that we shall need. These are based on properties of the Poisson kernel, which is, essentially, an Eisenstein series for the trivial group, and which we shall discuss first. The results of this section are mainly to be found in the thesis of M. Mandouvalos [8], at least in the case $N = 2$. See also [16].

The first result that we shall need is an asymptotic development. This will play a very important role in our later discussion. Before we formulate the result we need some new definitions. Let $\phi \in L^1(S^N, m)$ and let $\zeta \in S^N$ be a point so that ϕ is smooth in a neighborhood of ζ. Then we define

$$C_0(\phi, s, \zeta) = c_N(s)\phi(\zeta),$$

where

$$c_N(s) = \pi^{N/2} \cdot 2^{2s-N} \cdot \frac{\Gamma\left(\frac{s-N}{2}\right)}{\Gamma(s)}$$

and

$$D_0(\phi, s, \zeta) = \int_{S^N} \|\zeta - \eta\|^{-2s} \phi(\eta)\, dm(\eta),$$

which is to mean that $D_0(\phi, s, \zeta)$ is defined for $\mathrm{Re}(s) < N/2$ by this integral and is analytically continued to $\mathbb{C}$ as a meromorphic function (with poles of the first order at $N/2 + \mathbb{N}_0$, [6] I.3, II.2. Note that $D_0(\phi, s, \zeta)$ is also smooth in a neighborhood of ζ. We shall always apply this concept in cases where ϕ is smooth on a fixed open set in S^N and will be taken to be zero outside it.

Having defined C_0, D_0 we define C_j, D_j recursively by

$$C_k(\phi, s, \zeta) = (2s - N + k)^{-1} k^{-1} (N - s + k - 1)\left(N - s + k - \frac{(N+1)}{2}\right) \times C_{k-1}(\phi, s, \zeta) + (2s - N - k)^{-1} k^{-1} \Delta_0 C_{k-2}(\phi, s, \zeta)$$

and

$$D_k(\phi, s, \zeta) = (N - 2s + k)^{-1} \cdot k^{-1} \cdot (s + k - 1)\left(s + k - \frac{(N+1)}{2}\right) \times D_{k-1}(\phi, s, \zeta) + (N - 2s + k)^{-1} \cdot k^{-1} \Delta_0 D_{k-2}(\phi, s, \zeta),$$

where Δ_0 is the standard Laplacian on S^N. We shall define

$$C_k'(\phi, s, \zeta) = \frac{\partial}{\partial s} C_k(\phi, s, \zeta)$$

and

$$D_k'(\phi, s, \zeta) = \frac{\partial}{\partial s} D_k(\phi, s, \zeta).$$

We shall have to study the asymptotic behavior of

$$\int_{S^N} P(w, \zeta)^s \phi(\zeta)\, dm(\zeta) \qquad \text{for } w \text{ "near" } S^N.$$

We shall need also some variants on this. These results are very closely related to the corresponding hyperfunction expansion ([19] Section 6.3). A direct approach is to be found in [5].

Proposition 3.1. *Let $\phi \in L^1(S^N, m)$ and let $\zeta \in S^N$ be such that ϕ is smooth in a neighborhood of ζ. Then one has as $w \mapsto \zeta$, $w \in B^{N+1}$ locally uniformly in ζ for any $k \geq 0$ and s such that $s \notin N/2 + \mathbb{Z}$*

$$\begin{aligned}\int P(w, \eta)^s \phi(\eta)\, dm(\eta) = & \sum_{0 \leq j \leq k} (1 - \|w\|^2)^{N-s+j} C_j\left(\phi, s, \frac{w}{\|w\|}\right) \\ & + \sum_{0 \leq j \leq k} (1 - \|w\|^2)^{s+j} D_j\left(\phi, s, \frac{w}{\|w\|}\right) \\ & + O((1 - \|w\|^2)^{N-s+k+1} + (1 - \|w\|^2)^{s+k+1}).\end{aligned}$$

Moreover this asymptotic expansion can be differentiated formally with respect to s and w.

The other result that we shall need concerns the "resolvent kernel." We consider solutions of

$$\Delta F(L(w, w')) = -s(N-s)F(L(w, w')),$$

where Δ is computed with respect to the w variable. This yields the hypergeometric equation

$$x(1-x)F'' + (N+1)\left(\frac{1}{2} - x\right)F' - s(N-s)F = 0.$$

One solution is given by

$$r(w, w'; s) = \left(4 \cdot \left(s - \frac{N}{2}\right) c_N(s)\right)^{-1} L(w, w')^{-s}$$
$$\times {}_2F_1\left(s, s - \frac{(N-1)}{2}; 2s - (N-1); L(w, w')^{-1}\right),$$

which behaves as $L(w, w') \to \infty$ as

$$\left(4 \cdot \left(s - \frac{N}{2}\right) \cdot c_N(s)\right)^{-1} L(w, w')^{-s}$$

and as $w \to w'$ as ([24] p. 291)

$$\frac{\Gamma((N-1)/2)}{4\pi^{(N+1)/2}} (L(w, w') - 1)^{-(N-1)}.$$

Consequently we have that $r(w, w'; s) - r(w, w'; N - s)$ is regular along $w = w'$. We have now

Proposition 3.2. *Let s be such that* $\mathrm{Re}(s) \geq N/2$. *Then $r(w, w'; s)$ is the kernel of the resolvent operator $(-\Delta - s(N-s)I)^{-1}$. One has, for all s,*

$$\int_{S^N} P(w, \zeta)^s \cdot P(w', \zeta)^{N-s} \, dm(\zeta)$$

is equal to

$$4 \cdot c_N(s) c_N(N-s) \cdot \left(s - \frac{N}{2}\right) \cdot (r(w, w'; s) - r(w, w'; N-s)).$$

The determination of the resolvent kernel is a standard matter once one knows that it exists; for this see [1], [10], [12]. As we have already indicated $r(w, w'; s) - r(w, w'; N - s)$ is, up to a constant multiple, the only solution of $\Delta F = -s(N-s)F$ for generic s that is symmetric about w'. But $\int_{S^N} P(w, \zeta)^s \cdot P(w', \zeta)^{N-s} \, dm(\zeta)$ also has this property, and one determines the constant by evaluating both expressions along $w = w'$.

We need one further property of $r(w, w'; s)$. Let $H \subset \mathrm{Con}(N)$ be a hyperbolic subgroup and let Z_H be as in Section 1. Then we have

Proposition 3.3. *One has for* $\mathrm{Re}(s) > N/2$ *for any cyclic hyperbolic group H*

$$(2s - N) \int_{H \backslash B^{N+1}} \sum_{\substack{\gamma \neq I \\ \gamma \in H}} r(w, \gamma w; s) \, d\sigma(w) = \frac{-Z_H'(s)}{Z_H(s)}.$$

Proof. This proof is much easier to carry out in the upper half-space model of $\mathbb{H}^{N+1}$. The maps are given by

$$\alpha\colon B^{N+1} \to \mathbb{R}_+^\times \times \mathbb{R}^N;\ x \mapsto (1 + \|x\|^2 - 2x_1)^{-1} \times \left(\frac{1}{2}(1 - \|x\|^2), x_2, \ldots, x_{N+1}\right)$$

and

$$\alpha'\colon \mathbb{R}_+^\times \times \mathbb{R}^N \to B^{N+1};\ (y_0, y) \mapsto \left(\left(y_0 + \frac{1}{2}\right)^2 + \|y\|^2\right)^{-1} \times \left(y_0^2 + \|y\|^2 - \frac{1}{4}, y_1, \ldots, y_N\right).$$

One has

$$L(\alpha'(y_0, y), \alpha'(y_0', y')) = \frac{(y_0 + y_0')^2 + \|y - y'\|^2}{4y_0 y_0'}$$

and that $d\sigma$ pulls back to

$$2^{-(N+1)} \cdot y_0^{-(N+1)}\, dm(y)\, dy_0.$$

Also

$$P(\alpha'(y_0, y), \alpha'(0, z)) = \frac{y_0}{y_0^2 + \|y - z\|^2} \cdot 2\left(\frac{1}{4} + \|z\|^2\right).$$

If $A \in \mathrm{Con}(N)$ fixes $(1, 0, 0, \ldots, 0)$ and $(-1, 0, \ldots, 0)$ and corresponds to $\lambda_A \in \mathbb{R}_+^\times$ and $A_0 \in O(N)$ then

$$A\alpha'(y_0, y) = \alpha'(\lambda_A y_0, \lambda_A \cdot A_0(y)).$$

Now we have for $\mathrm{Re}(s) > \max(\mathrm{Re}(t), N - \mathrm{Re}(t))$

$$2^{-(N+1)} \iint r_s\left(\frac{(y_0 + y_0')^2}{4y_0 y_0'}\right) y_0'^{t-N-1}\, dy_0'\, dm(y') = (t \cdot (N - t) - s(N - s))^{-1} y_0^t,$$

where

$$r_s(L(w, w')) = r(w, w'; s).$$

This is equivalent to

$$2^{-(N+1)} \int_{\mathbb{R}_+} \int_{\mathbb{R}^N} r_s\left(\frac{1}{4} \cdot (1 + \|\eta\|^2) \cdot (y + y^{-1} + 2)\right) \times dm(\eta) y^{s-N/2} (y^{1/2} + y^{-1/2})^N y^{-1}\, dy = (t(N - t) - s(N - s))^{-1}.$$

This is a Mellin transform and we can invert it, obtaining

$$2^{-(N+1)}\int_{\mathbb{R}^N} r_s\left(\frac{1}{4}(1+\|\eta\|^2)(y+y^{-1}+2)\right)dm(\eta)\cdot(y^{1/2}+y^{-1/2})^N$$
$$=(2s-N)^{-1}\cdot\max(y,y^{-1})^{N/2-s}.$$

Now consider the group as being generated by $\gamma=(\lambda_A, A_0)$ with $\lambda_A>1$; we take as a fundamental domain

$$\{(y_0,y)\in\mathbb{R}_+^\times\times\mathbb{R}^N \,|\, 1\le y_0\le\lambda_A\}.$$

The contribution of γ^m is then

$$2^{-N-1}\int_1^{\lambda_A}\int_{\mathbb{R}^N} r_s\left(\frac{(y_0+\lambda_A^m y_0)^2+\|y-\lambda_A^m A_0^m(y)\|^2}{4\lambda_A^m y_0^2}\right)\frac{dy_0\,dm(y)}{y_0^{N+1}}.$$

Let

$$z=(1+\lambda_A^m)^{-1}y_0^{-1}\cdot(y-\lambda_A^m A_0^m(y));$$

the integral becomes

$$\log\lambda_A\cdot\int_{\mathbb{R}^N} r_s\left(\frac{1}{4}(1+\|z\|^2)(\lambda_A^m+\lambda_A^{-m}+2)\right)dm(z)$$
$$\cdot(\lambda_A^{m/2}+\lambda_A^{-m/2})^N\cdot\det(I-\lambda_A^m A_0^m)^{-1}\lambda_A^{mN/2}$$
$$=\log\lambda_A\cdot(2s-N)^{-1}\cdot\lambda_A^{1/2|m|(N-2s)}\lambda_A^{mN/2}|\det(I-\lambda_A^m A_0^m)|^{-1}$$
$$=\log\lambda_A\cdot(2s-N)^{-1}\cdot\lambda_A^{-|m|\cdot s}|\det(I-\lambda_A^{-m}A_0^m)|^{-1}.$$

Thus we have proved that

$$\int_{H\backslash B^{N+1}}\sum_{\substack{\gamma\neq I\\ \gamma\in H}} r_s(w,\gamma w;s)\,d\sigma(w)$$
$$=(2s-N)^{-1}\cdot\sum_{\substack{m\neq 0\\ m\in\mathbb{Z}}}\log\lambda_A\cdot\lambda_A^{-|m|s}|\det(I-\lambda_A^{-|m|}A_0^m)|^{-1}$$
$$=-2(2s-N)^{-1}\cdot\frac{d}{ds}\left\{\sum_{m>0}\frac{1}{m}\lambda_A^{-ms}|\det(I-\lambda_A^{-m}A_0^m)|^{-1}\right\}$$
$$=(2s-N)^{-1}\cdot\frac{d}{ds}\log Z_H(s)$$

as required.

Now let Γ be a convex cocompact subgroup of Con(N). We can define for $w \in B^{N+1}$, $\zeta \in \Omega(\Gamma)$, $s \in \mathbb{C}$, $\mathrm{Re}(s) > \delta(\Gamma)$ the Eisenstein series

$$E(w, \zeta, s) = \sum_{\gamma \in \Gamma} P(\gamma(w), \zeta)^s.$$

Let $\phi \in L(s)$; then

$$E(w, \phi, s) = \int_{\Gamma \backslash \Omega(\Gamma)} E(w, \zeta, s)\phi(\zeta)\, dm(\zeta)$$

is defined. In view of Proposition 2.2 we have for w approaching a point of $\Omega(\Gamma)$

$$E(w, \phi, s) \sim (1 - \|w\|^2)^{N-s} \cdot c_N(s)\phi\left(\frac{w}{\|w\|}\right) + (1 - \|w\|^2)^s S(s)(\phi)\left(\frac{w}{\|w\|}\right),$$

where $S(s)$ is the pseudodifferential operator with "kernel" $S(\zeta_1, \zeta_2; s)$ defined in Section 1. The kernel has a singularity along the diagonal that has to be interpreted in the same way as the analogous singularity of D_0 above.

Suppose now that $\delta(\Gamma) < N/2$. Then one can interpret the equation of Proposition 3.2 as giving

$$S(N - s)S(s) = c_N(s)c_N(N - s)\, \mathrm{Id}_{L(s)}$$

and

$$E(w, S(s)(\phi), N - s) = c_N(N - s)E(w, \phi, s)$$

for s satisfying $\delta(\Gamma) < \mathrm{Re}(s) < N - \delta(\Gamma)$.

Let now F be as in Proposition 2.2 and let F_0 be

$$F_0(\zeta_1, \zeta_2) = \sum_{\delta \in \Gamma} F(\delta(\zeta_1), \delta(\zeta_2)).$$

Then define

$$S^F(\zeta_1, \zeta_2; s) = \sum_{\gamma \in \Gamma} j(\gamma, \zeta_1)^s F_0(\gamma\zeta_1, \zeta_2)\|\gamma\zeta_1 - \zeta_2\|^{-2s}.$$

This is defined for all s and is the kernel of a pseudodifferential operator. There is a neighborhood U of the diagonal $\Delta(\Omega(\Gamma))$ in $\Omega(\Gamma) \times \Omega(\Gamma)$ so that $F_0(\zeta_1, \zeta_2) = 1$ for $(\zeta_1, \zeta_2) \in U$. It follows that (cf. [7]4(v), [9])

$$S^F(N - s)S^F(s) = c_N(s)c_N(N - s) \cdot \mathrm{Id}_{L(s)} + K^F(s),$$

where $K(s)$ represents a meromorphic family of operators with smooth kernels and holomorphic in $\mathbb{C} - (N/2 + \mathbb{Z})$. It follows from this that $S^F(s)^{-1}$ exists for almost all s. Moreover, $S(s)S^F(s)^{-1}$ is of the form $\mathrm{Id}_{L(s)} + \sigma^F(s)$ where σ^F is again a meromorphic family of operators with smooth kernels for $\mathrm{Re}(s) > \delta(\Gamma)$. Moreover, $S(s)$ and $S^F(s)$ are symmetric; thus for $\mathrm{Re}(s) > \delta(\Gamma)$

$$
\begin{aligned}
{}^t(S(N-s)S^F(N-s)^{-1})\cdot(S(s)\cdot S^F(s)^{-1}) \\
= I + c_N(s)^{-1}\cdot c_N(N-s)^{-1}\cdot K^F(s)
\end{aligned}
$$

and hence, as $S(N-s)S^F(N-s)^{-1}$ can be inverted by Fredholm theory, we obtain the meromorphic continuation of $S(s)S^F(s)^{-1}$ to $\mathbb{C} - E(N)$, $E(N) = (-\mathbb{N}_0) \cup (N + \mathbb{N}_0)$—and hence also that of $S(s)$. Note also that the Fredholm determinant of $S(s)S^F(s)^{-1}$ exists. These matters are discussed in the case $N = 1$ in [11].

This method of analytic continuation has the advantage of considerable simplicity, and of illustrating how one can approach the problem of regularizing the kernel of the $S(s)$. In [10] the analytic continuation of the $S(s)$ is proved, without any reference to an exceptional set such as $E(N)$. R. Melrose informs me that the method of [10] can also be applied when Γ is merely assumed to be geometrically finite. See also [16].

Let now for $\mathrm{Re}(s) > \delta(\Gamma)$

$$r_\Gamma(w, w'; s) = \sum_{\gamma\in\Gamma} r(w, \gamma w'; s).$$

From Proposition 3.2 we have for $\delta(\Gamma) < \mathrm{Re}(s) < N - \delta(\Gamma)$

$$
\begin{aligned}
\int_{\Gamma\backslash\Omega(\Gamma)} E(w, \zeta, s)E(w', \zeta, N-s)\, dm(\zeta) \\
= 4c_N(s)\cdot c_N(N-s)\left(s - \frac{N}{2}\right)(r_\Gamma(w, w'; s) - r_\Gamma(w, w'; N-s)).
\end{aligned}
$$

The contribution of the summands with $\gamma = I$ to the right-hand side is $2\cdot\pi^{(N+1)/2}\Gamma((N+1)/2)$. By Proposition 3.3 and Selberg's transformation one has for $\mathrm{Re}(s) > N/2$,

$$\int_{\Gamma\backslash F(A/Y)} (r_\Gamma(w, w; s) - r(w_0, w_0; s))\, d\sigma(w) \to -\frac{1}{2s-N}\frac{Z_\Gamma'(s)}{Z_\Gamma(s)}$$

as $Y \to \infty$.

Next, as $Y \to \infty$

$$r_\Gamma(r_{A/Y}(\eta), r_{A/Y}(\eta), s) - r(w_0, w_0; s)$$

has an asymptotic expansion in Y of the form

$$(1 - \|r_{A/Y}(\eta)\|^2)^{2s}\left(\sum_{\gamma \neq I} j(\gamma, \eta)^s \|\gamma\eta - \eta\|^{-2s}\right)\cdot(4\cdot(s - N/2))^{-1}\cdot c_N(s)^{-1}$$
$$+ (1 - \|r_{A/Y}(\eta)\|^2)^{2s+1}\cdot F_1(\eta, s)$$
$$+ \cdots$$

and consequently if $|s - N/2| < \frac{1}{2}$, $\mathrm{Re}(s) > \delta(\Gamma)$

$$\int_{\Gamma\setminus F(A/Y)} \{r_\Gamma(w, w; s) - r(w_0, w_0; s)\}\, d\sigma(w)$$
$$= a(s) + b_1(s)Y^{N-2s} + O(Y^{N-2\mathrm{Re}(s)-1}\log Y),$$

where a and b are holomorphic in s. Using the case $\mathrm{Re}(s) > N/2$ we see that

$$a(s) = -(2s - N)^{-1}\frac{Z'_\Gamma(s)}{Z_\Gamma(s)},$$

whereas the asymptotic expansion above shows that

$$b_1(s) = -\left(4^2\cdot\left(\frac{s - N}{2}\right)^2 c_N(s)\right)^{-1}$$
$$\cdot\int_{\Gamma\setminus\Omega(\Gamma)}\left(\sum_{\gamma\neq I}\|\eta - \gamma\eta\|^{-2s} j(\gamma, \eta)^s\right)(4A(\eta))^{2s-N}\, dm(\eta).$$

It now follows that as $Y \to \infty$, $N/2 < \mathrm{Re}(s) < N - \max(\frac{1}{2}, \delta(\Gamma))$

$$\int_{\Gamma\setminus F(A/Y)}\int_{\Gamma\setminus\Omega(\Gamma)} E(w, \zeta, s)E(w, \zeta, N - s)\, dm(\eta)\, d\sigma(w)$$
$$= \frac{2\cdot\pi^{(N+1)/2}}{\Gamma\left(\frac{N+1}{2}\right)\sigma\left(\Gamma\setminus F\left(\frac{A}{Y}\right)\right)} - 2c_N(s)c_N(N - s)\left\{\frac{Z'_\Gamma(s)}{Z_\Gamma(s)} + \frac{Z'_\Gamma(N - s)}{Z_\Gamma(N - s)}\right\}$$
$$+ b(s)Y^{2s-N} + O(1),$$

where

$$b(s) = \left(s - \frac{N}{2}\right)^{-1}\cdot c_N(s)\cdot\int_{\Gamma\setminus\Omega(\Gamma)}$$
$$\cdot\sum_{\gamma\neq I} j(\gamma, \eta)^{N-s}\|\gamma\eta - \eta\|^{-2(N-s)}(2A(\eta))^{N-2s}\, dm(\eta).$$

We can now apply Green's theorem to convert the integral over $\Gamma\setminus F(A/Y)$ to a "surface" integral over $\Gamma\setminus\partial F(A/Y)$, which itself is parameterized by $\Gamma\backslash\Omega(\Gamma)$ via $r_{A/Y}$ for Y large enough. Let dS be the "surface area element" along the outward normal on $\partial F(A/Y)$. For a holomorphic function g we define ∂g to be g', for notational reasons.

We define now

$$E_A(w, \zeta, s) = E(w, \zeta, s)A(\zeta)^s$$
$$S_A(\zeta, \eta; s) = S(\zeta, \eta; s)A(\zeta)^s A(\eta)^s;$$

these are Γ-invariant in ζ and η. Define also the Γ-invariant measure σ^A on $\Omega(\Gamma)$ by

$$d\sigma_A(\zeta) = A(\zeta)^{-N}\, dm(\zeta).$$

Our final result is:

Proposition 3.4. *With the notations above for s satisfying $N/2 <$ $\mathrm{Re}(s) < N\text{-}\max(\frac{1}{2}, \delta(\Gamma))$ one has as $Y \to \infty$*

$$\begin{aligned}
&\iint_{\Gamma\backslash\Omega(\Gamma)\times\Gamma\backslash\Omega(\Gamma)} (1-\|r_{A/Y}(\eta)\|^2)^{-N}\\
&\quad\cdot\{(1-\|r_{A/Y}(\eta)\|^2)\partial\nabla E_A(r_{A/Y}(\eta), \zeta, N-s)E_A(r_{A/Y}(\eta), \zeta, s)\\
&\quad-(1-\|r_{A/Y}(\eta)\|^2)\nabla E_A(r_{A/Y}(\eta), \zeta, s)\partial E_A(r_{A/Y}(\eta), \zeta, N-s)\}\\
&\quad\cdot dS(r_{A/Y}(\eta))\, d\sigma_A(\zeta)\\
&= -(2s-N)\sigma\left(\Gamma\backslash F\left(\frac{A}{Y}\right)\right)\cdot 2\pi^{(N+1)/2}/\Gamma\left(\frac{N+1}{2}\right)\\
&\quad+8c_N(s)c_N(N-s)\cdot\left(s-\frac{N}{2}\right)\left\{\frac{Z'_\Gamma(s)}{Z_\Gamma(s)}+\frac{Z'_\Gamma(N-s)}{Z_\Gamma(N-s)}\right\}\\
&\quad-2\left(s-\frac{N}{2}\right)b(s)Y^{2s-N}+O(1)
\end{aligned}$$

Proof. Consider first

$$(s(N-s)-t(N-t))\int_{\Gamma\backslash F(A/Y)} E_A(w, \zeta, s)E_A(w, \zeta, t)\, d\sigma(w);$$

this is equal to

$$\int_{\Gamma\backslash F(A/Y)} \{\Delta E(w, \zeta, t)\cdot E_A(w, \zeta, s) - E_A(w, \zeta, t)\Delta E_A(w, \zeta, s)\}\, d\sigma(w).$$

To this we can apply Green's theorem and obtain

$$\int_{\Gamma\backslash\partial F(A/Y)} (1-\|w\|^2)^{-N}\{(1-\|w\|^2)\nabla E_A(w,\zeta,t)\cdot E_A(w,\zeta,s) - (1-\|w\|^2)\nabla E_A(w,\zeta,s)\cdot E_A(w,\zeta,t)\}\cdot dS(w).$$

Differentiate this with respect to t and set $t = N - s$. We obtain

$$-(2s-N)\int_{\Gamma\backslash F(A/Y)} E_A(w,\zeta,s)\cdot E_A(w,\zeta,N-s)\,d\sigma(w)$$

$$= \int_{\Gamma\backslash\partial F(A/Y)} (1-\|w\|^2)^{-N}\{(1-\|w\|^2)\partial\nabla E_A(w,\zeta,N-s)E_A(w,\zeta,s) - (1-\|w\|^2)\nabla E_A(w,\zeta,s)\cdot\partial E_A(w,\zeta,N-s)\}\cdot dS(w),$$

from which the assertion follows.

One should regard this as a "Maass–Selberg" relation. It forms the basis of the further analysis of Eisenstein series, as we shall see in the next section.

4. The Asymptotic Analysis

Our next objective is to study the integral appearing in Proposition 3.4 using Proposition 3.1. For later use we need a version of the analysis that is a little more elaborate than absolutely necessary.

Proposition 4.1. *Let $F\colon \Omega(\Gamma)\times\Omega(\Gamma)\to\mathbb{R}$ be a function so that*

(i) $F(\zeta_1,\zeta_2)\geq 0$,

(ii) F is smooth of compact support, and

(iii) $\sum_{g_1,g_2\in\Gamma} F(g_1(\zeta_1),g_2(\zeta_2)) = 1.$

Let $T\subset\Gamma$ be a finite subset so that if $F(g_1(\zeta_1),g_2(\zeta_2))$ does not vanish on every sufficiently small neighborhood of the diagonal $\Omega(\Gamma)\times\Omega(\Gamma)$, then $g_1^{-1}g_2\in T$. Let

$$S_A^T(\zeta_1,\zeta_2;s) = \sum_{\gamma\in\Gamma-T} j(\gamma,\zeta_1)^s\|\gamma(\zeta_1)-\zeta_2\|^{-2s}A(\gamma(\zeta_1))^s A(\zeta_2)^s.$$

Let for s: $\delta(\Gamma) < \mathrm{Re}(s) < N/2$

$$c(T, F, s) = \iint F(\zeta, \eta) S_A^T(\zeta, \eta; s) S_A^T(\zeta, \eta; N - s)\, d\sigma_A(\zeta) \cdot d\sigma_A(\eta)$$

$$+ \sum_{\gamma \in T} \int D_A(F(\zeta, \cdot) S_A^{\{\gamma\}}(\zeta, \cdot; s), N - s; \gamma(\zeta))\, d\sigma_A(\zeta)$$

$$+ \sum_{\gamma \in T} \int D_A(F(\zeta, \cdot) S_A^T(\zeta, \cdot; N - s), s; \gamma(\zeta))\, d\sigma_A(\zeta),$$

where $D_A(\cdot, s; \eta)$ *is the distribution*

$$D_A(\phi, s; \eta) = \int_{S^N} \frac{A(\zeta)^s A(\eta)^s}{\|\zeta - \eta\|^{2s}} \phi(\zeta)\, d\sigma_A(\zeta)$$

to be understood in the same sense as above. Then $c(T, F, s)$ *does not depend on* s *and we write*

$$c(T, F, s) = c(T, F).$$

There exists $c'(T, F, A)$ *so that*

$$\pi^{(N+1)/2} \Gamma\left(\frac{N+1}{2}\right)^{-1} \sigma\left(\Gamma \backslash F\left(\frac{A}{Y}\right)\right)$$

$$= c'(T, F, A) + c(T, F) \log Y$$

$$- \iint \sum_{\gamma \in T} F(\zeta, \gamma\eta) \det\left[\frac{\partial r_{A/Y}(\eta)}{\partial \eta_1}, \ldots, \frac{\partial r_{A/Y}(\eta)}{\partial \eta_N},\right.$$

$$\left. \cdot\, \eta_{N+1}^{-1}(r_{A/Y}(\eta) + (r_{A/Y}(\eta) - \zeta) P(r_{A/Y}(\eta), \zeta)\right]$$

$$\cdot \log(P(r_{A/Y}(\eta), \zeta) A(\zeta)) \|r_{A/Y}(\eta) - \zeta\|^{-N} A(\zeta)^N A(\eta)^N$$

$$\cdot\, d\sigma_A(\zeta)\, d\sigma_A(\eta) + o(1).$$

For s *satisfying* $\delta(\Gamma) < \mathrm{Re}(s) < N/2$

$$c_N(s) c_N(N - s) \left\{\frac{Z_\Gamma'(s)}{Z_\Gamma(s)} + \frac{Z_\Gamma'(N - s)}{Z_\Gamma(N - s)}\right\} = c'(T, F, A) + c(T, F)$$

$$+ \iint F(\zeta, \eta) S_A^{T'}(\zeta, \eta; N - s) S_A^T(\zeta, \eta; s)\, d\sigma_A(\zeta)\, d\sigma_A(\eta)$$

$$+ \sum_{\gamma \in T} \int D_A'(F(\zeta, \cdot) S_\Omega^{\{\gamma\}}(\zeta, \cdot; s), \gamma(\zeta))\, d\sigma_A(\zeta)$$

$$+ \sum_{\gamma \in T} \int D_A(F(\zeta, \cdot) S_A^{T'}(\zeta, \cdot; N - s), s; \gamma(\zeta))\, d\sigma_A(\zeta).$$

We remark here that Condition (ii) on F can be relaxed considerably but we do not need that now.

Proof. The proof consists in studying the behavior for large Y of the integral appearing on the left-hand side of the equation of Proposition 3.4. This we write as

$$\iint_{S^N \times S^N} F(\eta, \zeta)(1 - \|r_{A/Y}(\eta)\|^2)^{-N} \det\Bigg[\frac{\partial r_{A/Y}(\eta)}{\partial \eta_1}, \ldots, \frac{\partial r_{A/Y}(\eta)}{\partial \eta_N},$$

$$\times\, \eta_{N+1}^{-1}((1 - \|r_{A/Y}(\eta)\|^2)\nabla\partial E_A(r_{A/Y}(\eta), \zeta, N-s)E_A(r_{A/Y}(\eta), \zeta, s)$$

$$- (1 - \|r_{A/Y}(\eta)\|^2)\nabla E_A(r_{A/Y}(\eta), \zeta, s)\partial E_A(r_{A/Y}(\eta), \zeta, N-s))\Bigg]$$

$$\times\, d\sigma_A(\zeta)\, d\sigma_A(\eta).$$

We define now for a finite subset T of Γ

$$E_A^T(w, \zeta, s) = E_A(w, \zeta, s) - \sum_{\gamma \in T} P(\gamma(w), \zeta)^s A(\zeta)^s.$$

Note that if $\|w\| \to 1$, $w \to \eta$ and if $\eta \notin (\Gamma - T)\{\zeta\}$, then

$$E_A^T(w, \zeta, s) = (1 - \|w\|^2)^s A(\eta)^{-s} S_A^T(\eta, \zeta; s) + O((1 - \|w\|^2)^{1 + \operatorname{Re}(s)})$$

locally uniformly. We substitute for $E(w, \zeta, s)$ and its derivatives

$$\sum_{\gamma \in T} P(\gamma(w), \zeta)^s A(\zeta)^s + E_A^T(w, \zeta, s)$$

and its derivatives in the integral above. This gives a sum of eight integrals. These we have to investigate individually. The first term arises from the term involving the $\sum_{\gamma_1 \in T} P(\gamma_1 w, \zeta)^s$ and the $\sum_{\gamma_2 \in T} P(\gamma_2 w, \zeta)^{N-s}$ (and their derivatives). Those with $\gamma_1 = \gamma_2$ yield

$$2(2s - N) \iint_{S^N \times S^N} F(\eta, \zeta) \sum_{\gamma \in T} \det\Bigg[\frac{\partial r_{A/Y}}{\partial \eta_1}, \ldots, \frac{\partial r_{A/Y}}{\partial \eta_N},$$

$$\eta_{N+1}^{-1}(r_{A/Y}(\eta) + r_{A/Y}(\eta)P(r_{A/Y}(\eta), \gamma^{-1}\zeta)\Bigg]$$

$$\cdot \log P(r_{A/Y}(\eta), \gamma^{-1}\zeta)A(\gamma^{-1}\zeta) \cdot \|r_{A/Y}(\eta) - \gamma^{-1}\zeta\|^{-N} A(\eta)^N A(\gamma^{-1}\zeta)^N$$

$$\cdot\, d\sigma_A(\zeta)\, d\sigma_A(\eta),$$

and this is equal to

$$2(2s-N)\iint\limits_{S^N\times S^N}\sum_{\gamma\in T}F(\eta,\gamma(\zeta))\det\left[\frac{\partial r_{A/Y}}{\partial\eta_1},\ldots,\frac{\partial r_{A/Y}}{\partial\eta_N},\right.$$

$$\left.\eta_{N+1}^{-1}(r_{A/Y}(\eta)+(r_{A/Y}(\eta)-\zeta)P(r_{A/Y}(\eta),\zeta)\right]$$

$$\times\log(P(r_{A/Y}(\eta),\zeta)A(\zeta))\|r_{A/Y}(\eta)-\zeta\|^{-N}\,dm(\zeta)\,dm(\eta). \tag{1}$$

The terms with $\gamma_1\neq\gamma_2$ can be summed as

$$\sum_{\gamma_2\in T}\ \sum_{\gamma_1\in T,\gamma_1\neq\gamma_2},$$

and this means that in those terms arising from a term of the form $P(\gamma_2 w,\zeta)^{N-s}A(\zeta)^{N-s}$ and $E_A^T(w,\zeta,s)$ that $E_A^T(w,\zeta,s)$ and its derivatives can be replaced by $E_A^{\{\gamma\}}(w,\zeta,s)$ or the appropriate derivatives.

Now consider those terms arising from the $E_A^T(w,\zeta,s)$ and $E_A^T(w,\zeta,N-s)$. These terms yield

$$2(2s-N)\iint\limits_{S^N\times S^N}F(\eta,\zeta)S_A^T(\eta,\zeta;N-s)S_A^T(\eta,\zeta;s)\,d\sigma_A(\zeta)$$

$$\times\,d\sigma_A(\eta)\log((4Y)^{-1})$$

$$-2\iint\limits_{S^N\times S^N}F(\eta,\zeta)S_A^T(\eta,\zeta;N-s)S_A^T(\eta,\zeta;s)\,d\sigma_A(\zeta)\,d\sigma_A(\eta)$$

$$+2(2s-N)\iint\limits_{S^N\times S^N}F(\eta,\zeta)S_A^{T'}(\eta,\zeta;N-s)S_A^T(\eta,\zeta;s)\,d\sigma_A(\zeta)\,d\sigma_A(\eta)$$

$$+O(Y^{-1}\cdot\log Y). \tag{2}$$

This deals with four of the eight terms. We now have to treat the four "mixed" terms, which we do using Proposition 3.1. The results are somewhat unpleasant. One finds first

$$\sum_{\gamma\in T}\frac{\partial}{\partial t}\iint F(\eta,\zeta)(1-\|r_{A/Y}(\eta)\|^2)^{-N}\det\left[\frac{\partial r_{A/Y}(\eta)}{\partial\eta_1},\ldots,\frac{\partial r_{A/Y}(\eta)}{\partial\eta_N},\right.$$

$$\left.\eta_{N+1}^{-1}((1-\|r_{A/Y}(\eta)\|^2)(\nabla P^t)(r_{A/Y}(\gamma(\eta),\zeta)A(\zeta)^t E_A^{\{\gamma\}}(r_{A/Y}(\eta),\zeta,s)\right]$$

$$\cdot A(\zeta)^N A(\eta)^N\,d\sigma_A(\zeta)\,d\sigma_A(\eta)|_{t=N-s}$$

$$= -2(N-s)\sum_{\gamma\in T}\int D'_A(F(\eta,\cdot)S_A^{\{\gamma\}}(\eta,\cdot;s), N-s;\gamma(\eta))\,d\sigma_A(\eta)$$

$$+ 2(N-s)\sum_{\gamma\in T}\int D_A(F(\eta,\cdot)S_A^{\{\gamma\}}(\eta,\cdot;s), N-s,\gamma(\eta))\,d\sigma_A(\eta)\cdot\log\left(\frac{Y}{4}\right)$$

$$- 2\sum_{\gamma\in T}\int D_A(F(\eta,\cdot)S_A^{\{\gamma\}}(\eta,\cdot;s), N-s,\gamma(\eta))\,d\sigma_A(\eta)$$

$$- 2s\sum_{\gamma\in T}\int C'_0(F(\eta,\cdot)S_A^{\{\gamma\}}(\eta,\cdot;s), N-s;\gamma(\eta))\,d\sigma_A(\eta)\left(\frac{Y}{4}\right)^{N-2s}$$

$$- 2s\sum_{\gamma\in T}\int C_0(F(\eta,\cdot)S_A^{\{\gamma\}}(\eta,\cdot;s), N-s;\gamma(\eta))\,d\sigma_A(\eta)\left(\frac{Y}{4}\right)^{N-2s}$$

$$\times\log\left(\frac{Y}{4}\right) + 2\sum_{\gamma\in T}\int C_0(F(\eta,\cdot)S_A^{\{\gamma\}}(\eta,\cdot;s), N-s;\gamma(\eta))\,d\sigma_A(\eta)$$

$$\times\left(\frac{Y}{4}\right)^{N-2s} + O((\log Y)(Y^{-1} + Y^{N-2\mathrm{Re}(s)-1})), \tag{3}$$

$$- \sum_{\gamma\in T}\frac{\partial}{\partial t}\iint_{S^N\times S^N} F(\eta,\zeta)(1-\|r_{A/Y}(\eta)\|^2)^{-N}$$

$$\times\det\left[\frac{\partial r_{A/Y}(\eta)}{\partial\eta_1},\ldots,\frac{\partial r_{A/Y}(\eta)}{\partial\eta_N},\right.$$

$$\left.\times\,\eta_{N+1}^{-1}((1-\|r_{A/Y}(\eta)\|^2\nabla E_A^{\{\gamma\}}(r_{A/Y}(\eta),\zeta,s)P^t(r_A(\gamma(\eta)),\zeta)\right]$$

$$\times A(\zeta)^t A(\eta)^N\,d\sigma_A(\zeta)\,d\sigma_A(\eta)|_{t=N-s}$$

$$= 2s\sum_{\gamma\in T}\int_{S^N} D'_A(F(\eta,\cdot)S_A^{\{\gamma\}}(\eta,\cdot;s), N-s;\gamma(\eta))\,d\sigma_A(\eta)$$

$$- 2s\sum_{\gamma\in T}\int_{S^N} D_A(F(\eta,\cdot)S_A^{\{\gamma\}}(\eta,\cdot;s), N-s;\gamma(\eta))\,d\sigma_A(\eta)\cdot\log\left(\frac{Y}{4}\right)$$

$$+ 2s\sum_{\gamma\in T}\int_{S^N} C'_0(F(\eta,\cdot)S_A^{\{\gamma\}}(\eta,\cdot;s), N-s;\gamma(\eta))\,d\sigma_A(\eta)\left(\frac{Y}{4}\right)^{N-2s}$$

$$+ 2s\sum_{\gamma\in T}\int_{S^N} C_0(F(\eta,\cdot)S_A^{\{\gamma\}}(\eta,\cdot;S), N-s;\gamma(\eta))\,d\sigma_A(\eta)\left(\frac{Y}{4}\right)^{N-2s}$$

$$\cdot\log\left(\frac{Y}{4}\right) + O(\log Y(Y^{-1} + Y^{N-2\mathrm{Re}(s)-1})). \tag{4}$$

$$\sum_{\gamma \in T} \frac{\partial}{\partial t} \iint\limits_{S^N \times S^N} F(\eta, \zeta)(1 - \|r_{A/Y}(\eta)\|^2)^{-N} \det\left[\frac{\partial r_{A/Y}(\eta)}{\partial \eta_1}, \ldots, \frac{\partial r_{A/Y}(\eta)}{\partial \eta_N}\right.$$

$$\left. \eta_{N+1}^{-1}(1 - \|r_{A/Y}(\eta)\|^2)\nabla E_A^T(r_{A/Y}(\eta), \zeta, t)P(r_{A/Y}(\gamma(\eta)), \zeta)^s\right]$$

$$\times A(\zeta)^s A(\eta)^N \, d\sigma_A(\zeta) \, d\sigma_A(\eta)|_{t=N-s}$$

$$= -2(N-s) \sum_{\gamma \in T} \iint\limits_{S^N} D_A(F(\eta, \cdot)S_A^{T\prime}(\eta, \cdot; N-s)s, \gamma(\eta)) \, d\sigma_A(\eta)$$

$$+ 2(N-s) \sum_{\gamma \in T} \int_{S^N} D_A(F(\eta, \cdot)S_A^T(\eta, \cdot; N-s), s; \gamma(\eta)) \, d\sigma_A(\eta) \log(\tfrac{Y}{4})$$

$$- 2 \sum_{\gamma \in T} \int_{S^N} D_A(F(\eta, \cdot)S_A^T(\eta, \cdot; N-s), s; \gamma(\eta)) \, d\sigma_A(\eta)$$

$$- 2(N-s) \sum_{\gamma \in T} \int_{S^N} C_0(F(\eta, \cdot)S_A^{T\prime}(\eta, \cdot; N-s), s; \gamma(\eta)) \, d\sigma_A(\eta)\left(\frac{Y}{4}\right)^{2s-N}$$

$$+ 2(N-s) \sum_{\gamma \in T} \int_{S^N} C_0(F(\eta, \cdot)S_A^T(\eta, \cdot; N-s, s; \gamma(\eta)) \, d\sigma_A(\eta)$$

$$\times \left(\frac{Y}{4}\right)^{2s-N} \log\left(\frac{Y}{4}\right)$$

$$- 2 \sum_{\gamma \in T} \int_{S^N} C_0(F(\eta, \cdot)S_A^T(\eta, \cdot; N-s), s; \gamma(\eta)) \, d\sigma_A(\eta) \cdot \left(\frac{Y}{4}\right)^{2s-N}$$

$$+ O(\log Y(Y^{-1} + Y^{2\mathrm{Re}(s) - N - 1})). \tag{5}$$

Finally,

$$- \sum_{\gamma \in T} \frac{\partial}{\partial t} \iint\limits_{S^N \times S^N} F(\eta, \zeta)(1 - \|r_{A/Y}(\eta)\|^2)^{-N} \det\left[\frac{\partial r_{A/Y}(\eta)}{\partial \eta_1}, \ldots, \frac{\partial r_{A/Y}(\eta)}{\partial \eta_N},\right.$$

$$\left. \eta_{N+1}^{-1}((1 - \|r_{A/Y}(\eta)\|^2)\nabla P^s(r_{A/Y}(\gamma(\eta)), \zeta)\partial E_A^0(r_{A/Y}(\eta), \zeta, N-s)\right]$$

$$\times A(\zeta)^s A(\eta)^N \cdot d\sigma_A(\zeta) \, d\sigma_A(\zeta) \, d\sigma_A(\eta)|_{t=N-s}$$

$$= 2s \int_{S^N} D_A(F(\eta, \cdot)S_A^{T'}(\eta, \cdot; N - s), s; \gamma(\eta))\, d\sigma_A(\eta)$$

$$- 2s \sum_{\gamma \in T} \int_{S^N} D_A(F(\eta, \cdot)S_A^{T}(\eta, \cdot; N - s), s; \gamma(\eta))\, d\sigma_A(\eta)\left(\frac{Y}{4}\right)$$

$$+ 2(N - s) \sum_{\gamma \in T} \int_{S^N} C_0(F(\eta, \cdot)S_A^{T}(\eta, \cdot; N - s), s; \gamma(\eta))\, d\sigma_A(\eta)\left(\frac{Y}{4}\right)^{2s - N}$$

$$+ 2(N - s) \sum_{\gamma \in T} \int_{S^N} C_0(F(\eta, \cdot)S_A^{T}(\eta, \cdot; N - s), s; \gamma(\eta))\, d\sigma_A(\eta) \cdot \left(\frac{Y}{4}\right)^{2s - N}$$

$$- 2(N - s) \sum_{\gamma \in T} \int_{S^N} C_0(F(\eta, \cdot)S_A^{T}(\eta, \cdot; N - s), s; \gamma(\eta))$$

$$\times\, d\sigma_A(\eta) \cdot \left(\frac{Y}{4}\right)^{2s - N} \log\left(\frac{Y}{4}\right)$$

$$+ O(\log Y \cdot (Y^{-1} + Y^{2\mathrm{Re}(s) - N - 1})). \tag{6}$$

Adding together (1) to (6) we see that for $s: \delta(\Gamma) < \mathrm{Re}(s) < N/2$ the integral is asymptotic to an expression of the form

$$2\beta(Y)(2s - N) + \alpha_0(s) + \alpha_1(s) \log Y + \alpha_2(s) Y^{N - 2s} + O(1)$$

(the coefficient of $Y^{N-2s} \log Y$ is identically 0). On the other hand, this is equal to

$$- 2(2s - N)\pi^{(N+1)/2}\Gamma\left(\frac{N+1}{2}\right)^{-1} \sigma\left(\Gamma \backslash F\left(\frac{A}{Y}\right)\right)$$

$$+ 4c_N(s) \cdot c_N(N - s)\left\{\frac{Z_\Gamma'(s)}{Z_\Gamma(s)} + \frac{Z_\Gamma'(N - s)}{Z_\Gamma(N - s)}\right\}(2s - N)$$

$$- 2\left(s - \frac{N}{2}\right) b(N - s) Y^{N - 2s} + O(1)$$

by Proposition 3.4. It turns out that the coefficients of Y^{N-2s} are identically equal so that one sees that $\alpha_1(s)$ has to be of the form $\alpha \cdot (2s - N)$, where α is a constant, and for some constant α' one has

$$-2\pi^{(N+1)/2}\Gamma\left(\frac{N+1}{2}\right)^{-1} \cdot \sigma\left(\Gamma \backslash F\left(\frac{A}{Y}\right)\right) = 2\beta(Y) + \alpha \cdot \log Y + \alpha' + O(1);$$

moreover, one has

$$\alpha_0(s) - \alpha'(2s - N) = 2c_N(s)c_N(N - s)\left\{\frac{Z'_\Gamma(s)}{Z_\Gamma(s)} + \frac{Z'_\Gamma(N - s)}{Z_\Gamma(N - s)}\right\}(2s - N).$$

In these expressions $\beta(Y)$ is the integral appearing in (1) and

$$\alpha_0(s) = 2(2s - N) \iint_{S^N \times S^N} F(\eta, \zeta) S_A^{T'}(\eta, \zeta; N - s) S_A^{T}(\eta, \zeta; s)\, d\sigma_A(\zeta)\, d\sigma_A(\eta)$$

$$+ \sum_{\gamma \in T} \int_{S^N} D'_A(F(\eta, \cdot) S_A^{\{\gamma\}}(\eta, \cdot; s), N - s, \gamma(\eta))\, d\sigma_A(h)$$

$$+ \sum_{\gamma \in T} \int_{S^N} D_A(F(\eta, \cdot) S_A^{T'}(\eta, \cdot; N - s), s; \gamma(\eta))\, d\sigma_A(\eta)$$

$$- 2 \cdot \iint_{S^N \times S^N} F(\eta, \zeta) S^T(\eta, \zeta; s) S^T(\eta, \zeta; N - s)\, d\sigma_A(\zeta)\, d\sigma_A(\eta)$$

$$+ \sum_{\gamma \in T} \int_{S^N} D_A(F(\eta, \cdot) S_A^{\{\gamma\}}(\eta, \cdot; s), N - s; \gamma(\eta))\, d\sigma_A(\eta)$$

$$+ \sum_{\gamma \in T} \int_{S^N} D_A(F(\eta, \cdot) S_A^{T}(\eta, \cdot; N - s), s; \gamma(\eta))\, d\sigma_A(\zeta)\, d\sigma_A(\eta)$$

$$+ \sum_{\gamma \in T} \int_{S^N} D_A(F(\eta, \cdot) S_A^{\{\gamma\}}(\eta, \cdot; s), N - s; \gamma(\eta))\, d\sigma_A(\eta)$$

$$+ \sum_{\gamma \in T} \int_{S^N} D_A(F(\eta, \cdot) S_A^{T}(\eta, \cdot; N - s), s; \gamma(\eta))\, d\sigma_A(\eta).$$

The conclusions of the proposition follow on rewriting these results.

Proposition 4.2. *There exist $V'_0(A), V_0(A), \ldots, V_N(A)$ so that*

$$\sigma\left(\Gamma \backslash F\left(\frac{A}{Y}\right)\right) = V_N(A) Y^N + \cdots + V_1(A) Y$$
$$+ V_0(A) \log Y + V'_0(A) + o(1)$$

as $Y \to \infty$.

Proof. For Y_1 sufficiently large $\Gamma\backslash(F(A/Y_2) - F(A/Y_1))$, $Y_2 > Y_1$ can be parametrized by $[Y_1, Y_2] \times \Gamma\backslash\Omega(\Gamma)$ using

$$(Y, \eta) \mapsto r_Y(\eta).$$

The result follows now on using the asymptotic expansion of $r_Y(\eta)$ in Y and evaluating the Jacobian of this map.

5. Main Theorem

Let S_A, S_A^F be as above. We shall regard these as the kernels of pseudo-differential operators defined on the space of smooth functions on $\Gamma\backslash\Omega(\Gamma)$ and taken with respect to σ_A. One should note that $S_A(s)$ and $S_A^F(s)$ have poles as meromorphic families of pseudodifferential operators at $N/2 + j, j \geq 0, j$ integral. In the region of absolute convergence of the series $S_A(s) - S_A^F(s)$ has a smooth kernel; moreover, in $\mathrm{Re}(s) \geq N/2$ the meromorphic family of operators

$$s \mapsto S_A(s)^{-1} - (c_N(s)c_N(N-s))^{-1} S_A^F(N-s)$$

is represented by a meromorphic family of smooth kernels, as one can verify by computing the symbol.

Our objective is now to express our previous results in more invariant terms than before. This we do by means of the following theorem.

Theorem 5.1. *Let Γ be a convex cocompact Kleinian group with $\delta(\Gamma) < N/2$. Then the function $s \mapsto Z'_\Gamma(s)/Z_\Gamma(s)$ can be analytically continued as a meromorphic function in $\mathbb{C}$. Suppose that F satisfies the condition of Proposition 2.2; then for a certain constant $\Delta(A)$ one has*

$$\frac{Z'_\Gamma(s)}{Z_\Gamma(s)} + \frac{Z'_\Gamma(N-s)}{Z_\Gamma(N-s)} = \frac{\Delta(A)}{c_N(s)c_N(N-s)}$$

$$+ \mathrm{Tr}\left(S_A(s)^{-1}S'_A(s) - \frac{1}{c_N(s)c_N(N-s)} S_A^F(N-s)S_A^{F\prime}(s)\right)$$

$$+ \frac{1}{c_N(s)c_N(N-s)} \iint (S_A^F(\zeta, \eta; N-s)S_A^{F\prime}(\zeta, \eta; s) - S'_A(\zeta, \eta; N))$$

$$\cdot\, d\sigma_A(\zeta)\, d\sigma_A(\eta).$$

Before we give the proof of this theorem we shall append some remarks. First of all, note that this theorem does *not* state that Z_Γ has an analytic continuation as a meromorphic function. It seems to be rather difficult to deduce this statement from the theorem. One would have to verify that the poles of the right-hand side of the functional equation are simple and have residues that are integral. There are two separate cases to be considered. If N is even then $(c_N(s)c_N(N-s))^{-1}$ is regular, and so the nature of $\Delta(A)$ plays no role. The formula of the theorem then makes it very plausible that the residues are of this form. There are, however, considerable technical difficulties in deducing this rigorously. If N is odd then $(c_N(s)c_N(N-s))^{-1}$ has poles at $s=-j (j \geq 0)$ and $s = N - j (j \geq 0)$ where j is integral. This is only possible if $\Delta(A)$ satisfies some arithmetical condition. In the case of cocompact groups, the integrality of the corresponding quantity is a consequence of the Gauss–Bonnet theorem; that the Euler characteristic is even corresponds to the "square" in the definition of Z_Γ. It is tempting to speculate that this is also the case here and that the residue of the expression on the right-hand side of the functional equation at $s = j$, $j \geq N$, j integral has a differential-geometric interpretation.

One should also remark here that if Γ is as above, it can also be regarded as a subgroup of $\mathrm{Con}(N+1)$. Denote this subgroup by Γ_0. Then it is easy to verify that

$$Z_{\Gamma_0}(s) = \prod_{j=0}^{\infty} Z_\Gamma(s+j).$$

Thus if we could verify the meromorphy of Z_{Γ_0} we could also verify that of Z_Γ and conversely. Thus we could, in principle, restrict our attention to the apparently simpler case of even N. There are also relationships between the corresponding S-matrices.

There is a quite different approach to the study of the meromorphy of the Selberg zeta-function. This is through the theory of the Ruelle zeta-function. We recall that if ρ is a representation class of $O(N)$ then we can define, following Fried [3], [4]

$$Z(s, \rho) = \prod_{\gamma} \det(1 - N(\gamma)^{-s}\rho(A(\gamma))),$$

where γ runs through a set of generators of representatives of conjugacy classes, of maximal hyperbolic subgroups, and $A(\gamma)$ is as in

Section 1. If V_j denotes the j-th symmetric power of the standard N-dimensional representation of $O(N)$ then we see that

$$Z(s) = \prod_{j=0}^{\infty} Z(s + \dim(V_j), V_j),$$

where V_0 denotes the trivial representation. It follows that the meromorphy of the $Z(s, \rho)$ for ρ of the form V_j, implies that of the Selberg zeta-function. Fried shows how to prove this using Ruelle's ideas [3], [4] (see [17], [18]) when Γ is cocompact. It seems quite likely that the same proof applies in the convex cocompact case, since an analogous Markov partition exists following work of Bowen, Rees, and Series —see [22] for details. The difficulty lies in the fact that one considers the flow restricted to a small subset of the sphere bundle of $\Gamma \backslash B^{N+1}$—the subset consists of the lifts of geodesics with both endpoints in the limit set—and this is not a real analytic manifold. However, it inherits almost a real-analytic structure from the sphere bundle of $\Gamma \backslash B^{N+1}$, and this has a real analytic Ansov foliation. One may therefore expect that Fried's proof can be extended, but it should be emphasized that this is merely a suggestion.

One attractive feature of such an interpretation is that it would yield another proof of the theorem of [14], according to which S has no poles on $\mathrm{Re}(s) = \delta(\Gamma)$ other than a simple one at $s = \delta(\Gamma)$. This is also a property of the Ruelle zeta-function, which is derived from a study of the Perron-Frobenius operator.

Note that the statement of the theorem essentially gives a meaning to "det S" modulo a power of the primitive of $(c_N(s)c_N(N-s))^{-1}$. One way of regarding the classical Selberg zeta-function is as "$\det((-\Delta - s(N-s)I)^{-1}$," although this formulation tends to hide its geometric significance. This formulation is less meaningful in our case. The "regularization" of the S-matrix given here seems to be of independent interest.

Proof. We now turn to the proof of Theorem 5.1. First of all we shall need some notation. Let X be a measure space and let μ be a measure on X. Let $f(s)$ be an analytic family of functions in $L^1(X, \mu)$ defined on a domain D in $\mathbb{C}$. Then $\int_X f(s)\, d\mu$ is also analytic in D, and we write $PV \int_X f(s)\, d\mu$ for the analytic continuation of $\int_X f(s)\, d\mu$ to some larger domain.

With this notation we can write the final equation of Proposition 4.1 as (with F as above and $T = \{I\}$)

$$c_N(s)c_N(N-s)\left\{\frac{Z'_\Gamma(s)}{Z_\Gamma(s)} + \frac{Z'_\Gamma(N-s)}{Z_\Gamma(N-s)}\right\} = c'(T, F, A) + c(T, F)$$

$$+ \iint F(\zeta, \eta)(S'_A(\zeta, \eta; N-s) - \|\zeta - \eta\|^{-2(N-s)} A(\zeta)^{N-s} A(\eta)^{N-s}$$

$$\times \log\left(\frac{A(\zeta)A(\eta)}{\|\zeta - \eta\|^2)}\right) \cdot (S_A(\zeta, \eta; s)$$

$$- A(\zeta)^s A(\eta)^s \|\zeta - \eta\|^{-2s}) \, d\sigma_A(\zeta) \, d\sigma_A(\eta)$$

$$+ PV \iint F(\zeta, \eta)\|\zeta - \eta\|^{-2(N-s)}(\log(A(\zeta)A(\eta)/\|\zeta - \eta\|^2))$$

$$\cdot A(\zeta)^{N-s} A(\eta)^{N-s}(S_A(\zeta, \eta; s)$$

$$- A(\zeta)^s A(\eta)^s \|\zeta - \eta\|^{-2s}) \, d\sigma_A(\zeta) \, d\sigma_A(\eta)$$

$$+ PV \iint F(\zeta, \eta)(S'_A(\zeta, \eta; N-s) - \|\zeta - \eta\|^{-2(N-s)}$$

$$\times A(\zeta)^{N-s} A(\eta)^{N-s}$$

$$\times \log(A(\zeta)A(\eta)/\|\zeta - \eta\|^2))A(\zeta)^s A(\eta)^s \|\zeta - \eta\|^{-2s} \, d\sigma_A(\zeta) \, d\sigma_A(\eta).$$

In this the first term is convergent in s: $\delta(\Gamma) < \mathrm{Re}(s) < N - \delta(\Gamma)$, the second in $N/2 < \mathrm{Re}(s) < N - \delta(\Gamma)$, and the third in $\delta(\Gamma) < \mathrm{Re}(s) < N/2$. In the region of convergence we can transform the integrals over $\Omega(\Gamma) \times \Omega(\Gamma)$ into integrals over $(\Gamma\backslash\Omega(\Gamma)) \times (\Gamma\backslash\Omega(\Gamma))$ on summing the integrand over $\Gamma \times \Gamma$. Using the fact that $S_A(\zeta, \eta; s)$ and $S'_A(\zeta, \eta; s)$ are $\Gamma \times \Gamma$–invariant we see that the right-hand side can be transformed into $c'(T, F, A) + c(T, F)$

$$+ \iint (S'_A(\zeta, \eta; N-s)S_A(\zeta, \eta; s) - S_A^{F'}(\zeta, \eta; N-s)S_A(\zeta, \eta; s)$$

$$- S'_A(\zeta, \eta; N-s)S_A^F(\zeta, \eta; N-s) + S_A^{F'}(\zeta, \eta; N)) \, d\sigma_A(\zeta) \, d\sigma_A(\eta)$$

$$+ PV \iint (S_A^{F'}(\zeta, \eta; N-s)S_A(\zeta, \eta; s) - S_A^{F'}(\zeta, \eta; N)) \, d\sigma_A(\zeta) \, d\sigma_A(\eta)$$

$$+ PV \iint S'_A(\zeta, \eta; N-s)S_A^F(\zeta, \eta; s) - S_A^{F'}(\zeta, \eta; N)) \, d\sigma_A(\zeta) \, d\sigma_A(\eta).$$

Next observe that

$$S_A^{F'}(\zeta, \eta; N) - S_A^{F'}(\zeta, \eta; N - s)S_A^F(\zeta, \eta; s)$$

is smooth and also that

$$S_A^{F'}(\zeta, \eta; N) - S'_A(\zeta, \eta; N)$$

is smooth. Thus we can rewrite this again as

$$\begin{aligned}
c'(T, F, A) + c(T, F) &+ \iint (S_A^{F'}(\zeta, \eta; N) - S'_A(\zeta, \eta; N))\, d\sigma_A(\zeta)\, d\sigma_A(\eta) \\
&+ \iint (S'_A(\zeta, \eta; N - s) - S_A^{F'}(\zeta, \eta; N - s)) \\
&\cdot (S_A(\zeta, \eta; s) - S_A^F(\zeta, \eta; s))\, d\sigma_A(\zeta)\, d\sigma_A(\eta) \\
&+ PV \iint S_A^{F'}(\zeta, \eta; N - s)(S_A(\zeta, \eta; s) - S_A^F(\zeta, \eta; s))\, d\sigma_A(\zeta)\, d\sigma_A(\eta) \\
&+ PV \iint (S'_A(\zeta, \eta; N - s) - S_A^{F'}(\zeta, \eta; s)\, d\sigma_A(\zeta)\, d\sigma_A(\eta) \\
&+ \iint (S_A^{F'}(\zeta, \eta; N - s)S_A^F(\zeta, \eta; N))\, d\sigma_A(\zeta)\, d\sigma_A(\eta).
\end{aligned}$$

The second integral here is

$$\mathrm{Tr}((S'_A(N - s) - S_A^{F'}(N - s)) \cdot (S_A(s) - S_A^F(s)));$$

the third is

$$\mathrm{Tr}(S_A^{F'}(N - s)(S_A(s) - S_A^F(s)));$$

and the fourth is

$$\mathrm{Tr}((S'_A(N - s) - S_A^{F'}(N - s))S_A(s)).$$

These three terms add together to give

$$\mathrm{Tr}(S'_A(N - s)S_A(s) - S_A^{F'}(N - s)S_A^F(s)).$$

When we replace s by $N - s$ and divide by $c_N(s)c_N(N - s)$ we obtain the "trace term" of the formula of the theorem. The final integral here coincides with the last term in the formula. Since if F_1 and F^* satisfy the condition of Proposition 2.2

$$S^F(\zeta, \eta; s) - S^{F^*}(\zeta, \eta; s)$$

is smooth, it is easy to see that the sum of the last two terms does not depend on F. Thus (as $T = \{I\}$)

$$\Delta(A) = c'(T, F, A) + c(T, F) + \iint (S_A^{F'}(\zeta, \eta; N) - S'_A(\zeta, \eta; N))\, d\sigma_A(\zeta)\, d\sigma_A(\eta)$$

also does not depend on F and is the $\Delta(A)$ of the theorem.

References

[1] Colin de Verdière, Y. "Théorie spectrale des surfaces de Riemann d'aire infinie," *Astérisque* **132** (1985), 259–275.

[2] Elstrodt, J. Grunewald, F. and Mennicke, J. "The Selberg zeta-function for cocompact discrete subgroups of PSL(2, $\mathbb{C}$)." In: *Elementary and Analytic Theory of Numbers*, Banach Centre Publications, 17, PWN, Warsaw (1985), 83–120.

[3] Fried, D. "Fuchsian groups and Reidemeister torsion," *Contemp. Math.* **53** (1986), 141–163.

[4] Fried, D. "The zeta functions of Ruelle and Selberg, I," *Ann. Scient. Ec. Norm. Sup.* **19** (1986), 491–517.

[5] Gaillard, P.-Y. "Transformation de Poisson de formes differentielles—le cas de l'espace hyperbolique," Genève (1985), preprint.

[6] Gelfand, I.M. and Shilov, G.E. *Generalized Functions*, Vol. 1. Academic Press, (1964) New York, London.

[7] Guillopé, L. "Sur la distribution des longeurs des géodésiques fermés d'une surface compacte à bord totallement géodésique," *Duke Math. J.* **53** (1986), 827–848.

[8] Mandouvalos, N. "The theory of Eisenstein series and spectral theory for Kleinian groups," Ph.D. Thesis, Cambridge (1984).

[9] Mandouvalos, N. "The theory of Eisenstein series for Kleinian groups," *IHES* (1984), preprint.

[10] Mazzeo, R.R. and Melrose, R.B. "Meromorphic extension of the resolvent on complete spaces with asymptotically constant negative curvature," *J. Functional Anal.* **75** (1987) 260–310.

[11] Patterson, S.J. "The Laplacian operator on a Riemannian surface,"
I. *Comp. Math.* **31** (1975), 83–107;
II. *Comp. Math.* **32** (1976), 71–112;
III. *Comp. Math.* **33** (1976), 227–259.

[12] Patterson, S.J. "Spectral theory and Fuchsian groups," *Math. Proc. Cambridge Phil. Soc.* **81** (1977), 59–75.

[13] Patterson, S.J. "Lectures on measures of limit sets of Kleinian groups." In: *Analytical and Geometrical Aspects of Hyperbolic Space*, D.B. Epstein, ed., LMS Lecture Notes, **111**, CUP (1987), 281–323.

[14] Patterson, S.J. "On a lattice-point in hyperbolic space and related questions in spectral theory," *Arkiv för Mat.*, **26** (1988) 167–172.

[15] Perry, P.A. "The Laplace operator on a hyperbolic manifold, I. Spectral theory and scattering theory," *J. Functional Anal.* **75** (1987) 161–187.

[16] Perry, P.A. "Eisenstein series and the scattering matrix for certain Kleinian groups," (1987), preprint.

[17] Ruelle, D. "Zeta-functions for expanding maps and Anosov flows," *Inv. Math.* **34** (1976), 231–242.

[18] Ruelle, D. "Repellers for real analytic maps," *Erg. Thy. Dyn. Systs.* **2** (1982), 99–107.

[19] Schlichtkrull, H. *Hyperfunctions and Harmonic Analysis on Symmetric Spaces*. Birkhäuser (1984), Boston, Basel, Stuttgart.

[20] Selberg, A. "Harmonic analysis and discontinuous groups in weakly symmetric Riemannian spaces with applications to Dirichlet series," *J. Indian Math. Soc.* **20** (1956), 47–87.

[21] Stein, E.M. *Singular Integrals and Differentiability Properties of Functions*. Princeton Univ. Press (1970), Princeton.

[22] Stratmann, B. "Ergodentheoretische Untersuchungen Klein'scher Gruppen." Diplomarbeit, Göttingen (1986).

[23] Thurston, W. "The geometry and topology of three-manifolds," *Lecture Notes*, Princeton.

[24] Whittaker, E.T. and Watson, G.N. *Modern Analysis*. Cambridge Univ. Press (1946), Cambridge.

24 Nonarithmetic Lattices in Lobachevsky Spaces of Arbitrary Dimension

*I. PIATETSKI–SHAPIRO**

The purpose of this paper is to present examples of nonarithmetic lattices in hyperbolic spaces of arbitrary dimension. The first such examples for dimension three were given by Makarov. In these examples the lattices are generated by reflections. Vinberg gave a general theory of groups generated by reflections in Lobachevsky spaces. Using this theory he gave examples of nonarithmetic lattices for dimensions four and five. Later it was proved that for dimension larger than 995 lattices generated by reflections cannot exist.

The existence of nonarithmetic lattices for other symmetric spaces of rank one is known only for spaces of Hermitian type of dimension two or three. These examples were obtained by Mostow using his theory of quasireflections.

Our construction of nonarithmetic lattices is based on the following: Assume that we have two compact hyperbolic manifolds, M_1, M_2 having isomorphic submanifolds of codimension one; denote each of them by M_0. Now we cut M_1 and M_2 along M_0, and we glue one part of M_1 with one part of M_2. This construction was essentially known to F. Klein.

* All results are joint with M. Gromov.

NUMBER THEORY,
TRACE FORMULAS and
DISCRETE GROUPS

ISBN 0-12-067570-6

Let us recall that a Lobachevsky space can be described as $SO(n, 1)/K$, where K is the maximal compact subgroup of $SO(n, 1)$. Let F be a totally real field of algebraic numbers of degree m over Q. Consider a quadratic form over F, which under one embedding of F in $\mathbb{R}$ becomes a form of type $(n, 1)$ and the other embeddings of F in $\mathbb{R}$ becomes a definite form. Consider the set Γ of matrices in $M_{n+1}(F)$ with integral entries. It is well known that Γ is a lattice in the corresponding Lobachevsky space. If $m > 1$ then this lattice has compact fundamental domain, and if $m = 1$ it is usually noncompact, but still has finite volume.

In order to construct two such manifolds that have isomorphic submanifolds of codimension one, we consider a form f defined over F of signature $(n, 1)$ under one embedding γ_0 of F in $\mathbb{R}$, and definite under the other embedding of F in $\mathbb{R}$. Consider now the following two forms, $f + ax_{n+2}^2$ and $f + bx_{n+2}^2$ where $a/b \notin (F^*)^2$. We assume, of course, that under γ_0 both forms are of signature $(n + 1, 1)$ and definite under the other embeddings of F in $\mathbb{R}$. It is clear that the resulting hyperbolic manifolds have isomorphic submanifolds of codimension one. Now we glue two parts, one of each manifold, along the submanifold. Call the new manifold a hybrid. Our basic result is that under the above assumptions the hybrid is a nonarithmetic manifold. The proof is based on the following: The fundamental group of the hybrid contains large subgroups of the fundamental groups of M_1 and M_2. Let L be a lattice in hyperbolic space. A subgroup $\Delta \in L$ is said to be a large subgroup of L if Δ is Zariski dense in the group G of motions of the hyperbolic space.

We can prove that if two arithmetic lattices of G have conjugate large subgroups, then they are commensurable. Using this result it is easy to prove that the lattice constructed above is nonarithmetic.

25 On Some Functions Connected with the Sieve

H.-E. RICHERT

1.

At the basis of sieve theory are estimations of the sifting function

$$S(\mathscr{A}, \mathscr{P}, z) := |\{a \in \mathscr{A};\, (a, P(z)) = 1\}| \qquad \left(\mathscr{P} CP,\, P(z) := \prod_{\substack{p<z \\ p\in\mathscr{P}}} p\right),$$

counting the number of elements of a sequence $\mathscr{A}$ of integers that are not divisible by the primes $p < z$, $p \in \mathscr{P}$. One generally defines for the subsequences

$$\mathscr{A}_d := \{a \in \mathscr{A};\, a \equiv 0 \bmod d\}, \qquad d \,|\, P(z),$$

successively, approximations of the form

$$|\mathscr{A}_d| = \frac{\omega(d)}{d} X + r_d, \tag{1}$$

with a multiplicative function $\omega(d)$ satisfying $0 \le \omega(p) < p$ such that r_d are in the nature of remainder terms, and we put

$$\omega(p) := 0,\, p \notin \mathscr{P}; \qquad V(z) := \prod_{p<z} \left(1 - \frac{\omega(p)}{p}\right).$$

NUMBER THEORY,
TRACE FORMULAS and
DISCRETE GROUPS

ISBN 0-12-067570-6

If we assume that there are constants $\kappa > 0$ and $A \geq 1$ such that

$$\left| \sum_{w_1 \leq p < w_2} \frac{\omega(p)}{p} \log p - \kappa \log \frac{w_2}{w_1} \right| \leq A \qquad \text{for all } w_2 > w_1 \geq 2$$

(or a similar—even a one-sided—condition), we can classify sieves with respect to κ, which is called their dimension.

Then the inequalities for the sifting function take the form

$$S(\mathscr{A}, \mathscr{P}, z) \lesseqgtr XV(z) \begin{Bmatrix} F_\kappa \\ f_\kappa \end{Bmatrix} \left(\frac{\log y}{\log z} \right) + o \Bigg\} + R_y. \tag{2}$$

Here, $F_\kappa(u)$, $f_\kappa(u)$ are functions depending on the dimension κ, o stands in place of an (less important) error term, R_y is formed by the r_d's of (1), and $y(\geq z)$ is a parameter that permits balancing between the precision of the first term and the size of the remainder R_y. Usually, y is chosen as large as possible, but such that R_y is still of lower order than the first term, resulting in

$$S(\mathscr{A}, \mathscr{P}, z) \lesseqgtr XV(z) \begin{Bmatrix} F_\kappa \\ f_\kappa \end{Bmatrix} \left(\frac{\log y}{\log z} \right) \pm \varepsilon \Bigg\}. \tag{3}$$

The problem of "best" results of the type (3) has to be classified with respect to the dimension κ. For $\kappa = 1$ we know that

$$F(u) := \frac{2e^\gamma}{u}, \qquad f(u) := 0, \qquad 0 < u \leq 2,$$

$$(uF(u))' = f(u-1), \qquad (uf(u))' = F(u-1), \qquad u \geq 2$$

(continuous at $u = 2$) are suitable functions, and A. Selberg gave the examples

$$\mathscr{A}_\nu := \{n \leq x, \Omega(n) \equiv \nu \bmod 2\}, \qquad \nu = 0, 1,$$

for which in (3), instead of the $\geq, \leq$ – signs, one has (asymptotically) equality. He also gave examples corresponding to the case $\kappa = \frac{1}{2}$ ([12]), thus proving that for $\kappa = 1$ and $\kappa = \frac{1}{2}$ an improvement of the above general sieve inequalities is impossible.

Many results of type (2) have been proved or improved by the following principle, first systematically used by A.A. Buchstab. The sifting function satisfies

$$S(\mathscr{A}, \mathscr{P}, z) = S(\mathscr{A}, \mathscr{P}, z_1) - \sum_{\substack{z_1 \leq p < z \\ p \in \mathscr{P}}} S(\mathscr{A}_p, \mathscr{P}, p) \qquad \text{for } z > z_1 \geq 2. \tag{4}$$

Hence, assuming that $S(\mathscr{A}, \mathscr{P}, z_1)$ is already known (e.g., by choosing z_1 sufficiently small and applying the so-called Fundamental Lemma (cf. [7], Chapter 7)), any upper bound for the terms in the sum of (4) gives for the left-hand side of (4) a (new) lower bound and vice versa.

In his important paper [9], Iwaniec solved the aforementioned problem for all dimensions $\kappa \leq 1$. His functions F_κ, f_κ have not been proved best in the preceding sense, i.e., by virtue of examples, however, they are "Buchstab-best," namely, they are invariant with respect to Buchstab's principle.

For dimensions $\kappa > 1$ Iwaniec also pointed out that Selberg's sieve yields results superior to his, and, together with van de Lune and te Riele [10], he provided numerical material and prepared the ground for solving the problem to find Buchstab-invariant functions in this case.

In a recent paper with H. Halberstam and H. Diamond ([3]), a result of the form (2) with "Buchstab-best" functions for $\kappa > 1$ is proved. It is based on [10] and on what we have called the Fundamental Sieve Identity ([8]), namely

$$\sum_{d|P(z)} \mu(d)\phi(d) = \sum_{d|P(z)} \mu(d)\chi(d)\phi(d) + \sum_{d|P(z)} \mu(d)\bar{\chi}(d) \sum_{d|P(p(d))} \mu(t)\phi(dt).$$

Here $\chi(d)$ and $\phi(d)$ are arbitrary arithmetic functions, $\chi(1) = 1$, and $\bar{\chi}(d)$ is defined by

$$\bar{\chi}(1) := 0; \qquad \bar{\chi}(d) := \chi\left(\frac{d}{p(d)}\right) - \chi(d), \qquad d > 1,$$

$p(d)$ denoting the smallest prime divisor of d.

2.

In Selberg's sieve the function $\sigma_\kappa(u)$, defined by

$$\sigma_\kappa(u) = \begin{cases} 0 & \text{for } u \leq 0 \\ \dfrac{u^\kappa}{A_\kappa} & \text{for } 0 \leq u \leq 2, \end{cases} \qquad A_\kappa := (2e^\gamma)^\kappa \Gamma(\kappa + 1),$$

$$(u^{-\kappa}\sigma_\kappa(u))' = -\kappa u^{-\kappa-1}\sigma_\kappa(u-2), \qquad u \geq 2, \tag{5}$$

and continuous at $u = 2$ (cf. [7]. Chapter 6) plays a fundamental role. In fact, the first consequence of the Λ^2-method is

$$F_\kappa(u) = \frac{1}{\sigma_\kappa(u)}$$

([2]), and so it serves—via Buchstab—as a starting point for a first lower bound and subsequent improvements ([2], [11], [7], [3]).

Starting from its definition (5), the behavior of $\sigma_\kappa(u)$ turns out to be rather complicated and often requires a manifold integration (cf. [4]). Moreover, more advanced investigations show that its behavior as a function of κ is most important, whereas the definition acts as if $\sigma_\kappa(u)$ were a function of u with a fixed parameter κ. The same problem arises with several related functions (see, e.g., $r_\kappa(u)$ below). With respect to κ, it is known ([7], p. 196) that $\sigma_\kappa(u)$ is a decreasing function of κ, and a more intricate property,

$$\sigma_\kappa(2\kappa) > \frac{1}{2} \tag{6}$$

has been conjectured for years.

3.

Let

$$\kappa > 1.$$

Following [2], it is more convenient to consider the function

$$j_\kappa(u) := \sigma_\kappa(2u),$$

for which (5) becomes

$$j_\kappa(u) := \begin{cases} 0 & \text{for } u \le 0 \\ \dfrac{u^\kappa}{B_\kappa} & \text{for } 0 \le u \le 1, \end{cases} \qquad B_\kappa := e^{\gamma\kappa}\Gamma(\kappa + 1),$$

and for all $u \in \mathbb{R}$

$$(u^{-\kappa} j_\kappa(u))' = -\kappa u^{-\kappa-1} j_\kappa(u-1)$$

or

$$u j_\kappa'(u) = \kappa(j_\kappa(u) - j_\kappa(u-1)) \tag{7}$$

(the $'$ always refers to the argument u), and we quote ([2]*)

$$j_\kappa(u) > 0,\ j_\kappa'(u) > 0,\ u > 0; \qquad \lim_{u \to \infty} j_\kappa(u) = 1, \tag{8}$$

$$u j_\kappa''(u) = (\kappa - 1) j_\kappa'(u) - \kappa j_\kappa'(u-1), \qquad u > 0. \tag{9}$$

* There the notation is $\tau_\kappa(u) = e^{\gamma\kappa}\Gamma(\kappa) j_\kappa'(u)$.

For $u > 0$, $j''_\kappa(u)$ is continuous in u and has a unique, simple zero,

$$u_2 = u_2(\kappa),$$

$$j''_\kappa(u_2) = 0, \qquad j''_\kappa(u)\begin{cases} >0 & \text{for } 0 < u < u_2 \\ <0 & \text{for } u > u_2, \end{cases} \tag{10}$$

$$uj'''_\kappa(u) = (\kappa - 2)j''_\kappa(u) - \kappa j''_\kappa(u-1) \begin{cases} u > 0 & \text{if } \kappa > 2 \\ u > 1 & \text{if } 1 < \kappa \le 2, \end{cases} \tag{11}$$

and

$$\max(1, \kappa - 1) < u_2(\kappa) < \kappa.$$

It is well known that for differential-difference equations the Laplace transform constitutes a most convenient tool, here we obtain the pair (cf. [2])

$$\begin{aligned} e^{-\kappa g(s)} &= \int_0^\infty e^{-su} j'_\kappa(u)\, du, \qquad s \ge 0, \\ j'_\kappa(u) &= \frac{1}{2\pi i}\int_{(\sigma)} e^{uz - \kappa g(z)}\, dz, \qquad \sigma \ge 0, \end{aligned} \tag{12}$$

where $g(z)$ denotes the entire function

$$g(z) := \sum_{\nu=1}^{\infty} \frac{(-1)^{\nu+1} z^\nu}{\nu!\,\nu}.$$

It is related to the exponential integral by

$$g(x) = \gamma + \log x + \int_x^\infty \frac{e^{-t}}{t}\, dt = \int_0^x \frac{1 - e^{-t}}{t}\, dt, \qquad x > 0, \tag{13}$$

(cf. [1], (5.1.11), (5.1.33)). Applying the inversion formula on the convolution $\{j'_\kappa * j'_\lambda\}$ leads at once to the symmetric formula

$$j'_{\kappa+\lambda}(u) = \int_0^u j'_\kappa(t) j'_\lambda(u - t)\, dt, \tag{14}$$

and after integration we have

$$j_{\kappa+\lambda}(u) = \int_0^u j'_\kappa(u - t) j_\lambda(t)\, dt, \qquad \kappa > 0, \lambda > 0. \tag{15}$$

For different proofs of (14) and (15) see [2], Theorem 2.2 and [7], (6.3.13).

Rewriting (15) as

$$j_\kappa(u) - j_{\kappa-\lambda}(u) = -\int_0^1 j'_{\kappa-\lambda}(u-t)(1-j_\lambda(t))\,dt$$
$$-\int_1^u j'_{\kappa-\lambda}(u-t)(1-j_\lambda(t))\,dt, \qquad 0<\lambda<\kappa-1,$$

the properties of $j_\kappa(u)$ and the fact that

$$B_\lambda = 1 + O(\lambda^2), \qquad \text{as } \lambda \to 0,$$

yield a first general κ-result, namely,

For $\kappa > 1$ and fixed $u \in \mathbb{R}$,

$$\frac{\partial}{\partial\kappa} j_\kappa(u) = -\int_0^1 j'_\kappa(u-t)\log\frac{1}{t}\,dt = -\int_{u-1}^u \frac{j_\kappa(u)-j_\kappa(t)}{u-t}\,dt. \tag{16}$$

The last formula follows merely by partial integration from the first equation.

In order to obtain a result for $j'_\kappa(u)$ corresponding to (16), one simply employs (7) yielding

$$u\frac{\partial}{\partial\kappa} j'_\kappa(u) = j_\kappa(u) - j_\kappa(u-1) + \kappa\left(\frac{\partial}{\partial\kappa} j_\kappa(u) - \frac{\partial}{\partial\kappa} j_\kappa(u-1)\right), \tag{17}$$

and by continuity

$$\frac{\partial}{\partial\kappa} j'_\kappa(u) = \left(\frac{\partial}{\partial\kappa} j_\kappa(u)\right)', \qquad u>0, \tag{18}$$

from which, by (16),

$$\frac{\partial}{\partial\kappa} j'_\kappa(u) = -\int_0^1 j''_\kappa(u-t)\log\frac{1}{t}\,dt = -\int_{u-1}^u \frac{j'_\kappa(u)-j'_\kappa(t)}{u-t}, \qquad u>0,$$

and this procedure may be continued for obtaining corresponding formulas for the higher derivatives of $j_\kappa(u)$, only the upcoming singularities have to be watched.

4.

These results already enable us to prove proper κ-results for $j_\kappa(u)$ and so for $\sigma_\kappa(u)$. As an example, consider $u_2 = u_2(\kappa)$ which, according to (10), is uniquely determined by the equation

$$j''_\kappa(u) = 0. \tag{19}$$

If we consider (19) as an equation $F(\kappa, u) = 0$ for the implicit function $u = u_2(\kappa)$, we obtain, after verification of sufficient conditions and some simplification by the differential-difference equation,

$u_2(\kappa)$ has a continuous derivative:

$$u_2'(\kappa) = \frac{u}{\kappa j''_\kappa(u-1)} \int_0^1 j''_\kappa(u-t)\frac{dt}{t}\bigg|_{u=u_2} \left(= -\frac{F_\kappa}{F_u}\bigg|_{u=u_2}\right).$$

Obviously, this method can be applied to other parameters connected with $j_\kappa(u)$, e.g., to the zeros of the higher derivatives of $j_\kappa(u)$ or to the (by (8) unique) solution $h_c(\kappa)$ of

$$j_\kappa(h_c(\kappa)) = c \qquad \text{for } 0 < c < 1.$$

5.

In order to attack the problem (6) one is lead to a study of

$$\frac{d}{d\kappa} j_\kappa(u(\kappa)) = u'(\kappa) j'_\kappa(u) + \frac{\partial}{\partial \kappa} j_\kappa(u)$$

for $u(\kappa) = a\kappa + b$ and $a = 1$. Then the monotonicity of $j_\kappa(\kappa + b)$ is clearly determined by the sign of

$$s_\kappa(u) := j'_\kappa(u) + \frac{\partial}{\partial \kappa} j_\kappa(u). \tag{20}$$

Applying the Mean Value Theorem to the last expression in (16) yields

$$s_\kappa(u) = j'_\kappa(u) - \int_{u-1}^{u} j'_\kappa(\zeta_t)\,dt, \qquad u - 1 < t < u, \tag{21}$$

and the monotonicity of j'_κ (see (10)) permits reading off at once from (21) that

$$s_\kappa(u)\begin{cases} >0 & \text{for } 0 < u \le u_2(\kappa) \\ <0 & \text{for } u \ge u_2(\kappa) + 1, \end{cases} \tag{22}$$

so that, by continuity, $s_\kappa(u)$ has at least a zero in $(u_2, u_2 + 1)$.

By (18), (17), (9) and (7) one obtains

$$us'_\kappa(u) = \left(\frac{u}{\kappa} - 1\right) j'_\kappa(u) + \kappa(s_\kappa(u) - s_\kappa(u-1)), \qquad u > 0, \tag{23}$$

from which one infers that, for $u \le \kappa$, $s_\kappa(u)$ has at most one, simple zero.

With the first formula of (16) we further have, from (20), in view of

$$\int_0^1 \log \frac{1}{t}\, dt = 1$$

and with partial integration

$$s_\kappa(u) = \int_0^1 (j'_\kappa(u) - j'_\kappa(u-t)) \log \frac{1}{t}\, dt = \int_0^1 K(t) j''_\kappa(u-t)\, dt, \qquad u > 0, \quad (24)$$

as well as

$$s'_\kappa(u) = \int_0^1 K(t) j'''_\kappa(u-t)\, dt, \qquad u > 1, \qquad (25)$$

where

$$K(t) := 1 - t + t\log t, \qquad K(1) = 0,\ K'(t) = \log t \le 0,$$
$$0 < t \le 1; \qquad K(0) := 1.$$

We now combine (23), (25), and (24) to obtain

$$\left(\frac{u}{\kappa} - 1\right) j'_\kappa(u) + 2s_\kappa(u) = us'_\kappa(u) - (\kappa - 2)s_\kappa(u) + \kappa s_\kappa(u-1)$$
$$= \int_0^1 K(t)\{uj'''_\kappa(u-t) - (\kappa-2)j''_\kappa(u-1) + \kappa j''_\kappa(u-1-t)\}\, dt$$
$$= \int_0^1 tK(t) j'''_\kappa(u-t)\, dt, \qquad (26)$$

the latter by using the differential-difference equation (11). Since $j'''_\kappa(v) < 0$ in $\kappa - 1 \le v \le u_2 + 1$, (26) proves, in view of (22), that

$$s_\kappa(u) < 0 \qquad \text{for } u \ge \kappa.$$

So altogether, for $u > 0$, $s_\kappa(u)$ has exactly one, simple zero ζ_κ, satisfying

$$u_2(\kappa) < \zeta_\kappa < \kappa. \qquad (27)$$

As to the lower bound in (27) we know for $\kappa > 1$ only that

$$u_2(\kappa) > \kappa - \frac{1}{2},$$

which leaves a gap of width $\frac{1}{2}$ for ζ_κ; however, a careful lower estimation of the last integral in (26) shows that

$$\zeta_\kappa > \kappa - 0.26 \qquad \text{for } 1 < \kappa \le 2.$$

Recalling the meaning of $s_\kappa(u)$, these results prove that for constant b

$$j_\kappa(\kappa + b) \downarrow \qquad \text{if } b \geq 0 \tag{28}$$

and

$$j_\kappa(\kappa - b) \uparrow \qquad \text{if } b \geq \frac{1}{2},$$

and

$$b \geq 0.26 \qquad \text{if } 1 < \kappa \leq 2\,(b \leq \kappa).$$

Similar to (12), Laplace transforms provide also the formula

$$1 - j_\kappa(u) = \frac{1}{2\pi i}\int_{(\sigma)} e^{uz}\frac{1 - e^{-\kappa g(z)}}{z}\,dz, \qquad \sigma \geq 0, u > 0,$$

and a study of this integral on the vertical line $\sigma = 0$ leads to the result

$$\lim_{\kappa \to \infty} j_\kappa(\kappa + c) = \frac{1}{2} \tag{29}$$

for any constant $c \in \mathbb{R}$.

In particular, for $c = 0$, (29) proves, together with (28), the conjecture (6), and we mention that for $u = a\kappa + b$, $a \neq 1$, (29) is simply supplemented by

$$\lim_{\kappa \to \infty} j_\kappa(a\kappa + b) = \begin{cases} 1 & \text{if } a > 1 \\ 0 & \text{if } 0 < a < 1. \end{cases}$$

6.

Let us finally consider one of the related functions, $r_\kappa(u)$, defined by

$$r_\kappa(u) := \int_0^\infty e^{-ux + \kappa g(x)}\,dx, \qquad u > 0 \tag{30}$$

(for $g(x)$ see (13)).

Iwaniec ([9], p. 180) has pointed out that in connection with a differential-difference equation

$$sp'(s) = -ap(s) - bp(s - 1),$$

it is advantageous to consider also the "conjugate" equation

$$(sq(s))' = aq(s) + bq(s + 1),$$

because then

$$sp(s)q(s) = b\int_{s-1}^{s} p(t)q(t+1)\,dt + c \qquad \text{for } s \geq s_0 \tag{31}$$

and some constant c.

$r_\kappa(u)$ satisfies the differential-difference equation

$$(ur_\kappa(u))' = \kappa r_\kappa(u+1) - \kappa r_\kappa(u). \tag{32}$$

Therefore, using the above language, since by (7),

$$j_\kappa(u) \text{ is a } \quad p \qquad \text{with } a = -\kappa,\ b = \kappa,$$

(32) tells us that $r_\kappa(u)$ is a q with the same pair (a, b) and thus a conjugate to $j_\kappa(u)$. Hence an application of (31) leads to

$$uj_\kappa(u)r_\kappa(u) = \kappa\int_{u-1}^{u} j_\kappa(t)r_\kappa(t+1)\,dt + 1, \qquad u > 0,$$

or, equivalently,

$$j_\kappa(u) = 1 - \kappa\int_{u-1}^{u} (j_\kappa(u) - j_\kappa(t))r_\kappa(t+1)\,dt, \qquad u > 0. \tag{33}$$

For $r_\kappa(u)$ the analogue of (16) is readily obtained from (30). Making use of the formula

$$g(x) = x\int_0^1 e^{-xt}\log\frac{1}{t}\,dt$$

leads to

$$\frac{\partial}{\partial\kappa} r_\kappa(u) = -\int_0^1 r_\kappa'(u+t)\log\frac{1}{t}\,dt = -\int_u^{u+1}\frac{r_\kappa(t) - r_\kappa(u)}{t-u}\,dt, \qquad u > 0.$$

Monotonicity results analogous to those we showed for $j_\kappa(a\kappa + b)$ can, in this case, be obtained in different ways. Instead of elaborating on those, we mention that by observing that the integrand in (33) is, for $u \geq \kappa + 1$, convex in $u - 1 \leq t \leq u$, one obtains the inequality

$$j_\kappa(u) > 1 - \frac{ur_\kappa(u)}{2} j_\kappa'(u) \qquad \text{for } u \geq \kappa + 1$$

(for $2 \leq \kappa \leq 8$ this holds true even for $u \geq \kappa$). This inequality can be used to derive that

$$j_\kappa(u) > 1 - (1 - j_\kappa(\kappa+1))e^{2(\kappa+1-u)}\left(\frac{u}{\kappa+1}\right)^{\kappa}\left(\frac{\mu u + 1}{\mu(\kappa+1)+1}\right)^{\kappa} \qquad \text{for } u \geq \kappa + 1, \tag{34}$$

where μ is any constant satisfying

$$\mu \geq \frac{1}{2u}(u - \kappa - 1 + \sqrt{(u - \kappa + 1)^2 + 4\kappa}.$$

(34) improves a previous estimate of this type of Ankeny–Onishi ([2], (2.8), cf. [6], Lemma 3).

Details of these results will appear in joint papers with F. Grupp ([5]) and with H. Diamond and H. Halberstam ([3]), respectively.

References

[1] Abramowitz, M. and Stegun, I. *Handbook of Mathematical Functions*. J. Wiley, New York (1972).

[2] Ankeny, N.C. and Onishi, H. "The general sieve," *Acta Arith.* **10** (1964), 31–62.

[3] Diamond, H., Halberstam, H., and Richert, H.-E. "Combinatorial sieves of dimension exceeding one. I, II," I: *J. Number Theory* **28** (1988), 306–346; II: to appear.

[4] Grupp, F. "On difference-differential equations in the theory of sieves," *J. Number Theory* **24** (1986), 154–172.

[5] Grupp, F. and Richert, H.-E. "Notes on functions connected with the sieve," *Analysis*, **8** (1988), 1–23.

[6] Hagedorn, H. W. "Sieve methods and polynomial sequences," *Acta Arith.* **28** (1975), 245–252.

[7] Halberstam, H. and Richert, H.-E. *Sieve Methods*. Academic Press, London (1974).

[8] Halberstam, H. and Richert, H.-E. "A weighted sieve of Greaves' type. I." Banach Center Publications, Warsaw, **17** (1985), 155–182.

[9] Iwaniec, H. "Rosser's sieve," *Acta Arith.* **36** (1980), 171–202.

[10] Iwaniec, H., van de Lune, J., and te Riele, H.J. "The limits of Buchstab's iteration sieve," *Indag. Math.* **42** (1980), 409–417.

[11] Porter, J. W. "On the non-linear sieve," *Acta Arith.* **29** (1976), 377–400.

[12] Selberg, A. "Sieve Methods," *Proc. Symp. Pure Math.* **20** (1971), 311–351.

26 Special Values of Selberg's Zeta-function*

P. SARNAK

In his 1956 paper on the Trace Formula [12] Selberg introduced a zeta-like function associated to a lattice $\Gamma \leq \mathrm{SL}(2, \mathbb{R})$. We assume that $\begin{pmatrix} -1 & 0 \\ 0 & -1 \end{pmatrix} \in \Gamma$. To define this function we need some notation. An element $\gamma \in \Gamma$ is called primitive (or prime) if it generates its centralizer in Γ. The norm $N(\gamma)$ of a hyperbolic member of $\mathrm{SL}(2, \mathbb{R})$ is λ^2 where γ is conjugate to $\begin{pmatrix} \lambda & \\ & \lambda^{-1} \end{pmatrix}$, $|\lambda| > 1$. For χ a unitary character of Γ the above zeta-function is defined to be

$$Z(s, \chi) = \prod_{\{\gamma\}} \prod_{k=0}^{\infty} (1 - \chi(\gamma) N(\gamma)^{-(s+k)}), \tag{1}$$

where the product runs over the conjugacy classes of primitive $\gamma \in \Gamma$ with trace $(\gamma) > 2$. The product converges absolutely for $\mathrm{Re}(s) > 1$ and extends to a meromorphic function in the plane with functional equation $s \to 1 - s$. As with the classical L-functions this function may be used to study the asymptotics of these "prime geodesics" [6]. The

* I would like to thank D. Bertrand, R. Phillips and H. Royden for their helpful insights and comments as well as Alvarez-Gaume', Moore, and Vafa, whose paper [1] motivated this paper.

ISBN 0-12-067570-6

above characters allow one to study the distribution of these primes in a given homology class [9]. One striking difference between $Z(s, \chi)$ and the classical L-functions $L(s, \chi)$ is that the analogue of the Riemann Hypothesis is easily established for $Z(s, \chi)$ (there may be some zeros on the real axis). In this note, we examine the special value of $Z(s, \chi)$ at $s = \frac{1}{2}$, the middle of the critical strip. Actually we do this only for odd characters, i.e., $\chi(-I) = -1$, for the even characters the value at $s = 1$ would be more appropriate, but the situation there is much more complicated. We therefore assume, unless otherwise indicated, that χ is odd. The importance of this special value for certain L-functions is well known. In our case, its importance lies in its relation to the Dirac operator and especially its determinant. These determinants have received much attention recently because of their appearance in quantum string theory [3, 1].

In order to describe our results we need to specify certain quantities more explicitly. Firstly, assume that $\bar{\Gamma} = \Gamma/\{\pm I\}$ is the fundamental group of a compact Riemann surface M of genus $g \geq 2$. Thus $M = \mathfrak{h}/\bar{\Gamma}$ where $\mathfrak{h}$ is the upper half-plane. Once and for all let $a_1, a_2, \ldots, a_g, b_1, \ldots, b_g$ be a canonical basis for the homology on M. Let τ be the corresponding period matrix of the normalized Abelian differentials of the first kind. Let $J(M)$ be the Jacobian of M, i.e., $J(M) = \mathbb{C}^g/L$ where $L = \{m + \tau n | m, n \in \mathbb{Z}^g\}$ and let ϕ be the corresponding map of M into $J(M)$ relative to some base point. Since $\bar{\Gamma}/[\bar{\Gamma}, \bar{\Gamma}] \cong \mathbb{Z}a_1 \oplus \mathbb{Z}a_2 \oplus \cdots \oplus \mathbb{Z}a_g \oplus \cdots \oplus \mathbb{Z}b_g$, we can parametrize the characters of $\bar{\Gamma}$ by $(\theta, \psi) \in \mathbb{R}^g/\mathbb{Z}^g \times \mathbb{R}^g/\mathbb{Z}^g$, where

$$
\begin{aligned}
\chi_{(\theta,\psi)}(\gamma) &= e^{2\pi i(\langle \pi_1(\gamma), \theta \rangle + \langle \pi_2(\gamma), \psi \rangle)} \\
\pi(\gamma) &= n_1 a_1 \oplus n_2 a_2 \oplus \cdots \oplus n_g a_g \oplus \cdots \oplus m_g b_g \\
\pi_1(\gamma) &= (n_1, n_2, \ldots, n_g) \\
\pi_2(\gamma) &= (m_1, m_2, \ldots, m_g). \qquad (2)
\end{aligned}
$$

By setting

$$u = \tau(-\theta) + \psi, \qquad (3)$$

we can parametrize the character group of $\bar{\Gamma}$ by $J(M)$ by letting χ_u correspond to $\chi_{(\theta,\psi)}$. This, of course, parametrizes the even characters of Γ. If v is any odd character (they exist) then we may parametrize all odd characters as $v\chi_u$ for $u \in J(M)$. In order to normalize this representation of all odd characters, we single out an odd character of order two as follows: There are 2^{2g} odd characters of order 2, which we refer to as spin structures. Corresponding to any odd character χ we consider automorphic forms on $\mathfrak{h}$ satisfying

$$f(\gamma z) = \chi(\gamma)(cz + d)f(z) \tag{4}$$

for $\gamma \in \Gamma$. The divisor of any meromorphic (nonzero) section of (4) defines a divisor class on M that we denote by B_χ. If v is a spin structure, it is clear that $2B_v$ is the canonical class W. On the other hand, it follows from Riemann's Vanishing Theorem [4] that $\phi(W) = -2K$, where K is Riemann's vector of constants. It follows that for any spin structure v, $2(\phi(B_v) + K) = 0$ in $J(M)$. In particular, there is a unique spin structure v_0 for which

$$\phi(B_{v_0}) = -K. \tag{5}$$

With this v_0 we parametrize the odd characters of Γ by λ_u, $u \in J(M)$, where

$$\lambda_u = v_0 \chi_u. \tag{6}$$

Finally, recall the definition of the theta function

$$\theta(u, \tau) = \sum_{n \in \mathbb{Z}^g} e^{\pi i n' \tau n + 2\pi i n' u} \tag{7}$$

for $u \in \mathbb{C}^g$.

Let

$$\begin{cases} Y = \mathrm{Im}(\tau) & \text{and} \\ u = \xi + i\eta \end{cases}. \tag{8}$$

Our main result is the following evaluation of $Z(s, \chi)$ at $s = \frac{1}{2}$, χ odd.

Theorem 1.

$$Z\left(\frac{1}{2}, \lambda_u\right) = ce^{-2\pi\eta' Y^{-1}\eta} |\theta(u, \tau)|^2$$

for some $c > 0$ independent of u.

Remark 1. (a) It does not seem possible to express c in terms of τ in an elementary way.

(b) An analogue of Theorem 1 for L-functions is the theorem of Waldpurger see [7].

(c) The above result fits nicely into the theory of heights and special values. In fact, the logarithm of either side is the Neron function on

$J(M)$ with respect to the θ divisor. This essentially will be the method of establishing the theorem.

The first step in proving Theorem 1 is to express $Z(\frac{1}{2}, \chi)$ in terms of determinants. Let $T^n(\chi)$ denote the automorphic forms on $\mathfrak{h}$ satisfying

$$f(\gamma z) = \chi(\gamma)(cz + d)^{2n} f(z). \tag{9}$$

The first-order operators (essentially the Maass operators) ∇_n and ∇^n are defined as follows

$$\nabla_n \colon T^n(\chi) \to T^{n-1}(\chi) \qquad \text{by } \nabla_n f = y^2 \frac{\partial f}{\partial \bar{z}} \tag{10}$$

and

$$\nabla^{n-1} \colon T^{n-1}(\chi) \to T^n(\chi) \qquad \text{by } \nabla^{n-1} = y^{-2n} \frac{\partial}{\partial z} y^{2n}. \tag{10'}$$

We may turn $T^n(\chi)$ into a Hilbert space by

$$(f, g)_{T^n} = \int_{\mathfrak{h}/\Gamma} f(z)\overline{g(z)} y^{2n} \frac{dx\, dy}{y^2}. \tag{11}$$

With this it is easily checked that

$$\nabla_n^* = -\nabla^{n-1}. \tag{12}$$

In particular $\nabla_{1/2}$ is called the Dirac operator and $\Delta_{1/2} = (\nabla_{1/2})^* \nabla_{1/2}$ is its corresponding Laplacian. Now given a positive elliptic operator such as Δ above, we may define its determinant by

$$\det \Delta = \prod_{j=0}^{\infty} \lambda_j, \tag{13}$$

where $\lambda_0 \leq \lambda_1 \ldots$ are its eigenvalues. This product must be regularized in some way. The method we have in mind here is

$$\det \Delta = 0 \qquad \text{if } \lambda_0 = 0,$$

while if $\lambda_0 > 0$, $\det \Delta = e^{-Z'(0)}$, where

$$Z(s) = \sum_{j=0}^{\infty} \lambda_j^{-s}. \tag{14}$$

That $Z(s)$ is regular at $s = 0$ is well known [13].

where

$$D_u\colon T^{1/2}(v) \to T^{-1/2}(v)$$
$$D_u^*\colon T^{-1/2}(v) \to T^{1/2}(v)$$

are defined by

$$D_u = y^2\left(2\pi i\bar{\mu} + \frac{\partial}{\partial \bar{z}}\right)$$

$$D_u^* = y^2\left(2\pi i\mu + \frac{\partial}{\partial z}\right)$$

$$-i\bar{\mu}\,\overline{dz} = \sum_{j=1}^{g}\sum_{k=1}^{g}(Y^{-1})_{jk}u_k\bar{\zeta}_j,$$

where $\zeta_1, \ldots, \zeta_g$ is the canonical basis of Abelian differentials of the first kind dual to $a_1, \ldots, b_g$. The point is that L_u is factored into an operator D_u that is clearly holomorphic in u and D_u^*, which is antiholomorphic. Furthermore, on $\mathbb{C}^g\setminus V$ both D_u and D_u^* have no kernel so that Quillen's method of holomorphic factorization of the determinant [10] leads to (17)(i).

To establish (17)(ii) or, more directly, Theorem 1, we rewrite (i) as

$$\partial\,\bar{\partial}(\log(e^{-(\pi/2)(u-\bar{u})'Y^{-1}(u-\bar{u})}\det \Delta_{1/2}(\lambda_u))) = 0 \tag{20}$$

for $u \in \mathbb{C}^g\setminus V$. From this it follows that locally for $u \in \mathbb{C}^g\setminus V$,

$$\det \Delta_{1/2}(\lambda_u)e^{-(\pi/2)(u-\bar{u})'Y^{-1}(u-\bar{u})} = |h(u)|^2 \tag{21}$$

for some holomorphic function $h(u)$. If we could show that $h(u)$ is in fact single valued and holomorphic in $\mathbb{C}^g$, then it follows easily from the fact that $\det \Delta_{1/2}(\lambda_u)$ is periodic by the lattice L that $h(u)$ is a θ function (see [1]). Since the divisor of h is V it would then follow that, in fact, h is a multiple of $\theta(u, \tau)$, and the theorem would follow.

So we let $h(u)$ be a function element as determined in (21). It clearly admits unrestricted analytic continuations to a multiple valued function on $\mathbb{C}^g\setminus V$. Let $R \subset V$ be the set of regular points of the variety V [5], i.e., those points $P \in V$ that have neighborhoods U that are biholomorphic to a neighborhood of 0 such that $V \cap U$ is mapped to $z_1 = 0$. It is known that $C = V\setminus R$ is of dimension $\leq g - 2$ [5].

Lemma 1. *$h(u)$ admits unrestricted analytic continuations in $\mathbb{C}^g\setminus C$.*

Once Lemma 1 is established, the theorem follows easily since $\mathbb{C}^g \setminus C$ is simply connected so that, in fact, $h(u)$ is single valued there. Since $h(u)$ is clearly locally bounded, it extends to a holomorphic function in $\mathbb{C}^g$ as needed. To establish Lemma 1 we need

Lemma 2. *$Z(\frac{1}{2}, \lambda_u)$ is real analytic in u.*

Proof. That $Z(s, \lambda_u)$ for $\mathrm{Re}(s) > 1$ is real analytic in u is clear from the definition in (1), the absolute convergence of the product and the easily established estimate $|(\pi_1(\gamma), \pi_2(\gamma))| \le \tilde{C} \log N(\gamma)$ for some $\tilde{C}$ independent of γ. To obtain the same fact for any s and, in particular, $s = \frac{1}{2}$, we use the functional equation as follows: Let $k(w)$ be an analytic function in the strip $-3 < \mathrm{Re}(w) < 3$ for which $k(0) = 1$ and which is rapidly decreasing as $|\mathrm{Im}(w)| \to \infty$. Consider

$$\frac{1}{2\pi i} \int_{2 - i\infty}^{2 + i\infty} Z\left(\frac{1}{2} + w, \lambda_u\right) \frac{k(w)}{w}\, dw.$$

On shifting the line of integration to $\mathrm{Re}(w) = -1$ we pick up the residue at 0, which is $Z(\frac{1}{2}, \lambda_u)$. Thus

$$Z\left(\frac{1}{2}, \lambda_u\right) = \frac{1}{2\pi i} \int_{2 - i\infty}^{2 + i\infty} Z\left(\frac{1}{2} + w, \lambda_u\right) \frac{k(w)}{w}\, dw + \frac{1}{2\pi i} \int_{-1 - i\infty}^{-1 + i\infty} Z\left(\frac{1}{2} + w, \lambda_u\right) \frac{k(w)}{w}\, dw.$$

Using the functional equation $Z(s) = \Lambda(s) Z(1 - s)$, where $\Lambda(s)$ is the appropriate ratio of double gamma functions, we have

$$Z\left(\frac{1}{2}, \lambda_u\right) = \frac{1}{2\pi i} \int_{2 - i\infty}^{2 + i\infty} Z\left(\frac{1}{2} + w, \lambda_u\right) \frac{k(w)}{w}\, dw + \frac{1}{2\pi i} \int_{-1 - i\infty}^{-1 + i\infty} \Lambda\left(\frac{1}{2} + w\right) Z\left(\frac{1}{2} - w, \lambda_u\right) \frac{k(w)}{w}\, dw.$$

Lemma 2 follows immediately from this representation.

Finally, we establish Lemma 1. Let $P \in V$ be a regular point. We must show that h is analytic at P. We can assume $P = 0$ and $V = \{z \mid z_1 = 0\}$ in a neighborhood U of 0. Our function h is multivalued on $U \setminus V$. The problem of multiple valuedness quickly reduces to one variable in this case, and the following lemma, together with Lemma 2, is what is needed.

Lemma 3. *Let $h(z)$ be a multiple-valued analytic function of one variable in $0 < |z| < 1$ for which $F(z) = |h(z)|^2$ is real analytic in $|z| < 1$. Then $h(z)$ is single valued and analytic in $|z| < 1$.*

Proof. It is easy to see that such an $F(z)$ is of the form $|z|^{\alpha}|g(z)|^2$, $\alpha \in \mathbb{R}$, where $g(z)$ is analytic in $|z| < 1$ and $g(0) \neq 0$. Since $F(z)$ is real analytic, it follows that α is an even non-negative integer and hence that $h(z) = c_1 z^{\alpha/2} g(z)$ is analytic in $|z| < 1$.

References

[1] Alvarez-Gaume', L., Moore, G., and Vafa, C. "Theta functions, modular invariance and strings," *Comm. Math. Phys.* **106** (1986), 1–40.

[2] Barnes, E.W. "The theory of the *G*-function," *Quarterly Journal of Math.* **31** (1900), 264–314.

[3] D'Hoker, E. and Phong, D.H. "Multiloop amplitudes for the bosonic Polyakov string," *Nucl. Phys.* **B 269** (1986), 205–234.

[4] Farkas, H. and Kra, I. "Riemann surfaces," *GTM* **71**, Springer Verlag (1980).

[5] Gunning, R. and Rossi, H. *Analytic Functions of Several Complex Variables.* Prentice Hall (1965).

[6] Hejhal, D.A. "The Selberg trace formula for PSL(2, ℝ)," *Lecture Notes in Math.* **1** (1976), 548.

[7] Kohnen, W. and Zagier, D. "Values of *L*-series of modular forms at the center of the critical strip," *Invent. Math.* **64** (1981), 175–198.

[8] Lang, S. *Fundamentals of Diophantine Geometry.* Springer Verlag (1983).

[9] Phillips, R. and Sarnak, P. "Geodesics in homology classes," *Duke Math. Jr.*, **55**, 2 (1987), 287–297.

[10] Quillen, D. "Determinant of Cauchy–Riemann operators on Riemann surfaces," *Funct. Anal.* **19** (1985), 37.

[11] Sarnak, P. "Determinants of Laplacians," *Comm. Math. Physics* **110** (1987), 113–120.

[12] Selberg, A. "Harmonic analysis and discontinuous groups in weakly symmetric Riemannian spaces with applications to Dirichlet series," *Jr. Ind. Math. Soc.* **20** (1956), 47–87.

[13] Singer, I. "Eigenvalues of the Laplacian and invariants of manifolds" *Int. Cong. of Math.* Vancouver (1974).

[14] Vignéras, M.F. "L'equation fonctionelle de la fonction zéta de Selberg do groupe modulare SL(2, ℤ)," Société Arithmétique de France, *Astérisque* **61** (1979), 235–249.

27 Sifting Problems, Sifting Density, and Sieves

ATLE SELBERG

1.

My aims in this lecture are several. First, to introduce some new terminology that clearly distinguishes between the problem, the sifting, and the tool or instrument, the sieve. Second, to indicate some generalizations of the theory sketched out in my lectures at Stony Brook 1969. Finally, I also present some new results concerning the sifting limit for high constant density.

All of this material will be more fully presented in a comprehensive account of my work on the theory of sieves that is under preparation.

2.

The original sifting problem can be stated as follows:

We have given an interval $\mathscr{I}_x$ of length x and a finite set of primes P, which we call the sifting range. The elements of P we denote by p, and with each p in P is associated a set of $u(p)$ distinct residue classes modulo p. We refer to $u(p)$ as the sifting density with respect to p.

The problem is now to estimate the number of integers remaining in $\mathscr{I}_x$ after we have excluded all those integers that belong to one of the $u(p)$ residue classes associated with some p in P.

NUMBER THEORY,
TRACE FORMULAS and
DISCRETE GROUPS

ISBN 0-12-067570-6

We shall here look at a more general formulation of the problem, which of course is modeled on the case of the interval $\mathscr{I}_x$ and the exclusion of certain residue classes for the primes p in P.

We assume that we have a set of non-negative numbers ("weights") w_n associated with each integer n and such that $\sum_n w_n = x$. We call this a weighted set and denote it by W_x. We also have a finite set of primes P, which we call the sifting range, and assume that with each p there is associated a *sifting property* $p \times n$ (read "p excludes n" or "p crosses out n"), which may be valid or not for each individual n. If for some n the relation $p \times n$ is false for all p in P, we write $P \overline{\times} n$ (read "P does not exclude n" or "P does not cross out n").

In this setting our sifting problem consists of estimating the quantity

$$M(W_x, P) = \sum_{P \overline{\times} n} w_n. \tag{2.1}$$

Some assumptions have to be made in order to say anything nontrivial, and these assumptions are modeled on the earlier problem for $\mathscr{I}_x$. First we introduce the notation (P) to denote the set of square-free non-negative integers all of whose prime factors belong to P. The elements of (P) we denote by the letters d or δ. If the relation $p \times n$ holds for all p that divide d, we write $d \times n$, (note that $1 \times n$ holds for all n), we assume

$$N_d = N_d(W_x, P) = \sum_{d \times n} w_n = \frac{u(d)}{d} x + R_d, \tag{2.2}$$

where $u(d)$ is a multiplicative function on (P) such that $0 < u(p) < p$,[1] and $|R_d| \leq u(d)$. We again call $u(p)$ the sifting density with respect to p.

It is easily seen that

$$\begin{aligned} M(W_x, P) &= \sum_d \mu(d) N_d(W_x, P) \\ &= x \sum_d \mu(d) \frac{u(d)}{d} + \sum_d \mu(d) R_d \\ &= x \prod_{p \in P} \left(1 - \frac{u(p)}{p}\right) + \sum_d \mu(d) R_d, \end{aligned} \tag{2.3}$$

[1] If $u(p) = p$, the problem is clearly trivial, and if $u(p) = 0$, we may omit p from the sifting range.

where $\mu(d)$ is the Möbius function. Unless P contains very few elements, this expression is useless since the term

$$\sum_d \mu(d)R_d$$

cannot be estimated well enough.

We shall write

$$E(W_x, P) = x \prod_{p \in P} \left(1 - \frac{u(p)}{p}\right) = xe(P) \tag{2.4}$$

and refer to $E(W_x, P)$ as the *expected value* of $M(W_x, P)$ and to $e(P)$ as the *expectation.*

3.

By a Λ-system $\Lambda(P)$ we shall understand a set of real numbers λ_d associated with each d in (P). If $\Lambda(P)$ for all d in (P) satisfies the condition

$$\theta_d = \sum_{\delta | d} \lambda_\delta \geq \sum_{\delta | d} \mu(\delta), \tag{3.1}$$

we call $\Lambda(P)$ an upper bound sieve and denote it by $\Lambda^+(P)$. If the $\Lambda(P)$ for all d in (P) satisfies the conditions

$$\theta_d = \sum_{\delta | d} \lambda_\delta \leq \sum_{\delta | d} \mu(\delta), \tag{3.1$'$}$$

for all d in (P), we refer to it as a lower bound sieve and denote it by $\Lambda^-(P)$.

Evidently, we have for a Λ^+ that

$$\begin{aligned} M(W_x, P) &\leq \sum_d \lambda_d N_d(W_x, P) \\ &= x \sum_d \lambda_d \frac{u(d)}{d} + \sum_d R_d \lambda_d \\ &\leq x \sum_d \lambda_d \frac{u(d)}{d} + \sum_d u(d) |\lambda_d|, \end{aligned} \tag{3.2}$$

and similarly for a $\Lambda^-(P)$ that

$$M(W_x, P) \geq x \sum_d \lambda_d \frac{u(d)}{d} - \sum_d u(d) |\lambda_d|. \tag{3.2$'$}$$

It can be shown, as indicated in my Stony Brook lectures, that by suitable choice of $\Lambda^+(P)$ and $\Lambda^-(P)$, (3.2) and (3.2′) actually give the best possible upper bound and best possible lower bound for $M(W_x, P)$ under the assumptions about W_x that we made above. In other words, if we denote by $M^+(W_x, P)$ the maximal value (it will be attained) of $M(W_x, P)$ and, similarly, by $M^-(W_x, P)$ the minimal value (also attained) with some fixed P and $u(p)$, we have that there exists at least one $\Lambda^+(P)$ such that

$$M^+(W_x, P) = x \sum_d \frac{u(d)}{d} \lambda_d + \sum_d u(d)|\lambda_d|, \tag{3.3}$$

and at least one $\Lambda^-(P)$ such that

$$M^-(W_x, P) = x \sum_d \frac{u(d)}{d} \lambda_d - \sum_d u(d)|\lambda_d|. \tag{3.3′}$$

We refer to these sieves as *optimal* for the given sifting problem. They are not necessarily unique. In the case of an upper bound sieve, it is easily seen that an optimal sieve must have $\lambda_1 = 1$, since otherwise we would by (3.1) have $\lambda_1 > 1$, and we would obtain a better upper bound sieve by dividing all λ_d by λ_1. In the case of a lower bound sieve, it is seen that an optimal sieve that gives a positive lower bound must have $0 < \lambda_1 \le 1$; if $\lambda_1 < 1$ we could divide all λ_d by λ_1 and would obtain a better lower bound. Thus, we may restrict ourselves always to $\Lambda(P)$ with $\lambda_1 = 1$, since the best nontrivial bounds can always be obtained by such sieves.

We shall introduce the notation $f(d)$ for the multiplicative function $d/u(d)$, and $f'(d)$ as the multiplicative function for which $f'(p) = f(p) - 1$.

By using the relation

$$\lambda_d = \sum_{\delta|d} \mu(\delta)\theta_{d/\delta},$$

we find

$$\sum_d \frac{u(d)}{d} \lambda_d = \prod_{p\in P}\left(1 - \frac{u(p)}{p}\right) \sum_{d\in(P)} \frac{\theta_d}{f'(d)} = e(P)T(\Lambda).$$

The expression $T(\Lambda)$ for a sieve thus measures the ratio between the *main term*

$$x \sum_d \frac{u(d)}{d} \lambda_d$$

and the *expected value* $E(W_x, P)$.

4.

Generally we will be concerned with situations where $P = P(x^{\alpha})$, the set of all primes $p < x^{\alpha}$, or $P = P(x^{\alpha}, x^{\alpha'})$, the set of all primes $x^{\alpha} \leq p < x^{\alpha'}$, and we ask what can be said about $M^{+}(W_x, P)$ and $M^{-}(W_x, P)$ as $x \to \infty$.

This requires, of course, some further assumptions about the $u(p)$. We shall assume that as $x \to \infty$

$$e(P) > x^{-\varepsilon} \qquad \text{if } x > x_0(\varepsilon) \tag{4.1}$$

holds for any fixed $\varepsilon > 0$. We shall assume

$$S(x^{\beta}) = \sum_{p < x^{\beta}} \frac{u(p)}{p} \log p = g(\beta) \log x + o(\log x), \tag{4.2}$$

where $g(\beta)$ is an increasing continuous function with $g(0) = 0$. For simplicity we shall assume (though less would do) that $g(\beta)$ has a continuous derivative $g'(\beta)$, except possibly at isolated points, and that $\sqrt{\beta} g'(\beta) \to 0$ as $\beta \to 0$. Where $g'(\beta)$ exists, we call it the sifting density around x^{β}. The case most considered in the literature is the case of constant sifting density k, when $g(\beta) = k\beta$ with some positive constant k.

Under these assumptions the following statements can be proved:

Theorem 1. *There exists a function $T_g^{+}(\alpha)$ such that*

$$\lim_{x \to \infty} \frac{M^{+}(W_x, P(x^{\alpha}))}{E(W_x, P(x^{\alpha}))} = T_g^{+}(\alpha) \tag{4.3}$$

and a function $T_g^{-}(\alpha)$ such that

$$\lim_{x \to \infty} \frac{M^{-}(W_x, P(x^{\alpha}))}{E(W_x, P(x^{\alpha}))} = T_g^{-}(\alpha). \tag{4.3$'$}$$

The function $T_g^{+}(\alpha)$ is continuous and increasing for $0 \leq \alpha$, and the function $T_g^{-}(\alpha)$ continuous and decreasing for $0 \leq \alpha < 1$. Both functions are effectively computable.

The proof is similar to that given in my Stony Brook lectures for Theorem 1 there. It also follows from the proof that in approaching the limits given by (4.3) and (4.3′) we may restrict ourselves to sieves $\Lambda(P(x^{\alpha}))$ such that the $\lambda_d = 0$ for $d > z$, with some $z = x^{1-\varepsilon}$ where $\varepsilon > 0$ is a function of x tending to zero as x tends to infinity, and where λ_d, for $d = p_1 p_2 \cdots p_r$, is of the form

$$\lambda_d = \ell_r\left(\frac{\log p_1}{\log z}, \frac{\log p_2}{\log z}, \ldots, \frac{\log p_r}{\log z}\right), \tag{4.4}$$

where $\ell_r(u_1, \ldots, u_r)$ is a continuous function of the r arguments for $u_1 + u_2 + \cdots + u_r \leq 1$.

An analogous theorem can be proved for the more general problem of sifting with weights. We assume that, instead of estimating the quantity

$$\sum_{P \overline{\times} n} w_n,$$

we wish to estimate a sum where we also count in n's for which a limited number of p, say $p_1 < \cdots < p_r$ with $r \leq r_0$, in P satisfy the relation $p_i \times n$ for $i = 1, \ldots, r$, while for all other p in P, we have $p \overline{\times} n$, and counting them with a weight factor of the form

$$\sigma_r\left(\frac{\log p_1}{\log x}, \frac{\log p_2}{\log x}, \ldots, \frac{\log p_r}{\log x}\right) \geq 0.$$

Defining $\Lambda_\sigma^+(P)$ to be a Λ for which, if $d = p_1 \cdots p_r$ with $r \leq r_0$, we have

$$\sum_{\delta | d} \lambda_\delta \geq \sigma_r\left(\frac{\log p_1}{\log x}, \frac{\log p_2}{\log x}, \ldots, \frac{\log p_r}{\log x}\right),$$

and if $r > r_0$, $\sum_{\delta | d} \lambda_\delta \geq 0$, and $\Lambda_\sigma^-(P)$ if we replace $\geq$ by $\leq$ in these inequalities, we can prove a result similar to Theorem 1, if we impose suitable conditions on $g(\beta)$ and the functions $\sigma_r(u_1, \ldots, u_r)$. For instance, if we require $\sigma_r(u_1, \ldots, u_r) = \mathcal{O}(u_1^\delta)$ (u_1 being the smallest argument) for some $\delta > 0$, and assume $g'(\beta)$ bounded uniformly in $0 \leq \beta \leq \alpha$, the corresponding theorem holds for sifting with weights of the form σ_r, and in approaching the respective limit functions, we may again restrict ourselves to sieves of the form (4.4).

While the case of a variable density function $g'(\beta)$ is theoretically interesting, it is unlikely to be of much interest when it comes to applications of sieve methods.

Returning to the function $T_g^-(\alpha)$ it is easily seen that for $\alpha > 1$ we have $T_g^-(\alpha) = 0$, while for α sufficiently small $T_g^-(\alpha)$ is positive, since $T_g^-(0) = 1$. Therefore, there exists an $0 < \alpha_g \leq 1$ such that $T_g^-(\alpha_g - \varepsilon) > 0$ and $T_g^-(\alpha_g + \varepsilon) = 0$ for any positive small ε. We may refer to α_g as the *sifting limit* of the problem. Up to now it has been studied only for the case of constant sifting density k, we shall in this case write α_k instead of α_g. It is known that for $0 \leq k \leq \frac{1}{2}$ we have $\alpha_k = 1$, and that for $k = 1$ we have $\alpha_k = \frac{1}{2}$. For no other value of k is the precise value of α_k known, for some small k we have both upper and lower bounds for α_k (the lower bounds probably being closer to the truth), and for large k some lower bounds are known.

Lower bounds for α_k are in general produced by constructing a special lower bound sieve to fit the problem, upper bounds for α_k, since they imply a statement about all lower bound sieves, involve producing a suitable W_x that gives a contradiction if we assume α_k is too large, or sometimes replacing the original sifting problem with another that is such that a positive lower bound for $M^-(W_x, P(x^\alpha))$ for the first problem implies a positive lower bound for the second problem, and finally constructing a W'_x for the second problem for which the result of the sifting is not positive.

5.

The specific methods so far utilized for constructing sieves can be listed as follows

a) The combinatorial sieve, developed first by Brun and perfected by Buchstab and Rosser, I shall here refer to it as the "B^2R sieve." For sifting problems with constant density 1, its theory was developed fully by Rosser in the case $k = 1$, for general k it was done by myself and independently by Iwaniec (his proofs are rather different from mine though the results, of course, are the same). For the combinatorial sieve the λ_d are either equal to $\mu(d)$ or to zero. I shall not give the rules for the choice here, they are found in the literature, for instance, in my Stony Brook lectures.

b) The process of Buchstab iteration, which I shall here define not in terms of Λ systems (though this is easily done), but rather in a way that is intuitively easier and corresponds more to Buchstab's original definition. We introduce the notation $W_x(p)$ to denote the weighted set formed by the weights w_n for which $p \times n$. Obviously if p does not divide d, we have

$$N_d(W_x(p)) = N_{dp}(W_x).$$

If now $\alpha > \beta \geq 0$, we have clearly

$$M(W_x, P(x^\alpha)) = M(W_x, P(x^\beta)) - \sum_{x^\beta \leq p < x^\alpha} M(W_x(p), P(p)), \qquad (5.1)$$

from which, for instance, we can conclude

$$M^-(W_x, P(x^\alpha)) \geq M^-(W_x, P(x^\beta)) - \sum_{x^\beta \leq p < x^\alpha} M^+(W_x(p), P(p)). \qquad (5.1')$$

Similarly

$$M^+(W_x, P(x^\alpha)) \le M^+(W_x, P(x^\beta)) - \sum_{x^\beta \le p < x^\alpha} M^-(W_x(p), P(p)). \quad (5.1'')$$

c) Multiplication of Λ-systems and sieves. If we define for two Λ-systems Λ' and Λ'' a new

$$\Lambda = \Lambda'\Lambda'' \quad (5.2)$$

by the rule

$$\lambda_d = \sum_{[d_1, d_2] = d} \lambda'_{d_1}\lambda''_{d_2}, \quad (5.3)$$

where $[d_1, d_2]$ denotes the least common multiple of d_1 and d_2, it is easily seen that

$$\theta_d = \sum_{\delta | d} \lambda_\delta = \sum_{\delta_1 | d} \lambda'_{\delta_1} \sum_{\delta_2 | d} \lambda''_{\delta_2} = \theta'_d \theta''_d. \quad (5.4)$$

Thus if $\Lambda' = \Lambda''$, θ_d becomes a square, so we clearly have an upper bound sieve Λ^+ if $\lambda'_1 = 1$. We write $\Lambda^+ = \Lambda'^2$. We can also construct new lower bound sieves in this way by multiplying a lower bound sieve Λ'^- with a Λ^2; the resulting $\Lambda^- = \Lambda'^-\Lambda^2$ may be better suited for the problem. In (5.3) it is seen that if the λ'_{d_1} vanish for $d_1 > z$ and the λ''_{d_2} for $d_2 > z$ then λ_d vanish for $d > z^2$, and if the λ'_d in Λ'^- vanish for $d > \xi$, those in $\Lambda'^-\Lambda^2$ vanish for $d > \xi z^2$. This puts some constraints on our choices since for the Λ^2 or the $\Lambda'^-\Lambda^2$ to be useful, we require $z^2 \le x^{1-\varepsilon}$ or $\xi z^2 \le x^{1-\varepsilon}$, respectively, with some ε going sufficiently slowly to zero. For the main terms corresponding to $\sum_d \lambda_d/f(d)$, we get, respectively, in the two cases, if we introduce the notation,

$$\frac{y_\rho}{f'(\rho)} = \mu(\rho) \sum_{\rho | d} \frac{\lambda_d}{f(d)}, \quad (5.5)$$

which implies

$$\frac{\lambda_d}{f(d)} = \mu(d) \sum_{d | \rho} \frac{y_\rho}{f'(\rho)}, \quad (5.5')$$

that for the Λ^2

$$\sum_{d_1, d_2} \frac{\lambda_{d_1}\lambda_{d_2}}{f([d_1, d_2])} = \sum_\rho \frac{y_\rho^2}{f'(\rho)}, \quad (5.6)$$

and for the $\Lambda'^{-}\Lambda^2$,

$$\sum_{d_1,d_2,d_3} \frac{\lambda'_{d_1}\lambda_{d_2}\lambda_{d_3}}{f([d_1,d_2,d_3])} = \sum_d \frac{\lambda'_d}{f(d)} \sum_{(\rho,d)=1} \frac{1}{f'(\rho)} \left\{ \sum_{\delta|d} \mu(\delta) y_{\rho\delta} \right\}^2. \tag{5.6'}$$

In (5.6) and (5.6′) the y_ρ are zero for $\rho > z$ and subject only to the constraint

$$\sum_\rho \frac{y_\rho}{f'(\rho)} = 1, \tag{5.7}$$

which follows from (5.5′) and the fact that $\lambda_1 = 1$.

In earlier attempts to find lower bounds for the sifting limit α_k for large k, only the B²R sieve and the B²R sieve combined with the Buchstab iteration (5.1′) using upper bounds obtained by the Λ^2 sieve (we shall refer to this as the B²RΛ^2 sieve) have been used so far. The sifting limit for the B²R sieve with constant sifting density k is usually denoted by β_k; it has been studied in great detail by K.M. Tsang, who has given an asymptotic expansion, but we shall here only quote the result

$$\beta_k \sim \frac{c}{k}, \tag{5.8}$$

where c is the real solution of the transcendental equation $ce^{1+c} = 1$; we have

$$c \approx \frac{1}{3.591\ldots}.$$

The B²RΛ^2 sieve gives a better lower bound for α_k; for large k it behaves like

$$\beta'_k \sim \frac{c'}{k}, \tag{5.9}$$

where $c' \approx 1/2.445\ldots$ is given by a transcendental expression which I will not quote here.

6.

Before continuing the study of α_k for large k, I wish to turn to a much simpler sifting problem, which, since it can be analyzed rather more fully, is quite instructive about how to proceed in the more general situation.

Let R be a positive integer. We assume we have a sifting range $P(x^{\alpha}, x^{\alpha'})$ with $1/R + 1 < \alpha < \alpha' < 1/R$, and we shall assume that the sifting densities satisfy the condition

$$\sum_{x^{\alpha} \le p < x^{\alpha'}} \frac{u^2(p)}{p^2} \to 0$$

as $x \to \infty$; furthermore, we write

$$v = \sum_{x^{\alpha} \le p < x^{\alpha'}} \frac{u(p)}{p}. \tag{6.1}$$

We observe that if d is the product of r primes from the range, then for $r \ge R + 1$

$$d > x^{(R+1)\alpha} > x^{1+\eta} \qquad \text{with some fixed } \eta > 0,$$

while if $r \le R$

$$d < x^{R\alpha'} < x^{1-\eta} \qquad \text{with some fixed } \eta > 0.$$

Thus we may choose $\lambda_d \ne 0$ only for the d which have $< R + 1$ prime factors. It is not difficult to show that we may, without loss, restrict ourselves to sieves where λ_d depends only on the number of prime factors d has, so we may write

$$\lambda_d = \lambda(r) \qquad \text{for } d = p_1 p_2 \cdots p_r \tag{6.2}$$

with $\lambda(0) = 1$ and $\lambda(r) = 0$ for $r > R$. We get

$$\theta_d = \sum_{\nu=0}^{R} \lambda(\nu)\binom{r}{\nu} = \theta(r).$$

$\theta(r)$ is therefore a polynomial in r of degree at most R. As $x \to \infty$ we get that the expression

$$\sum_d \frac{\lambda_d u(d)}{d}$$

tends to the limit

$$\sum_{\nu \le R} \frac{\lambda(\nu)}{\nu!} v^{\nu} = e^{-v} \sum_{r=0}^{\infty} \frac{\theta(r)}{r!} v^{r}. \tag{6.3}$$

It is possible to show that for the upper bound the optimal value is attained for a $\theta(r)$ of the form

$$\theta(r) = \prod_{i \le [R/2]} \left(1 - \frac{r}{\nu_i}\right)\left(1 - \frac{r}{1 + \nu_i}\right), \tag{6.4}$$

where the ν_i for $i = 1, \ldots, [R/2]$ are positive integers such that

$$\nu_{i+1} \geq \nu_i + 2. \tag{6.5}$$

Similarly for the lower bound sieve the optimal (or largest) value of (6.3) is attained by a sieve for which $\theta(r)$ has the form

$$\theta(r) = (1 - r) \prod_{i \leq [(R-1)/2]} \left(1 - \frac{r}{\nu_i}\right)\left(1 - \frac{r}{1 + \nu_i}\right), \tag{6.6}$$

where the ν_i satisfy the same inequalities (6.5) as before and the additional condition $\nu_1 \geq 2$.

In both cases it can be shown that the largest ν_i for the optimal sieve is bounded by $c(v + R)$ where c is a positive constant.

The combinatorial sieve in this setting is simply given by

$$\lambda(r) = \begin{cases} (-1)^r & \text{for } r \leq 2[R/2], \\ 0 & \text{for } r > 2[R/2], \end{cases} \tag{6.7}$$

for the upper bound case, and

$$\lambda(r) = \begin{cases} (-1)^r & \text{for } r \leq 2[(R-1)/2] + 1, \\ 0 & \text{for } r > 2[(R-1)/2 + 1], \end{cases} \tag{6.7$'$}$$

for the lower bound sieve.

It is easily seen that this corresponds to the forms

$$\theta(r) = \prod_{1 \leq \nu \leq 2[R/2]} \left(1 - \frac{r}{\nu}\right), \tag{6.8}$$

and

$$\theta(r) = \prod_{1 \leq \nu \leq 2[(R-1)/2]+1} \left(1 - \frac{r}{\nu}\right), \tag{6.8$'$}$$

respectively.

In the case of the Λ^2 sieve, this is seen to correspond to the form

$$\theta(r) = q^2(r), \tag{6.9}$$

where $q(r)$ is a polynomial of degree $[R/2]$ with $q(0) = 1$. It is not difficult to see that the optimal choice of $q(r)$ will be of the form

$$q(r) = \prod_{i \leq [R/2]} \left(1 - \frac{r}{\mu_i}\right), \tag{6.10}$$

where the μ_i are real numbers > 1, and this form of $\theta(r)$ seems well suited to mimic the behavior of (6.4), if μ_i is some number between ν_i

and $\nu_i + 1$, which may to some extent explain the effectiveness of the Λ^2 sieve in this case. The form of (6.6) makes it tempting to try the form

$$\theta(r) = (1 - r)q^2(r) \tag{6.11}$$

for the lower bound sieve in this case. This corresponds to the construction of a sieve $\Lambda'^{-}\Lambda^2$, where the sieve Λ'^{-} is defined as follows: $\lambda'_1 = 1$, $\lambda'_p = -1$ for p in P, and $\lambda'_d = 0$ for all other d.

For a given R we may ask: What is the least upper bound of those v for which the expression (6.3) can be made positive by a suitable choice of the $\lambda(r)$ in the lower bound sieve? We shall call this number v_R. For any given R this v_R can be found by trying out the various choices of the ν_i in (6.6). Since we can give an upper bound for the largest ν_i (for instance $\nu_i < 4R$ is easily proven, probably $\nu_i < 2R$ is true, but $\nu_i < \frac{3}{2}R$ false), this is a finite number of choices. For each choice, we look at the polynomial

$$\sum_{\nu \leq 2R' + 1} \frac{\lambda(\nu)}{\nu!} v^{\nu},$$

where we have written $R' = [(R - 1)/2]$, and determine its positive zero (there is only one). The largest of these zeroes will be v_R.

If we use the combinatorial sieve we get the polynomial

$$\begin{aligned}\sum_{\nu \leq 2R' + 1} \frac{(-1)^{\nu}}{\nu!} v^{\nu} &= e^{-v} - \frac{1}{R!}\int_0^v (v - t)^R e^{-t}\, dt \\ &= e^{-v} - \frac{v^R}{R!}\int_0^v \left(1 - \frac{t}{v}\right)^R e^{-t}\, dt \\ &= e^{-v} - \frac{v^R}{R!}\frac{v}{R + v + \eta},\end{aligned} \tag{6.12}$$

where $0 < \eta < 1$.

Denoting the positive root of the polynomial by v'_R and using Stirling's formula for $R!$, we easily get that

$$v'_R \sim cR \tag{6.13}$$

for large R, where c is the real constant satisfying the equation $ce^{1+c} = 1$, that is, the same constant as in (5.8), $c \approx 1/3.591\ldots$.

We may also see what one can get by combining the combinatorial sieve with Buchstab iteration, using upper bounds obtained by the Λ^2

sieve. We cannot really do this while remaining within the setup where $\lambda_d = \lambda(r)$ depending only on the number of prime factors in d, but after completing the construction one can replace the resulting sieve by one where $\lambda_d = \lambda(r)$, and which gives the same result. For this method we find that the lower bound vanishes for $v = v''_R$, where

$$v''_R \sim c''R, \tag{6.14}$$

where c'' is not the same constant as in (5.9) but a somewhat smaller

$$c'' = \frac{1}{2.88\ldots}.$$

We now use a sieve such as was suggested by the expression (6.11), a $\Lambda'^{-}\Lambda^2$ where Λ'^{-} is defined by $\lambda'(0) = 1$, $\lambda'(1) = -1$, and $\lambda'(\nu) = 0$ for $\nu > 1$, while $\lambda(0) = 1$, $\lambda(\nu) = 0$ for $\nu > R'$, $R' = [(R-1)/2]$, and $\lambda(\nu)$ for $\nu = 1, \ldots, R'$ are at our disposal to choose.

Writing

$$y_r = (-1)^r \sum_\nu \frac{\lambda(r+\nu)}{\nu!} v^\nu, \tag{6.15}$$

so that

$$\lambda(r) = (-1)^r \sum_\nu \frac{y_{r+\nu}}{\nu!} v^\nu, \tag{6.15$'$}$$

and $y_r = 0$ for $r > R'$, while

$$\sum_\nu \frac{y_\nu}{\nu!} v^\nu = \lambda(0) = 1, \tag{6.15$''$}$$

we find that the leading term (5.6′) as $x \to \infty$ tends to the limit

$$Q(y) = \sum_{r \le R'} \frac{v^r}{r!} y_r^2 - \sum_{r \le R'} \frac{v^{r+1}}{r!} (y_r - y_{r+1})^2. \tag{6.16}$$

If we write $y_r - y_{r+1} = \Delta_r$ we have

$$y_r = \sum_{j=r}^{R'} \Delta_j,$$

from which we get

$$y_r^2 \le (R' + 1 - r) \sum_{j=r}^{R'} \Delta_j^2, \tag{6.17}$$

where we can have equality only if all Δ_j are equal for $0 \le j \le R'$. We insert this upper bound for y_r^2 in (6.16) and get

$$Q(y) \le \sum_{r \le R'} \frac{v^r}{r!} (R' + 1 - r) \sum_{j=r}^{R'} \Delta_j^2 - \sum_{j \le R'} \frac{v^{j+1}}{j!} \Delta_j^2$$

$$= (R' + 1 - v) \sum_{j \le R'} \Delta_j^2 \sum_{r \le j} \frac{v^r}{r!}. \tag{6.18}$$

From (6.18) it is clear that $Q(y)$ cannot be positive for $v > R' + 1$ whatever our choice of y_r. On the other hand, $Q(y)$ is positive for $v < R' + 1$ and vanishes for $v = R' + 1 = [(R+1)/2]$ if we choose the Δ_j all equal so that $y_r = (R' + 1 - r)\Delta$ for $0 \le r \le R'$, where Δ is to be determined by (6.15″). This method thus leads to a v_R'''

$$v_R''' = \left[\frac{R+1}{2}\right]. \tag{6.19}$$

We clearly have $v_R \ge v_R'''$. Calculations show that $v_R = v_R'''$ for $R < 5$, and for the next few odd values it is only a little larger than v_R'''. It is tempting to guess that

$$v_R \sim \frac{R}{2} \tag{6.20}$$

as $R \to \infty$, but I cannot prove this. It is not hard to show, by looking at the expression

$$\sum_r \frac{\theta(r)}{r!} v^r, \tag{6.21}$$

that it will be negative for any $\theta(r)$ of the form (6.6) (and thus for any θ of degree at most R that satisfies $\theta(0) = 1$ and $\theta(r) \le 0$ for positive integers r) if $v \ge R > 1$. This implies

$$v_R < R \tag{6.22}$$

for $R > 1$. While it is possible to improve this upper bound somewhat, I see no way of showing that

$$\limsup_{R \to \infty} \frac{v_R}{R} < 1,$$

or even

$$\liminf_{R \to \infty} \frac{v_R}{R} < 1.$$

7.

We now return to the study of α_k for large k. The results in the previous section clearly indicate the direction to take. We shall, for simplicity, write $\xi = x^{\alpha}$, our sifting range will be $P(\xi)$ or $p < \xi$. We shall need some asymptotic results for sums of the form

$$\sum_{d < \xi^u} \frac{1}{f'(d)} \tag{7.1}$$

where the d again belongs to $(P(\xi))$. It is not hard to show that as $x \to \infty$, we have

$$\sum_{d < \xi^u} \frac{1}{f'(d)} \sim h_k(u) \prod_{p < \xi} \frac{1}{\left(1 - \dfrac{u(p)}{p}\right)} = \frac{h_k(u)}{e(P(\xi))}, \tag{7.2}$$

where $h_k(u)$ is a function of u that is increasing from 0 to 1 as u goes from 0 to ∞.

$h_k(u)$ satisfies a simple difference-differential equation, which can be explicitly solved. For our purposes it is more useful to look at $h_k'(u)$, and we can show that

$$h_k'(u) = \frac{1}{2\pi i} \int_{-i\infty}^{i\infty} \exp\left(uz - k \int_0^z \frac{1 - e^{-t}}{t} \, dt \right) dz. \tag{7.3}$$

We shall also write $k' = k - \frac{1}{3}$. For large k we can estimate the integral (7.3) by the method of steepest descent, if u is near to k', and show that for $|u - k'| < k^{3/5}$, we have

$$h_k'(u) = \frac{1}{\sqrt{k\pi}} e^{-(u - k')^2/k} \left(1 + \frac{4}{9} \frac{(u - k')^3}{k^2} + \mathcal{O}\left(\frac{1}{k} + \frac{(u - k')^4}{k^3} \right) \right). \tag{7.4}$$

For $|u - k'| \geq k^{3/5}$ we can show that

$$h_k'(u) \ll e^{-(u - k')^2/2k}. \tag{7.4$'$}$$

We apply a lower bound sieve to $\Lambda'^{-}\Lambda^2$ where Λ'^{-} is defined by $\lambda_1' = 1$, $\lambda_p' = -1$ for $p < \xi$ and $\lambda_d' = 0$ for all other d, for Λ we assume $\lambda_1 = 1$; $\lambda_d = 0$ for $d > z = \xi^u$, and the λ_d with $1 < d \leq z$ are to be chosen freely. Introducing, as in Section 5, the y_ρ defined by (5.5), we get for the leading term (5.6′) in this case

$$Q(y) = \sum_{\rho \leq z} \frac{y_\rho^2}{f'(\rho)} - \sum_{p \leq \xi} \frac{1}{f(p)} \sum_{\substack{\rho \leq z \\ p \nmid \rho}} \frac{1}{f'(\rho)} \{y_\rho - y_{\rho p}\}^2. \tag{7.5}$$

Here the y_ρ are zero for $\rho > z$ and otherwise only subject to the constraint (5.7). In order that the remainder term be of smaller order than the main term we need to assume that $z^2\xi$ is $< x^{1-\varepsilon}$ for some $\varepsilon > 0$ if $x > x_0(\varepsilon)$. This holds if $\alpha(1 + 2u) < 1$.

We make the assumption that

$$y_\rho = \sigma\left(\frac{\log \rho}{\log z}\right), \tag{7.6}$$

where $\sigma(t)$ is a bounded function defined for $t \geq 0$, continuous except possibly for $t = 1$, and such that $\sigma(t) = 0$ for $t > 1$. As $x \to \infty$ we can then show that $Q(y)$ is asymptotic to

$$Q(y) \sim Q^*(\sigma)\frac{1}{e(P(\xi))}, \tag{7.7}$$

where

$$Q^*(\sigma) = \int_0^u \sigma^2\left(\frac{v}{u}\right)h_k'(v)\,dv - k \iint\limits_{\substack{0<v<u\\0<t<1}} \left(\sigma\left(\frac{v}{u}\right) - \sigma\left(\frac{v+t}{u}\right)\right)^2 \frac{h_k'(v)}{t}\,dv\,dt. \tag{7.8}$$

The question is now: How small can we make u and still get $Q^*(\sigma)$ positive with some choice of σ?

For large k we can for a given choice of σ estimate these integrals using (7.4) and (7.4′) if u lies close to k'. Analogy with the expression (6.16) leads us first to try the function $\sigma(t) = 1 - t$, and we find indeed that for large k, $Q^*(\sigma) > 0$ for $u \geq k' + c_1$ for a fairly small positive constant c_1. If we replace this form of $\sigma(t)$ by the more general form

$$\sigma(t) = 1 - t + \frac{\mu}{k}, \tag{7.9}$$

where μ is a constant to be chosen later, we find that the optimal choice of μ is $\frac{1}{4}$ and that this choice gives that $Q^*(\sigma) > 0$ for

$$u \geq k - \frac{17}{72} - \frac{c_2}{\sqrt{k}}, \qquad k > k_0, \tag{7.10}$$

where c_2 is a positive constant. Finally we may add a third term to (7.9) and use a $\sigma(t)$ of the form

$$\sigma(t) = 1 - t + \frac{\mu}{k} + \delta(1-t)^2, \tag{7.11}$$

and try to choose both μ and δ in the optimal way. This leads only to a slight improvement in (7.10) since $\mu = \frac{1}{4}$ is still the optimal choice and the new term $\delta(1-t)^2$ only improves (7.10) to the extent of the terms of order $k^{-1/2}$ and smaller. But we do get that $Q^*(\sigma) > 0$ for $k > k_0$ for

$$u \geq k - \frac{17}{72} + \frac{c_3}{\sqrt{k}} \tag{7.12}$$

with a small positive constant c_3. Recalling the condition $\alpha(1+2u) < 1$, we get that we obtain a positive lower bound for $M^-(W_x, P(x^\alpha))$ of the correct order $E(W_x, P(x^\alpha))$ (or what is the same, of $T_k^-(\alpha)$) if

$$\alpha < \frac{1}{1+2u},$$

where u is given by (7.12) for k large enough, and in particular for

$$\alpha = \frac{1}{2k + \frac{19}{36}}.$$

Thus we have, for $k > k_0$,

$$\alpha_k > \frac{1}{2k + \frac{19}{36}}. \tag{7.13}$$

(7.13) probably is true for all $k > \frac{17}{72}$.

It should be mentioned that the calculation of Q^* for u near $k' = k - \frac{1}{3}$ is best carried out by computing at first the value for $u = k'$ and the first derivative with respect to u at $u = k'$. Finally one makes a good estimate of the second derivative in the region $|u - k'| \leq 1$, say.

It does not seem very likely that we can get beyond the constant $\frac{19}{36}$ in (7.13) by changing our choice of $\sigma(t)$. There remains, however, the possibility of changing our Λ'^- without changing the restriction $\lambda'_d = 0$ for $d > \xi$. For instance, one might attempt to replace our former Λ'^- with a more sophisticated combinatorial sieve, say a B^2R lower bound sieve, which retains the bound ξ. This would bring many more terms into the quadratic form $Q(y)$ (as (5.6′) shows) and corresponding terms into the $Q^*(\sigma)$ that would replace (7.8). These further terms contain higher powers of k as factors and make it obvious that a $\sigma(t)$ like (7.9) or (7.11) could no longer be used; we would have to make a completely new guess for $\sigma(t)$, and it is somewhat doubtful that it could lead to any improvement, while the calculations would be forbidding.

A better choice would be to pick Λ'^{-} as follows: $\lambda'_1 = 1$; $\lambda'_p = -1$ for $p < \xi$ as before, $\lambda'_{p_1p_2} = (4T-2)/T(T+1)$, $\lambda'_{p_1p_2p_3} = -6/T(T+1)$ for p_1, p_2, $p_3 < \xi^{1/3}$, while all other $\lambda'_d = 0$. Here T is a positive integer to be chosen later. This introduces only two more terms in $Q^*(\sigma)$. It is easily seen that by choosing $T = [\rho k]$ with a sufficiently large constant ρ we can always with our former choice of $\sigma(t)$ make the new $Q^*(\sigma)$ larger than the former $Q^*(\sigma)$. This must lead to some improvement in our lower bound for α_k, but the necessary calculations are rather involved because of the many terms entering and have not yet been completed.

It is possible to obtain an upper bound for α_k,

$$\alpha_k < \frac{e}{k} \tag{7.14}$$

for all k. This result, which is of course quite trivial for small k, is derived from (6.22), and the method used has severe limitations. It could not, in any case, give more than

$$\limsup_{k \to \infty} k\alpha_k \leq \frac{e}{2},$$

which would follow if we knew that

$$\lim_{R \to \infty} \frac{v_R}{R} = \frac{1}{2}.$$

It seems reasonable to conjecture that $k\alpha_k$ approaches a limit as $k \to \infty$; my guess is that this limit is actually $\frac{1}{2}$, and that for large k, $k\alpha_k > \frac{1}{2}$, so that the approach is from above.

References

[1] Stony Brook lectures. *AMS Proceedings of Symposia in Pure Mathematics.* **XX**, 311–351.

[2] Iwaniec, H. "Rosser's Sieve," *Acta Arithmetica.* **36** (1980), 171–202.

[3] Iwaniec, H. *Recent Progress in Analytic Number Theory*, Vol 1. Academic Press (1981), 203–230.

28 Remarks on the Sieving Limit of the Buchstab–Rosser Sieve*

KAI-MAN TSANG

I. Introduction

Let

$$\omega(z) = \int_0^z \frac{1 - e^{-u}}{u} \, du$$

and, for any $k > 0$, let

$$g_k(s) = \frac{-i}{2} \int_\Gamma t^{-2k} e^{st + k\omega(-t)} \, dt,$$

where Γ is any path of the shape

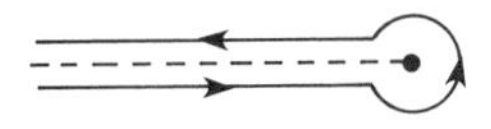

Let $\beta_k - 1$ be the largest positive zero of $g_k(s)$ (its existence will be shown in the sequel). β_k is the sieving limit of the Buchstab–Rosser

* The author wishes to thank Professors A. Selberg and H. Iwaniec for many valuable discussions on the materials in this paper. Thanks are also due to Professor H.E. Richert who has kindly informed the author a simple proof of the fact that $g_k(s) > 0$ when $s > \tau k$.

ISBN 0-12-067570-6

sieve with parameter k. Selberg [2] and Iwaniec [1] have given excellent expositions of this sieve. In [1, Theorem 6], Iwaniec proved that

$$\beta_k = \tau k + O(k^{2/3}),$$

as $k \to \infty$, where $\tau(\approx 3.5917\ldots)$ is the positive zero of the equation

$$\tau(\log \tau - 1) = 1. \tag{1}$$

By means of more careful analysis, we are able to obtain an asymptotic formula for β_k.

Theorem 1. *For any given $n \geq 3$, β_k has the following asymptotic expansion:*

$$\beta_k = \tau k + \tau_1 k^{1/3} + \tau_2 + \tau_3 k^{-1/3} + \tau_4 k^{-2/3} + \cdots + \tau_{n-2} k^{(4-n)/3} + O_n(k^{1-n/3}),$$

where the τ_i are numerical constants. In particular, τ is defined by (1) *and $\tau_1 \approx -2.6006\ldots$, $\tau_2 \approx 0.775622\ldots$, etc.*

Remark. The proof actually revealed that the several largest zeros of $g_k(s)$ are about $ck^{1/3}$ apart.

As an application, we prove in Section 6 the following

Theorem 2. *The largest value of β for which the Buchstab–Rosser sieve fails to converge is given by*

$$\beta^* = \beta_k - \frac{1}{\log \tau} + O(k^{-1/3}),$$

where β_k and τ are given in Theorem 1.

2. Some Preparations

Throughout the paper, we assume that k is sufficiently large. For $s > 0$, we write

$$\eta = \frac{s}{k}.$$

The proof of Theorem 1 consists of two parts: (i) proof that $g_k(s) \geq 0$ when $\eta \geq \tau$; (ii) demonstration of a sign change of $g_k(s)$ when $\eta < \tau$. In view of Iwaniec's result, we shall restrict our consideration to those η lying in the range

$$\tau - O(k^{-2/3}) \leq \eta \leq \tau + O(k^{-1/3}).$$

Let

$$\ell(z) = -2k \log z + sz + k\omega(-z) \tag{2}$$

with $-\pi < \arg z \leq \pi$. Then

$$\ell'(z) = k\left(\eta - \frac{1 + e^z}{z}\right), \tag{3}$$

$$\ell''(z) = -kz^{-2}(e^z(z-1) - 1), \tag{4}$$

$$\ell^{(3)}(z) = -kz^{-3}(e^z(z^2 - 2z + 2) + 2), \tag{5}$$

$$\ell^{(4)}(z) = -kz^{-4}(e^z(z^3 - 3z^2 + 6z - 6) - 6), \tag{6}$$

and, for $n \geq 2$,

$$\ell^{(n)}(z) = -kz^{-n}(e^z P_{n-1}(z) - (-1)^n(n-1)!), \tag{7}$$

where $P_{n-1}(z)$ is a monic polynomial of degree $n - 1$ whose coefficients do not depend on k.

There are two separate cases: (a) $\eta \geq \tau$ and (b) $\eta < \tau$. It is easily seen from the graphs of $\ell'(z)$ and $\ell''(z)$ that in case (a), $\ell'(z)$ has two positive zeros (repeated zeros if $\eta = \tau$). We use r in this case to denote the smaller one. In case (b), no such zero exists and we use r to denote, instead, the positive zero of $\ell''(z)$. Thus (cf. (3), (4), and (1)),

$$r = \begin{cases} \text{smallest positive zero of } \quad \eta = (1 + e^z)/z, & \text{in case (a)}, \\ \log \tau, & \text{in case (b)}. \end{cases} \tag{8}$$

So, in case (b), we have

$$e^r(r-1) = 1.$$

Put

$$v = |\eta - \tau| k, \tag{9}$$

and define

$$\delta = \begin{cases} \left.\begin{cases} k^{-1/4} & \text{for } 0 \leq v \leq k^{1/2}, \\ (kv)^{-1/6} & \text{for } k^{1/2} < v \leq k^{2/3}, \end{cases}\right\} & \text{in case (a)}, \\ \left(\dfrac{-\ell^{(3)}(r)}{6}\right)^{-1/4} = \left(\dfrac{k}{6r(r-1)}\right)^{-1/4} & \text{in case (b)}. \end{cases} \tag{10}$$

Clearly,

$$g_k(s) = \text{Im} \int_{\Gamma^+} e^{\ell(t)}\, dt,$$

where Γ^+ is the part of Γ that lies above the real axis. Let h and ϕ be determined by the equations (see the accompanying figure below)

$$h^2 = r^2 + \delta^2 - r\delta,$$

$$\sin \phi = \frac{\sqrt{3}\delta}{2h}.$$

Let

$$\Gamma_1 = \{xe^{\pi i} | x: h \to \infty\},$$

$$\Gamma_2 = \{he^{i\theta} | \theta: \phi \to \pi\},$$

$$\Gamma_3 = \{r + ye^{2\pi i/3} | y: 0 \to \delta\},$$

and choose $\Gamma^+ = \Gamma_1 \cup \Gamma_2 \cup \Gamma_3$.

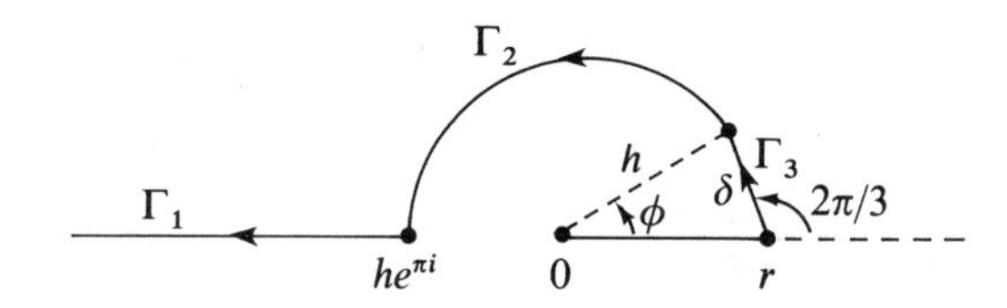

Plainly,

$$r - h = \frac{\delta}{2}(1 + O(\delta)) \tag{11}$$

and

$$\sin \frac{\phi}{2} = \frac{\sqrt{3}\delta}{4h}(1 + O(\delta^2)). \tag{12}$$

3. Some Lemmas

Lemma 1. *For any real θ and any $\rho > 0$, we have*

$$\text{Re}\, \omega(\rho e^{i\theta}) \le \omega(-\rho) + 2\rho e^{\rho} \cos^2 \frac{\theta}{2}.$$

Proof. From the definition of $\omega(z)$ in Section 1, we have

$$\omega(z) = -\sum_{j=1}^{\infty} \frac{(-z)^j}{j!\,j}. \tag{12a}$$

Therefore,

$$\omega(\rho e^{i\theta}) = -\sum_{j=1}^{\infty} \frac{\rho^j e^{i(\pi+\theta)j}}{j!\,j},$$

and

$$\begin{aligned}\operatorname{Re}\omega(\rho e^{i\theta}) &= -\sum_{j=1}^{\infty}\frac{\rho^j}{j!\,j} + 2\sum_{j=1}^{\infty}\frac{\rho^j}{j!\,j}\sin^2\!\left(\frac{\pi+\theta}{2}j\right)\\ &\leq \omega(-\rho) + 2\sin^2\!\left(\frac{\pi+\theta}{2}\right)\sum_{j=1}^{\infty}\frac{\rho^j}{(j-1)!}\\ &= \omega(-\rho) + 2\rho e^{\rho}\cos^2\frac{\theta}{2}.\end{aligned}$$

Lemma 2. *There exists a positive constant c such that*

$$\int_0^{\infty} e^{-x^2}\cos(Lx^3)\,dx > \frac{c}{1+L^{1/3}}$$

for every positive L.

Proof. Define temporarily

$$f(L) = \int_{-\infty}^{\infty} e^{-x^2 + iLx^3}\,dx.$$

Clearly, $f(L) \to \sqrt{\pi}$ as $L \to 0$. By a change of variable, we have

$$f(L) = 2\operatorname{Re}\int_0^{\infty} \exp(-x^2 e^{\pi i/3} - Lx^3)e^{\pi i/6}\,dx,$$

so, $f(L) \to L^{-1/3}\Gamma(\frac{1}{3})/\sqrt{3}$ as $L \to \infty$. In fact, one can obtain asymptotic formulas for $f(L)$ when $L \to 0$ and $L \to \infty$. It remains to show that $f(L) > 0$ for every $L > 0$. This can be seen by comparing the area of the regions enclosed by the curve $y = e^{-x^2}\cos(Lx^3)$ and the x-axis.

Lemma 3. *In case (a), that is $\eta \geq \tau$, we have*

$$e^r(1-r) + 1 = r(2\tau(\eta-\tau)\log\tau)^{1/2} + O(\eta-\tau).$$

Proof. According to (8),

$$\eta r = 1 + e^r. \tag{13}$$

Therefore, $e^r(1-r)+1=r(\eta-e^r)$, and we only need to show that

$$\eta - e^r = (2\tau(\eta-\tau)\log\tau)^{1/2} + O(\eta-\tau). \tag{14}$$

Let us write $r=f(\eta)$ and $g(\eta)=\eta-e^r=\eta-e^f$. Differentiating this equation with respect to η, we get

$$g'(\eta) = 1 - e^f f'.$$

Similarly, from (13) we have

$$f + \eta f' = e^f f' \tag{15}$$

These two equations combined give

$$g'(\eta) = 1 + \frac{fe^f}{g(\eta)},$$

that is,

$$g^2(\eta) = 2\int_\tau^\eta \{g(u) + f(u)e^{f(u)}\}\,du \tag{16}$$

($g(\tau)=0$ because $f(\tau)=\log\tau$).

Since $f(r)$ increases to $f(\tau)$ as η decreases towards τ,

$$\eta - \tau \ll \int_\tau^\eta f(u)e^{f(u)}\,du \ll \eta - \tau.$$

Substituting this into (16), we have

$$\eta - \tau \ll g^2(\eta) \ll \eta - \tau.$$

Put $P(u)=f(u)e^{f(u)}$. We deduce from (15) that

$$P'(u) = f'(1+f)e^f = -\frac{1+f}{g}P(u).$$

Thus,

$$\int_\tau^\eta f(u)e^{f(u)}\,du = P(\tau)(\eta-\tau) - \int_\tau^\eta (u-\eta)P'(u)\,du$$

$$= (\eta-\tau)\tau\log\tau + O((\eta-\tau)^{3/2}).$$

This together with (16) establishes (14).

Lemma 4. *We have*

$$\int_{\Gamma_1} e^{\ell(t)}\,dt \ll e^{\ell(r)-4k}.$$

Proof. The integral on the left is majorized by

$$\int_h^\infty t^{-2k} e^{-st + k\omega(t)}\, dt,$$

and this is

$$\begin{aligned}&\le h^{-2k} e^{k\omega(h)} \int_h^\infty e^{-st}\, dt \ll h^{-2k} e^{-sh + k\omega(h)}\\ &= \exp(-2k \log h - sh + k\omega(h)).\end{aligned}$$

In view of (2) and (11),

$$\begin{aligned}&-2k \log h - sh + k\omega(h) \le \ell(r) - 2sr + k(\omega(r) - \omega(-r)) + O(\delta k)\\ &= \ell(r) - 2k\eta r + 2k\left(r + \frac{r^3}{3!\,3} + \frac{r^5}{5!\,5} + \cdots\right) + O(\delta k)\\ &\le \ell(r) - 4k.\end{aligned}$$

Lemma 5. *There exists a positive constant c such that*

$$\int_{\Gamma_2} e^{\ell(t)}\, dt \ll e^{\ell(r) - ck^{1/4}}.$$

Proof. By Lemma 1,

$$\begin{aligned}\operatorname{Re} \ell(he^{i\theta}) &\le -2k \log h + sh \cos\theta + k\omega(-h) + 2khe^h \sin^2\frac{\theta}{2}\\ &= \ell(h) - 2hk(\eta - e^h)\sin^2\frac{\theta}{2}.\end{aligned}$$

Write $\sigma = 2hk(\eta - e^h)$. By (11),

$$\begin{aligned}\sigma &= 2hk(\eta - e^r + (r - h)e^r + O(\delta^2))\\ &= 2hk\left(\eta - e^r + \frac{\delta}{2} e^r + O(\delta^2)\right) > 0. \qquad (17)\end{aligned}$$

Therefore,

$$\begin{aligned}\int_{\Gamma_2} e^{\ell(t)}\, dt &\ll e^{\ell(h)} \int_\phi^\pi \exp\left(-\sigma \sin^2\frac{\theta}{2}\right) d\theta\\ &\ll \left(\sigma \sin\frac{\phi}{2}\right)^{-1} \exp\left(\ell(h) - \sigma \sin^2\frac{\phi}{2}\right). \qquad (18)\end{aligned}$$

Using Taylor expansion, (6) and (11), we have

$$\ell(h) = \ell(r) + (h - r)\ell'(r) + (h - r)^2 \frac{\ell''(r)}{2} + (h - r)^3 \frac{\ell^{(3)}(r)}{6} + O(k\delta^4).$$

We now consider the two cases (a) and (b) separately.

In case (a), $\ell'(r) = 0$. So, in view of (4), (5), and (10), we have

$$\ell(h) = \ell(r) + \frac{k\delta^2}{8r^2}(e^r(1 - r) + 1)(1 + O(\delta))$$

$$+ \frac{k\delta^3}{48r^3}(e^r(r^2 - 2r + 2) + 2)(1 + O(\delta)) + O(1). \tag{19}$$

Using (17) and (12), we get

$$\ell(h) - \sigma \sin^2 \frac{\phi}{2} = \ell(r) + \frac{k\delta^2}{4r^2}(e^r(r - 1) - 1)$$

$$+ \frac{k\delta^3}{48r^3}(e^r(-8r^2 - 2r + 2) + 2)$$

$$+ O\{k\delta^4 + k\delta^3(\eta - e^r) + k\delta^3(e^r(1 - r) + 1)\}.$$

By Lemma 3, (14), (9), and (10), the O-term is

$$\ll k\delta^4 + \delta^3\sqrt{kv} \ll 1.$$

Thus, using Lemma 3 again, there exists $c > 0$ such that

$$\ell(h) - \sigma \sin^2 \frac{\phi}{2} = \ell(r) - \frac{k\delta^2}{4r^2}\left(1 - \frac{\delta}{6r}\right)(e^r(1 - r) + 1) - \frac{k\delta^3 e^r}{6r} + O(1)$$

$$\leq \ell(r) - c\delta^2\sqrt{kv} - ck\delta^3 + O(1)$$

$$\leq \ell(r) - ck^{1/4}. \tag{20}$$

In case (b), we have $r = \log \tau$ so that $\ell''(r) = 0$. In place of (19), we have

$$\ell(h) = \ell(r) + k(r - h)(\tau - \eta) + \frac{k(r - h)^3}{6r^3}(e^r(r^2 - 2r + 2) + 2) + O(1)$$

$$= \ell(r) + \frac{k}{2}\delta(\tau - \eta) + \frac{k\delta^3}{48r(r - 1)} + O(1 + \delta^2 k(\tau - \eta))$$

$$= \ell(r) + \frac{k\delta^3}{48r(r - 1)} + O(k^{1/12}).$$

Thus, there exists $c > 0$ such that

$$\ell(h) - \sigma \sin^2 \frac{\phi}{2} = \ell(r) + \frac{k\delta^3}{48r(r-1)} - \frac{3\tau k\delta^3}{16h} + O(k^{1/12})$$
$$\leq \ell(r) - \frac{k\delta^3}{6r(r-1)}$$
$$\leq \ell(r) - ck^{1/4}.$$

This together with (20) and (18) completes the proof of the lemma.

4. Proof that $g_k(s) > 0$ for $\eta \geq \tau$

Lemma 6. *In case $\eta \geq \tau$*

$$\operatorname{Im} \int_{\Gamma_3} e^{\ell(t)}\, dt = (1 + O(k^{-1/3}))e^{\ell(r)} \int_0^{\infty} \exp\left\{\frac{-\ell''(r)}{2} y^2\right\}$$
$$\times \cos\left\{\frac{-\ell^{(3)}(r)}{6} y^3\right\} dy$$
$$\gg e^{\ell(r)} k^{-1/3}.$$

Proof. For simplicity, we write

$$v = e^{2\pi i/3}.$$

By Taylor's expansion at r, we have

$$\ell(r + yv) = \ell(r) + \ell'(r)yv + \frac{\ell''(r)}{2}(yv)^2 + \frac{\ell^{(3)}(r)}{6}(yv)^3 + O(ky^4).$$

Since $\eta \geq \tau$, $\ell'(r) = 0$. Therefore,

$$\int_{\Gamma_3} e^{\ell(t)}\, dt = e^{\ell(r)} \int_0^{\delta} \exp\left\{\frac{\ell''(r)}{2}(yv)^2 + \frac{\ell^{(3)}(r)}{6}(yv)^3\right\}(1 + O(ky^4))\, d(yv).$$

Note that $\ell''(r) > 0$ and $\ell^{(3)}(r) < 0$. Hence

$$\operatorname{Im} \int_{\Gamma_3} e^{\ell(t)}\, dt = e^{\ell(r)} \operatorname{Im} \int_0^{\infty} \exp\left\{\frac{\ell''(r)}{2}(yi)^2 + \frac{\ell^{(3)}(r)}{6}(yi)^3\right\} d(yi)$$
$$+ O\left(e^{\ell(r)} \int_{\delta}^{\infty} \exp\left\{\frac{-\ell''(r)}{4} y^2 + \frac{\ell^{(3)}(r)}{6} y^3\right\} dy\right)$$
$$+ O\left(e^{\ell(r)} k \int_0^{\delta} y^4 \exp\left\{\frac{-\ell''(r)}{4} y^2 + \frac{\ell^{(3)}(r)}{6} y^3\right\} dy\right). \quad (21)$$

The main term here, by Lemma 2, is

$$\gg e^{\ell(r)}\{\sqrt{\ell''(r)} + (-\ell^{(3)}(r))^{1/3}\}^{-1} \gg e^{\ell(r)}k^{-1/3}. \tag{22}$$

On the other hand, by Lemma 3,

$$\ell''(r) = \frac{k}{r}\{(2\tau(\eta-\tau)\log\tau)^{1/2} + O(\eta-\tau)\} \approx \text{constant} \times \sqrt{kv}$$

and, by (5),

$$-\ell^{(3)}(r) \approx \text{constant} \times k.$$

Hence, in view of (10),

$$\max\{\ell''(r)\delta^2, -\ell^{(3)}(r)\delta^3\} \approx \max(\delta^2\sqrt{kv}, k\delta^3) \gg k^{1/4}.$$

So, the first O-term in (21) is negligible. The second O-term is

$$\ll e^{\ell(r)}k\min\{(\ell''(r))^{-5/2}, (-\ell^{(3)}(r))^{-5/3}\}$$
$$\ll e^{\ell(r)}k^{-1/3}\{\sqrt{\ell''(r)} + (-\ell^{(3)}(r))^{1/3}\}^{-1}.$$

Our lemma follows readily from this and (22).

Combining now the estimates in Lemmas 4, 5, and 6, we have, for some $c > 0$,

$$g_k(s) > ck^{-1/3}e^{\ell(r)},$$

whenever $\eta \geq \tau$. Thus $g_k(s)$ has no zeros on the right side of τk.

5. Proof that $g_k(s)$ changes sign for $\eta \approx \tau - O(k^{-2/3})$

We now consider case (b), that is, when

$$\tau - O(k^{-2/3}) \leq \eta < \tau. \tag{23}$$

Let $\mathscr{L}(\alpha)$ denote the path $\{xe^{2\pi i/3} | x: 0 \to \alpha\}$.

Lemma 7. *The integral*

$$I(u) = \operatorname{Im}\int_{\mathscr{L}(\infty)} e^{-uz-z^3}\,dz, \qquad u \geq 0$$

has the following two properties:

(i) $I(0) > 0$

(ii) *$I(u)$ has infinitely many positive zeros and they are all simple.*

Proof. We can write

$$I(u) = \int_0^\infty \cos(x^3 - ux)\, dx,$$

which is an Airy's integral. It is well known that (see Watson [3, p. 190])

$$I(u) = \frac{\pi}{3}\sqrt{\frac{u}{3}}\left[J_{1/3}\left(\frac{2u\sqrt{u}}{3\sqrt{3}}\right) + J_{-1/3}\left(\frac{2u\sqrt{u}}{3\sqrt{3}}\right)\right],$$

for $u \geq 0$, where $J_{\pm 1/3}$ are Bessel functions. The asymptotic expansions of $J_{\pm 1/3}$ give immediately that $I(0) = \frac{\pi}{3}/\Gamma(\frac{2}{3}) > 0$ and

$$I(u) = \left(\frac{\pi}{\sqrt{3u}}\right)^{1/2}\left\{\cos\left(\frac{2u\sqrt{u}}{3\sqrt{3}} - \frac{\pi}{4}\right) + O(u^{-3/2})\right\}$$

as $u \to +\infty$. This shows that $I(u)$ has infinitely many positive zeros, and all the large ones are simple and about $cu^{-1/2}$ apart. The fact that all the positive zeros are simple follows easily from the well-known differential equation [3, p. 189]

$$I''(u) + \frac{u}{3} I(u) = 0, \tag{24}$$

for $u \geq 0$.

We now proceed to the estimation of the integral

$$\operatorname{Im} \int_{\Gamma_3} e^{\ell(t)}\, dt = \operatorname{Im} \int_{\mathscr{L}(\delta)} e^{\ell(r+y)}\, dy.$$

For a given integer $n \geq 3$, since $\ell''(r) = 0$, we have

$$\ell(r+y) = \ell(r) + \ell'(r)y + \frac{\ell^{(3)}(r)}{6} y^3 + H(y) + O(k|y|^{n+1}),$$

where

$$H(y) = \frac{\ell^{(4)}(r)}{4!} y^4 + \cdots + \frac{\ell^{(n)}(r)}{n!} y^n.$$

We have used the fact that $\ell^{(n+1)}(z) \ll k$ for any $z \in r + \mathscr{L}(\delta)$ (see (7)). From (3) and (5), we have

$$\ell'(r) = k(\eta - \tau) = -v < 0$$

and

$$\ell^{(3)}(r) = \frac{-k}{r(r-1)} < 0.$$

Let

$$\gamma_3 = \frac{1}{6r(r-1)}$$

and

$$\gamma_m = \frac{\ell^{(m)}(r)}{m!\,k} \qquad \text{for } m = 4, 5, \ldots, n,$$

then, $\gamma_m = O(1)$. We have

$$\operatorname{Im} \int_{\Gamma_3} e^{\ell(t)}\, dt = e^{\ell(r)} \operatorname{Im} \int_{\mathscr{L}(\delta)} \exp\{-vy - k\gamma_3 y^3 + H(y)\}\, dy$$

$$+ O\left\{e^{\ell(r)} \int_0^{\delta} ky^{n+1} \exp\left\{\frac{v}{2} y - k\gamma_3 y^3 + O(ky^4)\right\} dy\right\}.$$

In view of (23), the O-term here is

$$\ll e^{\ell(r)} k^{1-(n+2)/3} \int_0^{\infty} y^{n+1} \exp\left\{\frac{v}{2} k^{-1/3} y - \gamma_3 y^3 + O(1)\right\} dy$$

$$\ll e^{\ell(r)} k^{-(n-1)/3}.$$

Let

$$\xi = (k\gamma_3)^{-1/3}.$$

ξ is close to zero because we have assumed throughout that k is large. Then, in view of (10),

$$\operatorname{Im} \int_{\Gamma_3} e^{\ell(t)}\, dt = \xi e^{\ell(r)} \operatorname{Im} \int_{\mathscr{L}(\xi^{-1/4})} \exp\{-v\xi y - y^3 + \xi y^4 P(\xi y)\}\, dy$$

$$+ O(e^{\ell(r)} \xi^{n-1}), \tag{25}$$

where

$$P(z) = (\gamma_4 + \gamma_5 z + \cdots + \gamma_n z^{n-4})\gamma_3^{-1}.$$

Note that $\xi \approx k^{-1/3}$, $\xi^{-1/4} \approx k^{1/12}$, and $0 < v\xi \ll 1$, by (23). Consider the function

$$V(q, \xi) = \int_{\mathscr{L}(\xi^{-1/4})} \exp\{-qy - y^3 + \xi y^4 P(\xi y)\}\, dy$$

of the two variables q and ξ in the ranges $0 \le q \ll 1$ and $\xi \to 0^+$.

Clearly, $V(q, \xi)$ is a C^∞-function in each of the variables and

$$V(q, \xi) = \int_{\mathscr{L}(\xi^{-1/4})} e^{-qy - y^3}(1 + O(\xi |y|^4))\, dy$$

$$= \int_{\mathscr{L}(\infty)} e^{-qy - y^3}\, dy + O(\xi).$$

By Lemma 7(i),

$$\text{Im } V(q, \xi) = I(q) + O(\xi),$$

which is positive when both ξ and q are sufficiently close to zero. In view of the assertion (ii) of the same lemma, we may let $q_0 = q_0(\xi)$ be the smallest positive zero of Im $V(q, \xi)$ for each given small ξ. Let q^* be the smallest positive zero of $I(u)$, then

$$\lim_{\xi \to 0^+} q_0(\xi) = q^*.$$

Let

$$F(q, \xi) = \text{Im } V(q, \xi).$$

For each q close to q_0,

$$F(q, \xi) = F(q_0, \xi) + F_1(\tilde{q}, \xi)(q - q_0)$$

$$= F_1(\tilde{q}, \xi)(q - q_0), \tag{26}$$

where $\tilde{q} \in (q_0, q)$. Now

$$F_1(\tilde{q}, \xi) = \text{Im} \int_{\mathscr{L}(\xi^{-1/4})} -y \exp\{-\tilde{q}y - y^3 + \xi y^4 P(\xi y)\}\, dy$$

$$= \text{Im} \int_{\mathscr{L}(\infty)} -y \exp\{-\tilde{q}y - y^3\}\, dy + O(\xi).$$

This integral is actually equal to $I'(\tilde{q})$. Since $I(0) > 0$ and q^* is the smallest positive zero of $I(u)$, $I'(q^*)$ is negative. Thus, when q is sufficiently close to q_0, which is also close to q^*, we have $F_1(\tilde{q}, \xi) < 0$; in particular, it is nonzero. Collecting the estimates in (25) and Lemmas 4 and 5, we have that

$$g_k(s) = \xi e^{\ell(r)} F(v\xi, \xi) + O(\xi^{n-1} e^{\ell(r)}). \tag{27}$$

Let L be large and fixed. We see from (26) that the two numbers

$$q_1 = q_0 - \frac{L\xi^{n-2}}{|I'(q^*)|} = q_0 - O(\xi^{n-2})$$

and

$$q_2 = q_0 + \frac{L\xi^{n-2}}{|I'(q^*)|} = q_0 + O(\xi^{n-2})$$

satisfy

$$F(q_1, \xi) = -F_1(\tilde{q}, \xi)\frac{L\xi^{n-2}}{|I'(q^*)|} > \frac{L}{2}\xi^{n-2}$$

and

$$F(q_2, \xi) = F_1(\tilde{q}, \xi)\frac{L\xi^{n-2}}{|I'(q^*)|} < -\frac{L}{2}\xi^{n-2}.$$

It then follows from (27) that, for suitably chosen L, $g_k(s)$ changes sign when $v\xi$ goes from q_1 to q_2 and does not change sign for $0 \le v\xi < q_1$. Thus,

$$\begin{aligned}\beta_k &= \eta k + 1 = \tau k - v + 1 = \tau k - \xi^{-1}q_0 + 1 + O(\xi^{n-3}) \\ &= \tau k + 1 - (k\gamma_3)^{1/3}q_0 + O(k^{1-(n/3)}). \qquad (28)\end{aligned}$$

So, to complete the proof of Theorem 1, it remains to develop $q_0 = q_0(\xi)$ into an asymptotic power series. To this end, we differentiate implicitly with respect to ξ the equation $F(q_0, \xi) = 0$. This gives

$$F_1(q_0, \xi)q_0'(\xi) + F_2(q_0, \xi) = 0.$$

The key point here is whether $F_1(q_0, \xi) = 0$. However,

$$\begin{aligned}F_1(q_0, \xi) &= \operatorname{Im}\int_{\mathscr{L}(\infty)} -y\exp(-q_0 y - y^3)\,dy + O(\xi) \\ &= I'(q_0) + O(\xi) \to I'(q^*)\end{aligned}$$

as $\xi \to 0^+$. Since $I'(q^*) < 0$, we have $F_1(q_0, \xi) < 0$ and

$$q_0'(\xi) = \frac{-F_2(q_0, \xi)}{F_1(q_0, \xi)}. \qquad (29)$$

Employing our familiar techniques, we have

$$
\begin{aligned}
F_2(q_0, \xi) &= \operatorname{Im} \int_{\mathscr{L}(\xi^{-1/4})} \exp\{-q_0 y - y^3 + \xi y^4 P(\xi y)\} \\
&\qquad \times \{y^4 P(\xi y) + \xi y^5 P'(\xi y)\}\, dy \\
&\qquad - \frac{1}{4}\xi^{-5/4} \operatorname{Im}\{\exp\{-q_0 \xi^{-1/4} e^{2\pi i/3} \\
&\qquad - \xi^{-3/4} + e^{2\pi i/3} P(\xi^{3/4} e^{2\pi i/3})\} e^{2\pi i/3}\} \\
&= \operatorname{Im} \int_{\mathscr{L}(\xi^{-1/4})} \exp\{-q_0 y - y^3 + O(\xi |y|^4)\} \left\{\frac{\gamma_4}{\gamma_3} y^4 + O(\xi |y|^5)\right\} dy \\
&\qquad + O\{\exp(q_0 \xi^{-1/4} - \xi^{-3/4})\} \\
&= \frac{\gamma_4}{\gamma_3} \operatorname{Im} \int_{\mathscr{L}(\infty)} y^4 \exp(-q_0 y - y^3)\, dy + O(\xi) \\
&= \frac{\gamma_4}{\gamma_3} I^{(4)}(q_0) + O(\xi).
\end{aligned}
$$

By (24),

$$
I^{(4)}(q_0) = \frac{1}{9} q_0^2 I(q_0) - \frac{2}{3} I'(q_0).
$$

Thus,

$$
q_0'(\xi) = \left\{\frac{\gamma_4}{\gamma_3}\left(\frac{2}{3} I'(q_0) - \frac{1}{9} q_0^2 I(q_0)\right) + O(\xi)\right\}\{I'(q_0) + O(\xi)\}^{-1} \to \frac{2\gamma_4}{3\gamma_3}
$$

as $\xi \to 0^+$. Since $F_1(q_0, \xi) \neq 0$, higher derivatives of $q_0(\xi)$ can be obtained by successive differentiation of equation (29). Thus, if $\xi > \xi_1 > 0$, we have

$$
\begin{aligned}
q_0(\xi) &= q_0(\xi_1) + (\xi - \xi_1) q_0'(\xi_1) + \frac{(\xi - \xi_1)^2}{2} q_0''(\xi_1) + \cdots \\
&\qquad + \frac{(\xi - \xi_1)^{n-3}}{(n-3)!} q_0^{(n-3)}(\xi_1) + O((\xi - \xi_1)^{n-2} |q_0^{(n-2)}(u)|),
\end{aligned}
$$

where $u \in (\xi_1, \xi)$. Letting $\xi_1 \to 0^+$, this becomes

$$
\begin{aligned}
q_0(\xi) &= q_0(0^+) + \xi q_0'(0^+) + \frac{\xi^2}{2} q_0''(0^+) + \cdots \\
&\qquad + \frac{\xi^{n-3}}{(n-3)!} q_0^{(n-3)}(0^+) + O(\xi^{n-2}) \\
&= q^* + \frac{2\gamma_4}{3\gamma_3}\xi + \frac{q_0''(0^+)}{2}\xi^2 + \cdots + \frac{q_0^{(n-3)}(0^+)}{(n-3)!}\xi^{n-3} + O(\xi^{n-2}).
\end{aligned}
$$

Substituting this into (28), we obtain the desired asymptotic formula for β_k.

6. Estimation of β*

The sieving functions $F(s)$ and $f(s)$ corresponding to the Buchstab–Rosser sieve with parameter k satisfy the differential-difference equations

$$\begin{cases} (s^k F(s))' = ks^{k-1} f(s-1) & \text{for } s > \beta + 1, \\ (s^k f(s))' = ks^{k-1} F(s-1) & \text{for } s > \beta, \end{cases}$$

and

$$\begin{cases} s^k F(s) = A & \text{for } s \le \beta + 1, \\ s^k f(s) = B & \text{for } s \le \beta. \end{cases}$$

(see equations (1.8) and (1.9) in [1]). Here A and B are determined by the pair of linear equations on p. 196 of [1]. Solving this pair of equations, we have

$$f(\beta) = \beta^{-k} B = \frac{2g_k(\beta - 1)}{\beta D_k(\beta)}, \tag{30}$$

where

$$D_k(\beta) = g_k(\beta - 1) h_k(\beta) + g_k(\beta) h_k(\beta - 1) \tag{31}$$

and

$$h_k(\beta) = \int_0^\infty e^{-\beta t - k\omega(t)}\, dt.$$

Since $h_k(\beta) > 0$, $D_k(\beta) > 0$ for all $\beta \ge \beta_k$.

The optimal choice of β is β_k, which corresponds to the point where $f(\beta)$ changes from positive to negative. Thus, $1/\beta_k$ represents the maximum amount of sieving that can be done by this sieve. As β decreases further, it will reach $\beta^* = \beta^*(k)$, the largest zero of $D_k(\beta)$. Equation (30) shows that $f(\beta)$ is a holomorphic function of k for $\beta > \beta^*$ and $f(\beta) \to -\infty$ as $\beta \to \beta^*$. Thus, β^* is the lower limit of those β for which the Buchstab–Rosser sieve converges.

To prove Theorem 2, we assume k is large and consider those β lying in the range $\beta_k - 1 < \beta < \beta_k$. By Theorem 1,

$$\beta = \tau k + \tau_1 k^{1/3} + O(1). \tag{32}$$

Lemma 8. *For $s = \beta$ or $\beta - 1$, we have*

$$h_k(s) = \frac{1}{(\tau + 1)k} + O(k^{-5/3}). \tag{33}$$

Proof. By (12a), $t - \omega(t) = O(t^2)$ for all real t. Writing $\alpha = k^{-1/2}$, since $\beta \sim \tau k$, we have

$$\begin{aligned} h_k(\beta) &= \int_0^\alpha e^{-(\beta + k)t + k(t - \omega(t))}\,dt + \int_\alpha^\infty e^{-\beta t - k\omega(t)}\,dt \\ &= \int_0^\alpha e^{-(\beta + k)t}(1 + O(kt^2))\,dt + O\left(\int_\alpha^\infty e^{-\beta t}\,dt\right) \\ &= \int_0^\alpha e^{-(\beta + k)t}\,dt + O\left(k\int_0^\alpha t^2 e^{-(\beta + k)t}\,dt\right) + O(\beta^{-1}e^{-\beta\alpha}) \\ &= \frac{1}{\beta + k} + O(k^{-2}). \end{aligned}$$

By (32), $\dfrac{1}{\beta + k} = \dfrac{1}{(\tau + 1)k} + O(k^{-5/3})$. This proves Lemma 8.

With $n = 4$, equation (27) yields

$$g_k(\beta) = \xi e^{\ell(r)}F((k\tau - \beta)\xi, \xi) + O(\xi^3 e^{\ell(r)}). \tag{34}$$

Let

$$\sigma = 1 - \frac{1}{r} = 1 - \frac{1}{\log \tau}. \tag{35}$$

Since $e^r(r - 1) = 1$, we have

$$\sigma(1 + e^{-r}) = e^{-r}. \tag{36}$$

Let L_1 be a large positive number that is independent of k and let

$$q_3 = k\tau - q_0\xi^{-1} + \sigma - L_1\xi,$$

where q_0 is defined in Section 5. Then

$$(k\tau - q_3)\xi - q_0 = -\sigma\xi + L_1\xi^2.$$

Using Taylor's expansion at q_0,

$$F((k\tau - q_3)\xi, \xi) = F_1(q_0, \xi)(-\sigma\xi + L_1\xi^2) + O(\xi^2).$$

Substituting this into (34), we have

$$g_k(q_3) = \xi^2 e^{\ell(r)}F_1(q_0, \xi)(-\sigma + L_1\xi) + O(\xi^3 e^{\ell(r)}).$$

Similarly,

$$g_k(q_3 - 1) = \xi e^{\ell(r)} e^{-r} F((k\tau - q_3 + 1)\xi, \xi) + O(\xi^3 e^{\ell(r)})$$
$$= \xi^2 e^{\ell(r)} e^{-r} F_1(q_0, \xi)(1 - \sigma + L_1 \xi) + O(\xi^3 e^{\ell(r)}).$$

Putting these and (33) into (31), we find that

$$D_k(q_3) = \frac{\xi^2}{(\tau + 1)k} e^{\ell(r)} F_1(q_0, \xi)(-\sigma - e^{-r}(\sigma - 1)$$
$$+ (e^{-r} + 1)L_1 \xi) + O(\xi^3 k^{-1} e^{\ell(r)})$$
$$= \frac{\xi^3}{(\tau + 1)k} e^{\ell(r)} F_1(q_0, \xi)(e^{-r} + 1)L_1 + O(\xi^3 k^{-1} e^{\ell(r)}),$$

by (36). Since $F_1(q_0, \xi) < 0$, we have $D_k(q_3) < 0$ when L_1 is sufficiently large. Taking

$$q_4 = k\tau - q_0 \xi^{-1} + \sigma + L_1 \xi,$$

we show in the same way that

$$D_k(q_4) = \frac{-\xi^3}{(\tau + 1)k} e^{\ell(r)} F_1(q_0, \xi)(e^{-r} + 1)L_1 + O(\xi^3 k^{-1} e^{\ell(r)}) > 0$$

when L_1 is sufficiently large. Hence $q_3 < \beta^* < q_4$. We deduce from (35) and (28) that

$$\beta^* = k\tau - q_0 \xi^{-1} + \sigma + O(\xi)$$
$$= \beta_k - \frac{1}{\log \tau} + O(\xi).$$

This proves Theorem 2.

References

[1] Iwaniec, H. "*Rosser's sieve*," *Acta Arithmetica.* **36** (1980), 171–202.

[2] Selberg, A. "*Sieve methods*," *Proc. Sympos. Pure Math.* **20** (1971), 311–351.

[3] Watson, G.N. *Theory of Bessel Functions*, 2nd ed. Camb. Univ. Press, London, New York (1962).

29 Recent Work on Waring's Problem

R. C. VAUGHAN

1. Introduction

I would like to take this opportunity to give a brief account of some very recent work on Waring's problem. Although much of what I shall say is special to that problem, there is one aspect of the work which may have consequences in other areas of analytic number theory.

2. Waring's Problem

The central problem in Waring's problem today is the determination of $G(k)$, i.e., the smallest number s such that every *sufficiently large* natural number is a sum of at most s positive k-th powers. The work I shall describe leads to improvements in the known upper bounds for $G(k)$ for all $k \geq 5$. For instance, for smaller k it is now possible to obtain

k	4*	5	6	7	8	9
$G(k) \leq$	12	19	29	41	58	75

This may be compared with the earlier results of Vaughan [1986a,b]

$G(k) \leq$	13	21	31	45	62	82

NUMBER THEORY,
TRACE FORMULAS and
DISCRETE GROUPS

ISBN 0-12-067570-6

Here 4* indicates that an obviously necessary local condition has to be included for solubility.

For large k it is now possible to obtain

$$G(k) < k\left(2\log k + 2\log\log k + 2 + 2\log 2 + O\left(\frac{\log\log k}{\log k}\right)\right).$$

This may be compared with Vinogradov's estimate [1959], valid only for $k > 170{,}000$,

$$G(k) < k(2\log k + 4\log\log k + 2\log\log\log k + 13),$$

which makes heavy use of the Vinogradov Mean Value Theorem.

Another question in which interest has been shown, especially in connection with cubes, is that of obtaining lower estimates for

$$\mathcal{N}_k(N) = \operatorname{card}\{n \le N: n = x_1^k + x_2^k + x_3^k,\ x_i \ge 0\}$$

of the kind

$$\mathcal{N}_k(N) \gg N^{\alpha_k - \varepsilon} \qquad (N \ge 3).$$

It is now possible to obtain

$$\alpha_3 = \frac{11}{12}, \qquad \alpha_k = \frac{3}{k} - \frac{2}{k(2k-3+\sqrt{8+(2k-3)^2})} \qquad (k \ge 4).$$

This may be compared with the earlier results

$$\alpha_3 = \frac{19}{21} \qquad \text{(Vaughan 1986c)},$$

$$\alpha_4 = \frac{19}{28} \qquad \text{(Davenport 1939)},$$

$$\alpha_5 = \frac{5}{9}, \qquad \alpha_6 = \frac{59}{126} \qquad \text{(Davenport 1942)},$$

$$\alpha_7 = \frac{65}{161}, \qquad \alpha_8 = \frac{77}{216} \qquad \text{(Davenport's methods)},$$

$$\alpha_k = \frac{3}{k} - \frac{1}{k^2} - \frac{2}{k^3} \qquad (k \ge 9) \qquad \text{(Davenport and Erdős 1939)}.$$

and the conditional estimate of Hooley 1986

$$\alpha_3 = \frac{18}{19}.$$

3. Diagonal Forms

The method is also applicable to diagonal forms. Consider the equation

$$c_1x_1^k + \cdots + c_sx_s^k = 0. \tag{1}$$

We say that (1) satisfies the *local condition* if for every natural number q the equation (1) has a solution modulo q with $(x_j, q) = 1$ for some j. Then, following Davenport and Lewis [1963], we define $G^*(k)$ to be the least number t such that the equation (1) has a nontrivial integer solution whenever

(i) $s \geq t$,
(ii) it has a real solution, and
(iii) the local condition is satisfied.

They instituted a programme to show that

$$G^*(k) \leq k^2 + 1 \qquad \text{for all } k \geq 4.$$

The bound $k^2 + 1$ is of particular interest since, as Davenport and Lewis show, when $k + 1$ is prime and $s = k^2$ there are $c_1, \ldots, c_s$ and a prime p for which (1) has no nontrivial p-adic solution.

It is now possible to complete this programme and, in particular, to settle the stubborn case $k = 10$. Here is a list of upper bounds for $G^*(k)$ which it is now possible to obtain. The same bounds also hold for $G(k)$.

k	10	11	12	13	14	15	16	17	18	19	20
$G^*(k) \leq$	93	109	125	141	156	171	187	202	217	232	248

This may be compared with the earlier bounds of Thanigasalam [1982, 1985].

$G^*(k) \leq$	103	119	134	150	165	181	197	213	229	245	262

4. An Exponential Sum

A theorem of a more technical nature, but which indicates that the underlying techniques might nevertheless have wider applicability than those described so far, concerns the exponential sum $S(\alpha)$ defined by taking

$$\mathscr{A}(P, R) = \{n \leq P : p \mid n \Rightarrow p \leq R\},$$
$$S(\alpha) = \sum_{x \in \mathscr{A}(P,R)} e(\alpha x^k).$$

Theorem. *Suppose that* $0 < \delta < 1/2k$,

$$\mathfrak{m} = \left\{\alpha \in \mathbb{R}: \left|\alpha - \frac{a}{q}\right| \le q^{-1}P^{1/2+\delta k - k} \Rightarrow q > P^{1/2+\delta k}\right\}$$

and

$$\rho(k) = \max_{\substack{s \in \mathbb{N} \\ s \ge 2}} \frac{1}{4s}\left(1 - (k-2)\left(1 - \frac{1}{k}\right)^{s-2}\right).$$

Then for each positive number ε *there is a positive number* $\eta = \eta(\varepsilon)$ *such that whenever* $2 \le R \le P^{\eta}$ *one has*

$$\sup_{\mathfrak{m}} |S(\alpha)| \ll P^{1+\varepsilon}(P^{-\delta} + P^{-\rho(k)})$$

Note that

$$4\rho(k)k \log k \sim 1 \qquad \text{as } k \to \infty$$

and

$$\operatorname{card}(\mathscr{A}(P, P^{\eta})) \sim c_{\eta}P \qquad \text{as } P \to \infty.$$

Presumably the best possible exponent in the theorem is

$$1 - \frac{1}{k}.$$

The bound given by the theorem may be compared with that obtaining from Vinogradov's Mean Value Theorem for the classical generating function in Waring's problem, namely,

$$\sup_{\mathfrak{m}} \left| \sum_{x \le P} e(\alpha x^k) \right| \ll P^{1-\sigma(k)+\varepsilon}$$

with

$$4\sigma(k)k^2 \log k \sim 1 \qquad \text{as } k \to \infty.$$

Note that the exponent 2 is replaced by 1 in the theorem.

A refined version of the above theorem gives an exponent $\rho(k)$ that is superior to that provided by Weyl's inequality or Vinogradov's Mean Value Theorem whenever $k \ge 8$.

5. Description of the Method

After the seminal work of Hardy and Littlewood on additive number theory, and Waring's problem in particular (see Hardy [1966]), the best upper bounds for $G(k)$ when $k \geq 4$ have been based, in essence, on prior estimates for the number of solutions of auxiliary equations of the form

$$x_1^k + \cdots + x_s^k = y_1^k + \cdots + y_s^k, \tag{2}$$

with the x_j and y_j lying in ranges of the kind

$$P_j < x_j < 2P_j,\ P_j < y_j < 2P_j,$$

where $P_1 \geq P_2 \geq \cdots$. The use of *diminishing ranges* in this context was refined and perfected by Davenport (see Davenport [1977]) and Vinogradov (see Vinogradov [1984] and Vaughan [1981]), and the earlier work of Thanigasalam [1985] and Vaughan [1986a,b] is largely based on a variant of a special case of a theorem of Davenport [1942].

The use of diminishing ranges in (2), whilst conferring a number of benefits, has one serious drawback, namely that the homogeneity of (2) is lost.

The underlying theme of the new method is the conservation of homogeneity in equations such as (2). Thus (2) is considered with $x_j \in \mathscr{A}$, $y_j \in \mathscr{A}$ where $\mathscr{A}$ is a fairly dense subset of $[1, P] \cap \mathbb{Z}$. For a suitable $\mathscr{A}$ we relate the number of such solutions to the number of solutions of

$$x^k + m^k(z_2^k + \cdots + z_{s-1}^k) = y^k + m^k(t_2^k + \cdots + t_{s-1}^k), \tag{3}$$

with $x \leq P$, $y \leq P$, $M < m \leq M'$, $z_j \in \mathscr{B}$, $t_j \in \mathscr{B}$ where $\mathscr{B}$ has similar properties to $\mathscr{A}$, but $\mathscr{B} \subset [1, P/M] \cap \mathbb{Z}$. Then by the use of ideas stemming from the diminishing range circle of ideas combined with Hölder's inequality and the homogeneity of

$$z_2^k + \cdots + z_{s-1}^k - t_2^k - \cdots - t_{s-1}^k,$$

we are able to estimate the number of solutions of (3) in terms of the number of solutions of (2) with s replaced by a number not exceeding s and with $\mathscr{A}$ replaced by $\mathscr{B}$. This enables an iterative procedure of an entirely new kind to be created. In a certain sense this does for a single equation what the arguments underlying the proof of Vinogradov's Mean Value Theorem do for the corresponding system of equations.

It transpires that our technique puts no serious obstacle in the way of methods that have been developed in the context of diminishing ranges. Thus the technique has great flexibility.

An important role is played throughout this work by the set $\mathscr{A}(P, R)$ of natural numbers not exceeding P with no prime factor exceeding R. Other sets could be substituted in some of the arguments, but no alternate seems to provide the same general degree of flexibility.

6. A Fundamental Lemma

Let me now give a few brief details. Let

$$\mathscr{A}(P, R) = \{n: n \le P, p|n \Rightarrow p \le R\}, \tag{4}$$

let $S_s(P, R)$ denote the number of solutions of

$$x_1^k + \cdots + x_s^k = y_1^k + \cdots + y_s^k \tag{5}$$

with

$$x_j \in \mathscr{A}(P, R), y_j \in \mathscr{A}(P, R), \tag{6}$$

and for a given real number θ with $0 < \theta < 1$ let $T_s(P, R, \theta)$ denote the number of solutions of

$$x^k + m^k(x_1^k + \cdots + x_{s-1}^k) = y^k + m^k(y_1^k + \cdots + y_{s-1}^k) \tag{7}$$

with

$$x \le P, y \le P, x \equiv y \pmod{m^k}, \qquad P^\theta < m \le \min(P, P^\theta R), \tag{8}$$

$$x_j \in \mathscr{A}(P^{1-\theta}, R), y_j \in \mathscr{A}(P^{1-\theta}, R). \tag{9}$$

The core of the method is a lemma that relates S_s to T_s.

Lemma. *Let $\theta = \theta(s, k)$ satisfy $0 < \theta < 1$ and suppose that*

$$1 \le D \le P.$$

Then

$$S_s(P, R) \ll \left(\sum_{d > D}\left(S_s\left(\frac{P}{d}, R\right)\right)^{1/s}\right)^s + S_s(D^{1-\theta}P^\theta, R)$$
$$+ P^\varepsilon\left(\sum_{d \le D}\left(\left(\frac{P}{d}\right)^\theta R\right)^{2-3/s}\left(T_s\left(\frac{P}{d}, R, \theta\right)\right)^{1/s}\right)^s.$$

When $s > k$ and R is not too small by comparison with P we expect that $S_s(P, R) \gg P^\sigma$ and $T_s(P, R, \theta) \gg P^\tau$ with $\sigma > s$, $\tau > s$. Thus for a suitable choice of D the first two terms on the right of the above inequality can be expected to be small compared with the left-hand side, and the third term will be dominated by the term in the sum with $d = 1$. Thus, in principle, the lemma says that either

$$S_s(P, R) \ll P^s$$

or

$$S_s(P, R) \ll (P^\theta R)^{2s-3} T_s(P, R, \theta).$$

From the lemma we can establish an iterative procedure by estimating

$$T_s(P, R, \theta)$$

in terms of

$$S_s(P^{1-\theta}, R) \qquad \text{and} \qquad S_{s-1}(P^{1-\theta}, R).$$

There is an immensity of technical detail, since it is possible to utilise all the diminishing range weaponry in the process. Unfortunately, I do not have sufficient time to go into details.

References

Balasubramanian, R. and Mozzochi, C.J. "An improved upper bound for $G(k)$ in Waring's problem for relatively small k," *Acta Arith.* **43** (1984), 283–285.

de Bruijn, N.G. "The asymptotic behaviour of a function occurring in the theory of primes," *J. Indian Math. Soc.* (N.S.) **15** (1951a), 25–32.

de Bruijn, N.G. "On the number of positive integers $\leq x$ and free of prime factors $>y$," *Nederl. Acad. Wetensch. Proc. Ser. A.* **54** (1951b), 50–60.

Chen, J.-R. "On Waring's problem for n-th powers," *Acta Math. Sinica.* **8** (1958), 253–257, translated in *Chin. Math. Acta* **8** (1966), 849–853.

Davenport, H. "On Waring's problem for fourth powers," *Ann. of Math.* **40** (1939), 731–747.

Davenport, H. "On sums of positive integral kth powers," *Amer. J. Math.* **64** (1942), 189–198.

Davenport, H. *The Collected Works of Harold Davenport*, vol. III, Birch, B.J. Halberstam, H. and Rogers, C.A. eds. Academic Press (1977).

Davenport, H. *Multiplicative Number Theory*, 2nd edition. Springer-Verlag (1980).

Davenport, H. and Erdős, P. "On sums of positive integral kth powers," *Ann. Math.* **40** (1939), 533–536.

Davenport, H. and Lewis, D.J. "Homogeneous additive equations," *Proc. R. Soc. London* **274A** (1963), 443–460.

Estermann, T. "Einige Sätze über quadratfreie Zahlen," *Math. Annalen.* **105** (1931), 653–662.

Halberstam, H. and Richert, H.-E. *Sieve Methods*. Academic Press (1974).

Hardy, G.H. and Littlewood, J.E. *Collected Papers of G.H. Hardy, Including Joint Papers with J.E. Littlewood and Others*, vol. I. London Mathematical Society, Clarendon Press (1966).

Hooley, C. "On Waring's problem," *Acta Math.* **57** (1986), 49–97.

Linnik, J.V. "On the representation of large numbers as sums of seven cubes," *Doklady Akad. Nauk SSSR.* **35** (1942), 162; *Mat. Sbornik.* **12** (1943), 218–224.

Thanigasalam, K. "On Waring's problem," *Acta Arith.* **38** (1980), 141–155.

Thanigasalam, K. "Some new estimates for $G(k)$ in Waring's problem," *Acta Arith.* **42** (1982), 73–78.

Thanigasalam, K. "Improvement on Davenport's iterative method and new results in additive number theory, I & II. Proof that $G(5) \leq 22$," *Acta Arith.* **46** (1985), 1–31, 91–112.

Vaughan, R.C. *The Hardy-Littlewood Method.* Cambridge Univ. Press (1981).

Vaughan, R.C. "Sums of three positive cubes," *Bull. London Math. Soc.* **17** (1985), 17–20.

Vaughan, R.C. "On Waring's problem for smaller exponents," *Proc. London Math. Soc.* (3), **52** (1986a), 445–463.

Vaughan, R.C. "On Waring's problem for sixth powers," *J. London Math. Soc.* (2), **33** (1986b), 227–236.

Vaughan, R.C. "On Waring's problem for cubes," *J. für die Reine und Angewandte Mathematik.* **365** (1986c), 122–170.

Vaughan, R.C. "On Waring's problem for smaller exponents II," *Mathematika.* **33** (1986d), 6–22.

Vinogradov, I.M. "The method of trigonometrical sums in the theory of numbers," *Trudy Matem. Instituta im V.A. Steklov.* **23** (1947), 1–109.

Vinogradov, I.M. *Selected Works*, Springer-Verlag (1984).

Watson, G.L. "A proof of the seven cube theorem," *J. London Math. Soc.* **26** (1951), 153–156.